# Lecture Notes in Computer Science　11554

*Commenced Publication in 1973*
Founding and Former Series Editors:
Gerhard Goos, Juris Hartmanis, and Jan van Leeuwen

More information about this series at http://www.springer.com/series/7407

Huchuan Lu · Huajin Tang ·
Zhanshan Wang (Eds.)

# Advances in Neural Networks – ISNN 2019

16th International Symposium on Neural Networks, ISNN 2019
Moscow, Russia, July 10–12, 2019
Proceedings, Part I

Springer

*Editors*
Huchuan Lu
Dalian University of Technology
Dalian, China

Huajin Tang
Sichuan University
Chengdu, China

Zhanshan Wang
Northeastern University
Shenyang, China

ISSN 0302-9743     ISSN 1611-3349   (electronic)
Lecture Notes in Computer Science
ISBN 978-3-030-22795-1     ISBN 978-3-030-22796-8   (eBook)
https://doi.org/10.1007/978-3-030-22796-8

LNCS Sublibrary: SL1 – Theoretical Computer Science and General Issues

This Springer imprint is published by the registered company Springer Nature Switzerland AG
The registered company address is: Gewerbestrasse 11, 6330 Cham, Switzerland

# Preface

This volume contains the papers presented at ISNN 2019: the 16th International Symposium on Neural Networks held during July 10–12, 2019, in Moscow. Located by the Moskva River, Moscow is the capital and largest city in Russia with a population of over 13 million. Thanks to the success of the previous events, ISNN has become a well-established series of popular and high-quality conferences on the theory and methodology of neural networks and their applications. ISNN 2019 aimed to provide a high-level international forum for scientists, engineers, and educators to present the state of the art of neural network research and applications in related fields. The symposium also featured plenary speeches given by world renowned scholars, regular sessions with a broad coverage, and special sessions focusing on popular topics.

This year, the symposium received more submissions than previous years. Each submission was reviewed by at least two, and on average, 4.5 Program Committee members. After the rigorous peer reviews, the committee decided to accept 111 papers for publication in the *Lecture Notes in Computer Science* (LNCS) proceedings. These papers cover many topics of neural network-related research including learning system, graph model, adversarial learning, time series analysis, dynamic prediction, uncertain estimation, model optimization, clustering, game theory, stability analysis, control method, industrial application, image recognition, scene understanding, biomedical engineering, hardware. In addition to the contributed papers, the ISNN 2019 technical program included three keynotes and plenary speeches by renowned scholars: Prof. Andrzej Cichocki (IEEE Fellow, Skolkovo Institute of Science and Technology, Moscow, Russia), Prof. Yaochu Jin (IEEE Fellow, University of Surrey, Guildford, UK), and Prof. Nikhil R. Pal (IEEE Fellow, Indian Statistical Institute, Calcutta, India).

Many organizations and volunteers made great contributions toward the success of this symposium. We would like to express our sincere gratitude to Skolkovo Institute of Science and Technology and City University of Hong Kong for their sponsorship, the International Neural Network Society, Asian Pacific Neural Network Society, Polish Neural Network Society, and Russian Neural Network Society for their technical co-sponsorship. We would also like to sincerely thank all the committee members for their great efforts in organizing the symposium. Special thanks to the Program Committee members and reviewers whose insightful reviews and timely feedback ensured the high quality of the accepted papers and the smooth flow of the symposium. We would also like to thank Springer for their cooperation in publishing the proceedings in the prestigious LNCS series. Finally, we would like to thank all the speakers, authors, and participants for their support.

June 2019

Huchuan Lu
Huajin Tang
Zhanshan Wang

# Organization

## Program Committee

| | |
|---|---|
| Ping An | Shanghai University, China |
| Sabri Arik | Istanbul University, Turkey |
| Xiaochun Cao | Institute of Information Engineering, Chinese Academy of Sciences, China |
| Wenliang Chen | Soochow University, China |
| Long Cheng | Institute of Automation, Chinese Academy of Sciences, China |
| Zheru Chi | The Hong Kong Polytechnic University, China |
| Yang Cong | Chinese Academy of Science, China |
| Jose Alfredo Ferreira Costa | Federal University of Rio Grande do Norte, Brazil |
| Ruxandra Liana Costea | Polytechnic University of Bucharest, Romania |
| Chaoran Cui | Shandong University of Finance and Economics, China |
| Cheng Deng | Xidian University, China |
| Lei Deng | University of California, Santa Barbara, USA |
| Jing Dong | National Laboratory of Pattern Recognition, China |
| Minghui Dong | Institute for Infocomm Research, Singapore |
| Jianchao Fan | National Marine Environmental Monitoring Center, China |
| Xin Fan | Dalian University of Technology, China |
| Liang Feng | Chongqing University, China |
| Wai-Keung Fung | Robert Gordon University, UK |
| Shaobing Gao | Sichuan University, China |
| Shenghua Gao | ShanghaiTech University, China |
| Shiming Ge | Chinese Academy of Sciences, China |
| Tianyu Geng | Sichuan University, China |
| Xin Geng | Southeast University, China |
| Xiaofeng Gong | Dalian University of Technology, China |
| Jie Gui | Chinese Academy of Sciences, China |
| Chengan Guo | Dalian University of Technology, China |
| Ping Guo | Beijing Normal University, China |
| Zhenhua Guo | Tsinghua University, China |
| Zhenyuan Guo | Hunan University, China |
| Zhishan Guo | University of Central Florida, USA |
| Huiguang He | Institute of Automation, Chinese Academy of Sciences, China |
| Ran He | National Laboratory of Pattern Recognition, China |
| Jin Hu | Chongqing Jiaotong University, China |
| Jinglu Hu | Waseda University, Japan |

| | |
|---|---|
| Qinglai Wei | Institute of Automation, Chinese Academy of Sciences, China |
| Huaqiang Wu | Tsinghua University, China |
| Yong Xia | Northwestern Polytechnical University, China |
| Xiurui Xie | Institute for Infocomm Research, Singapore |
| Shaofu Yang | Southeast University, China |
| Mao Ye | University of Electronic Science and Technology of China, China |
| Qiang Yu | Tianjin University, China |
| Shan Yu | Institute of Automation, Chinese Academy of Sciences, China |
| Wen Yu | Cinvestav, Mexico |
| Zhaofei Yu | Peking University, China |
| Zhi-Hui Zhan | South China University of Technology, China |
| Jie Zhang | Newcastle University, UK |
| Malu Zhang | National University of Singapore, Singapore |
| Nian Zhang | University of the District of Columbia, USA |
| Tielin Zhang | Institute of Automation, Chinese Academy of Sciences, China |
| Xiaoting Zhang | Boston University, USA |
| Xuetao Zhang | Xi'an Jiaotong University, China |
| Yongqing Zhang | Chengdu University of Information Technology, China |
| Zhaoxiang Zhang | Institute of Automation, Chinese Academy of Sciences, China |
| Wenda Zhao | Dalian University of Technology, China |
| Chengqing Zong | Institute of Automation, Chinese Academy of Sciences, China |

## Additional Reviewers

Bai, Ao
Bao, Gang
Bhuiyan, Ashik Ahmed
Bian, Jiang
Cai, Qing
Che, Hangjun
Chen, Boyu
Chen, Haoran
Chen, Junya
Chen, Xiaofeng
Chen, Yi
Chen, Yuanyuan
Chen, Zhaodong
Cui, Hengfei
Cui, Lili

Dang, Suogui
Deng, Xiaodan
Dong, Jiahua
Dong, Junfei
Fan, Huijie
Fan, Xiaofei
Fang, Xiaomeng
Gu, Peng
Gu, Pengjie
Guo, Zhenyuan
Han, Kezhen
Harsha Vardhan Reddy
    Velmula, Harsha
Hu, Jin
Hu, Zhanhao

Huang, Haoyu
Huang, Jiangshuai
Huang, Li
Huang, Yaoting
Huang, Zitian
Huimin, Xu
Jia, Haozhe
Jiang, Peilin
Jiang, Xinyu
Ju, Xiping
Kong, Wanzeng
Li, Cong
Li, Dengju
Li, Jiaxin
Li, Lixiang

Li, Lukai
Li, Zhe
Liang, Hongjing
Liang, Ling
Liang, Yudong
Lin, Jilan
Liu, Faqiang
Liu, Jiayuan
Liu, Kang
Liu, Lining
Liu, Liu
Liu, Lu
Liu, Mingkai
Liu, Sichao
Liu, Yingying
Liu, Yongsheng
Lu, Bao-Liang
Lu, Junwei
Luo, Xiaoling
Luo, Xuefang
Ma, Xin
Ma, Yuhao
Mao, Xin
Mikheev, Aleksandr
Mohammed, Mahmoud
Niu, Dong
Niu, Weina
Pan, Zihan
Pang, Dong
Peng, Jianxin
Qu, Zheng
Ren, He
Ren, Yanhao
Rong, Nannan

Shan, Qihe
Shi, Zhan
Song, Shiming
Sun, Pengfei
Sun, Xiaoxuan
Tan, Guoqiang
Tang, Chufeng
Tang, Tang
Tang, Tianqi
Tian, Yufeng
Tian, Zhiqiang
Vaidhun, Sudharsan
Wan, Fukang
Wang, Bangyan
Wang, Dingheng
Wang, Hongfeng
Wang, Jiru
Wang, Jue
Wang, Lin
Wang, Songsheng
Wang, Xiaoping
Wang, Yaoyuan
Wang, Yixuan
Wang, Yuan
Wang, Yulong
Wang, Zengyun
Wei, Wang
Wen, Shiping
Wen, Xianglan
Wen, Yinlei
Wu, Jian
Wu, Jibin
Wu, Yanming
Wu, Yicheng

Wu, Yining
Wu, Yujie
Xie, Xinfeng
Xie, Xiurui
Xie, Yutong
Xing, Yanli
Xu, Qi
Xu, Yang
Xujun, Yang
Yan, Mingyu
Yang, Yukuan
Yang, Zeheng
Yang, Zhaoxu
Yang, Zheyu
Yao, Xianshuang
Yao, Yanli
Yu, Xue
Yu, Yikuan
Yuan, Mengwen
Zeng, Zhigang
Zhan, Qiugang
Zhang, Haodong
Zhang, Houzhan
Zhang, Jianpeng
Zhang, Malu
Zhang, Shizhou
Zhang, Xin
Zhang, Yun
Zhao, Bo
Zheng, Hao
Zheng, Wei-Long
Zhou, Bo
Zhou, Yue
Zhu, Mingchao

# Contents – Part I

**Time Series Analysis, Dynamic Prediction, and Uncertain Estimation**

## Model Optimization, Bayesian Learning, and Clustering

# Contents – Part II

## Signal Processing, Industrial Application, and Data Generation

## Image Recognition, Scene Understanding, and Video Analysis

**Bio-signal, Biomedical Engineering, and Hardware**

# Learning System, Graph Model, and Adversarial Learning

# Fast Training of Deep LSTM Networks

Wen Yu[1(✉)], Xiaoou Li[2], and Jesus Gonzalez[1]

[1] Departamento de Control Automático,
CINVESTAV-IPN (National Polytechnic Institute), Mexico City, Mexico
yuw@ctrl.cinvestav.mx
[2] Departamento de Computación,
CINVESTAV-IPN (National Polytechnic Institute), Mexico City, Mexico

**Abstract.** Deep recurrent neural networks (RNN), such as LSTM, have many advantages over forward networks. However, the LSTM training method, such as backward propagation through time (BPTT), is really slow.

In this paper, by separating the LSTM cell into forward and recurrent substructures, we propose a much simpler and faster training method than the BPTT. The deep LSTM is modified by combining the deep RNN with the multilayer perceptron (MLP). The simulation results show that our fast training method for LSTM is better than BPTT for LSTM.

## 1 Introduction

For many time series, the current data values depend on their past data, such as sentences and sound waves. Recurrent neural networks (RNNs) have similar property. The propagation backwards through time (BPTT) is an effective training method for RNN. However, BPTT training has many problems, such as gradient loss and slow convergence. The long-term memory network (LSTM) in [6] is very popular RNN in recent years. The deep LSTM is one of the most important deep learning models. By using the gate units, LSTM avoids the problem of gradient degradation. But LSTM still needs a lot more computing time.

Deep learning models augment the hidden layers of neural networks instead of the neuronal nodes, see [5]. This idea can successfully avoid the local minimums in [12] and the problem of determining the structure in [1]. Deep LSTM has been widely applied in many areas, especially in time series modeling, including speech recognition, natural language processing and sequence prediction in [3]. Simplified LSTM, like GRU in [2], is also very effective for modeling time series.

In addition to time series modeling, LSTM can also be applied for modeling nonlinear systems. In [22], LSTM is the basic sub-model of multiple models for unknown dynamic systems. In [25], the deep LSTM is regrouped as a dynamic model. However, these LSTM still use BPTT, they have the problem of slow learning.

In this paper, the slow training problem of LSTM networks is solved by combining the stable training of RNN with the feedforward NN, as in [13–15]. Each

© Springer Nature Switzerland AG 2019
H. Lu et al. (Eds.): ISNN 2019, LNCS 11554, pp. 3–10, 2019.
https://doi.org/10.1007/978-3-030-22796-8_1

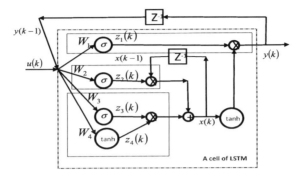

**Fig. 1.** One block of LSTM

block of LSTM includes feedforward and recurrent sub-networks. The training error is back propagated to all LSTM blocks. The results of the simulation show that the new LSTM training model proposed is much better than the BPTT method.

## 2   LSTM as a Dynamic System

Theoretically, the RNN can model any time series, regardless of how long the current status depends on their previous information. In practice, it cannot, because the information between the relevant and place becomes smaller and smaller. LSTM uses the "gate" technique to let useful information pass. So it has capable of handling "long-term dependencies".

In this session, we transform the classical LSTM cell into the form of a dynamic system. We first re-define the three types of gate of LSTM as follows, see Fig. 1.

In the forgetting process,

$$z_2 (k) = \varphi (W_2 (k) [y (k-1), u (k)])$$

where $u(k)$ is the input, $y(k-1)$ is the output, $\varphi$ is a sigmoid function, $\varphi = 1$ represents "keep this", $\varphi = 0$ represents "get rid of this" $z_2 (k)$ is the input to the cell state $x_{k-1}$, $W_2 (k)$ is the weight. From the input $u(k)$ to the hidden layer, the state is

$$z_3 (k) = \varphi (W_3 (k) [y (k-1), u (k)])$$
$$z_4 (k) = \tanh (W_4 (k) [y (k-1), u (k)])$$

where $z_4 (k)$ and $z_3 (k)$ are inner states. The output needs the cell state as

$$z_1 (k) = \varphi (W_1 [y (k-1), u (k)])$$
$$y (k) = z_1 (k) \tanh (x (k))$$

here $\varphi$ select the passing state. For the cell state $x (k)$

$$x (k) = z_2 (k) x (k-1) + z_3 (k) z_4 (k)$$

here

$$x(k) = \varphi(W_2(k)[\hat{y}(k-1), u(k)]) x(k-1)$$
$$+\varphi(W_3(k)[\hat{y}(k-1), u(k)]) \tanh(W_4(k)[\hat{y}_{k-1}, u_k]) \tag{1}$$
$$\hat{y}(k) = \varphi(W_1(k)[\hat{y}(k-1), u(k)]) \tanh(x(k))$$

When we use the peephole connections definition, the LSTM cell model (1) becomes

$$z_2(k) = \varphi(W_2(k)[x(k-1), y(k-1), u(k)])$$
$$z_3(k) = \varphi(W_3(k)[x(k-1), y(k-1), u(k)])$$
$$z_1(k) = \varphi(W_1(k)[x(k-1), y(k-1), u(k)])$$

where $z_4(k)$ is the same as LSTM. The coupled forget and input gates are used. Only new values without forgetting are sent to the state $x(k)$

$$x(k) = z_2(k) x(k-1) + (1 - z_2(k)) z_4(k).$$

The GRU combines the forget gate and the input gate into a single "update gate".

(1) Forget gate:

$$z_2(k) = \varphi(W_2(k)[y(k-1), u(k)]).$$

(2) Input gate:

$$z_3(k) = \varphi(W_3(k)[y(k-1), u(k)])$$
$$z_5(k) = \tanh(W_4(k)[z_3(k) y(k-1)]).$$

The cell state and hidden state are merged as

$$y(k) = y(k-1) + z_2(k)[z_5(k) - y(k-1)]. \tag{2}$$

The GRU cell (2), shown in Fig. 2 is simpler than LSTM cell (1).

## 3  Fast Training of LSTM

The dynamic nonlinear system has the following form

$$d(k) = \Psi[d(k-1), d(k-2)\cdots, u(k), \cdots, u(k - n_u)] \tag{3}$$

where $\Psi(\cdot)$ is an unknown nonlinear difference equation representing the plant dynamics, $u(k)$ and $d(k)$ are input and output.

To model the plant (3), we use the following neural networks

$$y(k) = N[y(k-1), y(k-2), \cdots, u(k), u(k-1), \cdots] \tag{4}$$

where $y(k)$ is the output of the neural networks. We use the LSTMs shown in Fig. 4 to model the nonlinear plant (3). The identification error of the last block is defined as

$$e(k) = y(k) - d(k)$$

where $e(k)$ is the identification error.

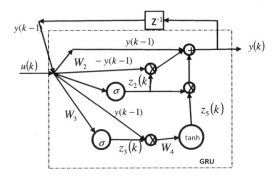

**Fig. 2.** The cell of GRU

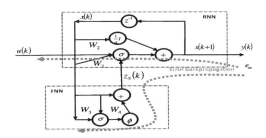

**Fig. 3.** LSTM block separated to FNN and RNN

We first study the fast learning for one block of the GRU. The output of this block is $d(k) = x(k+1)$, the input is $u(k)$. So the dynamic of the GRU (2) is

$$x(k+1) = x(k) + \varphi(W_2(k)[x(k), u(k)])$$
$$\times [\tanh(W_4(k)[\varphi(W_3(k)[x(k), u(k)])x(k)]) - x(k)]$$

The GRU block in Fig. 2 can be transformed into Fig. 3. So GRU can be regarded as a recurrent neural network (RNN) with a feedforward neural networks (FNN)

$$\begin{aligned}
\text{RNN:} \quad & x(k+1) = Ax(k) - \varphi[W_2 x(k) + W_2 u(k)]z_5(k) \\
\text{FNN:} \quad & z_5(k) = \phi(W_3\varphi[W_4 x(k)]x(k)) + x(k)
\end{aligned} \tag{5}$$

where $A = I$, $z_5(k)$ is the virtual input from FNN to RNN.

The deep LSTM for the nonlinear system modeling is shown in Fig. 4. There are $p$ hidden layers, each layer has $q$ LSTM blocks. The last layer is a multilayer perception (MLP).

To train the whole deep LSTM, we define the index as

$$J = \frac{1}{2}e^2 \quad e = y - d$$

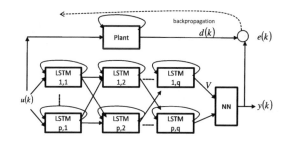

**Fig. 4.** Deep LSTM for nonlinear system modeling

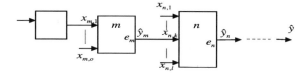

**Fig. 5.** Error back-propagation

Because the last layer is MLP, the weights in this layer are updated by the gradient descent algorithm

$$V(k+1) = V(k) - \eta y_{i,q}(k) e(k) \tag{6}$$

where $y_{i,q}$, $i = 1 \cdots p$, are the outputs of the last layer, $\eta$ is the positive learning rate, $V(k)$ is the weight matrix in the last layer.

The training error should be sent to the output point of each block. If the error is back propagated from the block $n$ to the block $m$, as in Fig. 5, the error in the output of the block $m$ is

$$e_m = \frac{\partial \varphi_n}{\partial t} W_n e_n \tag{7}$$

where $e_n$ is the training error in the output of the block $n$, $\varphi_n$ is the total nonlinae functions of the block $n$, $W_n$ is the weight matrix of the block $n$.

All LSTM blocks have the same structure as in Fig. 3, so only discuss how to train this block. For the block $m$, the training error is $e_m$.

The weights in RNN part are trained as

$$W_2(k+1) = W_2(k) - \eta_k \varphi'[x(k), u(k)]^T z_5(k) e_m(k) \tag{8}$$

where $\eta_k$ satisfies

$$\eta_k = \begin{cases} \dfrac{\eta}{1 + \|\varphi'x(k) z_5(k)\|^2 + \|\varphi'u(k) z_5(k)\|^2} & \text{if } a\|e_m(k+1)\| \geq \|e_m(k)\| \\ 0 & \text{if } a\|e_m(k+1)\| < \|e_m(k)\| \end{cases}$$

$0 < \eta \leq 1$, $a$ the eigenvalue of $A$.

The weights in FNN part are trained by the backpropagation method

$$W_3 (k+1) = W_3 (k) - \eta_k e_m (k) W_4 (k) x (k)$$
$$W_4 (k+1) = W_4 (k) - \eta_k e_m (k) W_3 (k) \varphi' x (k)$$

$$(9)$$

where $\eta_k = \dfrac{\eta}{1 + \|W_4 (k) x (k)\|^2 + \|W_3 (k) \varphi' x (k)\|^2}, 0 < \eta \leq 1.$

## 4   Simulations

We use the deep LSTM networks to model the transonic unsteady aerodynamic system with the proposed fast training method. The aerodynamic system under transonic is shown in Fig. 6. This system can be described as

$$D = F (M_\infty^*, \alpha_0^*, h, \beta)$$

where $D = \begin{bmatrix} d_1 \\ d_2 \end{bmatrix}$, it has 4 input and 2 output. One flight condition is considered $M$ and $\alpha$ are fixed, $M_\infty^*$ and $\alpha_0^*$ are constants, $\{M_\infty, \alpha_0\}$ are selected 20 conditions. The deep LSTM is

$$Y = NN (M_\infty^*, \alpha_0^*, h, \beta)$$

The input is $(M_\infty^*, \alpha_0^*, h, \beta)$, there are 20 groups $(M_\infty^*, \alpha_0^*, h, \beta)$. The output is $Y = \begin{bmatrix} y_1 \\ y_2 \end{bmatrix}$. The testing results are shown in Fig. 6. To compare with MLP, we use different $p$ and $q$ for the deep LSTMs, the results are shown in Table 1 (Fig. 7).

**Table 1.** Squared error of different neural model $(\times 10^{-3})$

| Layers | ML testing | LSTM testing | GRU testing |
|---|---|---|---|
| $p = 4, q = 4$ | 17.3 | 9.16 | 11.22 |
| $p = 8, q = 8$ | 14.72 | 8.32 | 10.81 |
| $p = 4, q = 20$ | 11.14 | 8.01 | 8.17 |
| $p = 8, q = 40$ | 10.51 | 7.81 | 8.01 |

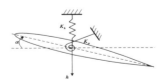

**Fig. 6.** Aerodynamic

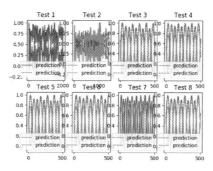

**Fig. 7.** Testing results in different condition

## 5 Conclusion

In this paper, we propose the deep LSTM, which takes advantages of multilayer perceptrons and LSTM, for dynamic system modeling when only the test input $u$ is available. To avoid using the complex training method for recurrent neural networks, backpropagation through time, we give stable learning algorithm for LSTM cell. Three nonlinear systems are used to compare with MLP and normal LSTM. The results show the proposed deep LSTM is better than the other existed neural models for the simulation mode.

## References

1. Bengio, Y., Lamblin, P., Popovici, D., Larochelle, H.: Greedy layer-wise training of deep networks. In: Advances in Neural Information Processing Systems (NIPS 2006), pp. 153–160 (2007)
2. Chung, J., Gulcehre, C., Cho, K., Bengio, Y.: Empirical Evaluation of Gated Recurrent Neural Networks on Sequence Modeling. arXiv:1412.3555 [cs.NE] (2014)
3. Graves, A., Mohamed, A., Hinton, G.: Speech Recognition with Deep Recurrent Neural Networks. arXiv:1303.5778 (2013)
4. Hirose, N., Tajima, R.: Modeling of rolling friction by recurrent neural network using LSTM. In: 2017 IEEE International Conference on Robotics and Automation (ICRA), Singapore, pp. 6471–6475 (2017)
5. Hinton, G., Osindero, S., Teh, Y.: A fast learning algorithm for deep belief nets. Neural Comput. **18**(7), 1–6 (2006)
6. Hochreiter, S., Schmidhuber, J.: Long short-term memory. Neural Comput. **9**(8), 1735–1780 (1997)
7. Box, G., Jenkins, G., Reinsel, G.: Time Series Analysis: Forecasting and Control, 4th edn. Wiley, Hoboken (2008)
8. Kumar, A., Chandel, Y.: Solar radiation prediction using artificial neural network techniques: a review. Renew. Sustain. Energy Rev. **33**(2), 772–781 (2014)
9. Kingma, P., Ba, J.: Adam: A Method for Stochastic Optimization. arXiv:1412.6980 [cs.LG] (2014)
10. Narendra, K., Parthasarathy, K.: Gradient methods for optimization of dynamical systems containing neural networks. IEEE Trans. Neural Netw. **2**(2), 252–262 (1991)

11. Krizhevsky, A., Sutskever, I., Hinton, G.: ImageNet classification with deep convolutional neural networks. In: NIPS, pp. 12–19 (2012)
12. LeCun, Y., Bottou, L., Bengio, Y., Haffne, P.: Gradient based learning applied to document recognition. Proc. IEEE **86**(11), 2278–2324 (1998)
13. Cao, W., Wang, X., Ming, Z., Gao, J.: A review on neural networks with random weights. Neurocomputing **275**(2), 278–287 (2018)
14. Wang, X., Musa, A.: Advances in neural network based learning. Int. J. Mach. Learn. Cybern. **5**(1), 1–2 (2014)
15. Wang, X., Cao, W.: Non-iterative approaches in training feed-forward neural networks and their applications. Soft Comput. **22**(11), 3473–3476 (2018)
16. Ljung, L.: System Identification-Theory for User. Prentice Hall, Englewood Cliffs (1987)
17. Nelles, O.: Nonlinear System Identification: From Classical Approaches to Neural Networks and Fuzzy Models. Springer, Heidelberg (2013)
18. Ogunmolu, O., Gu, X., Jiang, S., Gans, N.: Nonlinear Systems Identification Using Deep Dynamic Neural Networks. arXiv:1610.01439v1 [cs.NE] (2016)
19. Schoukens, J., Schoukens, J., Ljung, L.: Wiener-Hammerstein benchmark. In: 15th IFAC Symposium on System Identification, pp. 1–6 (2009)
20. Srivastava, N., Hinton, G., Krizhevsky, A., Sutskever, I., Salakhutdinov, R.: Dropout: a simple way to prevent neural networks from overfitting. J. Mach. Learn. Res. **15**(1), 1929–1958 (2014)
21. Sugeno, M., Yasukawa, T.: A fuzzy logic based approach to qualitative modeling. IEEE Trans. Fuzzy Syst. **5**(1), 7–31 (1993)
22. Wang, Y.: A new concept using LSTM neural networks for dynamic system identification. In: 2017 American Control Conference, Seattle, USA, pp. 5324–5329 (2017)
23. Wang, L., Langari, R.: Complex systems modeling via fuzzy logic. IEEE Trans. Syst. Man Cybern. **26**(1), 100–106 (1996)
24. Yu, W.: Nonlinear system identification using discrete-time recurrent neural networks with stable learning algorithms. Inf. Sci. **58**(1), 131–147 (2004)
25. Yu, W., Li, X.: Discrete-time neuro identification without robust modification. IEE Proc. Control Theory Appl. **150**(3), 311–316 (2003)
26. Yu, W., Rubio, J.: Recurrent neural networks training with stable bounding ellipsoid algorithm. IEEE Trans. Neural Netw. **20**(6), 983–991 (2009)

# Bus Arrival Time Prediction with LSTM Neural Network

Anton Agafonov$^{(\boxtimes)}$ and Alexander Yumaganov

Samara National Research University,
34, Moskovskoye shosse, Samara 443086, Russia
ant.agafonov@gmail.com, yumagan@gmail.com

**Abstract.** Arrival time is a key aspect of passenger information systems. Provision of accurate bus arrival information is essential for delivering an attractive service and necessary to passengers for reducing their waiting time and bus stops and choosing alternative routes. Recently, the same information is used in smart-phone trip planners. In this paper, we explore an LSTM neural network model for bus arrival time prediction. We take into account heterogeneous information about the transport situation, directly or indirectly affecting the prediction travel time. We evaluate the proposed models with bus operation data from Samara, Russia. Evaluation results show that the proposed model outperforms some typical prediction algorithms.

**Keywords:** Arrival time prediction · Artificial neural network ·
Long short-term memory · Intelligent transportation systems

## 1 Introduction

Public transport is an important part of the transport system. For passengers, providing accurate real-time information about arrival and departure time of public transport is a key aspect of intelligent transportation systems. Nowadays, widely used electronic boards at bus stops or smart-phone applications are considered as a standard way to display such information.

Arrival time at stops can be considered as stochastic, since it depends on many factors, such as travel time of road segments, dwell time at stops and delays at intersections, which can fluctuate spatially and temporally. In addition, traffic congestion, incidents, and weather conditions may cause additional delays in arrival time. Considering these facts, developing a model that can take into account various space-time factors and predict arrival time at stops with high accuracy in real time is a difficult task.

In recent years, many research works focused on the arrival time prediction problem. Despite the popularity of this problem, many papers consider only a small number of features to describe spatial and temporal characteristics of the road situation (for example, speed on the current and previous road segments).

© Springer Nature Switzerland AG 2019
H. Lu et al. (Eds.): ISNN 2019, LNCS 11554, pp. 11–18, 2019.
https://doi.org/10.1007/978-3-030-22796-8_2

In addition, the comparison of algorithms is carried out on different data sets, so, authors cannot conclude, that one model always outperforms to others.

In this paper, we propose to use a recurrent neural network with long short-term memory units (LSTM network) for short-term bus arrival time prediction based on heterogeneous information describing the current and historical transport situation.

The rest of the paper is organized as follows: in the next section, we review related work. In Sect. 3, we present an overview of our approach. Section 4 presents a case study and experimental results of the proposed model with bus operation data from Samara, Russia. Finally, we conclude our work in Sect. 5.

## 2   Related Work

In the past decade, bus arrival time prediction and bus link travel time prediction problems have attracted the interest of many researchers in the transportation area.

Early approaches for travel time prediction used historical average models [15]. Such models can only be used when the road situation is stable; the accuracy is degraded in the case of traffic congestion or accidents. Linear regression models [2,11] determine a dependent variable from a set of independent variables. Regression models assume independence among various factors, which is often impractical. These models have low prediction accuracy, but still widely used in the industry because of their simplicity. Recent research uses these models only for comparison purposes.

In papers [3,14] authors used k-nearest neighbors non-parametric regression models. However, the requirement of a large sample size imposes a restriction on the use of these methods in real time. In [17] authors designed a clustering approach to estimate the distribution of travel time on each link.

Kalman filter based models [4,6] allow estimating the future values based on a series of stochastic measurements over time, containing statistical noise and other inaccuracies. However, the models are limited in representing complex non-linear spatially-temporal relations. Time-series models [21] can also be used for predicting travel times; however, time lags in the historical traffic patterns can lead to inaccurate prediction results.

Recently, in different research was shown, that machine learning methods, such as artificial neural networks (ANN) and support vector machine (SVM), outperform other algorithms. ANN models [5,12,16] are one of the most commonly used models for arrival time prediction. ANN models capable of simulating a complex nonlinear relationship between the travel times and independent variables that characterize the traffic situation. In several works [18,20] was shown that SVM provides a similar prediction accuracy compared to ANN models, but these models have high computational cost. To improve the prediction accuracy, some authors propose hybrid models that combine parametric and non-parametric methods [1,19].

Nowadays, with the progress of machine learning technology, many researchers focus on deep learning models. LSTM model for highway travel time prediction was used in [8,9]. In [13] authors present method for bus travel time prediction using a convolutional LSTM neural network. However, as input data, only link travel times was used.

In this paper, we propose an LSTM neural network model that combines different factors describing the transport situation.

## 3   Methodology

### 3.1   Problem Formulation

Let $S$ denotes the set of stops, $R$ denotes the set of bus routes. The bus arrival time prediction problem can be formulated as follows:

$$t_j^{arr} = t_i^{dep} + T_{ij}^{travel}, \tag{1}$$

where $t_j^{arr}$ denotes the arrival time at stop $j \in S$ $j \in S$, $t_i^{dep}$ denotes the departure time from stop $i$, $T_{ij}^{travel}$ denotes the travel time between stops $i$ and $j$. Then, the prediction of bus arrival time $t_j^{arr}$ at a certain stop $j$ is equivalent to the prediction of bus travel time $T_{ij}^{travel}$.

### 3.2   Input Data

Let the objective bus vehicle run the route $r \in R$. To estimate the travel time $T_{ij}^{travel}$ of the route link between stops $i$ and $j$, the different factors that directly or indirectly affect the target value can be used. In this paper, we propose to use the following heterogeneous information describing the transport situation:

(1) The day of the week $day$ and the time of the day $time$.
(2) The travel speed $v_{i-1,i}$ of the objective bus on the previous route link. This value can show the road congestion degree on the route link that is close to the targeted one.
(3) The headway $h^r$ between the preceding bus vehicle with the same route $r$ and the objective bus.
(4) The travel time $T_{ij}^{m,r}$ of the preceding bus vehicle $m$ with the same route $r$.
(5) The weighted travel time $\tilde{T}_{ij}^r$ of preceding bus vehicles with the same route:

$$\tilde{T}_{ij}^r = \frac{\sum_{k \in N_r} \omega\left(t - t_i^{dep,k}\right) T_{ij}^{travel,k}}{\sum_{k \in N_r} \omega\left(t - t_i^{dep,k}\right)}, \tag{2}$$

where $t$ denotes the current time, $t_i^{dep,k}$ denotes the departure time of the bus vehicle $k$ from the stop $i$, $T_{ij}^{travel,k}$ denotes the travel time of the bus vehicle $k$ between stops $i$ and $j$, $N_r$ denotes the number of bus vehicles with the route $r$, $\omega(t) = \exp\left(-\alpha t\right), \quad t \leq \Delta_{max}$, is a kernel function, $\Delta_{max}$ is a maximum time interval.

(6) The headway $h^{any}$ between the preceding bus vehicle with any route and the objective bus.
(7) The travel time $T_{ij}^{m,any}$ of the preceding bus vehicle $m$ with any route.
(8) The weighted travel time $\tilde{T}_{ij}^{any}$ of preceding bus vehicles with any route.
(9) The historical average travel time $T_{ij}^{hist}(t)$ of bus vehicles with any route at time interval $t$.
(10) The historical average travel time $T_{ij}^{flow}(t)$ of traffic flow that shows historical traffic pattern at time interval $t$.
(11) The number of bus vehicles $c_{ij}$ on the targeted route link. A large number of vehicles on the link can cause an additional dwell time at the nearest stop.

The feature vector, describing a transport situation on the route segment between stops $i$ and $j$ for the bus vehicle with the route $r$, has the following:

$$s_{i,j} = \left(day, time, v_{i-1,i}, h^r, T_{ij}^{m,r}, \tilde{T}_{ij}^r, h^{any}, T_{ij}^{m,any}, \tilde{T}_{ij}^{any}, T_{ij}^{hist}, T^{flow}, c_{ij}\right) \quad (3)$$

### 3.3 Proposed Model

To predict the bus travel time of each transport segment to the end of the route, we propose to use a long short-term memory (LSTM) neural network that is a special type of recurrent neural network (RNN). LSTM network was proposed in [10] and in contrast to RNN it is capable to deal with long-term dependencies. It is achieved by its ability to pass a cell state from previous time step to next one, and to control the information flow in LSTM cell by three gates: input gate, forget gate and output gate. The structure of LSTM cell is shown in Fig. 1.

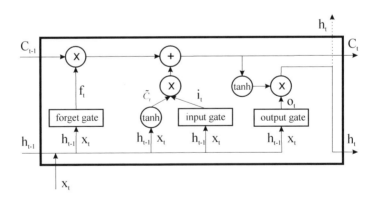

**Fig. 1.** Structure of LSTM cell

Let at time step $t$ the input is $x_t$, input gate state is $i_t$, forget gate state is $f_t$, output gate state is $o_t$, cell output state is $C_t$, cell input state is $\tilde{C}_t$, and layer output state is $h_t$. The output state $h_t$ computes as follow:

$$i_t = \sigma(W_{ix}x_t + W_{ih}h_{t-1} + b_i), \quad f_t = \sigma(W_{fx}x_t + W_{fh}h_{t-1} + b_f),$$
$$o_t = \sigma(W_{ox}x_t + W_{oh}h_{t-1} + b_o), \quad \tilde{C}_t = \tanh(W_{Cx}x_t + W_{Ch}h_{t-1} + b_C), \quad (4)$$
$$C_t = i_t * \tilde{C}_t + f_t * C_{t-1}, \quad h_t = o_t * \tanh(C_t),$$

where $W_{ix}, W_{fx}, W_{ox}, W_{Cx}$ are weight coefficients connecting $x_t$ to three gates and $\tilde{C}_t$; $W_{ih}, W_{fh}, W_{oh}, W_{Ch}$ are weight coefficients connecting $h_{t-1}$ to three gates and $\tilde{C}_t$; $b_i, b_f, b_o, b_C$ are bias values of three gates and $\tilde{C}_t$; $\sigma(x) = \frac{1}{1+\exp(-x)}$; $\tanh(x) = \frac{\exp(x)-\exp(-x)}{\exp(x)+\exp(-x)}$.

The output states $C_t$ and $h_t$ are used as the input data to the next LSTM cell.

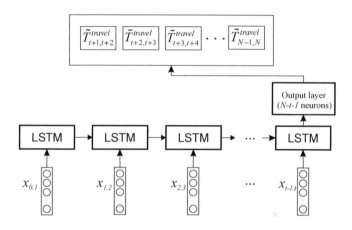

**Fig. 2.** Structure of LSTM network

For travel time prediction purposes, we use an LSTM based neural network, presented in Fig. 2. Let the analyzed public transport route consist of $N$ stops. Then there are $N-1$ route links for selected route and for each of them we build a separated LSTM network since the size of input and output data is different for each route link. Each LSTM network predicts the travel time between remaining bus stops using information related to passed bus stops. For the bus vehicle on the link $i$ $(0 < i < N-1)$ the input of corresponding LSTM network is the historical data $(x_{0,1}, x_{1,2}, ..., x_{i-1,i})$, and the output is predicted travel times $(\tilde{T}^{travel}_{i+1,i+2}, \cdots, \tilde{T}^{travel}_{N-1,N})$. One step feature vector $x_{i,i+1}$ describes observed travel time between stops $i$ and $i+1$ and the transport situation (2) on the whole bus route at the moment when the bus arrives at stop $i+1$. This complex description is used because the LSTM network allows to take into account the impact of the transport situation on remote route links in the long term forecast. The output data describes a travel time between each of the remaining bus stops.

The proposed LSTM networks have been developed with Keras [7], an open source neural network library. Every LSTM model consists of one LSTM layer with 64 units. The last time step output is connected with the output layer of a specific number of neurons. This number differs for each of the models dedicated to analyzed route and corresponds to the number of remaining bus stops to the final one. The Adam method was used as the optimizer; the mean absolute error was used as the loss function.

## 4    Experiments

The proposed method was evaluated on the dataset with bus operation data from Samara, Russia. For the experimental analysis, we chose one bus route with 30 route links and a total length of about 17 km. The bus GPS trajectories was processed and converted into travel times of route links. The dataset consists of travel time observations in the period of September 2018. The dataset was grouped by buses runs on the route and all obtained runs were divided into two parts: training set (80%) and test set (20%).

We compare the proposed model with the artificial neural network model (one hidden layer, 24 hidden neurons) and the linear regression model.

In order to evaluate the performance of the LSTM neural network and all the considered baseline algorithms, we use two standard metrics: mean absolute error (MAE) and mean absolute percentage error (MAPE) that can be formalized as:

$$\mathrm{MAE} = \frac{1}{n} \sum_{t=1}^{n} |T^{travel} - \tilde{T}^{travel}|, \tag{5}$$

$$\mathrm{MAPE} = \frac{1}{n} \sum_{t=1}^{n} \frac{|T^{travel} - \tilde{T}^{travel}|}{T^{travel}} \times 100\%, \tag{6}$$

where $T^{travel}$ is the true travel time, $\tilde{T}^{travel}$ is the predicted travel time, $n$ is the total number of test samples.

Table 1 shows the overall performance of the proposed models and baseline methods.

**Table 1.** Results of the proposed and the baseline models

|                   | MAE (seconds) | MAPE (%) |
|-------------------|---------------|----------|
| LSTM network      | 26.25         | 20.11    |
| ANN               | 34.186        | 28.32    |
| Linear regression | 42.77         | 37.52    |

At the next step, we compare the performance of the LSTM model and baseline models for the arrival time prediction at all remaining stops on the route.

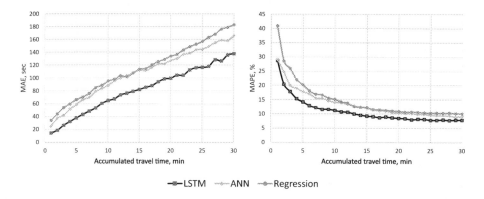

**Fig. 3.** Comparison of the predicted and measured accumulated travel time by the MAE and MAPE criteria

This is equal to predicting and accumulating the travel time of all route links. The MAE and MAPE results for the accumulated travel time are illustrated in Fig. 3.

Experiments show that the proposed method has a higher prediction accuracy than the baseline models.

Finally, we evaluate the computation time of the proposed model. Since the prediction method must be run each time, when the vehicle sent its GPS coordinates to the server, the computation time of the method is a critical consideration. Using laptop computer (Intel Core i5-3740 3.20 GHz, 8 GB RAM), the processing of one vehicle takes 2 ms in average, so, the LSTM model can be used for arrival time prediction in a large city in real time.

## 5   Conclusion

This paper proposed a multi-output model for bus arrival time prediction that uses LSTM (long short-term memory) neural network. The proposed model combines different factors describing the transport situation.

Our experimental results with bus operation data from Samara, Russia, demonstrate that the proposed model outperforms other baseline methods, such as artificial neural network or linear regression methods. The proposed model has a high prediction accuracy and reasonable computation time, sufficient for real-time prediction.

The possible direction of further research including works on choosing best neural network topology. Furthermore, we would like to validate the proposed model on the extended dataset containing records from multiple bus routes.

**Acknowledgments.** The work was supported by the Ministry of Science and Higher Education of the Russian Federation (unique project identifier RFMEFI57518X0177).

# References

1. Agafonov, A., Myasnikov, V.: An adaptive algorithm for public transport arrival time prediction based on hierarhical regression, vol. 2015-October, pp. 2776–2781 (2015)
2. Agafonov, A., Sergeev, A., Chernov, A.: Forecasting of the motion parameters of city transport by satellite monitoring data. Comput. Opt. **36**(3), 453–458 (2012)
3. Chang, H., Park, D., Lee, S., Lee, H., Baek, S.: Dynamic multi-interval bus travel time prediction using bus transit data. Transportmetrica **6**(1), 19–38 (2010)
4. Chen, M., Liu, X., Xia, J., Chien, S.: A dynamic bus-arrival time prediction model based on APC data. Comput.-Aided Civil Infrastruct. Eng. **19**(5), 364–376 (2004)
5. Chien, S.J., Ding, Y., Wei, C.: Dynamic bus arrival time prediction with artificial neural networks. J. Transp. Eng. **128**(5), 429–438 (2002)
6. Chien, S.J., Kuchipudi, C.: Dynamic travel time prediction with real-time and historic data. J. Transp. Eng. **129**(6), 608–616 (2003)
7. Chollet, F., et al.: Keras (2015). https://keras.io
8. Du, S., Li, T., Gong, X., Yu, Z., Huang, Y., Horng, S.J.: A Hybrid Method for Traffic Flow Forecasting Using Multimodal Deep Learning. arXiv e-prints arXiv:1803.02099, March 2018
9. Duan, Y., Lv, Y., Wang, F.Y.: Travel time prediction with LSTM neural network, pp. 1053–1058 (2016)
10. Hochreiter, S., Schmidhuber, J.: Long short-term memory. Neural Comput. **9**(8), 1735–1780 (1997)
11. Jeong, R., Rilett, L.: Prediction model of bus arrival time for real-time applications. Transp. Res. Rec. **1927**, 195–204 (2005)
12. Kumar, V., Kumar, B., Vanajakshi, L., Subramanian, S.: Comparison of model based and machine learning approaches for bus arrival time prediction, p. 14 (2014)
13. Petersen, N., Rodrigues, F., Pereira, F.: Multi-output bus travel time prediction with convolutional LSTM neural network. Expert Syst. Appl. **120**, 426–435 (2019)
14. Smith, B., Williams, B., Oswald, R.: Comparison of parametric and nonparametric models for traffic flow forecasting. Transp. Res. Part C: Emerg. Technol. **10**(4), 303–321 (2002)
15. Sun, D., Luo, H., Fu, L., Liu, W., Liao, X., Zhao, M.: Predicting bus arrival time on the basis of global positioning system data. Transp. Res. Rec. **2034**, 62–72 (2007)
16. Van Hinsbergen, C., Van Lint, J., Van Zuylen, H.: Bayesian committee of neural networks to predict travel times with confidence intervals. Transp. Res. Part C: Emerg. Technol. **17**(5), 498–509 (2009)
17. Xu, H., Ying, J.: Bus arrival time prediction with real-time and historic data. Cluster Comput. **20**(4), 3099–3106 (2017)
18. Yang, M., Chen, C., Wang, L., Yan, X., Zhou, L.: Bus arrival time prediction using support vector machine with genetic algorithm. Neural Netw. World **26**(3), 205–217 (2016)
19. Yu, B., Yang, Z.Z., Chen, K., Yu, B.: Hybrid model for prediction of bus arrival times at next station. J. Adv. Transp. **44**(3), 193–204 (2010)
20. Yu, B., Yang, Z.Z., Yao, B.Z.: Bus arrival time prediction using support vector machines. J. Intell. Transp. Syst.: Technol. Plan. Oper. **10**(4), 151–158 (2006)
21. Zhu, T., Ma, F., Ma, T., Li, C.: The prediction of bus arrival time using global positioning system data and dynamic traffic information (2011)

# From Differential Equations to Multilayer Neural Network Models

Tatiana T. Kaverzneva[1], Galina F. Malykhina[1,2(✉)],
and Dmitriy A. Tarkhov[1]

[1] Peter the Great St. Petersburg Polytechnic University, Saint Petersburg, Russia
g_f_malychina@mail.ru
[2] Russian State Scientific Center for Robotics and Technical Cybernetics,
Saint Petersburg, Russia

**Abstract.** A method for constructing multilayer neural network approximations of solutions of differential equations, based on the finite difference method, is proposed. The advantage of the method is the possibility of obtaining a neural network model of arbitrarily high accuracy without a time-consuming learning procedure. The solution is given by an analytical expression, explicitly including the parameters of the problem. The resulting neural network can, if necessary, be retrained according to the usual algorithm. The method is illustrated by the example of solving a particular ordinary second-order differential equation.

**Keywords:** The differential equation · Multilayer approximate solution ·
Neural network model · Deformation · Elastic thread

## 1 Introduction

All over the world, accidents of various scale, large and small, with and without loss of human life, occur every day. Collapse of the roofs of buildings and structures, cable breaks in elevator shafts, plane crashes, avalanches and other typical emergencies often take human lifes. One of the reasons for such accidents is that it is difficult to predict the technical condition of a particular object using its model, built on the basis of general physical laws due to a variety of factors affecting it. For example, the strength characteristics of various materials used in technology are influenced by such difficult-to-calculate factors as temperature differences, humidity and environmental aggressiveness, accumulated fatigue, type of load application (dynamic or static), and many others. Experimental study of the influence of such factors on a particular object with the subsequent refinement of its model in accordance with the data of an experiment or monitoring can significantly increase the reliability of predicting the behavior of the state of the studied systems.

To solve this problem, it is tempting to use neural networks because of their adaptability and resistance to errors in the data [1]. In this case, it is undesirable to completely abandon the information on the physical laws that describe the operation of the object. Here we see two possibilities. First, it is possible to construct an approximate neural network solution of the corresponding differential equation (system of equations) with boundary conditions, subsequently adapting it to the measurement

© Springer Nature Switzerland AG 2019
H. Lu et al. (Eds.): ISNN 2019, LNCS 11554, pp. 19–27, 2019.
https://doi.org/10.1007/978-3-030-22796-8_3

data. Secondly, you can immediately build a neural network solution using all the available information in the form of differential equations, boundary conditions, data, etc. Formalization and references are given in the next section.

However, currently used neural network methods have significant drawbacks, such as a long training procedure and, as a rule, the single-layer nature of the neural network (with one hidden layer), built on methods similar to FEM [2–19]. This article proposes our new approach, which allows you to quickly form a fairly accurate multilayer neural network solution of a differential equation. In contrast to the previous approaches, the method is not based on FEM, but on the finite difference method. The strength of the method is the automatic inclusion in the final formula of the parameters of the problem, which allows it to avoid multiple re-decisions if necessary to investigate the effect of parameters on the result. The solution obtained in this way is illustrated by a simple practical example.

The methods presented in this article can be used to model the dynamics and describe other processes in complex technical objects. This is especially important for development an individual model of a specific object, taking into account its unique features.

## 2   Materials and Methods

In addition to FEM, another basic method used to solve boundary-value problems (ordinary and partial derivatives) is the finite-difference method. This method consists in replacing the derivatives by approximating their difference ratios. As a result, the differential equation turns into a system of ordinary equations. Until recently, it was assumed that only numerical approximations could be obtained in this way. As one of the advantages of neural network modeling over the finite-difference method, we repeatedly noted the fact that the neural network approach allows one to obtain a solution analytically in the form of an explicit formula [10–17]. However, in [20] we showed that, using the finite difference method, we can obtain analytically defined approximations. Let us explain what has been said on the example of ordinary differential equations.

Consider the Cauchy problem for a system of ordinary differential equations

$$\begin{cases} \mathbf{y}'(x) = \mathbf{f}(x, \mathbf{y}(x)), \\ \quad \mathbf{y}(x_0) = \mathbf{y}_0 \end{cases} \tag{1}$$

on the interval $D = [x_0; x_0 + a]$. Here $x \in D \subset \mathbb{R}$, $\mathbf{y} \in \mathbb{R}^d$, $\mathbf{f} : \mathbb{R}^{d+1} \to \mathbb{R}^d$.

The classic Euler method is to split the interval $D$ into $n$ parts: $x_0 < x_1 < \ldots < x_k < x_{k+1} < \ldots < x_n = x_0 + a$, and apply an iterative formula

$$\mathbf{y}_{k+1} = \mathbf{y}_k + h_k \mathbf{f}(x_k, \mathbf{y}_k), \tag{2}$$

where $h_k = x_{k+1} - x_k$; $\mathbf{y}_k$ – approximation to the exact value of the desired solution $\mathbf{y}(x_k)$.

The estimate of the resulting approximations is known in the form $\|\mathbf{y}(x_k) - \mathbf{y}_k\| \leq C \max(h_k)$ where the constant $C$ depends on the estimates of the function $\mathbf{f}$ and its derivatives in the region in which the solution is located [21].

We propose to construct an approximate solution of problem (1) using formula (2) on an interval $\tilde{D} = [x_0, x]$ with a variable upper limit $x \in [x_0, x_0 + a]$. In this case $h_k = h_k(x)$, $\mathbf{y}_k = \mathbf{y}_k(x)$, is determined by the initial conditions $\mathbf{y}_0(x) = \mathbf{y}_0$. As approximate solution of the Eq. (1) is proposed to use $\mathbf{y}_n(x)$.

The simplest version of the algorithm is obtained by uniformly partitioning the interval with the step $h_k(x) = (x - x_0)/n$. As a result, we obtain a multilayer recurrent functional formula of the form

$$\mathbf{y}_{k+1}(x) = \mathbf{y}_k(x) + \frac{x - x_0}{n} \mathbf{f}\left(x_0 + \frac{k(x - x_0)}{n}, \mathbf{y}_k(x)\right) \tag{3}$$

wherein $\mathbf{y}_0(x) = \mathbf{y}_0$.

Note that the initial values $\mathbf{y}_0$ enter the expression for $\mathbf{y}_n(x)$ as parameters. In addition, if the function $\mathbf{f}$ depends on some parameters, these parameters will be included in the expression for $\mathbf{y}_n(x)$.

Note that if the function $\mathbf{f}$ in (1) is neural network, then formula (2) allows us to obtain a multilayer approximation of the solution. If $\mathbf{f}$ it is not a neural network, then it can be brought closer by a neural network from the corresponding class [1] and again get a multi-layer neural network solution.

It is known that the Euler method has low accuracy. It can be replaced by more accurate methods, for example, the Runge-Kutta method. The usual assessment of the accuracy of the original classical methods allow us to present convenient estimation of the accuracy of the approximations obtained. If the accuracy of the solution obtained in this way is insufficient, the corresponding neural networks can be extended using conventional methods [10–19].

The resulting solution in the form of a multilayer neural network function can be implemented as a neurochip to speed up computations.

## 3  Calculation

We will demonstrate the proposed methodology on a model task for calculating the dynamics of a person's fall with a rescue rope from a height. At the first stage, the body falls freely, and at the second stage it is braked by a cable with some force $F(x - 1)$. As a result, the equation of motion in the second stage is

$$\ddot{x} = G(x) = g - \frac{1}{m}F(x - l). \tag{4}$$

here $l$ is the length of the cable, $m$ is the body mass, $g$ is the acceleration of free fall.

On the basis of our experiments [22], it was possible to establish that the load at which the rupture occurs is practically independent of the sample length and the rate of increase of the tension force. Using neural network type dependency

$$F(x) = c_1 th[a_1(x - x_{c_1})] + c_2 th[a_2(x - x_{c_2})] \tag{5}$$

in [22], it was possible to quite accurately approximate the dependence of the elongation on the load. The weights of the network (5) were sought from the condition of the minimum of the error functional $\sum_{i=1}^{m} (F(x_i) - F_i)^2$. Here $x_i$ is the experimental elongation of the sample, and $F_i$ is the corresponding tensile force. The neural network approximation thus constructed for the dependence of force on elongation was

$$F(x) = 9.54 \tanh[0.0064(-280 + x)] + 12.3 \tanh[0.015(59 + x)]. \tag{6}$$

This result is consistent with theoretical studies [23–27]. The error of formula (6) relative to the experimental data was less than 1%.

The dependence (6) is substituted into the equation of motion (4). We obtain a problem of type (2) with the right part as the output of the neural network. Thus, for its approximate solution one can use the multilayer recurrent functional approximation described above (3). It remains only to choose a specific iterative method.

As a starting method, we have considered several options.

This is the classic Euler method and its two improved variations [23]

$$x_{k+1}(t) = x_{k-1}(t) + 2hG(x_k(t)), \tag{7}$$

$$x_{k+1}(t) = x_k(t) + h\left[G(x_k(t)) + \frac{h}{2}\left(G'_x(x_k(t)) + G'_y(x_k(t))G(x_k(t))\right)\right] \tag{8}$$

In addition, we used a special, more exact for Sturmer [23] method for the type of equation under consideration, based on the recurrent formula

$$x_{k+1}(t) = 2x_k(t) - x_{k-1}(t) + h^2 G(x_k(t)) \, h = t/n. \tag{9}$$

Note that in all cases, the desired solution $x_n(t)$ can be considered the output of a multi-layer neural network of direct propagation with hidden layers [1].

For convenience, we introduce a $p$-parameter depending on the mass. Then by replacing Eq. (4) we transform to the form:

$$\ddot{x} = p - F(x - l). \tag{10}$$

We present some calculation results for the above-mentioned methods for different values of the number $n$ of layers of an approximate solution in Table 1.

**Table 1.** The results of solving Eq. (10) using formula (3) for different methods and the number of layers

| Characteristics | n = 4 | n = 6 | n = 8 | n = 10 | n = 12 |
|---|---|---|---|---|---|
| Eulerian method | | | | | |
| Maximum relative error | 2.28317 | 1.54788 | 1.15885 | 0.918093 | 0.744812 |
| Mean square error. | 0.692613 | 0.487335 | 0.373508 | 0.298992 | 0.243284 |
| Computing time | 1.064 | 2.095 | 8.142 | 40.125 | 215.141 |
| Improved Eulerian method | | | | | |
| Maximum relative error | 1.09217 | 0.0463169 | 0.160146 | 0.0407372 | 0.0392853 |
| Mean square error. | 0.296747 | 0.0227818 | 0.0382626 | 0.0120647 | 0.00993122 |
| Computing time | 0.751 | 2.313 | 4.094 | 12. | 45.032 |
| Revised Eulerian method | | | | | |
| Maximum relative error | 0.205114 | 0.0794412 | 0.0293257 | - | - |
| Mean square error. | 0.0507965 | 0.0331571 | 0.0134021 | - | - |
| Computing time | 2.106 | 114.393 | 713.034 | - | - |
| Störmer method | | | | | |
| Maximum relative error | 0.159023 | 0.0392627 | 0.0269455 | 0.0163259 | - |
| Mean square error. | 0.0381197 | 0.00988471 | 0.00641539 | 0.00405411 | - |
| Computing time | 5.626 | 57.109 | 238.187 | 2692.31 | - |

Table 1 shows that you can get an approximate solution of sufficient accuracy without using the neural network training procedure.

The curves exemplifying this approximate solution of the computation problem are shown on the Fig. 1. Function, based on Störmer method and approximate solution of equation $x_2(t)$ (10) was constructed using the intrinsic function of Mathematica-10 with the parameters $l = 30; p = 10$.

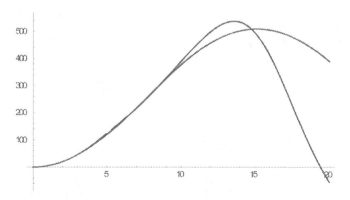

**Fig. 1.** Function, based on Störmer method and approximate solution of equation $x_2(t)$ (10), obtained with Mathematica 10 when $l = 30; p = 10$;

From the graphs it is clear that the accuracy of the approximation is insufficient. For $n = 3$, we obtain a significantly more accurate approximation (Fig. 2):

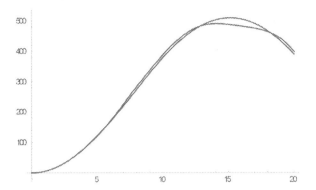

**Fig. 2.** Function $x_3(t)$, based on Störmer method and approximate solution of Eq. (10), obtained with Mathematica 10 when $l = 30; p = 10$;

As we see from Table 1, with increasing n the accuracy increases.

## 4    Results and Discussion

The results show the efficiency of suggested method. Well known theorems on the accuracy of iterative methods, for example, the Störmer method, allow us to obtain an estimate of the difference between the approximate and exact solution of the Eq. (10) $|x_n(t) - x(t)| \leq C \frac{t^2}{n^2}$.

In spite of the fact that in the considered problem we obtained a rather high accuracy of the result $x_n(t)$, without using additional selection of parameters, in some cases this may be necessary. For example, formulas with a large number of layers may be too cumbersome or require a lot of time to calculate.

Thus, the accuracy can be improved by replacing all or certain of the numerical coefficients in the formula for an approximate solution $x_n(t)$ with parameters and selecting them, minimizing the error functional, which for the problem in question has the following form

$$\sum_{j=1}^{M} (\ddot{x}(t_j) - G(x(t_j)))^2 + \delta x^2(0). \tag{11}$$

Here $\{t_j\}_{j=1}^{M}$ are test points on the interval [0, T]. If we want to clarify the solution for a certain range of parameters $l_1 < l < l_2; p_1 < p < p_2$, then functional (11) must be replaced by the following

$$\sum_{j=1}^{M} (\ddot{x}(t_j, p_j, l_j) - G(x(t_j, p_j, l_j)))^2 + \delta x^2(0, p_j, l_j). \tag{12}$$

Here $\{p_j, l_j\}_{j=1}^{M}$ are test points from the area of change of parameters of interest. The regeneration of test points is carried out as described above.

Another direction of the refinement of the solution is obtained by optimizing the choice of steps $h_k(t)$ in formulas of type (3).

Of great interest is the comparative testing of methods for constructing neural networks, given in this paper, and classical teaching methods based on the reverse propagation of error. There are two main features here. The first is to train the neural network of the same structure as the one constructed above using the back propagation method. At the same time, the choice of the initial weights of the network has a strong influence on the speed and the result of training. We tried to use as initial weights the numbers that we obtain using the methods we proposed in this paper. This significantly reduces the training time, but to reduce the error significantly faster simply increase the number $n$.

Comparative testing of the methods discussed in the article and other known methods of forming the structure and determining the weights of the neural network for different neural network architectures will be more objective, but such a study is beyond the scope of this work.

In [28–30], examples of problems with real measurements for which the constructed models more accurately reflect measurement data than exact solutions of initial boundary value problems are given.

Using the method of lines, you can solve problems for partial differential equations in a similar way.

## 5 Conclusions

The work has taken the first steps in the development of methods for constructing multilayer neural network solutions of differential equations based on finite difference methods. We hope that they will soon be developed in many directions and will allow to solve a wide range of practically interesting problems.

**Acknowledgments.** The article was prepared on the basis of scientific research carried out with the financial support of the Russian Science Foundation grant (project No. 18-19-00474).

## References

1. Haykin, S.: Neural Networks: A Comprehensive Foundation. Prentice Hall, Upper Saddle River (1999)
2. Lagaris, I.E., Likas, A., Fotiadis, D.I.: Artificial neural networks for solving ordinary and partial differential equations. IEEE Trans. Neural Netw. 9(5), 987–1000 (1998)

3. Dissanayake, M.W.M.G., Phan-Thien, N.: Neural-network-based approximations for solving partial differential equations. Commun. Numer. Methods Eng. **10**(3), 195–201 (1994)
4. Fasshauer, G.E.: Solving differential equations with radial basis functions: multilevel methods and smoothing. Adv. Comput. Math. **11**, 139–159 (1999)
5. Fornberg, B., Larsson, E.A.: Numerical study of some radial basis function based solution methods for elliptic PDEs. Comput. Math. Appl. **46**, 891–902 (2003)
6. Galperin, E., Pan, Z., Zheng, Q.: Application of global optimization to implicit solution of partial differential equations. Comput. Math. Appl. **25**(10/11), 119–124 (1993)
7. Galperin, E., Zheng, Q.: Solution and control of PDE via global optimization methods. Comput. Math. Appl. **25**(10/11), 103–118 (1993)
8. Sharan, M., Kansa, E.J., Gupta, S.: Application of the multiquadric method to the numerical solution of elliptic partial differential equations. Appl. Math. Comput. **84**, 275–302 (1997)
9. Vasilyev, A., Tarkhov, D., Guschin, G.: Neural networks method in pressure gauge modeling. In: Proceedings of the 10th IMEKO TC7 International Symposium on Advances of Measurement Science, Saint-Petersburg, Russia, vol. 2, pp. 275–279 (2004)
10. Tarkhov, D., Vasilyev, A.: New neural network technique to the numerical solution of mathematical physics problems. I: simple problems. Opt. Mem. Neural Netw. (Inf. Opt.) **14**(1), 59–72 (2005)
11. Tarkhov, D., Vasilyev, A.: New neural network technique to the numerical solution of mathematical physics problems. II: complicated and nonstandard problems. Opt. Mem. Neural Netw. (Inf. Opt.) **14**(2), 97–122 (2005)
12. Vasilyev, A., Tarkhov, D.: Mathematical models of complex systems on the basis of artificial neural networks. Nonlinear Phenom. Complex Syst. **17**(2), 327–335 (2014)
13. Shemyakina, T.A., Tarkhov, D.A., Vasilyev, A.N.: Neural network technique for processes modeling in porous catalyst and chemical reactor. In: Cheng, L., Liu, Q., Ronzhin, A. (eds.) ISNN 2016. LNCS, vol. 9719, pp. 547–554. Springer, Cham (2016). https://doi.org/10.1007/978-3-319-40663-3_63
14. Gorbachenko, V.I., Lazovskaya, T.V., Tarkhov, D.A., Vasilyev, A.N., Zhukov, M.V.: Neural network technique in some inverse problems of mathematical physics. In: Cheng, L., Liu, Q., Ronzhin, A. (eds.) ISNN 2016. LNCS, vol. 9719, pp. 310–316. Springer, Cham (2016). https://doi.org/10.1007/978-3-319-40663-3_36
15. Budkina, E.M., Kuznetsov, E.B., Lazovskaya, T.V., Leonov, S.S., Tarkhov, D.A., Vasilyev, A.N.: Neural network technique in boundary value problems for ordinary differential equations. In: Cheng, L., Liu, Q., Ronzhin, A. (eds.) ISNN 2016. LNCS, vol. 9719, pp. 277–283. Springer, Cham (2016). https://doi.org/10.1007/978-3-319-40663-3_32
16. Lozhkina, O., Lozhkin, V., Nevmerzhitsky, N., Tarkhov, D., Vasilyev, A.: Motor transport related harmful PM2.5 and PM10: from on road measurements to the modelling of air pollution by neural network approach on street and urban level. J. Phys.: Conf. Ser. **772** (2016). http://iopscience.iop.org/article/10.1088/1742-6596/772/1/012031
17. Kaverzneva, T., Lazovskaya, T., Tarkhov, D., Vasilyev, A.: Neural network modeling of air pollution in tunnels according to indirect measurements. J. Phys.: Conf. Ser. **772** (2016). http://iopscience.iop.org/article/10.1088/1742-6596/772/1/012035
18. Lazovskaya, T.V., Tarkhov, D.A., Vasilyev, A.N.: Parametric neural network modeling in engineering. Recent Patents Eng. **11**(1), 10–15 (2017)
19. Antonov, V., Tarkhov, D., Vasilyev, A.: Unified approach to constructing the neural network models of real objects. Part 1. Math. Models Methods Appl. Sci. **41**(18), 9244–9251 (2018)
20. Lazovskaya, T., Tarkhov, D.: Multilayer neural network models based on grid methods. In: IOP Conference Series: Materials Science and Engineering, vol. 158 (2016). http://iopscience.iop.org/article/10.1088/1757-899X/158/1/01206

21. Hairer, E., Norsett, S.P., Wanner, G.: Solving Ordinary Differential Equations I: Nonstiff Problem. Springer, Berlin (1987). https://doi.org/10.1007/978-3-662-12607-3
22. Bolgov, I., Kaverzneva, T., Kolesova, S., Lazovskaya, T., Stolyarov, O., Tarkhov, D.: Neural network model of rupture conditions for elastic material sample based on measurements at static loading under different strain rates. J. Phys.: Conf. Ser. **772** (2016). http://iopscience.iop.org/article/10.1088/1742-6596/772/1/012032
23. Aranda-Iglesias, D., Vadillo, G., Rodríguez-Martínez, J.A., Volokh, K.Y.: Modeling deformation and failure of elastomers at high strain rates. Mech. Mater. **104**, 85–92 (2017)
24. Zéhil, G.-P., Gavin, H.P.: Unified constitutive modeling of rubber-like materials under diverse loading conditions. Int. J. Eng. Sci. **62**, 90–105 (2013)
25. Hearle, J.W.S.: One-Dimensional Textiles. Handbook of Technical Textiles. Elsevier, Amsterdam (2016)
26. McKenna, H.A., Hearle, J.W.S., O'Hear, N.: Handbook of Fibre Rope Technology. Handbook of Fibre Rope Technology. Elsevier, Amsterdam (2004)
27. Weller, S.D., Johanning, L., Davies, P., Banfield, S.J.: Synthetic mooring ropes for marine renewable energy applications. Renew. Energy **83**, 1268–1278 (2015)
28. Vasilyev, A.N., Tarkhov, D.A., Tereshin, V.A., Berminova, M.S., Galyautdinova, A.R.: Semi-empirical neural network model of real thread sagging. In: Kryzhanovsky, B., Dunin-Barkowski, W., Redko, V. (eds.) NEUROINFORMATICS 2017. SCI, vol. 736, pp. 138–144. Springer, Cham (2018). https://doi.org/10.1007/978-3-319-66604-4_21
29. Zulkarnay, I.U., Kaverzneva, T.T., Tarkhov, D.A., Tereshin, V.A., Vinokhodov, T.V., Kapitsin, D.R.: A two-layer semi empirical model of nonlinear bending of the cantilevered beam. J. Phys.: Conf. Ser. **1044** (2018). https://iopscience.iop.org/article/10.1088/1742-6596/1044/1/012005/pdf
30. Bortkovskaya, M.R., et al.: Modeling of the membrane bending with multilayer semi-empirical models based on experimental data. In: Proceedings of the 2nd International Scientific Conference "Convergent Cognitive Information Technologies" (Convergent'2017) (2017). http://ceur-ws.org/Vol-2064/paper18.pdf

# Projectional Learning Laws for Differential Neural Networks Based on Double-Averaged Sub-Gradient Descent Technique

Isaac Chairez[1], Alexander Poznyak[2(✉)], Alexander Nazin[3], and Tatyana Poznyak[4]

[1] Biprocesses Departament, UPIBI-Instituto Politecnico Nacional, Mexico City, Mexico
[2] Automatic Control Department, CINVESTAV-IPN, Mexico City, Mexico
apoznyak@ctrl.cinvestav.mx
[3] Trapeznikov Institute of Control Sciences Russian Academy of Sciences, Moscow, Russia
[4] SEPI, ESIQIE-Instituto Politecnico Nacional, Mexico City, Mexico

**Abstract.** A new method to design learning laws for neural networks with continuous dynamics is proposed in this study. The learning method is based on the so-called double-averaged descendant technique (DAS-GDT), which is a variant of the gradient-descendant method. The learning law implements a double averaged algorithm which filters the effect of uncertainties of the states, which are continuously measurable. The learning law overcomes the classical assumption on the strict convexity of the functional with respect to the weights. The photocatalytic ozonation process of a single contaminant is estimated using the learning law design proposed in this study.

**Keywords:** Differential neural networks ·
Double-averaged subgradient · Optimization · Projection ·
Ozonation processes

## 1 Introduction

The problem of designing artificial learning methods (ALM) has received renewed attention because the increasing number of artificial intelligence applications [11]. Learning-based models typically have parameters (weights and biases) as well as a cost functional aimed to evaluate the goodness of a particular set of parameters. Mainly, the learning problems consist of finding the set of weights for the model which implies the minimization of the cost functional.

Most of the learning algorithms applies diverse forms of parametric estimation methods. The gradient descent method (GDM) has been the basis of

© Springer Nature Switzerland AG 2019
H. Lu et al. (Eds.): ISNN 2019, LNCS 11554, pp. 28–38, 2019.
https://doi.org/10.1007/978-3-030-22796-8_4

many renamed learning algorithms such as the back-propagation, Levenberg-Marquardt and many others. The application of the GDM implies the proposition of an optimization method, which introduces the feasible cost functional depending on the identification error. Most of the existing theory on the design of learning methods consider that such functional is strictly convex with respect to the parameters in the approximate models. However, this is a strong assumption which is rarely satisfied a-priori because the multivariable nature of the functional. Sometimes, the application of the projected GDM may help in solving the parameter identification problem with non-strict convex functional, which is dimension independent free (a remarkable characteristic when dealing with function approximation based on artificial neural networks) [7].

This study uses the so-called *mirror descent algorithm* (MDA) (proposed by Nemirovsky and Yudin) which may solve convex (non-strict) optimization problems [9,10]. This method has shown an efficient estimate which is mildly dependent on the decision variables dimension [2]. Therefore, it is suitable to solve large scale optimization problems such as the case of estimating the weights of an artificial neural network. Moreover, this method may produce better rates of convergence for the optimization problem.

The application of the DASGDT for a special class of functional describing the quality of approximation (in weak sense) of a neural network with continuous dynamics (usually called differential neural network or DNN) is the main contribution of this study. Moreover, the introduction of a box-type projection algorithm allowed to estimate the time-dependent weights of the DNN. The weights estimation algorithm proved the asymptotic convergence of the weights to their true values.

This paper is organized as follows. Section 2 introduces the problem formulation where the class of uncertain nonlinear systems approximated by neural networks, the DNN structure and the main assumptions are detailed. Section 3 formulates the DNN identifier structure and the main result of this study dealing with the application of DASGDT to solve the design of the learning laws for the DNN identifier. Section 4 presents the numerical example that illustrates the implementation of the learning law based on the DASGDT. Section 5 closes the study with some final remarks.

## 2    Problem Formulation

### 2.1    Class of Uncertain Systems and Main Assumptions

The class of nonlinear systems with uncertain structure considered in this study satisfies:

$$\frac{d}{dt}x(t) = f(x(t), u(t), t), \quad x(0) = x_0, \quad \|x_0\| < \infty \tag{1}$$

In (1), the variable $x \in \mathbb{R}^n$ is the state vector and let $X$ be such that $x \in Int\{X\} \subset \mathbb{R}^n$ ($Int\{X\}$ means the interior of set $X$). The system (1) has bounded trajectories so that $\|x(t)\| < x^+$, $\forall t \geq 0$, where $x^+$ corresponds to the maximum radius of set $X$.

The bounded and piecewise continuous (with respect to time) signal $u \in \mathbb{R}^m$ ($m < n$) is referred to as the external and given control function that belongs to the following admissible set

$$U_{adm} = \left\{ u : \|u\|^2 \leq u_0 + u_1\|x\|^2, \ u_0, u_1 \in \mathbb{R}^+ \right\} \qquad (2)$$

This admissible set allows to introduce linear feedbacks or bounded as the sliding mode controllers. The nonlinear uncertain vector function $f : \mathbb{R}^n \times \mathbb{R}^m \to \mathbb{R}^n$ describes the system dynamics and satisfies the following inequality.

$$\|f(x,u,t)\|^2_{A_\xi} \leq f_0 + f_1\|x\|^2 \ , \ \forall t \geq 0 \ \ \gamma_0, \gamma_1 \in \mathbb{R}^+ \ \ \ 0 < \Lambda_\xi = \Lambda_\xi^\top \in \mathbb{R}^{n \times n} \qquad (3)$$

This inequality justifies the approximation of $f$ by a neural network structure based on sigmoid functions as proposed by Cybenko [4].

## 2.2   Dynamic Neural Network Approximation

Taking into account the approximation properties of neural networks, the nonlinear function $f(x,u)$ in (1) always could be presented (based on the Stone-Weirsstrass and the Kolmogorov theorems [3]) as the composition of a nominal system $f_0(x, u \mid W_0) : \mathbb{R}^n \times \mathbb{R}^m \to \mathbb{R}^n$ and their corresponding modeling error term $\tilde{f}(x, u \mid W_0) : \mathbb{R}^n \times \mathbb{R}^m \to \mathbb{R}^n$. Therefore the following equivalence is always valid:

$$f(x(t), u(t)) = f_0\left(x(t), u(t) \mid W_0\right) + \tilde{f}\left(x(t), u(t) \mid W_0\right) \qquad (4)$$

where $W_0$ is known as the best attainable weights of the approximated neural network. The so-called nominal dynamics can be freely selected according to a predefined methodology (in this case, it has been used a neural network with continuous dynamics). The set of bounded parameters $W_0$ should be adjusted to obtain the best possible matching between the nominal $f_0(x, u \mid W_0)$ and the nonlinear dynamics $f(x,u)$.

Considering that the nonlinear dynamics $f(x,u)$ is locally-Lipschitz and under the class of admissible controls $u(t)$, the following upper bound for the error modeling $\tilde{f}(x, u \mid W_0)$ can be obtained:

$$\|\tilde{f}(x, u \mid W_0)\|^2_{A_f} \leq \tilde{f}_0 + \tilde{f}_1\|x\|^2 \quad \tilde{f}_0 \in \mathbb{R}^+, \tilde{f}_1 \in \mathbb{R}^+ \qquad (5)$$

According to the DNN theory, [6,13], the nominal dynamics can be selected as:

$$f_0(x, u \mid W_0) = Ax + W_0 \Psi(x, u) \qquad (6)$$

where $W_0 = \begin{bmatrix} W_{0,1} & W_{0,2} \end{bmatrix}$, $\Psi(x, u) = \begin{bmatrix} \Psi_1^\top(x) & (\Psi_2(x)u)^\top \end{bmatrix}^\top$. The matrices in the previous equation are $A \in \mathbb{R}^{n \times n}$, $W_{0,1} \in \mathbb{R}^{n \times l}$, $W_{0,2} \in \mathbb{R}^{n \times s}$, $\Psi_1(\cdot) \in \mathbb{R}^l$ and $\Psi_2(\cdot) \in \mathbb{R}^{s \times m}$ ($p = l + s$). In general, the weights $W_{0,1}$ and $W_{0,2}$ are unknown but bounded and they may depend on time.

The corresponding activation functions are represented by $\Psi_1 : \mathbb{R}^n \to \mathbb{R}^l$ and $\Psi_2 : \mathbb{R}^n \to \mathbb{R}^{s \times m}$. The components of the activation vector fields $\Psi_1$ and $\Psi_2$ are

$$\Psi_{1r}(x) = a_r \left(1 + b_r e^{\left(-c_r^\top x\right)}\right)^{-1} \quad \Psi_{2r,p}(x) = a_{r,p} \left(1 + b_{r,p} e^{\left(-c_{r,p}^\top x\right)}\right)^{-1} \qquad (7)$$

The parameters $a_r$, $a_{r,p}$, $b_r$ and $b_{r,p}$ are positive scalars while $c_r \in \mathbb{R}^n$ and $c_{r,p} \in \mathbb{R}^n$ are vectors of gains that shall be selected during the training process. Certainly, these functions satisfy the following sector conditions:

$$\|\Psi_1(z_1) - \Psi_1(z_2)\|^2_{\Lambda_{\Psi_1}} \leq L_{\Psi_1} \|z_1 - z_2\|^2$$

$$\|\Psi_2(z_1)u - \Psi_2(z_2)u\|^2_{\Lambda_{\Psi_2}} \leq L_{\Psi_2} \|z_1 - z_2\|^2 \|u\|^2$$

$$(8)$$

where $L_{\Psi_1}$ and $L_{\Psi_2}$ are positive constants while $\Lambda_{\Psi_1}$ and $\Lambda_{\Psi_2}$ are positive definite matrices of appropriate dimensions. Here $z_1, z_2 \in X \subseteq \mathbb{R}^n$ and their functions are globally bounded in $\mathbb{R}^n$, that is $\|\Psi_1(z_1)\| \leq L^+_{\Psi_1}$   $\|\Psi_2(z_1)u\| \leq L^+_{\Psi_2}\|u\|$.

## 2.3    Problem Formulation

Let consider the equivalent representation of the DNN (6)

$$f_0(x, u \mid W_0) = Ax + \bar{\Psi}(x,u)W_{0,v} \tag{9}$$

with $\bar{\Psi}(x,u) = I_n \otimes \Psi^{\mathsf{T}}(x,u)$ is the so-called extended activation matrix. Here the symbol $\otimes$ represents the Kronecker operator [12]. The symbol $I_n$ represents the identity matrix of dimensions $n \times n$. The vector $W_{0,v} \in \mathbb{R}^{np}$ is the vectorized form of the matrix $W_0$ that is

$$W_{0,v} = vec(W_0) \tag{10}$$

where $vec$ represents the vectorizing operator, i.e.:

$$W_{0,v} = \begin{bmatrix} W^{\mathsf{T}}_{0,1} & W^{\mathsf{T}}_{0,1} & W^{\mathsf{T}}_{0,3} & \cdots & W^{\mathsf{T}}_{0,n} \end{bmatrix}^{\mathsf{T}} \tag{11}$$

Here $W_{0,j}$ is the j-th column of $W_0$.

**Definition 1.** *Commonly, $W_{0,v}$ can be obtained as the solution that minimizes the distance $d(x)$ (over the set $X$ that introduces the local solution of the approximation). In this study, an averaged distance is proposed as part of the optimization problem, that is:*

$$W^*_{0,v}(t) = \operatorname*{argmin}_{W_{0,v} \in \mathbb{R}^{np}} J(t)$$

$$J(t, W_{0,v}(t)) = \frac{1}{t+\varepsilon} \int\limits_{\tau=0}^{t} \|\Delta(\tau)\|^2 \, d\tau, \;\; \varepsilon > 0 \tag{12}$$

*where $\Delta \in \mathbb{R}^n$ is the identification error dynamics, that is*

$$\Delta(t) = x(t) - \int\limits_{\tau=0}^{t} \left( Ax(\tau) + \bar{\Psi}(x(\tau), u(\tau))W_{0,v}(\tau) \right) d\tau$$

The approximate structure proposed in (6) has been tested in [13] to design adaptive observers and controllers. In (6), the set of parameters $W_0$ is assumed to be unknown and bounded. Lets consider the following assumptions for the design of the DNN learning laws:

**Assumption 1**: The vector $x(t)$ is available (physically measurable) at any time $t \geq 0$;

**Assumption 2**: The loss functional $J(t)$ is not completely known, convex (not obligatory, strongly convex), differentiable for almost all $x \in \mathbb{R}^n$ and $\nabla J(t)$ with respect to $W_{0,v}$ is supposed to be measurable at any point $x(t)$.

**Assumption 3**: The minimum of the loss functional $J(t)$ with respect to $W_{0,v}$ exists for each time, namely

$$W_{0,v}^*(t) = \underset{W_{0,v} \in \mathcal{W}_{0,adm} \subset \mathbb{R}^{np}}{\mathrm{argmin}} J(t, W_{0,v}(t))$$

$$\mathcal{W}_{0,adm} = \left\{ W_{0,v} = [W_{0,v,i}]_{i=\overline{1,np}}, \ W_{0,v,i} \in \left[ W_i^-, W_i^+ \right] \right\}$$

where $W_{0,v,i}$ is the i-th component of $W_{0,v}$ (The $\mathcal{W}_{0,adm}$ defines a box-type class of restrictions for $W_{0,v}$), so that $J^*(t, W_{0,v}(t)) = \underset{W_{0,v} \in \mathbb{R}^{np}}{\min} J(t, W_{0,v}(t)) > -\infty$ with a fixed $t$.

## 3   Identification Problem Based on DNN

Lets try to find the solution of the optimization problem (12) applying the extended variation of DASGDT. Then, the learning method to adjust the weights of the DNN structure in (6) is presented as follows

$$\mu(t) \frac{d}{dt} \hat{W}_{0,v}(t) + W_{0,v}(t) = \varsigma(t), \quad \varsigma \in \mathbb{R}^{np}$$

$$\frac{d}{dt} \varsigma(t) = -\nabla_W J(t, W_{0,v}(t)), \quad \varsigma(0) = 0 \tag{13}$$

$$\hat{W}_{0,v}(0) = 0, \ W_{0,v}(0) = \varsigma(0) \ - \varepsilon \frac{d}{dt} \hat{W}_{0,v}(0)$$

Here $\hat{W}_{0,v}$ is the projected vector of weights to the admissible set of weights $\mathcal{W}_{0,adm}$. The projection algorithm satisfies $W_{0,v}^{(i)}(t) = \left[ \hat{W}_{0,v}^{(i)}(t) \right]_{W_i^-}^{W_i^+}$, $i = 1, ..., np$. In (13) $\varsigma$ is the auxiliary variable inserted in integral form to the learning algorithm. The time dependent function $\mu(t) = t + \varepsilon$, $\varepsilon > 0$. As usual, $\nabla_W J$ represents the gradient with respect to $W_{0,v}$ of $J$. Notice here that a double integration process of the identifier error yields the estimation of the associated weights. This procedure yields to obtain a smooth filtering of the functional gradient with respect to the DNN weights.

The parameter $\nabla_W J$ in (13) is a convex, continuously differentiable functional $J : \mathbb{R}^+ \times \mathbb{R}^{np} \to \mathbb{R}^+$ having the following conjugate (Legendre-Fenchel transformation) [9].

$$\bar{H}(W_{0,v}(t)) = \underset{\varsigma \in \mathbb{R}^n}{\sup} \left\{ \langle \varsigma, W_{0,v}(t) \rangle - \frac{\|\varsigma\|^2}{2} \right\} \tag{14}$$

Based on the arguments proposed by [9], the algorithm presented in (13) is called the Method of Inertial Mirror Descent (MIDM). Assume that there exists a solution $W_{0,v}(t)$, $t \geq 0$ to the system of differential equations (13), namely $W_{0,v}^*$. Then, consider the functional

$$H(W_{0,v}^*(t)) = \frac{\|\varsigma\|^2}{2} - \langle \varsigma(t), W_{0,v}^*(t) \rangle, \quad \varsigma \in \mathbb{R}^{np}$$

Lets try to prove that a fixed-point for $H(W_{0,v}^*(t))$ at each time exists. The solution of this problem is gotten by the application of the second stability method of Lyapunov [5]. Lets take the time-derivative of $H(W_{0,v}^*(t))$

$$\frac{d}{dt}H(W_{0,v}^*(t)) = \left\langle \frac{d}{dt}\varsigma(t), \varsigma(t) \right\rangle - W_{0,v}^*(t) \right\rangle \tag{15}$$

Considering the algorithm proposed in (13), $\frac{d}{dt}H^*(t)$ can be calculated as follows

$$\frac{d}{dt}H^*(W_{0,v}(t)) = \left\langle -\nabla J(t, W_{0,v}(t)), \left( \mu(t)\frac{d}{dt}\hat{W}_{0,v}(t) + W_{0,v}(t) - W_{0,v}^*(t) \right) \right\rangle \tag{16}$$

Based on the convexity of $J(t)$ (non-strict), the expression in (16) satisfies

$$\frac{d}{dt}H^*(W_{0,v}(t)) \leq \left[ J(t, W_{0,v}^*(t)) - J(t, W_{0,v}(t)) \right] - \mu(t)\frac{d}{dt}\left[ J(t, W_{0,v}(t)) - J(t, W_{0,v}^*(t)) \right] \tag{17}$$

The integration (by-parts) of (17) on the interval $[0,t]$ with the assumption that $H^*(0) = 0$, yields

$$\int_{\tau=0}^{t} \left[ J(\tau, W_{0,v}(\tau)) - J(\tau, W_{0,v}^*(\tau)) \right] d\tau \leq -\mu(t)\left[ J(t, W_{0,v}(\tau)) - J(t, W_{0,v}^*) \right]_{\tau=0}^{t} -$$

$$H^*(W_{0,v}(t)) + \int_{\tau=0}^{t} \left[ J(t, W_{0,v}(\tau)) - J(t, W_{0,v}^*) \right] \left[ \frac{d}{dt}\mu(t) \right] dt \tag{18}$$

Considering the expression in (14) we obtain

$$\int_{\tau=0}^{t} \left[ J(\tau, W_{0,v}(\tau)) - J(\tau, W_{0,v}^*(\tau)) \right] d\tau \leq -H^*(W_{0,v}(t)) -$$

$$\mu(t)\left[ J(t, W_{0,v}(t)) - J(t, W_{0,v}^*) \right] + \mu(0)\left[ J(0, W_{0,v}(0)) - J(0, W_{0,v}^*) \right] +$$

$$\sup_{s \in [0,t]}\left[ \frac{d}{dt}\mu(s) \right] \int_{\tau=0}^{t} \left[ J(\tau, W_{0,v}(\tau)) - J(\tau, W_{0,v}^*(\tau)) \right] d\tau$$

Since $\dot{\mu}(t) = 1$ and in view of the relation (14), $-H^*(W_{0,v}(t)) \le \bar{H}(W_{0,v}^*(t))$. In consequence:

$$J(t, W_{0,v}(t)) - J(t, W_{0,v}^*(t)) \le \frac{\bar{H}(W_{0,v}^*(t))}{t + \varepsilon}$$

Noticing that

$$\bar{H}(W_{0,v}^*(t)) = \sup_{\varsigma \in \mathbb{R}^n} \left\{ \langle \varsigma, W_{0,v}^*(t) \rangle - \frac{\|\varsigma\|^2}{2} \right\} = \|W_{0,v}^*(t)\|^2$$

Consider the admissible set for the weights

$$\|W_{0,v}^*(t)\|^2 \le \sum_{i=1}^{np} \left( \max\{|W_i^-|, |W_i^+|\} \right)^2$$

This inequality implies that

$$J(t, W_{0,v}(t)) - J(t, W_{0,v}^*(t)) \le \frac{\sum_{i=1}^{np} \left( \max\{|W_i^-|, |W_i^+|\} \right)^2}{(t + \varepsilon)}$$

In consequence, $\lim_{t \to \infty} \left( J(t, W_{0,v}(t)) - J(t, W_{0,v}^*(t)) \right) = 0$.

## 4     Numerical Simulations: Photocatalytic Ozonation System

The proposed learning method for DNN was tested on a particular class of system describing the complex interaction between ozone and catalysts that may be activated by external photons. This system is known as photocatalytic ozonation, which has been applied for removing a large diversity of toxic, health-threatening contaminants [14].

Recent environmental concerns have led to the development of new treatment methods such as advanced oxidation processes (AOPs) [8]. The AOPs are physic-chemical methods capable of changing the chemical structure of the pollutant resulted from the species generation with a high oxidizing power [1]. Among the AOPs, the photocatalytic ozonation takes the advantages of both processes: ozonation and photocatalysis. An extension to the classical ozonation mathematical model [14], taking into account the presence of intermediate and final products, must consider the effect of catalyst with irradiation. The proposed model has the following mathematical structure:

$$\frac{d}{dt} c_t^g(t) = -\frac{[u_1(t) - u_2(t)]}{V_g} + k_{sat}(Q_{\max} - Q(t)) - Q(t) \sum_{i=1}^{N} k_i c_i(t) \tag{19}$$

$$\frac{d}{dt}Q(t) = k_{sat}(Q_{\max} - Q(t)) - Q(t)\sum_{i=1}^{N} k_i c_i(t) - k_{OH,F}(t)Q(t) \qquad (20)$$

$$\frac{d}{dt}c_i(t) = -k_{i,O_3}V_{liq}^{-1}c_i(t)Q(t) - k_{i,OH}c_i(t)c_{OH\cdot}(t) \qquad (21)$$

$$\frac{d}{dt}c_{OH\cdot}(t) = k_{OH,F}(t)V_{liq}^{-1}c_{OH\cdot}(t)Q(t) - \sum_{i=1}^{N} k_{i,O_3}c_i(t)c_{OH\cdot}(t) \qquad (22)$$

The new variables and parameters included in the model that consider the intermediates and final products are: $c_0^g$, $c_t^g$ are the initial and current ozone concentrations in the gas phase $(mole \cdot L^{-1})$, $W_g$ is the gas flow-rate $(L \cdot s^{-1})$, $V_g$ is the volume of the gas phase$(L)$, $Q_t$ is the current ozone amount in liquid phase $(mole)$, $V_{liq}$ is the volume of the liquid phase$(L)$, $Q_{\max}$ is the maximum amount of ozone in the saturation state of the liquid phase at a fixed temperature $(mole)$, $k_{sat}$ is the saturation constant of ozone in water $(s^{-1})$, $c_i$ is the contaminant concentration in the reactor $([mole \cdot L^{-1}])$. The components associated to the catalytic effect are: $c_{OH\cdot}$ is the concentration of OH radicals formed by the catalyst activity in the ozonation, $k_{OH,F}$ is the reaction rate constant characterizing the formation of $OH\cdot$ from $OH^-$ reacting with the dissolved ozone and modulated by photons $[L \cdot mole^{-1} \cdot s^{-1}]$; $k_{i,O_3}$ is the reaction rate constant characterizing the decomposition of compound $c_i$ by ozone $([L \cdot mole^{-1} \cdot s^{-1}])$, $k_{i,OH}$ is the reaction rate constant characterizing the decomposition of compound $c_i$ by $OH\cdot$ $([L \cdot mole^{-1} \cdot s^{-1}])$. Notice that the photocatalytic effect can be characterized by the new time-varying reaction rate parameter of the $OH\cdot$ formation $(k_{F,OH}(t))$ as follows

$$k_{OH,F}(t) = \tilde{k}_{F,OH}\frac{e^{-a(\lambda(t)-\lambda_0)^2}}{1 + be^{-c(I(t)-I_0)}},$$

where $\tilde{k}_{F,OH}$ is the reaction rate constant of the $OH\cdot$ formation at $\lambda(t) = \lambda_0$ and $I(t) \to \infty$, $\lambda$, is the current wavelength of the light used to stimulate the catalyst $(nm)$, $\lambda_0$ is the characteristic wavelength of the catalyst $(nm)$, $a$ is a constant characterizing the variation of the gain with respect to the wavelength $(m^{-2})$, $I$ is the current irradiation of the light applied over the catalyst $(E \cdot m^{-2} \cdot s^{-1})$, $I_0$ is the irradiation corresponding to the light saturation of the catalyst $(E \cdot m^{-2} \cdot s^{-1})$, $b$ is a characteristic parameter that defines the saturation property of the catalyst with respect to the light intensity $(E^{-1} \cdot m^2 \cdot s^1)$.

The numerical simulations were made in Matlab/Simulink. The integration algorithm was ODE-4 with fixed integration step of 0.0001 s. For comparison purposes, a class of Hopfield DNN identified the time variation of the photocatalytic ozonation system. The trajectories of the Hopfield network exhibited the regular oscillatory states which are commonly observed for such networks.

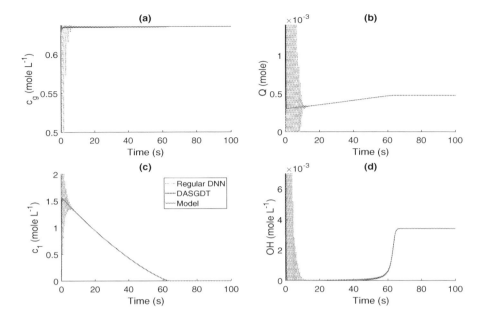

**Fig. 1.** States comparison of the catalytic model as well as the estimates by the DNN identifier with the MDA method aimed to estimate the learning law.

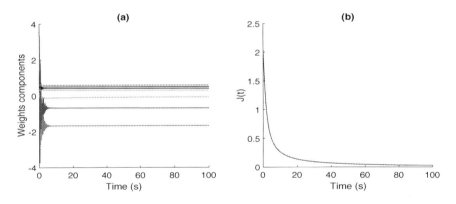

**Fig. 2.** Weights variations obtained with the MDA method (a) and the performance index $J(t)$ (b).

Figure 1 shows the time variation of the states included in the photocatalytic ozonation (19–22). This figure shows the time dependence of the estimated states obtained by the DNN identifier using the learning laws based on the DASGDT method. The weights bounds for the projection algorithm were fixed to $W_i^+ = 10000$, $W_i^- = -10000 \ \forall i \in [1, np]$. The trajectories produced by the DASGDT show smaller oscillations during the analyzed period of 100 s. Notice also that oscillations stop after 1.0 s.

Figure 2-a demonstrates the variations of some components of the weights estimated by the algorithm proposed in (13). The oscillations obtained during the estimation of the states are consequence of the large weights variations observed in the first 10 s. Figure 2-b depicts the variation of the performance index $J$. This variable has the expected variation tending to the zero asymptotically.

## 5   Conclusions

This study considers the design of a new learning law for DNN based on the application of a new variant of the GDM named DASGDT. The explicit idea of how implementing the DASGDT in continuous NN is described, which clarifies how to define the structure of the DASGDT method. The strict analysis for obtaining the upper bound of the method and to justify the asymptotic convergence of the performance index. The proposed method opens new researching opportunity to design new learning laws for a wide diversity of NN, even if they do not have continuous dynamics or they are not deterministic (stochastic DASGDT).

## References

1. Andreozzi, R., Caprio, V., Insola, A., Marotta, R.: Advanced oxidation processes (AOP) for water purification and recovery. Catal. Today **53**(1), 51–59 (1999)
2. Beck, A., Teboulle, M.: Mirror descent and nonlinear projected subgradient methods for convex optimization. Oper. Res. Lett. **31**(3), 167–175 (2003)
3. Cotter, N.: The stone-weierstrass theorem and its application to neural networks. IEEE Trans. Neural Netw. **1**, 290–295 (1990)
4. Cybenko, G.: Approximation by superpositions of sigmoidal function. Math. Control Sig. Syst. **1989**(2), 303–314 (1989)
5. Haddad, W., Chellaboina, V.: Nonlinear Dynamical Systems and Control. Princeton University Press, Princeton (2008)
6. Haykin, S.: Neural Networks and Learning Machines. Prentice Hall, Upper Saddle River (2009)
7. Kim, C.T., Lee, J.J.: Training two-layered feedforward networks with variable projection method. IEEE Trans. Neural Netw. **19**(2), 371–375 (2008)
8. Malato, S., Oller, I., Fernández-Ibánez, P., Fuerhacker, M.: Technologies for advanced wastewater treatment in the mediterranean region. In: Barcelá, D., Petrovic, M. (eds.) Waste Water Treatment and Reuse in the Mediterranean Region, pp. 1–28. Springer, Heidelberg (2010). https://doi.org/10.1007/698_2010_59
9. Nazin, A.V.: Algorithms of inertial mirror descent in convex problems of stochastic optimization. Autom. Remote Control **79**(1), 78–88 (2018)
10. Nemirovsky, A., Yudin, D.: Problem Complexity and Optimization Method Efficiency. Nauka, Moscow (1979)
11. Pérez-Sánchez, B., Fontenla-Romero, O., Guijarro-Berdiñas, B.: A review of adaptive online learning for artificial neural networks. Artif. Intell. Rev. **49**(2), 281–299 (2018)
12. Poznyak, A.: Advanced Mathematical Tools for Automatic Control Engineers: Volume 1: Deterministic Systems. Elsevier Science, Amsterdam (2008)

13. Poznyak, A., Sanchez, E., Yu, W.: Differential Neural Networks for Robust Nonlinear Control (Identification, State Estimation and Trajectory Tracking). World Scientific, Singapore (2001)
14. Poznyak, T., Chairez, I., Poznyak, A.: Ozonation and Biodegradation in Environmental Engineering. Dynamic Neural Network Approach. Elsevier, Amsterdam (2019)

# Better Performance of Memristive Convolutional Neural Network Due to Stochastic Memristors

Kechuan Wu[1,2], Xiaoping Wang[1,2(✉)], and Mian Li[1,2]

[1] School of Artificial Intelligence and Automation,
Huazhong University of Science and Technology, Wuhan, China
wangxiaoping@hust.edu.cn

[2] Key Laboratory of Image Processing and Intelligent Control of Education Ministry
of China, Wuhan 430074, People's Republic of China

**Abstract.** Convolutional Neural Network (CNN) has gotten admirable performance in the domain of image recognition. Nevertheless, the training of CNN on CPU or GPU is energy-intensive and time-consuming. Memristor crossbar is an alternative of the specific chip for CNN application. But it is hard to tune the memristor to certain conductance precisely. This work simulates the performance change of memristor-based CNN when memristor is with stochasticity. The simulation results demonstrate that stochastic memristor-based CNN performs better on CIFAR-10 dataset when memristive stochasticity is low. This is an encouragement for the engineer of memristor crossbar chip and edge computing application.

**Keywords:** Stochastic memristor · Convolutional neural network · Dataset noise

## 1 Introduction

In most scenarios, CNN and its variations are the most effective methods for computer vision applications [1]. However, using current general-purpose computers for training is extremely time-consuming and energy-intensive [2]. Modern society is developing rapidly, and people are increasingly demanding functions for electronic products such as cameras, mobile phones, and devices in Internet of Thing. Therefore, artificial neural network applications are playing an increasingly important role. But for portable devices such as cell phones, the cost of using neural network applications is very high. This situation may be changed using new processing devices. Memristor [3,4] is a possible choice.

This work was supported by the National Natural Science Foundation of China under grant 61876209 and the National Key R&D Program of China under Grant 2017YFC1501301.

H. Lu et al. (Eds.): ISNN 2019, LNCS 11554, pp. 39–47, 2019.
https://doi.org/10.1007/978-3-030-22796-8_5

Memristors have many advantages such as simple structure, easy for integration, low power, good compatibility with CMOS [5], and fast state switching [6]. Multilayer Perceptron based on memristors can greatly reduce area and power consumption [7]. Many CNN architecture is realized by memristors. Zeng *et al.* proposed an architecture based on binary memristors for image convolution in CNN [8]. Liu *et al.* proposed a 3D CNN based on memristors for classifying the behaviors of human in video [9]. Yakopcic *et al.* proposed an extremely parallel memristor crossbar for CNN [10]. However, the memristive devices' errors [11,12] impact the CNN accuracy, which is rarely been studied. Several noise sources have been presented in [11]. The memristor model will be too complex if we consider all the noise sources. The directly relevant sources mostly can be modeled by a zero-mean Gaussian distribution. Besides, the distribution of memductance adjustment errors conforms to a Gaussian distribution in [13]. The effect of the memristive stochasticity on classification accuracy of CNN is unknown.

In this paper, we study how the stochasticity in memristors impacts the performance of a CNN. The simulation is on CIFAR-10 dataset [14]. The rest of this paper is organized as follows. Section 2 introduces the background of memristor model and network architecture. Section 3 describes the memristive crossbar implementation of CNN. Simulation results are shown in Sect. 4. Lastly, the conclusion is presented in Sect. 5.

## 2   Background

### 2.1   Memristor

The memristor was postulated by L. O. Chua from symmetry arguments in 1971 [3]. Since it was produced by HP laboratory in 2008 [4], it is regarded as one of the next generations of computing devices. A voltage-controlled memristive system, is generally represented by

$$i(t) = G(w(t))v(t), \tag{1}$$

$$\dot{w}(t) = f(v, w, t), \tag{2}$$

where $i$ is the current flowing through the memristor, and $v$ is the voltage across the memristor. $G$ is the conductance of the memristor. $w$ is an internal variable. A memristor is fabricated in nanometer size and can be easily integrated. The conductance of memristors can be tuned to a range of continuous value, which means memristors can be treated as weights in the artificial neural network. Memristors maintain conductance after power off, which makes memristive neural network efficient.

### 2.2   CNN Architecture

The CNN is usually designed to imitate human visual processing. Compared with fully connected neural networks, CNN has fewer parameters and is easier

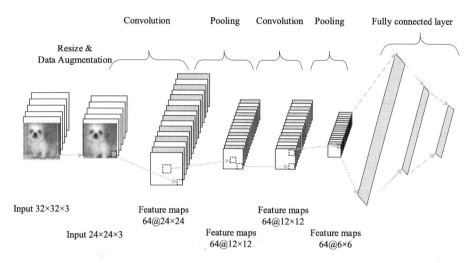

Convolution    Pooling   Convolution   Pooling   Fully connected layer

Resize &
Data Augmentation

Input 32×32×3

Feature maps
64@24×24

Input 24×24×3

Feature maps
64@12×12

Feature maps
64@12×12

Feature maps
64@6×6

**Fig. 1.** CNN block diagram. The upper is the name of the operations, and the lower is the data format.

to escape from vanishing gradients problem [15]. Figure 1 shows the main process of a CNN application. Input data is at the leftmost part of this figure, which is resized and augmented before adding to the network. The actions of data augmentation can be randomly flipping the pictures and randomly cropping the pictures, which force CNN to learn more essential feature.

The CNN contains three different layers: convolution layer, pooling layer, and fully connected layer. Convolution layer is the most important layer of a CNN. This architecture makes network concentrate on low-level features in the previous layer and assemble low-level feature into high-level features in the latter layers. The goal of pooling layers is to subsample the feature maps. This will reduce the computation burden. Convolution and pooling layer is used for feature exaction. Fully connect layer is used for classification like the situation in multilayer perceptron.

## 2.3   Training

To compare with the stochastic-memristor based CNN, the CNN was trained and tested firstly in software. The dataset used in this paper is CIFAR-10. The CIFAR-10 dataset contains 60000 $32 \times 32$ color images in 10 classes. The classes in the dataset are airplane, automobile, bird, cat, deer, dog, frog, horse, ship, and truck. Each class has 6000 images. There are 50000 images for training and 10000 images for test.

The training step is 100000. Each step has a batch of 128 images. The loss minimization is shown in Fig. 2. The training steps update CNN parameters. The CNN parameters (weights and kernel values) could be programmed into memristive crossbar after training [10] or updated in memristive crossbar directly [7].

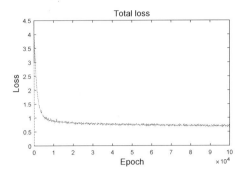

**Fig. 2.** Loss curve of the CNN in Fig. 1 implemented on software.

The following work describes the simulation of CNN implemented by stochastic memristor crossbar.

## 3  Memristor Crossbar Implementation

### 3.1  Stochastic Memristor

Memristors are a nanoscale device. Though the conductance of memristors is continuous, we can not set the conductance to the desired value precisely at one step. Feedback write scheme is used in [16,17]. In on-line neural network applications, the feedback write scheme will consume a lot of time in the training process. This paper simulated the stochastic memristor, which can not be programmed to desired value precisely, based CNN's performance. And the resulting classification accuracy under different dataset noise level is compared with that of the software approach.

There are many reasons why a memristor cannot be programmed to the desired value precisely. Feinberg *et al.* listed lots of them [11]. Mostly, the distribution of noise source, such as shot noise, thermal noise, conforms to Gaussian distribution. Thus we add a Gaussian variable to the model of the stochastic memristor, i.e.,

$$i(t) = G(w(t))(1 + s(t))v(t), \tag{3}$$

where s(t) is a Gaussian variable with a mean of zero and a standard deviation of $\sigma$. Every time the weights changes, the value of $s$ changes to simulate the stochasticity of memristors. Note that the random number seed in each memristor model is different.

### 3.2  Memristive Crossbar Structure

The memristors in crossbar are used to represent weights of neural networks. Thus the different memristive crossbar structure does not affect the classification result of CNN. Several memristive crossbar structures have been proposed,

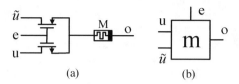

**Fig. 3.** (a) Structure of a single 2T1M and (b) its schematic diagram.

such as fully memristive crossbar [18], two transistors and a memristor crossbar (2T1M) [7], and so on.

In this paper, the 2T1M structure crossbar is elaborated. A single 2T1M structure is shown in Fig. 3(a), where $\tilde{u}$ is the inverse voltage of $u$. The $e$ port controls the states of the two MOSFETs. The upper MOSFET is a P-channel MOSFET, and the lower one is N-channel MOSFET. For the convenience of the following description, Fig. 3(a) is simplified as Fig. 3(b).

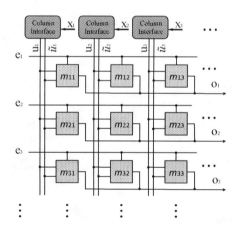

**Fig. 4.** The structure of 2T1M crossbar. The $m_{ij}$ cell is a 2T1M structure shown in Fig. 3(b). $e_i$ determines the states of MOSFETs in row $i$. $x_j$ determines the speed of conductance change in column $j$. The output currents can be sensed at $o$.

The structure of memristive crossbar is shown in Fig. 4. $m_{ij}$ represents the 2T1M structure in Fig. 3(b) at row $i$, column $j$. The amplitude of $e_i$ should be large enough to keep one of the two MOSFETs open. $e_i$ controls the states of MOSFETs in the whole row. That is to say, $e_i$ determines whether the conductance increases or decreases. The column interface converts a number $(x_j)$ to voltage ($u$ and $\tilde{u}$). $x_j$ determines the speed of the conductance change at the $j$-th column. During the weight update phase, The values of $e$ and $x$ are gotten using the BackPropagation algorithm. During the feedforward phase, the memristive conductance should be read to realize the computation of matrix-vector multiplication. In the read phase, $e$ ports stay positive for a little while, and then stays negative for the same amount of time. The time should be long enough to make sure that the current at port $o_i$ is correctly outputted, and not so long that the conductance changes too much.

The fully connected layer and convolution layer both can be implemented on memristive crossbar [10]. If we implement the CNN structure in Fig. 1 on Simulink or SPICE, it costs a lot of computer resources, especially training CNN online. Since the memristors in crossbar represent the weights in CNN, the stochasticity of the memristors is equivalent to the stochasticity of the weights in CNN. Thus we did this simulation on software.

**Table 1.** CNN parameters in simulation.

| | |
|---|---|
| Filter of 1st convolution layer | [5, 5, 3, 64] |
| Filter of 2nd convolution layer | [5, 5, 64, 64] |
| Strides of convolution layer | [1, 1] |
| Padding | SAME (zero padding) |
| Activation | ReLU |
| Strides of pooling layer | [2, 2] |
| Fully connected layer size | $384 \times 192 \times 10$ |

In the simulation on software, the hyper-parameters of the CNN is shown in Table 1. The numbers of convolution layer filter parameters stand for [filter height, filter width, input channels, output channels], respectively. The activation for all layers is ReLU. The strides of the two pooling layers are the same. The CNN structure has three fully connected layers, the number of neurons in these three layers is 384, 192, 10.

## 4   Simulation Results

We programmed some tensorflow code to get the results of the CNN shown in Fig. 1 on CIFAR-10 dataset. In memristive CNN, the conductance of memristors represents the weights. The stochasticity of memristors can be taken as weights' noise in the tensorflow model. The value of the random number changes during each weight update to simulate the stochasticity in memristive conductance update.

On the original dataset, the result of different weights' randomness ($\sigma = [0, 0.1, 0.2, 0.3]$) is shown in Table 2. When the memristive stochastic level is 0, the CNN's weights are not multiplied by Gaussian variables. When the memristive stochastic level is lower than 0.3, the classification accuracy is slightly increased. If the memristive stochastic level is higher, the classification accuracy goes down.

The results of different weights' randomness ($\sigma = [0, 0.1, 0.2, 0.3]$) is got. At a specific weights' randomness, the dataset was added with different levels of Gaussian noise. After the CNN is trained, it predicted the samples in the test set

**Table 2.** The classification accuracy on original dataset.

| Memristive stochastic level ($\sigma$) | Classification accuracy |
| --- | --- |
| 0 | 86% |
| 0.1 | 86.4% |
| 0.2 | 86.5% |
| 0.3 | 86.4% |
| 0.4 | 85.9% |
| 0.5 | 85.6% |

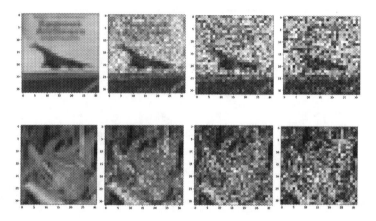

**Fig. 5.** Sample pictures with noise. From left to right, the dataset noise is 0, 0.2, 0.4, 0.6. The label of the pictures in the upper row is the *bird*. The label of the pictures in the lower row is the *frog*.

which has the same level of Gaussian noise in the train set. The effect of adding different levels of Gaussian noise is shown in Fig. 5. The test process is done ten times to get a reliable classification accuracy because of the randomness of the test set.

The results of classification accuracy are shown in Fig. 6. In the box plot, the x-axis is the value of $\sigma$ which indicates the randomness of weights. Outliers (red plus sign) is plotted as individual points. The median accuracy is the horizontal line in each subplot. The maximum and minimum is the highest value and the lowest value of the vertical line in each subplot. From Fig. 6, we can know that the accuracy is higher when weight randomness level is lower than 0.2. When the weight randomness level is higher than 0.2, the accuracy becomes lower. These results show that the stochasticity of memristors enhances the performance of memristive CNN.

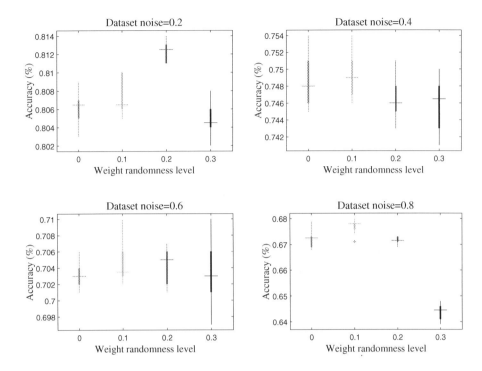

**Fig. 6.** Classification accuracy when dataset is added different levels of noise.

## 5    Conclusion

In this paper, we simulates the impact of memristive stochasticity to memristor-based CNN. It is found that the stochasticity of memristors enhances the CNN classification accuracy when weight randomness is at a low level (lower than 0.2). This is a great encouragement to the CNN applications in Internet of things, edge computing and memristive neural network chips.

## References

1. LeCun, Y., Bengio, Y., Hinton, G.: Deep learning. Nature **521**(7553), 436–444 (2015)
2. Jouppi, N.P., et al.: In-datacenter performance analysis of a tensor processing unit. In: 2017 ACM/IEEE 44th Annual International Symposium on Computer Architecture (ISCA), Toronto, pp. 1–12. IEEE (2017)
3. Chua, L.: Memristor-the missing circuit element. IEEE Trans. Circ. Theory **18**(5), 507–519 (1971)
4. Strukov, D.B., Snider, G.S., Stewart, D.R., Williams, R.S.: The missing memristor found. Nature **453**(7191), 80 (2008)
5. Xia, Q., et al.: Memristor-CMOS hybrid integrated circuits for reconfigurable logic. Nano Lett. **9**(10), 3640–3645 (2009)

6. Yang, J.J., et al.: High switching endurance in $TaO_x$ memristive devices. Appl. Phys. Lett. **97**(23), 232102 (2010)
7. Soudry, D., Di Castro, D., Gal, A., Kolodny, A., Kvatinsky, S.: Memristor-based multilayer neural networks with online gradient descent training. IEEE Trans. Neural Netw. Learn. Syst. **26**(10), 2408–2421 (2015)
8. Zeng, X., Wen, S., Zeng, Z., Huang, T.: Design of memristor-based image convolution calculation in convolutional neural network. Neural Comput. Appl. **30**(2), 503–508 (2018)
9. Liu, J., Li, Z., Tang, Y., Hu, W., Wu, J.: 3D convolutional neural network based on memristor for video recognition. Pattern Recogn. Lett. (2018). https://doi.org/10.1016/j.patrec.2018.12.005
10. Yakopcic, C., Alom, M.Z., Taha, T.M.: Extremely parallel memristor crossbar architecture for convolutional neural network implementation. In: 2017 International Joint Conference on Neural Networks (IJCNN), Alaska, pp. 1696–1703. IEEE (2017)
11. Feinberg, B., Wang, S., Ipek, E.: Making memristive neural network accelerators reliable. In: 2018 IEEE International Symposium on High Performance Computer Architecture (HPCA), Vienna, pp. 52–65. IEEE (2018)
12. Naous, R., Al-Shedivat, M., Salama, K.N.: Stochasticity modeling in memristors. IEEE Trans. Nanotechnol. **15**(1), 15–28 (2016)
13. Li, C., et al.: Analogue signal and image processing with large memristor crossbars. Nat. Electron. **1**(1), 52 (2018)
14. Krizhevsky, A., Hinton, G.: Learning multiple layers of features from tiny images. Technical report, vol. 1, no. 4, p. 7 (2009)
15. Glorot, X., Bengio, Y.: Understanding the difficulty of training deep feedforward neural networks. In: Proceedings of the 13th International Conference on Artificial Intelligence and Statistics, pp. 249–256 (2010)
16. Gao, L., Chen, P.Y., Yu, S.: Programming protocol optimization for analog weight tuning in resistive memories. IEEE Electron Device Lett. **36**(11), 1157–1159 (2015)
17. Merced-Grafals, E.J., Davila, N., Ge, N., Williams, R.S., Strachan, J.P.: Repeatable, accurate, and high speed multi-level programming of memristor 1T1R arrays for power efficient analog computing applications. Nanotechnology **27**(36), 365202 (2016)
18. Zhang, Y., Wang, X., Friedman, E.G.: Memristor-based circuit design for multilayer neural networks. IEEE Transactions on Circuits and Systems I: Regular Papers **65**(2), 677–686 (2018)

# Unsupervised Feature Selection Using RBF Autoencoder

Ling Yu, Zhen Zhang, Xuetao Xie, Hua Chen, and Jian Wang[✉]

College of Science, China University of Petroleum, Qingdao, China
ylingupc@163.com, zhang_zhen1995@163.com, xuetao_xie@163.com
{chenhua,wangjiannl}@upc.edu.cn
http://cilab.sci.upc.edu.cn/

**Abstract.** In this paper, a novel learning approach to solve unsupervised feature selection in high-dimensional data is proposed, namely Radial Basis Function Autoencoder feature selection (RAFS). This method based on autoencoder uses the radial basis function to achieve mapping instead of the weight. We also consider penalty to give a powerful constraint on redundant features. In extensive experiments, our method shows its outperformance in fair comparison with several other methods.

**Keywords:** Unsupervised · Feature selection · Radial basis function · Autoencoder · Penalty

## 1 Introduction

As an effective preprocessing method of data analysis, dimensionality reduction plays an indispensable role in many areas such as genetic engineering [1], text classification [2] and disease diagnosis [3]. It not only makes the learning model less design and decision-making cost, but also often improves the performance of the classifier. Often, Dimensionality reduction technology can be divided into two classes: feature extraction [4,5] and feature selection. Feature extraction means generating the transformed features which associate the original features. Feature selection focuses on finding the most informative features to construct a feature subset, which can preserve the intrinsic structure of the original data in lower dimension. In addition, as the selected features have clear connections to the original ones, feature selection has better interpretability than feature extraction.

According to whether the information of label is used in task or not, feature selection is categorized mainly as supervised [6,7], semi-supervised [8–10] and

This work was supported in part by the National Natural Science Foundation of China under Grant 6130507, in part by the Natural Science Foundation of Shandong Province under Grant ZR2015AL014 and Grant ZR201709220208, and in part by the Fundamental Research Funds for the Central Universities under Grant 15CX08011A and Grant 18CX02036A.

H. Lu et al. (Eds.): ISNN 2019, LNCS 11554, pp. 48–57, 2019.
https://doi.org/10.1007/978-3-030-22796-8_6

unsupervised [17,20]. Due to the class labels, supervised and semi-supervised feature selection, developed to exploit both labeled and unlabeled data simultaneously, are much easier to find the optimal objective function of the learning model and obtain a widely research development [11–14]. However, More high dimensionality data sets generated in practical problems are part-unlabeld even all-unlabeled. Accordingly, unsupervised feature selection has more challenge and practicability. Our paper focuses on it. Without label information, the fundamental issue in unsupervised feature selection is how to model the mainfold geometry structure of the whole feature set and produce a faithful feature subset which preserves the intrinsic structure accurately [15,16]. Several forms to characterize its structure have been proposed, including but not limited to the K-nearest neighbor (KNN) graph [17], Locally Linear Embedding [18] and Sammom's error [19]. On a novel view point, Han et al. [20] proposed a new method AEFS, based on the autoencoder and the group lasso regulation, demonstrating excellent performance.

In recent years, radial basis function neural network (RBFNN) has attracted extensive interests for a wide range of applications [21–24], because of the unusual property of radial basis function. In RBFNN, Radial basis activation functions for the units provide the network with the capability of forming more complex nonlinear mapping in the input space, which allows these data vectors to be mapped to higher dimensional space. As a result, Linear inseparable problem in low dimensional space can become linear separable. In our paper, we leveraged the ability of high-dimensional mapping to evaluate the quality of feature subset. As the salient features of RBFNN, article [25] pointed out that using simple properties of the basis functions and it is shown that a neural network with a single layer of hidden units of Gaussian type is a universal approximator. RBFNN requires besides less computation time for the learning [26], and a more compact topology than other neural networks [27]. Under the circumstances, we can utilize these advantages to deal with unsupervised feature selection problems.

Before the selection, we need to make clear which feature should to selected prior. Following paper [28], features can be classify into four categories: essential features, bad features, indifferent features and redundant features. Almost all of research only notice the first three kinds but skip redundant features and these are all favorable and dependent on each other. So we just need selected some of them.

Inspired by what mentioned above, in our paper, we perform AutoEncoder and RBF simultaneously to select the discriminative features for unsupervised learning, meanwhile, consider the control of redundant features. Experimental results that conducted on different datasets show the superiority of the proposed method.

## 2    The Proposed Method

In this section, we first summarize some methods that are the foundation of the new algorithm, including Autoencoder and RBF. Then the network structure for unsupervised feature selection is ascertained. Finally, we formulate the objective function and the optimization problem of our method.

## 2.1   Theoretical Foundation

Autoencoder is a type of artificial neural network used to learn efficient data codings in an unsupervised manner. Architecturally, the typical autoencoder generally is similar to the single layer perceptron. Mapping from input layer to hidden layer aims to get a lower-dimensional representation of the original data. This process is called encoding. Then transporting from hidden layer to output layer wants to copy its input to its output. This process is named decoding. In general, the network tends to approximate a function to make its output close to its input. Han et al. [20] further considers weight restrictions based on autoencoder to achieve unsupervised feature selection.

RBF is a special function whose value depends only on the distance from the origin, so that $\Phi(\mathbf{x}) = \Phi(\|\mathbf{x}\|)$. Usually, the Gaussian function is preferred among all possible radial basis functions due to the fact that it is factorizable [29]. Hence

$$\Phi(\mathbf{x}) = e^{-\frac{\|\mathbf{x}-c\|^2}{2\sigma^2}} \tag{1}$$

where $c$ and $\sigma$ are the center and the width of the Gaussian kernel, respectively.

Through the mapping of RBF, the network can be regarded as synthesizing an approximation of a set of multidimensional functions, to find a suitable mapping between a given set of patterns and their corresponding classes.

## 2.2   Objective Function

Based on the characteristics of Autoencoder and RBF, we combined them to get a special network structure, named RBFAE, then proposed a novel unsupervised feature selector (RAFS) to learn the optimal feature subset. RBFAE not only has the unsupervised learning ability of AutoEncoder, but also has the high-dimensional mapping power of RBF.

Now, assume that the origin data is $X \in \mathbb{R}^{p \times n}$, where $x_j$ is the $j$th column vector of $X$, i.e. the $j$th samples. In RBF, the number of Gaussian kernel is $q$, the centers are $\mathbf{c} = (c_1, c_2, ..., c_q) \in \mathbb{R}^{p \times q}$; the widths are $\boldsymbol{\sigma} = (\sigma_1, \sigma_2, ..., \sigma_q)$. So the output of the hidden layer is $\mathbf{H} = (H_{ij})_{q \times n}$. Where

$$H_{ij} = e^{\frac{-\|\mathbf{x}_j - \mathbf{c}_i\|^2}{2\sigma_i^2}} \tag{2}$$

Then the network output is $Y = WH$ and $W \in \mathbb{R}^{p \times q}$ is the weight applied to the hidden layer and the output layer. If this is the case, we can use the least square loss as the error function:

$$E = \frac{1}{2n} \sum_{j=1}^{n} \|WH_j - x_j\|^2 \tag{3}$$

Where $H_j$ is the hidden layer output of the $j$th samples.

To measure each feature how useful, some methods, based on neural network, used the Lass or group-Lasso to constrain the weight between the input layer and the hidden layer. Now we use the Guassian kernel function instead of the weight, so we adopt another way and associate gate function with each feature in the input layer, $f(\beta) = (f(\beta_1), f(\beta_2), ..., f(\beta_p))^T$, $\beta_k$ is called the gate parameter. So the output of the hidden layer will become

$$H_{ij} = e^{\frac{-\|(x_j - c_i) \circ f(\beta)\|^2}{2\sigma_i^2}} \tag{4}$$

where $\circ$ represents element-wise product of two vectors. In this paper, our choice of gate function is

$$f(\beta_k) = e^{-\beta_k^2} \tag{5}$$

The vaule of gate function is limited to $[0, 1]$. When $f(\beta_k) = 0$, the $k$th component of the norm, i.e. $(x_{kj} - c_{ki}) \cdot f(\beta_k)$ becomes zero. It means the $k$th feature of the $j$th sample data has no influence on the output of the $i$th node in the hidden layer. On the contrary, the computing will be regular when $f(\beta_k) = 1$.

## 2.3   Structure Determination

The determination of RBFAE network structure design includes the appropriate number and location of the centroids $c$, the widths $\sigma$ of the Gaussian kernels and the weights $w$ linking from the hidden layer to the output layer.

Early time, Powell [30] proposed a pure way, assigning all samples data as centers for strict interpolation in multidimensional space. This way always brings huge computation and even unpractical in high-dimensional data. Mody [26] used the input clustering strategy and regarded the cluster centers as the kernel centers by the K-means method which is facile to the unsupervised learning problems with good performance. So the number of the hidden units(the centers) is equal to the $k$ value of the K-means method. If the value of $k$ is large enough, the K-means method can satisfy the other requirement of Cover's theorem, which means that the dimensionality of the hidden layer is high enough. We therefore conclude that the K-means algorithm is indeed computationally powerful enough to transform a set of nonlinearly separable patterns into separable ones in accordance with this theorem [31]. In this paper, we use the K-means algorithm to confirm the centers location.

Now, we need consider the widths. One choice [32] is that fixing them as follow:

$$\sigma_k = \frac{d_{max}}{\sqrt{2q}}, \forall k = 1, 2, ..., q \tag{6}$$

$$d_{max} = \max_{i,j \leq q} \|c_i - c_j\|_2 \tag{7}$$

where the $d_{max}$ is the maximum distance between any two center points. Nevertheless, As for all Gaussian kernels, the same width is not appropriate to describe their respective curve characteristics and cover the sample data, which is often non-uniform distribution in most real-life problems. So it is better to assign a specific width to each Gaussian kernel. Moody [26] proposed to compute the width factors $\sigma_j$ (the radius of kernel $j$) by the $k$-nearest neighbours heuristic:

$$\sigma_j = \frac{1}{k} \left( \sum_{i=1}^{k} \|c_i - c_j\|_2 \right)^{\frac{1}{2}} \tag{8}$$

where $k$ is the adjustable number of the nearest neighbours to the center $c_j$. In our paper, we set $k = 2$.

Linear output layer and radial basis hidden layer structure of RBFN provide the possibility of learning the connection weights efficiently without local minima problem in a hierarchical procedure so that the linear weights are learned after determining the centers by a clustering process. The final layer of RBF do not use activation function and it rather linearly combines the output of the previous neuron. We exploited the pseudo-inverse method to compute the weights and the formulation is as follow:

$$\mathbf{W} = \mathbf{X} \cdot \mathbf{H}^{-1} \tag{9}$$

It should be pointed out that the pseudo-inverse method to confirm the weights maybe have some computational difficulty in ultra-high dimensional space. The conventional gradient descent method is a good alternative although it need some extra iteration time. We prefer to focus on the error difference of network before and after feature selection, not the error minimization in structure determination stage.

### 2.4  Optimization Learning

In an attempt to further constrain, the penalty is naturally considered. The first penalty, also called sparse item, aims to make the gate value of each feature close to zero or one and avoid too many partially open gates produced. The specific formula as follow:

$$PF_1 = \frac{1}{p} \sum_{i=1}^{p} e^{-\beta_i^2} (1 - e^{-\beta_i^2}) \tag{10}$$

Now to control redundancy, we add anther penalty which can avoid selection of dependent features. If some features are useful but dependent on each other, these features can be named redundant features which are not necessary, and only some are needed to solve the problem.

$$PF_2 = \frac{1}{p(p-1)} \sum_{i=1}^{p} e^{-\beta_i^2} \sum_{j \neq i} \rho_{ij}^2 e^{-\beta_j^2} \tag{11}$$

In (11), $\rho_{ij}$ is a measure of dependency between $x_i$ and $x_j$. In this paper, we use the Pearson's correlation to control linear dependency among features. If it is nonlinear, mutual information will be a good choice. According the penalty function method, we can get the objective function as:

$$TE = E + \mu PF_1 + \nu PF_2 \tag{12}$$

where $\mu$ and $\nu$ are the parameters to control the severity of the penalty.

To begin the learning, the modulators initialization is needed. We set that the random initial value $\beta_i$ between $[1, 1.01]$ to limit the initial gate function value of each feature between $[0.36, 0.37]$. It guarantees that the gates are partly opened at the beginning of learning and every feature has adequate opportunity to be selected or not. Then, the iterative optimization of the value of $\beta$ is done by a gradient descent technique, using the following equations.

$$\frac{\partial E}{\partial \beta_k} = \frac{f(\beta_k) f'(\beta_k)}{2n} \sum_{j=1}^{n} \left[ (x_{kj} - c^k) \circ \frac{1}{\sigma} \right]^2 \cdot \left[ (W^T x_j - W^T W H_j) \circ H_j \right] \tag{13}$$

$$\frac{\partial PF_1}{\partial \beta_k} = \frac{-2\beta_k e^{-\beta_k^2}}{p} (1 - 2e^{-\beta_k^2}) \tag{14}$$

$$\frac{\partial PF_2}{\partial \beta_k} = \frac{-4\beta_k e^{-\beta_k^2}}{p(p-1)} \sum_{i=1, i \neq k}^{p} e^{-\beta_i^2} \rho_{ik}^2 \tag{15}$$

$$\frac{\partial TE}{\partial \beta_k} = \frac{\partial E}{\partial \beta_k} + \lambda_1 \frac{\partial PF_1}{\partial \beta_k} + \lambda_2 \frac{\partial PF_2}{\partial \beta_k} \tag{16}$$

$$\beta_k^{new} = \beta_k^{old} - \lambda \cdot \frac{\frac{\partial TE}{\partial \beta_k^{old}}}{\sqrt{\sum_{i=1}^{p} \left( \frac{\partial TE}{\partial \beta_i^{old}} \right)^2}} \tag{17}$$

It must be noted that $c^k$ is the $k$th row vector of the centers $c$ in (13), and $\lambda$ is the learning coefficient in (17).

## 3   Experiments

The experiments are conducted on 6 classification data sets with very different dimensionality, as summarized in Table 1. In order to find the internal structure of data sets, we just used the original data sets and did not normalize.

To validate the effectiveness of our method RAFS, we compare it with these existing unsupervised feature selection methods.

**AllFea**: All original features are adopted.
**LS**: Features are selected using Laplacian score [17].
**NDFS**: Features are selected using nonnegative spectral analysis which joint the cluster labels and feature selection matrix [33].

---

**Algorithm 1.** The Optimization Algorithm of RAFS

---

**Input:** Dataset:$X \in \mathbb{R}^{p \times n}$; Penalty factors:$\mu, \nu$; learning coefficient:$\lambda$;The number of selected features:$d$.

**Output:** Subset of selected features.

1: Set $\beta_k \ (k = 1, ..., p) = 0$, i.e. all features join into the network and no feature selection.

2: Use K-means algorithm to calculate the centers location $c$.

3: Calculate the width of each kernel $\sigma$ by Eq.(8).

4: Calculate the weight $W$ by Eq.(9).

5: Randomly initialize $\beta_k$ between $[1, 1.01]$.

6: **if** Convergence criterion not satisfied **then**

7:     Compute the total error $(TE)$ by Eq.(3)-(5),(10)-(12).

8:     Compute the new values $\beta_k^{new}$ by Eq.(13)-(17) and get the new error $TE^{new}$

9:     **while** $TE^{new} > TE$ **do**

10:         $\lambda \leftarrow 0.9 \times \lambda$

11:         recompute $\beta_k^{new}$ and $TE^{new}$.

12:     $TE \leftarrow TE^{new}$ , $\beta_k \leftarrow \beta_k^{new}$.

13: Compute the value of gate function $f(\beta_k)$ and sort them in descending order, the top $d$ ranked features would be selected.

---

**Table 1.** Summary of data sets

| Data sets | Samples | Features | Classes |
|---|---|---|---|
| warPIE10P | 210 | 2420 | 10 |
| lung discrete | 73 | 325 | 7 |
| Isolet | 1560 | 617 | 26 |
| madlon | 2600 | 500 | 2 |
| ORL | 400 | 1024 | 40 |
| JAFFE | 213 | 676 | 10 |

**RSFS**: Which jointly improves the robustness of graph embedding and sparse spectral regression [34].

**AEFS**: Which is a embedded feature selection method based on autoencoder and group lasso penalty [20].

Now, the specify parameter setting for each method need to be given. For fair comparison, we set the size of the neighbors $k = 5$ for all data sets. The number of selected features as $50, 100, 150, ..., 300$; As for some personal settings of each method, we followed the author's setting in the original papers. In NDFS, the orthogonality condition parameter is fixed, $\gamma = 10^8$; the parameters are tuned from $\{10^{-6}, 10^{-4}, ..., 10^6\}$. In AEFS, the activation functions are $\sigma_1(X) = 1/(1 + e^{-X})$, $\sigma_2(X) = X$; the number of hidden layer nodes is $\{2^7, 2^8, 2^9, 2^{10}\}$; the parameters range is $\{10^{-3}, 10^{-2}, ..., 10^3\}$. In RAFS, the size of hidden layer is $\{2^2, 2^3, 2^4, 2^5, 2^6\}$, the parameters range is $\{10^{-3}, 10^{-2}, ..., 10^3\}$.

**Table 2.** Clustering results (ACC%±std) of different feature selection methods. The best results are highlighted on bold.

| Dataset | AllFea | LS | NDFS | RSFS | AEFS | RAFS |
|---|---|---|---|---|---|---|
| warPIE10P | $27.0 \pm 1.9$ | $46.4 \pm 2.0$ | $29.4 \pm 2.0$ | $36.0 \pm 3.1$ | $51.3 \pm 4.8$ | $\mathbf{54.3 \pm 4.5}$ |
| lung discrete | $63.3 \pm 5.4$ | $69.7 \pm 6.1$ | $69.2 \pm 8.6$ | $71.9 \pm 5.7$ | $72.5 \pm 6.3$ | $\mathbf{73.2 \pm 8.0}$ |
| Isolet | $59.4 \pm 2.5$ | $49.3 \pm 2.0$ | $59.4 \pm 2.0$ | $62.4 \pm 3.5$ | $61.4 \pm 3.8$ | $\mathbf{63.1 \pm 4.3}$ |
| madlon | $50.4 \pm 0.1$ | $57.4 \pm 0.1$ | $61.5 \pm 0.0$ | $61.7 \pm 0.0$ | $60.4 \pm 0.8$ | $\mathbf{61.8 \pm 0.1}$ |
| ORL | $50.3 \pm 2.8$ | $48.7 \pm 3.2$ | $52.5 \pm 3.3$ | $53.1 \pm 2.6$ | $53.9 \pm 2.7$ | $\mathbf{63.4 \pm 3.8}$ |
| JAFFE | $72.2 \pm 6.4$ | $74.0 \pm 7.0$ | $80.8 \pm 8.5$ | $63.4 \pm 3.5$ | $74.2 \pm 7.6$ | $\mathbf{81.8 \pm 7.4}$ |

**Table 3.** Clustering results (NMI%±std) of different feature selection methods. The best results are highlighted on bold.

| Dataset | AllFea | LS | NDFS | RSFS | AEFS | RAFS |
|---|---|---|---|---|---|---|
| warPIE10P | $28.1 \pm 3.6$ | $49.5 \pm 1.9$ | $27.4 \pm 2.4$ | $37.3 \pm 3.1$ | $55.4 \pm 3.1$ | $\mathbf{57.6 \pm 3.9}$ |
| lung discrete | $62.8 \pm 3.5$ | $66.9 \pm 4.8$ | $68.7 \pm 5.7$ | $71.5 \pm 3.6$ | $68.6 \pm 4.9$ | $\mathbf{72.3 \pm 6.6}$ |
| Isolet | $73.8 \pm 2.3$ | $65.6 \pm 0.7$ | $74.4 \pm 1.5$ | $\mathbf{78.7 \pm 1.2}$ | $75.0 \pm 1.7$ | $76.5 \pm 1.4$ |
| madlon | $1.8 \pm 0.1$ | $1.6 \pm 0.1$ | $3.8 \pm 0.0$ | $4.0 \pm 0.0$ | $3.5 \pm 0.1$ | $\mathbf{4.1 \pm 0.1}$ |
| ORL | $73.7 \pm 1.5$ | $71.1 \pm 1.4$ | $74.0 \pm 1.7$ | $74.2 \pm 1.3$ | $75.1 \pm 1.7$ | $\mathbf{75.8 \pm 1.4}$ |
| JAFFE | $81.0 \pm 4.0$ | $79.0 \pm 2.7$ | $87.9 \pm 5.6$ | $78.7 \pm 1.2$ | $79.9 \pm 3.8$ | $\mathbf{89.3 \pm 3.6}$ |

In evaluation metrics of feature subsets, we use Clustering Accuracy (ACC) and Normalized Mutual Information (NMI) to compare the performance with all above methods. In both metrics, higher value means more excellent performance. For the selected features, the K-means algorithm is repeat 20 times with random initialization and the average value and standard will be recorded.

From the results in Tables 2 and 3, we can observe that feature selection can get a subset which has less features but more satisfying clustering and classification performance. Obviously, the research for unsupervised feature selection is valuable and effective. We also get that our proposed method RAFS shows its outperformance in fair comparison with several other methods. For most data sets, RAFS is better than other methods.

## 4   Conclusion

In this study, a novel unsupervised feature selection scheme, RAFS has been proposed. In this method, radial basis function replaces the weight to activate autoencoder, gate function shows the situation of feature selection and control redundancy is considered. We give a specific learning process to make RAFS be applicative for unsupervised feature selection tasks in high-dimensional space. In the experiment, the efficiency and effectiveness of RAFS is validated by conducting several typical data sets.

In future work, there remains several facets to be further improved, including the determination of the kernel centers and the iterative optimization algorithm.

## References

1. Ding, C., Peng, H.: Minimum redundancy feature selection from microarray gene expression data. J. Bioinform. Comput. Biol. **3**(02), 185–205 (2003)
2. Forman, G.: An extensive empirical study of feature selection metrics for text classification. J. Mach. Learn. Res. **3**, 1289–1305 (2003)
3. Akay, M.F.: Support vector machines combined with feature selection for breast cancer diagnosis. Expert Syst. Appl. **36**(2–part–P2), 3240–3247 (2009)
4. Ding, S., Zhu, H., Jia, W., Su, C.: A survey on feature extraction for pattern recognition. Artif. Intell. Rev. **37**(3), 169–180 (2012)
5. Wiatowski, T., Bolcskei, H.: A mathematical theory of deep convolutional neural networks for feature extraction. IEEE Trans. Inf. Theory **1**(1), 1845–1866 (2017)
6. Basu, T., Murthy, C.A.: Effective text classification by a supervised feature selection approach. In: 12th IEEE International Conference on Data Mining Workshops, Brussels, pp. 918–925. IEEE Press (2013)
7. Chakraborty, R., Pal, N.R.: Feature selection using a neural framework with controlled redundancy. IEEE Trans. Neural Netw. Learn. Syst. **26**(1), 35–50 (2015)
8. Xu, Z., King, I., Lyu, R.T., Jin, R.: Discriminative semi-supervised feature selection via manifold regularization. IEEE Trans. Neural Netw. **21**(7), 1033–1047 (2010)
9. Zhao, J.: Locality sensitive semi-supervised feature selection. Neurocomputing **71**(10), 1842–1849 (2008)
10. Kalakech, M., Biela, P., Macaire, L., Hamad, D.: Constraint scores for semi-supervised feature selection: a comparative study. Pattern Recogn. Lett. **32**(5), 656–665 (2011)
11. Peña, J.M., Nilsson, R.: On the complexity of discrete feature selection for optimal classification. IEEE Trans. Pattern Anal. Mach. Intell. **32**, 1517–1522 (2010)
12. Sotoca, J.M., Pla, F.: Supervised feature selection by clustering using conditional mutual information-based distances. Pattern Recogn. **43**(6), 2068–2081 (2010)
13. Guyon, I., Weston, J., Barnhill, S., Vapnik, V.: Gene selection for cancer classification using support vector machines. Mach. Learn. **46**(1–3), 389–422 (2002)
14. Zhou, X., Tuck, D.P.: MSVM-RFE: extensions of SVM-RFE for multiclass gene selection on DNA microarray data. Bioinformatics **23**(9), 1106–1114 (2007)
15. Guyon, I.: Pattern classification. Pattern Anal. Appl. **44**(1), 87–87 (1998)
16. He, X.: Locality Preserving Projections. University of Chicago, Chicago (2005)
17. He, X., Cai, D., Niyogi, P.: Laplacian score for feature selection. In: 19th Advances in Neural Information Processing Systems, Vancouver, pp. 507–514. NIPS Press (2005)
18. Roweis, S.T., Saul, L.K.: Nonlinear dimensionality reduction by locally linear embedding. Science **290**(5500), 2323–2326 (2000)
19. Banerjee, M., Pal, N.: Unsupervised feature selection with controlled redundancy (UFeSCoR). IEEE Trans. Knowl. Data Eng. **27**(12), 3390–3403 (2015)
20. Han, K., Wang, Y., Zhang, C., Li, C., Xu, C.: Autoencoder inspired unsupervised feature selection. In: 43th IEEE International Conference on Acoustics, Speech and Signal Processing, Calgary. IEEE Press (2018)
21. Meng, K., Dong, Z.Y., Wang, D.H., Wong, K.P.: A self-adaptive RBF neural network classifier for transformer fault analysis. IEEE Trans. Power Syst. **25**(3), 1350–1360 (2010)

22. Han, H.G., Qiao, J.F., Chen, Q.L.: Model predictive control of dissolved oxygen concentration based on a self-organizing RBF neural network. Control Eng. Pract. **20**(4), 465–476 (2012)
23. Wilamowski, B., Cecati, C., Kolbusz, J., Rozycki, P., Siano, P.: A novel RBF training algorithm for short-term electric load forecasting and comparative studies. IEEE Trans. Ind. Electron. **62**(10), 6519–6529 (2015)
24. Baghaee, H.R., Mirsalim, M., Gharehpetian, G.B., Talebi, H.A.: Application of RBF neural networks and unscented transformation in probabilistic power-flow of microgrids including correlated wind/pv units and plug-in hybrid electric vehicles. Simul. Model. Pract. Theory **72**, 51–68 (2017)
25. Hartman, E.J., Keeler, J.D., Kowalski, J.M.: Layered neural networks with gaussian hidden units as universal approximations. Neural Comput. **2**(2), 210–215 (1990)
26. Moody, J., Darken, C.J.: Fast learning in networks of locally-tuned processing units. Neural Comput. **1**(2), 281–294 (1989)
27. Lee, S.: A Gaussian potential function network with hierarchically self-organizing learning. Neural Netw. **4**(2), 207–224 (1991)
28. Chakraborty, R., Pal, N.R.: Feature selection using a neural framework with controlled redundancy. IEEE Trans. Neural Netw. Learn. Syst. **26**(1), 35–50 (2014)
29. Meng, J.E., Wu, S., Lu, J.: Face recognition with radial basis function (RBF) neural networks. IEEE Trans Neural Netw. **13**(3), 697–710 (2002)
30. Micchelli, C.A.: Interpolation of scattered data: distance matrices and conditionally positive definite functions. Constr. Approx. **2**(1), 11–22 (1986)
31. Haykin, S.: Neural networks and learning machines. McMaster University, Ontario (2009)
32. Haykin, S.: Neural Networks A Comprehensive Foundation. McMaster University, Ontario (1994)
33. Li, Z., Yang, Y., Liu, J., Zhou, X., Lu, H.: Unsupervised feature selection using nonnegative spectral analysis. In: 26th AAAI Conference on Artificial Intelligence, Toronto, pp. 1026–1032. AAAI Press (2012)
34. Shi, L., Du, L., Shen, Y.D.: Robust spectral learning for unsupervised feature selection. In: 14th IEEE International Conference on Data Mining, Shenzhen, pp. 977–982. IEEE Press (2014)

# A Two-Teacher Framework for Knowledge Distillation

Xingjian Chen[1], Jianbo Su[1(✉)], and Jun Zhang[2]

[1] Research Center of Intelligent Robotics, Shanghai Jiao Tong University,
Shanghai, China
{chenxj,jbsu}@sjtu.edu.cn
[2] UM-SJTU Joint Institute, Shanghai Jiao Tong University, Shanghai, China

**Abstract.** Knowledge distillation aims at transferring knowledge from
a teacher network to a student network. Commonly, the teacher network
has high capacity, while the student network is compact and can be
deployed to embedded systems. However, existing distillation methods
use only one teacher to guide the student network, and there is no guarantee that the knowledge is sufficiently transferred to the student network. Thus, we propose a novel framework to improve the performance
of the student network. This framework consists of two teacher networks
trained with different strategies, one is trained strictly to guide the student network to learn sophisticated features, and the other is trained
loosely to guide the student network to learn general decision based on
learned features. We perform extensive experiments on two standard
image classification datasets: CIFAR-10 and CIFAR-100. And results
demonstrate that the proposed framework can significantly improve the
classification accuracy of a student network.

**Keywords:** Knowledge distillation · Convolutional neural networks ·
Adversarial learning · Image classification

## 1 Introduction

In recent years, deep neural networks (DNN) have achieved impressive success
in image classification [1–3], object detection [4–6], semantic segmentation [7,8],
etc. One of the reasons for DNN's success is attributed to the increasing depth
and huge amounts of parameters of neural networks. However, heavy computation and memory cost makes it difficult to deploy these cumbersome models on
resource-constrained platforms like autonomous cars, smart phones, humanoid
robots, etc. Therefore, knowledge distillation [9] is introduced as a method to
compress the knowledge in a cumbersome model into a single small model. But
the small model often has a worse performance when compared with the cumbersome model. Thus, great efforts have been made to make the performance
of the small student network comparable to that of the large teacher network,
and it is of great importance when considering the ever increasing demand from
industry that requires neural network models to work in real time.

© Springer Nature Switzerland AG 2019
H. Lu et al. (Eds.): ISNN 2019, LNCS 11554, pp. 58–66, 2019.
https://doi.org/10.1007/978-3-030-22796-8_7

To approximate the performance of the high-capacity teacher network, there have been several attempts in the literature to improve the student network's performance via knowledge transfer. Existing approaches can be divided into two categories according to what knowledge to transfer. One line of research takes the output probability distribution as knowledge [9,10], therefore the student network is forced to mimic the output probability distribution of the teacher network. Despite its simplicity and effectiveness, softmax loss function is a must for both teacher and student networks, which constrained its application scenario. Another direction takes the intermediate representations as knowledge, representative works include [11–13]. In [11], the knowledge is defined as the inner product between features from two layers, and it proved to be effective. In [12], attention maps of convolutional neural networks are considered as important knowledge representations. In [13], related experiments suggest that the activation boundaries formed by hidden neurons are suitable for knowledge transfer.

Recent breakthrough on knowledge distillation can be summarized as follows. First of all, adversarial learning was introduced to the knowledge distillation [10, 14,15], which greatly boosted the performance of student networks. Then Yang et al. [16] proposed a new strategy for training the same network in generation, and the descendant network trained with this strategy incredibly outperformed its ancestor.

Although recent research on knowledge distillation has significantly improved the classification accuracy of the student network, *there are still some limitations*: (1) Adversarial based methods mainly take output probability distribution [10,14] or coarse high-level feature maps [15] as input for the discriminator, which can not take full advantage of previous research on intermediate knowledge representations; (2) Yang et al. [16] found a more effective training method for the teacher network to give better output distribution guidance, but for intermediate representation guidance, its performance has not been generally validated so far; (3) Furthermore, previous studies employ only one teacher network trained with specific strategy to give two different kinds of guidance. The simple mode of processing deserves more careful consideration.

In order to address the aforementioned three problems, this paper proposes a framework containing two teacher networks trained with different strategies. *The first teacher network* is trained with traditional method, which gives intermediate representation guidance to the student network. Different from other methods [12,13,17,18] which align intermediate representations between the teacher network and the student network through $L_2$ loss, this paper introduces adversarial learning to make the alignment, and our experiment results show its superior performance. *The second teacher network* gives the output probability distribution guidance to the student network, and its training follows the strategy of [16]. Similar to other knowledge transfer methods [12,13], the proposed framework is verified on Wide Residual Network (WRN) [19] with different settings and VGG network [3], and extensive comparative experiments are conducted on CIFAR-10 and CIFAR-100 datasets [20]. Experiment results indicate that middle-level (intermediate representation) guidance and high-level (output distribution) guidance may need different training strategies for the teacher network.

## 2   The Proposed Two-Teacher Framework

Inspired by recent progress [10, 15, 16] on knowledge distillation, a two-teacher framework is proposed to better transfer knowledge from teacher networks to the student network. As depicted in Fig. 1, Teacher Network 2 (TN2) can give better output distribution guidance to the compact student network, but it may not give good intermediate representation guidance due to its specific objective function. In order to provide elaborate intermediate representation guidance, Teacher Network 1 (TN1) is trained in traditional way to ensure the effectiveness of intermediate representation transferring.

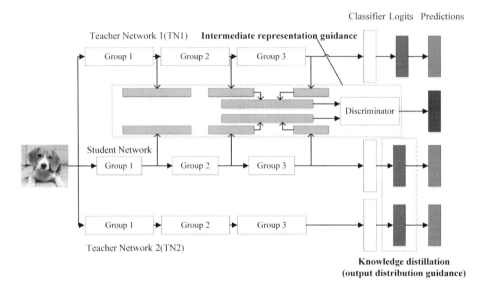

**Fig. 1.** The architecture of the proposed framework. Layers in every network are divided into 3 groups. Teacher Network 1 is responsible for knowledge embedded in output distribution, while Teacher Network 2 gives careful guidance to knowledge resided in intermediate representations.

### 2.1   Intermediate Representation Guidance

Many intermediate representations are proposed to represent the knowledge learned by neural networks, such as FSP matrix [11], activation boundaries

[13], attention maps [12], etc. However almost all intermediate representations between the teacher network and the student network are aligned by $L_2$ loss:

$$\mathcal{L}_2 = \sum_{j \in \mathcal{I}} \left\| \frac{R_S^j}{\left\| R_S^j \right\|_2} - \frac{R_T^j}{\left\| R_T^j \right\|_2} \right\|_2, \tag{1}$$

where $R_S^j$ and $R_T^j$ are respectively the $j$-th pair of intermediate representation for student and teacher networks in vectorized form.

In order to better transfer knowledge resided in intermediate representations, a discriminator is introduced to help to guide the student network. The discriminator aims to distinguish intermediate representations of the student network from those of the teacher network, while the student network learns to generate intermediate representations to fool the discriminator. To make the adversarial training process more stable, we adopt the conditional adversarial network [21] strategy in [10]. In this case, discriminator not only predicts the source of the input intermediate representations, but also predicts class labels.

Assume we have N training samples, and $(\boldsymbol{x}_i, \boldsymbol{y}_i)$ refers to the $i$-th instance, where $\boldsymbol{x}_i$ represents the input image, and $\boldsymbol{y}_i$ is a one-hot vector, which stands for the corresponding class label. If the number of classes is $K$, the output of discriminator $D(\cdot)$ is a $(K + 2)$-dimensional vector. For student network, the objective function involved in adversarial training can be divided into two parts. The first component $\mathcal{L}_{ad}$ is short for the standard adversarial loss in adversarial learning framework, and the second component $\mathcal{L}_{classification}$ refers to the ordinary classification loss. Hence, the objective function can be formulated as:

$$\begin{aligned} \mathcal{L}_A &= \mathcal{L}_{ad} + \mathcal{L}_{classification} \\ &= \frac{1}{N} \sum_{i=1}^{N} (log D(R_{si})_1 + log D(R_{ti})_2) \\ &\quad - \frac{1}{N} \sum_{i=1}^{N} \sum_{j=1}^{K} (y_i^j log D(R_{si})_{j+2} + y_i^j log D(R_{ti})_{j+2}) \end{aligned} \tag{2}$$

where $R_{si}$ and $R_{ti}$ represent intermediate representations of student network and teacher network given the $i$-th training sample, and $D(\cdot)_j$ represents the $j$-th element of $D(\cdot)$. The first and second elements of $D(\cdot)$ denote the probability that the input intermediate representation comes from the student network and the teacher network respectively. Moreover, according to the theory of adversarial training, the student network aims to learn intermediate representations similar to the teacher network's, while the discriminator tries to distinguish these two kinds of representations. Since the discriminator's objective is opposite to the student network's, its objective function can be defined as:

$$\mathcal{L}_{Discriminator} = -\mathcal{L}_{ad} + \mathcal{L}_{classification}. \tag{3}$$

## 2.2   Output Distribution Guidance

For image classification problem, the output of neural networks is a probability distribution $q = (q_1, ..., q_K)$, which is generated by applying softmax function on logits $z = (z_1, ..., z_K)$. Previous study [9] has demonstrated that rich information resides in logits or output probability distribution, and it is reasonable to use logits to transfer knowledge.

$$q_i = \frac{e^{z_i/T}}{\sum_{j=1}^{K} e^{z_j/T}}. \tag{4}$$

In Eq. 4, temperature $T$ is often set to 1, but a higher temperature leads to a softer probability distribution over classes. To distill knowledge embedded in output probability distribution, cross-entropy loss between the teacher network's output distribution $q^t$ and the student network's output distribution $q^s$ is used as an extra constraint, denoted as $\mathcal{L}_{KD}$.

$$\mathcal{L}_{KD} = \mathcal{H}(q^t, q^s) = -\sum_{i=1}^{K} q_i^t log q_i^s. \tag{5}$$

The teacher network is typically trained with the standard cross-entropy loss, however recent research on training deep neural networks in generations has demonstrated that a good teacher network should make the knowledge embedded in output distribution easier to transfer, not just aim to have a high accuracy for itself. Inspired by this idea, the new training strategy in [16] is introduced to Teacher Network 2 in our framework, which is responsible for output distribution guidance. The objective function of this strategy can be expressed as:

$$\mathcal{L}^T = \frac{1}{N} \sum_{i=1}^{N} \{-\lambda y_i^T log f_t(x_i) + (1 - \lambda)[f_{e1} - \frac{1}{M - 1} \sum_{j=2}^{M} f_{ej}]\}, \tag{6}$$

where $M$ is a fixed integer not greater than $K$, $f_t$ represents the corresponding function of the teacher network, and $f_{ej}$ is short for the $j$-th largest element of $f_t(x)$.

## 2.3   Model Training and Deployment

To combine different guidance from two teacher networks, we obtain the overall loss function for the student network during knowledge distillation:

$$\mathcal{L}_{Student} = \alpha \mathcal{L}_{CE} + (1 - \alpha)\mathcal{L}_{KD} + \beta \mathcal{L}_2 + \gamma \mathcal{L}_A, \tag{7}$$

where $\mathcal{L}_{CE}$ is the standard cross-entropy loss for image classification tasks, and $\alpha$, $\beta$, $\gamma$ are hyper-parameters. During training process, two teacher networks are fixed, and the discriminator and student network are optimized alternatively based on Eqs. 3 and 7 until the process converges. Once the framework is trained, it is convenient to obtain the compact student network by simply removing two teacher networks and the discriminator.

# 3    Experiment Results

In this section, we implement our method on two image classification datasets, CIFAR-10 and CIFAR-100. The CIFAR dataset consists of 50k training and 10k testing color images, and the image size is $32 \times 32$. CIFAR-10 contains 10 classes, while CIFAR-100 is a fine-grained visual classification dataset with 100 classes. First of all, adequate comparison is made between the proposed method and two representative distillation methods: knowledge distillation (KD) and attention transfer (AT) in Sect. 3.1. Then, ablation study is conducted to verify the necessity of two teacher networks with different training strategies in Sect. 3.2.

The neural networks employed in our experiment are Wide Residual Network (WRN) [19] and slightly modified VGG [3]. WRN is a variant of the widely acknowledged Resnet [1]. A specific setting of WRN can be denoted as WRN-$x$-$y$, where $x$ represents the depth, and $y$ is width, which represents the multiplication factor of channels. VGG's performance is worse than models like WRN and Resnet, but it is widely used due to its simple structure and convenient implementation, thus we also conduct some experiments on it.

## 3.1    Evaluation of the Proposed Method

Beside teacher-student combinations in [12], the proposed method is also verified on slightly modified VGG-13 models. To get the teacher network (VGG-T), we removed the last two convolutional layers of VGG-13, and the student network (VGG-S) only has half of convolutional layers compared with VGG-T. For knowledge distillation (KD), we set the temperature parameter T in Eq. 4 to 4 like [9,12], and hyper-parameters in Eq. 7 are set $\alpha = 0.9$, $\beta = 0$, and $\gamma = 0$, as in [12]. For attention transfer (AT), the settings of hyper-parameters are $\alpha = 1$, $\beta = 1e + 3$, and $\gamma = 0$. As for our method, we train Teacher Network 1 with the standard cross-entropy loss, but the objective function for Teacher Network 2 is Eq. 6, where M is set to 5 empirically. Moreover, we set $\alpha = 0.9$, $\beta = 1e + 3$ and $\gamma = 0.1$. Our experiments are repeated five times, and the median results are summarized in Tables 1 and 2.

**Table 1.** Error rate (%) of various methods on CIFAR-10 dataset. Our goal is to decrease the error rate of *Student*, and different algorithms to realize this goal are compared, including *AT*, *TN1*, *KD*, *TN2* and *Ours*. Note that *Ours* ia a combination of *TN1* and *TN2*.

| Teacher/student network | Teacher | Student | AT [12] | TN1 | KD [9] | TN2 | Ours |
|---|---|---|---|---|---|---|---|
| VGG-T/VGG-S | 7.84 | 9.87 | 8.61 | 8.39 | 9.50 | 9.12 | **8.04** |
| WRN-16-2/WRN-16-1 | 6.31 | 8.77 | 7.93 | 7.22 | 7.41 | 7.42 | **6.92** |
| WRN-40-1/WRN-16-1 | 6.58 | 8.77 | 8.25 | 7.72 | 8.09 | 7.95 | **7.44** |
| WRN-40-2/WRN-16-2 | 5.23 | 6.31 | 5.85 | **5.22** | 6.08 | 5.52 | 5.26 |

**Table 2.** Error rate (%) of various methods on CIFAR-100 dataset. The meaning of each column is the same as that in Table 1

| Teacher/student network | Teacher | Student | AT [12] | TN1 | KD [9] | TN2 | Ours |
|---|---|---|---|---|---|---|---|
| VGG-T/VGG-S | 28.13 | 30.52 | 29.38 | 29.23 | 30.09 | 29.63 | **29.19** |
| WRN-16-2/WRN-16-1 | 27.43 | 33.29 | 33.30 | 32.69 | 32.29 | 31.06 | **30.83** |
| WRN-40-1/WRN-16-1 | 29.47 | 33.29 | 33.30 | 32.66 | 32.89 | 32.58 | **31.62** |
| WRN-40-2/WRN-16-2 | 23.95 | 27.43 | 27.18 | 26.49 | 25.72 | 25.65 | **25.40** |

From the top row of the table, we can see many algorithms for knowledge transfer, including attention transfer (AT), knowledge distillation (KD) and the proposed two-teacher framework (Ours) which contains two teacher networks, TN1 and TN2. As can be seen, our method achieved the best performance when compared with AT and KD. Notably, from *comparison between column AT and TN1*, it is evident that the introduction of adversarial training for intermediate representations like attention maps can boost the performance. Moreover, quantitative results also demonstrated that the novel training strategy can improve the performance of KD when *comparing column KD and TN2*.

## 3.2    Ablation Study

Three experiments are done to validate the necessity of having two teacher networks with different training strategies. First of all, the teacher network is trained with the standard cross-entropy loss function, and it gives both output distribution and intermediate representation guidance to the student network (Strategy1). The second experiment is similar to the first one, but the teacher network is trained with Eq. 6 (Strategy2). For the last experiment, the teacher network in experiment1 and experiment2 are utilized to give output distribution and intermediate representation guidance respectively (Our Framework). The comparison results are shown in Table 3.

**Table 3.** Error rate(%) for ablation results on CIFAR-10 and CIFAR-100 datasets, *Strategy1* and *Strategy2* use only one teacher network to give two different guidance to the student network, while teacher networks in *Our Framework* are specially trained for two different guidance respectively.

| Dataset | Teacher/student network | Strategy1 | Strategy2 | Our framework |
|---|---|---|---|---|
| CIFAR-10 | WRN-16-2/WRN-16-1 | 7.24 | 7.04 | 6.92 |
| | WRN-40-1/WRN-16-1 | 7.77 | 7.69 | 7.44 |
| | WRN-40-2/WRN-16-2 | 5.58 | 5.32 | 5.26 |
| CIFAR-100 | WRN-16-2/WRN-16-1 | 31.60 | 31.25 | 30.83 |
| | WRN-40-1/WRN-16-1 | 32.09 | 31.68 | 31.62 |
| | WRN-40-2/WRN-16-2 | 25.95 | 25.96 | 25.40 |

Obviously, our two-teacher framework outperforms the first two strategies, which validates the necessity of two teacher networks. It also indicates that it is important to choose appropriate training strategy for the teacher network when it involves different kinds of knowledge to be transferred. Additionally, from comparison between Strategy2 and Our Framework, we draw the conclusion that teacher network trained with Strategy2 may not suitable for intermediate representation guidance.

## 4   Conclusions

A two-teacher framework has been proposed to distill knowledge to a compact student network. This framework is orthogonal to previous research on what knowledge to transfer, hence it can further improve the performance of the compact neural network. To verify the effectiveness of our method, Wide Residual Network and VGG network are used in our experiment. And all experimental results on CIFAR-10 and CIFAR-100 datasets demonstrate that our framework can significantly improve the classification accuracy of the student network. In addition, the two-teacher framework provide researchers with a new perspective on how to utilize different knowledge, and the training strategy for the teacher network may attract more attention.

**Acknowledgments.** This paper was partially financially supported by National Natural Science Foundation of China under grants 61533012 and 91748120.

## References

1. He, K., Zhang, X., Ren, S., Sun, J.: Deep residual learning for image recognition. In: Proceedings of the IEEE Conference on Computer Vision and Pattern Recognition, pp. 770–778 (2016)
2. Krizhevsky, A., Sutskever, I., Hinton, G.E.: ImageNet classification with deep convolutional neural networks. In: Advances in Neural Information Processing Systems, pp. 1097–1105 (2012)
3. Simonyan, K., Zisserman, A.: Very deep convolutional networks for large-scale image recognition. arXiv preprint arXiv:1409.1556 (2014)
4. Girshick, R., Donahue, J., Darrell, T., Malik, J.: Rich feature hierarchies for accurate object detection and semantic segmentation. In: Proceedings of the IEEE Conference on Computer Vision and Pattern Recognition, pp. 580–587 (2014)
5. Liu, W., et al.: SSD: single shot MultiBox detector. In: Leibe, B., Matas, J., Sebe, N., Welling, M. (eds.) ECCV 2016. LNCS, vol. 9905, pp. 21–37. Springer, Cham (2016). https://doi.org/10.1007/978-3-319-46448-0_2
6. Redmon, J., Divvala, S., Girshick, R., Farhadi, A.: You only look once: unified, real-time object detection. In: Proceedings of the IEEE Conference on Computer Vision and Pattern Recognition, pp. 779–788 (2016)
7. Long, J., Shelhamer, E., Darrell, T.: Fully convolutional networks for semantic segmentation. In: Proceedings of the IEEE Conference on Computer Vision and Pattern Recognition, pp. 3431–3440 (2015)

8. Zhao, H., Shi, J., Qi, X., Wang, X., Jia, J.: Pyramid scene parsing network. In: IEEE Conference on Computer Vision and Pattern Recognition (CVPR), pp. 2881–2890 (2017)
9. Hinton, G., Vinyals, O., Dean, J.: Distilling the knowledge in a neural network. arXiv preprint arXiv:1503.02531 (2015)
10. Xu, Z., Hsu, Y.C., Huang, J.: Training shallow and thin networks for acceleration via knowledge distillation with conditional adversarial networks. In: BMVC (2018)
11. Yim, J., Joo, D., Bae, J., Kim, J.: A gift from knowledge distillation: fast optimization, network minimization and transfer learning. In: 2017 IEEE Conference on Computer Vision and Pattern Recognition (CVPR), pp. 7130–7138. IEEE (2017)
12. Zagoruyko, S., Komodakis, N.: Paying more attention to attention: improving the performance of convolutional neural networks via attention transfer. In: ICLR (2017)
13. Heo, B., Lee, M., Yun, S., Choi, J.Y.: Knowledge transfer via distillation of activation boundaries formed by hidden neurons. arXiv preprint arXiv:1811.03233 (2018)
14. Wang, X., Zhang, R., Sun, Y., Qi, J.: KDGAN: knowledge distillation with generative adversarial networks. In: Advances in Neural Information Processing Systems, pp. 783–794 (2018)
15. Liu, P., Liu, W., Ma, H., Mei, T., Seok, M.: KTAN: knowledge transfer adversarial network. arXiv preprint arXiv:1810.08126v11 (2018)
16. Yang, C., Xie, L., Qiao, S., Yuille, A.: Knowledge distillation in generations: more tolerant teachers educate better students. arXiv preprint arXiv:1805.05551 (2018)
17. Lee, S.H., Kim, D.H., Song, B.C.: Self-supervised knowledge distillation using singular value decomposition. In: Ferrari, V., Hebert, M., Sminchisescu, C., Weiss, Y. (eds.) ECCV 2018. LNCS, vol. 11210, pp. 339–354. Springer, Cham (2018). https://doi.org/10.1007/978-3-030-01231-1_21
18. Romero, A., Ballas, N., Kahou, S.E., Chassang, A., Gatta, C., Bengio, Y.: FitNets: hints for thin deep nets. In: ICLR (2015)
19. Zagoruyko, S., Komodakis, N.: Wide residual networks. In: BMVC (2016)
20. Krizhevsky, A., Hinton, G.: Learning multiple layers of features from tiny images (2009)
21. Mirza, M., Osindero, S.: Conditional generative adversarial nets. arXiv preprint arXiv:1411.1784 (2014)

# The Neural Network for Online Learning Task Without Manual Feature Extraction

Yuriy Fedorenko, Valeriy Chernenkiy, and Yuriy Gapanyuk[✉]

Bauman Moscow State Technical University, Moscow, Russia
gapyu@bmstu.ru

**Abstract.** The article is devoted to the problem of feature extraction in online learning tasks. In many cases, the proper feature extraction is very time-consuming. Currently, in some cases, this problem is successfully solved by deep neural networks. However, deep models are computationally expensive and so hardly applicable for online learning tasks which require frequent updating of the model. This paper proposes the lightweight neural net architecture that can be learned in online mode and doesn't require complex hand-crafted features. The small sample processing time distinguishes the proposed model from more complex deep neural networks. The architecture and learning process of the proposed model are discussed in detail. The special attention is paid to fast software implementation. On benchmarks, we show that developed implementation processes one sample several times faster than implementations on the base of deep learning frameworks. The conducted experiments on CTR prediction task show that the proposed neural net with raw features gives the same performance as the logistic regression model with handcrafted features. For a clear description of the proposed architecture, we use the metagraph approach.

**Keywords:** Feature extraction · Online learning · Sparse neural network · CTR prediction · Logistic loss · ROC curve · Metagraph

## 1 Introduction

It is known that feature extraction is the crucial stage of input data processing. Sometimes the feature extraction task divides into two parts [1]: feature construction and feature selection. The goal of the first step is to obtain many features from the described data. On the second step, the most informative features are selected. To select informative features information gain or chi-squared test are often used [2]. The disadvantage of such approaches is that they consider features separately. However, the feature may be not informative by itself but informative jointly with other features. Therefore, there are methods to determine relevant feature sets, for example, minimum redundancy – maximum relevance approach (mRMR) [3]. The common drawback of such methods is that they require manual work from the researcher. Besides feature selection task may arise again and again if data rapidly changes over time.

The more modern approach is the representation learning in which deep learning is succeeded. The essential advantage of deep models is that they form features

© Springer Nature Switzerland AG 2019
H. Lu et al. (Eds.): ISNN 2019, LNCS 11554, pp. 67–76, 2019.
https://doi.org/10.1007/978-3-030-22796-8_8

automatically during training. They simplify the work of researchers and give better result in comparison with algorithms requiring manual feature selection [4].

However, such models are hardly applicable in high loaded systems where response time requirements are very strict. Also, it is impossible to provide often (for example, once per hour) update of such models because their retraining is computationally expensive. In online learning tasks complex models often lose to the simple ones because simple models can be updated more frequently. But they require manual features selection that takes a lot of researcher's time. In this article, we propose the neural network model which on the one hand is simple enough to be updated regularly during training and on the other hand is complex enough to implement the feature selection process automatically.

## 2   The Problem Definition

For definiteness, we developed a model on the example of CTR prediction task. However, it can be used for other online regression tasks such as stock movement prediction, malware prediction, goods recommendations, etc. The advertising system has many banners. The goal of the system is to show each user the most relevant banners to maximize proceeds (in most cases money is paid not for banner show but for active user action such as click, install the mobile app, purchase of goods, etc.).

Firstly, the system determines the set of banners which can be shown to the user (on the base of targeting and other restrictions). For this set of banners, the model predicts the probabilities of user actions. The banner with max probability is shown to the user. This is a predict regime. All features and user reaction on showed banners are written in the binary log. In the training regime model reads this log and calculate predictions. Then on the base of the difference between real and predicted values the model is updated. It is shown schematically in Fig. 1.

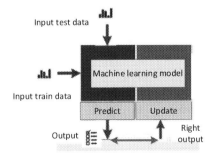

**Fig. 1.** Online learning task

The described system is dynamic. Some interests of users may change every day or even several hours. Also, advertising banners are very volatile: thousands of banners are created each hour and thousands of banners stop showing. So, we need to train model in online mode, and we want to do it without manual feature selection.

## 3   The Proposed Architecture

Unlike usual perceptron the proposed neural network is sparse: hidden layer neurons are connected with one or two inputs responsible for the concrete feature. The schema of this neural net is presented in Fig. 2.

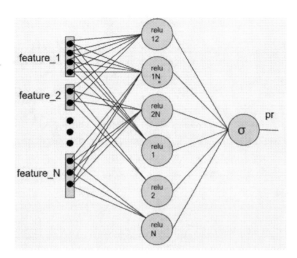

**Fig. 2.** The proposed neural network architecture

The input of this neural net is raw features of user and banner, output – probability that this user clicks (or perform another action) on this banner. Examples of user features are age, sex, region, list of visited websites last week, etc. Examples of banner features are image, text, target website or mobile app, etc.

The problem of simple logistic regression is that it is linear and so works poor with these raw features. To achieve an acceptable result, it is necessary to combine raw features in groups and hashing it. Determining exactly how to combine features requires exhaustive search that computationally very expensive (complexity grows exponentially on the number of features) or intellectual search which wastes the time of experts of domain area.

The proposed neural network is designed in such a way to consider all two-feature combinations. During training the weights of hidden layer corresponding relevant combinations obtain more high values than weights corresponding poorly relevant or not relevant combinations. So, the model implements feature extraction by itself. It is possible to use combinations with any number of features, but already addition of three features combination improves result only a few while significantly increases computational complexity. Addition of combination with a greater number of features only decreases model performance metrics. So, for CTR prediction task on considered dataset it is meaningless, but for other tasks, it may be useful.

For fitting the neural network, the logistic loss function is used [5]. The model is trained with stochastic gradient descent optimization (SGD) algorithm. Also, there can be used its modifications such as SGD with Momentum, RMSprop or Adam [6]. The training procedure of proposed architecture in more details is discussed in [7].

## 4 The Metagraph Representation of the Proposed Architecture

For a clear description of the proposed architecture, we use the metagraph approach. Metagraph is a kind of complex graph model: $MG = \langle V, MV, E \rangle$, where $MG$ – metagraph; $V$ – set of metagraph vertices; $MV$ – set of metagraph metavertices; $E$ – set of metagraph edges.

It is the metavertex that is the distinguishing feature of the model. The metavertex may include inner vertices, edges, and metavertices. From the general system theory point of view, metavertex is a special case of manifestation of emergence principle which means that metavertex with its private attributes, inner vertices, metavertices, and edges became whole that cannot be separated into its component parts.

For metagraph transformation, the metagraph agents are proposed. There are two kinds of metagraph agents: the metagraph function agent $(ag^F)$ and the metagraph rule agent $(ag^R)$.

The metagraph function agent $(ag^F)$ serves as a function with input and output parameter in the form of metagraph: $ag^F = \langle MG_{IN}, MG_{OUT}, AST \rangle$, where $MG_{IN}$ – input parameter metagraph; $MG_{OUT}$ – output parameter metagraph; $AST$ – abstract syntax tree of metagraph function agent in the form of metagraph.

The metagraph rule agent $(ag^R)$ uses a rule-based approach: $ag^R = \langle MG, R, AG^{ST} \rangle$, $R = \{r_i\}$, $r_i : MG_j \rightarrow OP^{MG}$, where $MG$ – working metagraph, a metagraph on the basis of which the rules of agent are performed; $R$ – set of rules $r_i$; $AG^{ST}$ – start condition (metagraph fragment for start rule check or start rule); $MG_j$ – a metagraph fragment on the basis of which the rule is performed; $OP^{MG}$ – set of actions performed on metagraph.

According to our paper [12], we can describe neural network operation using metagraph rule agents which are shown in Fig. 3.

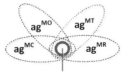

The metagraph representation of a neural network

**Fig. 3.** The structure of metagraph rule agents for neural network operation representation

In order to provide a neural network operation, the following agents are used:

- $ag^{MC}$ – the agent responsible for the creation of the network;
- $ag^{MO}$ – the agent responsible for the modification of the network;
- $ag^{MT}$ – the agent responsible for the training of the network;
- $ag^{MR}$ – the agent responsible for the execution of the network.

In Fig. 3 the agents are shown as metavertices by dotted-line ovals.

The network-creating agent ($ag^{MC}$) implements the rules of creating an original neural network topology. The agent holds both the rules of creating separate neurons and rules of connecting neurons into a network. In particular, the agent generates abstract syntactic trees of metagraph function agents.

The network-modification agent ($ag^{MO}$) holds the rules of modification the network topology in the process of operation. It is especially important for neural networks with variable topology such as SOINN.

The network-training agent ($ag^{MT}$) implements a particular training algorithm. As a result of training, the changed weights are written in the metagraph representation of the neural network. It is possible to implement a few training algorithms by using different sets of rules for agent $ag^{MT}$.

The network-executing agent ($ag^{MR}$) is responsible for the start and operation of the trained neural network.

The agents can work separately or jointly which may be especially important in the case of variable topologies. For example, when a SOINN network is trained, agent $ag^{MT}$ can call the rules of agent $ag^{MO}$ to change the network topology in the process of learning.

In fact, each agent uses its rules to implement a specific program "machine". The use of the metagraph approach allows us to implement the "multi-machine" principle: a few agents having different goals implement different operations on the same data structure.

In our paper [12] the metagraph representation of perceptron neural network was proposed. In our paper [13], the metagraph approach was applied to the analysis of regularization in deep neural networks. Using the results of these papers, the description of the proposed architecture is very straightforward and represented in Fig. 4.

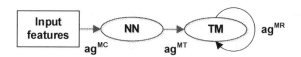

**Fig. 4.** The metagraph representation of the proposed architecture

In the creation mode, the metagraph representation of a neural network is created using metagraph agent $ag^{MC}$.

The advantage of the metagraph approach is that according to modeling tasks the complexity of created neuron structure can be various. In the simplest case, the neuron may be considered as a node with activation function. In more complex cases the neuron may be represented as a nested metagraph, which contains metavertices with complex activation function addressing neuron structure.

The key feature of the proposed neural network architecture is its sparsity which means the absence of some connections in comparison with the fully connected layer.

Since the structure of the neural network is generated by the metagraph agent, the input features are automatically converted into the corresponding sparse fragments of the neural network.

At the end of the creation mode, the "Neural Network" (NN) structure is created. It may be a flat graph of nodes (neurons) connected with edges. But also, nodes may be represented as complex metavertex and neurons of each layer of the network may also be combined into metavertex.

In the training mode, the "Training Metagraph" (TM) is created. TM structure is isomorphic to the NN structure. For each node $NN_i^n$ in NN, the corresponding metavertex $TM_i^n$ in TM is created. And for each edge $NN_i^e$ in NN, the corresponding edge $TM_i^e$ in TM is created. For the TM creation, the agent $ag^{MT}$ is used. Agent $ag^{MT}$ implements a particular training algorithm. As a result of training, the changed weights are written to the $TM_i^n$.

The agent $ag^{MR}$ is used for execution of the TM model.

In the proposed case the $ag^{MO}$ is not used because it is not necessary to change the network topology in the process of operation.

Thus, the metagraph approach allows representing the sparse neural network based on the input feature set.

## 5   The Software Implementation of the Proposed Architecture

When developing software implementation, special attention was paid to its performance. The sparsity of neural network connections may be implemented with deep learning frameworks but specific implementation optimized under the concrete architecture allows significantly increase performance as presented on benchmarks below.

The neural net architecture is statically set and stored into a two-dimensional array. The input data is a set of $L$ features each of which has $N$ different values. So, the input space is divided into $L$ ranges where each range corresponds to one feature. For this architecture, the number of hidden layer neurons is $L * (L + 1)/2$. In the developed implementation, only nonzero coefficients are stored. Due to fixed architecture, it is possible to predetermine the size of such a matrix with nonzero coefficients and create the matrix of neural network connections. In this matrix, the list of affected hidden layer neurons corresponds to each feature on the input layer.

As a result, for each input layer neuron, we can get the numbers of the connected hidden layer neurons. Since each neuron of the hidden layer is connected to exactly $L$ neurons of the hidden layer, weights of the hidden layer neurons can be stored in a matrix with size $N \cdot L \times L$. In case of a fully connected network of similar size, to store hidden layer weights $L/2$ times larger matrix would be required.

The summation for hidden layer neurons is presented in Fig. 5. The key feature of this process is that it is proceeded by input layer neurons. First, contributions to the hidden layer neurons are calculated from the first feature, then from the second feature and etc. Such way of summation allows getting acceleration due to the local location of the components (weights) in the memory.

We compare the performance of three implementations. The first one has been described above in details. The second one is based on the Lasagne library [8]. Lasagne

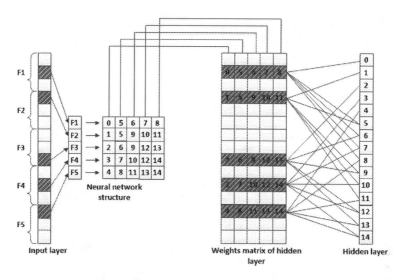

**Fig. 5.** Summation for hidden layer neurons

is a high-level library to build and train neural networks in Theano. Theano supports different formats of sparse matrices, so it is necessary only to write custom layer with sparse input. The third implementation was written on the base of the popular Pytorch library [9]. On the moment of conducting experiments, the last stable version of library 0.4.1 (July 2018) didn't support the automatic calculation of sparse gradients for summation or multiplication of matrices. So, implementation on the base of Pytorch uses dense tensors instead of sparse. The benchmarks are presented in Fig. 6.

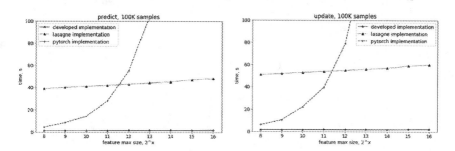

**Fig. 6.** Benchmarks of different implementations of proposed neural network

By x-axis feature, max size is plotted (all features have the same size). Because some features have large values (for example, banner id) they are mapped into fixed size vector using hashing trick. The size of such vector may be different and it is the variable parameter by the x-axis. By y-axis, the time of predict and update functions for 100 K samples are plotted. The figure shows that developed implementation is 20–30 times faster than implementation on the base of Lasagne. The Pytorch implementation shows exponential growth depending on the feature max size because it uses dense tensors. Thus, the described implementation is profitable.

## 6  The Experiments

The experimental analysis has been conducted on the CTR prediction. It is a regression problem. The model performance is estimated by several metrics. The first one is the cross-entropy or logistic loss value (logloss). This function has near to zero values if predicted, and real output differs slightly. So, it characterizes not only ranking quality but also adjacency of real and output values. The second metric is the area under the ROC curve. It characterizes only ranking quality. ROC curve allows well estimating the model quality in the case of unbalanced classes. The CTR prediction is such a case because the number of clicks is significantly less than the number of shows. Also, the area under ROC may be interpreted as the fraction of properly ranged pairs, i.e. fraction of pairs (A, B) with $p_A > p_B$ for which $\hat{p}_A > \hat{p}_B$ [10].

The first series of experiments is devoted to training models in single pass mode. Only a single pass by train set was done while model learning. Simple SGD was used as optimization algorithms. The four models were compared: the logistic regression model with raw features (lr-simple-features), the logistic regression with handcraft features (lr-handraft) and neural network models with raw features (with ReLU and hyperbolic tangent activations – nn-relu and nn-tanh accordingly). The results are presented in Fig. 7. It is worth noting that logistic loss values are normed on CTR entropy (1):

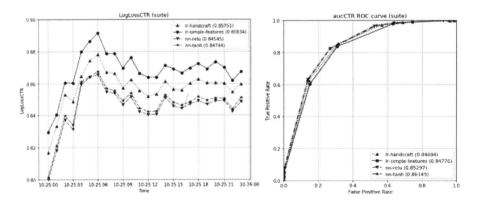

**Fig. 7.** Training models in single pass mode

$$norm = CTR \cdot \ln(CTR) + (1 - CTR) \cdot \ln(1 - CTR) \qquad (1)$$

Researchers from Facebook [11] state that such normalization serves to compare logloss values from different test sets. It is not difficult to see that by metrics values neural network model has better performance compared to not only logistic regression with raw features but also to logistic regression with handcrafted features. Also, it is worth noting that in such mode model isn't overfit (because the model has only single pass by data) and it is no need to apply regularization techniques.

In Fig. 8 experiments with multiple passes and early stopping technique are presented. It this case models were trained until the error value on validation set began to grow. For validation set, 10% data from train set were taken. Despite the decreasing of the training set, the result was improved due to preventing overfitting.

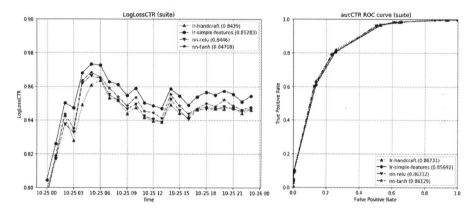

**Fig. 8.** Multiple pass training with early stopping

It is not difficult to see that with the proper organization of multiple passes by data the logistic regression improves its performance metrics, unlike the neural network. In result, it achieves a slightly better performance than the neural network. So, it may conclude that when training with the vanilla SGD the logistic regression model with handcrafted features is operated the same way as the proposed neural network with a simple listing of features. The neural network is trained faster, but the logistic regression may be trained slightly better with multiple passes. However, even with the same result, the neural network has a great advantage because it is not needed in manually selecting configurations of features.

## 7   Conclusion

So, the main contribution and innovation of the article are the proposed neural net architecture for solving online learning tasks without handcrafted features and fast software implementation for this architecture. The proposed neural network and the logistic regression model with handcrafted features work about equal in performance. Generally, the main advantage of the proposed model is the automatic feature selection which simplifies the researcher's work. Due to optimized custom implementation, the proposed neural net operates 20–30 times faster than analogous neural net implemented on the base of deep learning frameworks and only 20–30% slower than simple logistic regression with handcrafted features. Therefore, we call the proposed architecture "lightweight". So, the proposed neural net with custom implementation may be applied to high loaded real-world problems. The metagraph approach helps to represent the proposed architecture at a high level of abstraction.

# References

1. Guyon, I., Gunn, S., Nikravesh, M., Zadeh, L.A. (eds.): Feature Extraction: Foundations and Applications. Studies in Fuzziness and Soft Computing. Springer, Heidelberg (2008)
2. Manning, C.D., Raghavan, P., Schutze, H.: Introduction to Information Retrieval. University Press, Cambridge (2010)
3. Peng, H., Long, F., Ding, C.: Feature selection based on mutual information: criteria of max-dependency, max-relevance, and min-redundancy. IEEE Trans. Pattern Anal. Mach. Intell. **27**, 1226–1238 (2005)
4. LeCun, Y., Bengio, Y., Hinton, G.: Deep learning. Nature **521**, 436–444 (2015)
5. Bishop, C.: Pattern Recognition and Machine Learning. Springer, Heidelberg (2006). https://doi.org/10.1007/978-1-4615-7566-5
6. Ruder, S.: An Overview of Gradient Descent Optimization Algorithms (2016)
7. http://ruder.io/optimizing-gradient-descent/
8. Fedorenko, Y.S., Gapanyuk, Y.E.: The neural network with automatic feature selection for solving problems with categorical variables. In: Kryzhanovsky, B., Dunin-Barkowski, W., Redko, V., Tiumentsev, Y. (eds.) NEUROINFORMATICS 2018. SCI, vol. 799, pp. 129–135. Springer, Cham (2019). https://doi.org/10.1007/978-3-030-01328-8_13
9. Lasagne Documentation (2019). https://lasagne.readthedocs.io/en/latest/
10. PyTorch Documentation (2019). https://pytorch.org/docs/stable/index.html
11. Obuchowski, N.A., Bullen, J.A.: Receiver operating characteristic (ROC) curves: review of methods with applications in diagnostic medicine. Phys. Med. Biol. **63**, 1361–1367 (2018)
12. He, X., et al.: Practical lessons from predicting clicks on ads at Facebook. In: Eighth International Workshop on Data Mining for Online Advertising, ADKDD 2014, pp. 1–9. ACM, New York (2014)
13. Fedorenko, Y.S., Gapanyuk, Y.E.: Multilevel neural net adaptive models using the metagraph approach. Opt. Mem. Neural Netw. **25**(4), 228–235 (2016). https://doi.org/10.3103/S1060992X16040020
14. Fedorenko, Y.S., Gapanyuk, Y.E., Minakova, S.V.: The analysis of regularization in deep neural networks using metagraph approach. In: Kryzhanovsky, B., Dunin-Barkowski, W., Redko, V. (eds.) Advances in Neural Computation, Machine Learning, and Cognitive Research, pp. 3–8. Springer International Publishing, Cham (2018). https://doi.org/10.1007/978-3-319-66604-4_1

# Graph Convolution and Self Attention Based Non-maximum Suppression

Zhe Qiu and Xiaodong Gu[✉]

Department of Electronic Engineering, Fudan University, Shanghai, China
xdgu@fudan.edu.cn

**Abstract.** Non-maximum suppression is an integral and last part of object detection. Traditional NMS algorithm sorts the detection boxes according to their class scores. The detection boxes with maximum score are always selected while all other boxes with a sufficient overlap with the preserved boxes are discarded. This strategy is simple and effective. However, there still need some improvements in this process because the algorithm makes a 'hard' decision (accept or reject) for each box. In this paper, we formulate the non-maximum suppression as a rescoring process and construct a network called NmsNet which utilizes graph convolution and self attention mechanism to predict each box as an object or redundant one. We evaluate our method on the VOC2007 dataset. The experimental results show that our method achieves a higher MAP compared with the traditional greedy NMS and the Soft NMS.

**Keywords:** Graph convolution · Self attention · Non-maximum suppression

## 1 Introduction

Object detection is a key problem in the field of computer vision and has many comprehensive applications, such as self-driving car, robotics and video surveillance. Generally, almost all object detectors are consist of three modules: (1) generating a set of candidate windows. The typical methods include selective search [1], region proposal network [2] and so on. (2) refining these bounding boxes using regressor and classifier, (3) removing the accurate but redundant boxes. The last step is usually called non-maximum suppression (NMS). To illustrate the importance of NMS, in Table 1, we report the MAP of two detectors, namely, Faster R-CNN [2] and SSD [3], with and without NMS. Note that the Faster R-CNN was trained on VOC2007 trainval set while the SSD300 was trained on VOC2007 and 2012. It is clear that the NMS gives a huge improvement to each detector (0.460 for Faster R-CNN and 0.461 for SSD300). This evaluation reveals that the NMS is indispensable and has a strong impact on the final result.

Traditional NMS algorithm usually ranks all bounding boxes according to their class scores and then makes a 'hard' decision (accept or reject) for each

© Springer Nature Switzerland AG 2019
H. Lu et al. (Eds.): ISNN 2019, LNCS 11554, pp. 77–86, 2019.
https://doi.org/10.1007/978-3-030-22796-8_9

**Input**: $B = \{b_1,..,b_N\}, S = \{s_1,..,s_N\}, Th$
  $B$:list of detection boxes
  $S$: corresponding detection scores
  $Th$: NMS threshold
**Begin**:
  $D \leftarrow \emptyset$
  **while** $B \neq \emptyset$ **do**
   $m \leftarrow argmax\ S$
   $M \leftarrow b_m$
   $D \leftarrow D \cup M, \quad B \leftarrow B - M$
   **for** $b_i$ in $B$ **do**
    **if** $iou(M, b_i) \geq Th$ **then**
     $B \leftarrow B - b_i; \quad S \leftarrow S - s_i$
    **end**
   **end**
  **end**
  **return** $D, S$
**end**

Remove redundant boxes

**Fig. 1.** The pseudo code of traditional NMS algorithm.

**Table 1.** MAP of faster R-CNN and SSD with and without NMS.

| MAP | Detector | NMS |
|---|---|---|
| 0.752 | Faster R-CNN | Yes |
| 0.292 | Faster R-CNN | No |
| 0.777 | SSD300 | Yes |
| 0.316 | SSD300 | No |

box using a iou threshold. Figure 1 shows the details of this process. The NMS algorithm starts with a set of bounding boxes $B$ and their corresponding scores $S$ and then selects the box $M$ with the maximum score. After that, the algorithm removes the redundant boxes which have a significant iou (higher than the threshold) with the box $M$. This process continues until all boxes are selected or removed. Inspite of the great success of this traditional NMS algorithm, it has some problems. The major issue is that the algorithm makes a 'hard' decision (accept or reject) for each box which means it sets the score of neighboring detections to zero. Thus, if an object presents with a high iou (higher than threshold), it may be removed incorrectly and this would lead to a loss of average precision. We illustrate this problem in an intuitive fashion in Fig. 2. Furthermore, it is almost impossible to deal with a variety of situations with only one NMS threshold. In other words, the value of the threshold is difficult to determine.

In contrast to traditional greedy NMS, some methods [4,5] formulate the non-maximum suppression as a rescoring process. In these algorithms, the scores of neighboring boxes are not set to zero but decayed with a iou function which decreases as the rise of iou. In this paper, we follow these settings and propose an end-to-end network which is constructed from graph convolution and self attention to rescore each bounding box. In other words, our network learns how

**Fig. 2.** In this case, the detector predicts two confident person detections which are shown in dashed and solid line. The dashed box (with score 0.8) has a significant overlap with the solid one (with score 0.9). Thus, the dashed box will be suppressed incorrectly in the process of traditional NMS algorithm. The better way may be preserving the dashed box but decaying its score to a lower one such as 0.4.

to suppress redundant boxes. We describe the details of our method in the Sect. 3 and report the evaluation results in Sect. 4. Finally, the conclusion of this paper can be found in Sect. 5.

## 2 Related Work

The greedy NMS algorithm has been a integral part of detection tasks for many years and has been applied to some state-of-art detectors [2,3]. However, as we described above, this greedy algorithm has some flaws. In terms of these problems, some improved methods have been proposed in recent years. Here, we list two representative methods [4,5] (Soft NMS and GossipNet) which have motivated us to develop our network.

### 2.1 Soft NMS

The Soft NMS [4] revisits the greedy NMS algorithm in detail and rewrites the removing step as a rescoring function which is shown in Eq. (1).

$$s_i = \begin{cases} s_i, & iou(M, b_i) < N_t \\ 0, & iou(M, b_i) \geq N_t \end{cases} \tag{1}$$

Note that the denotation of each element above is similar to the one in Fig. 1. Therefore, the greedy NMS algorithm sets a hard threshold ($N_t$) to make a decision (accept or reject) for each neighborhood of $M$. As we described above, decaying the scores of neighboring boxes seems to be a feasible improvements. Hence, the Soft NMS formulates this rule as follows.

$$s_i = s_i e^{-\frac{iou(M,b_i)^2}{\sigma}} \tag{2}$$

The main idea behind this equation is that the algorithm applys a gaussian penalty function to each box and the higher the iou value, the more penalty will be applied to the score of box $b_i$. It is worth noting that the Soft NMS is also a greedy algorithm. It does not find the globally optimal solution and performs the penalty in a greedy style. More details of Soft NMS can be found in [4].

### 2.2  GossipNet

Similar to us, the GossipNet [5] also uses a neural network to rescore the detection boxes. The main components of the net are pair wise computations and fully connected layers. To get the feature of box $b_i$, the GossipNet collects handcrafted features of all neighboring boxes which have a significant overlap (IOU >0.2) with $b_i$ and then performs a global max-pooling over these related features. After that, the fully connected layers are used to map these features to a high level. In contrast to GossipNet, our method does not have any handcrafted steps. We get the related features of each box by stacking some graph convolutional layers. Furthermore, we also append a self attention module to these graph convolutional layers to extract more semantic information. We describe the details of our end-to-end NmsNet in next section.

## 3   Our NmsNet

After introducing the motivations of our design, this section presents the details of our NmsNet. The NmsNet is stacked by several identical building blocks and each block is composed of 4 graph convolutional layers, 1 self attention module and 1 fully connected layer. Figure 3 illustrates the architecture of our NmsNet. In our design, each detection box with corresponding score is firstly mapped to a feature vector by an encoding module. Then, these low level features are fed to several identical blocks. Note that to accelerate the training procedure, we perform layer normalization [6] for each minibatch before graph convolution and self attention. After processing of the base network, we use a matching module with ground truth boxes to generate training label for each detection box. Finally, a sigmoid cross entropy loss is computed for the entire network. We illustrate the main components of our NmsNet as follows.

### 3.1  Encoding Module

As we analysed above, for each box, the non-maximum suppression is related to three factors including the position in the entire image, the iou with adjacent boxes and the corresponding scores. Based on this analysis, we encode the

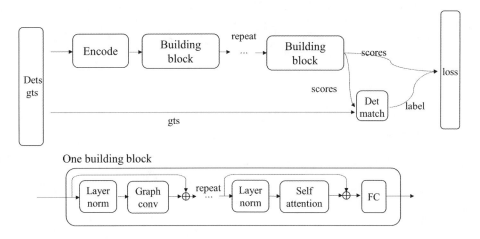

**Fig. 3.** The architecture of our NmsNet.

detection boxes as following equations.

$$f_i = (\frac{xmin_i}{W}, \frac{ymin_i}{H}, \frac{xmax_i - xmin_i}{W}, \frac{ymax_i - ymin_i}{H}, log(\frac{w_i}{h_i})) \qquad (3)$$

$$f_{ij} = (\frac{2\Delta xmin_{ij}}{w_i + w_j}, \frac{2\Delta ymin_{ij}}{h_i + h_j}, log(\frac{w_i}{w_j}), log(\frac{h_i}{h_j}), log(\frac{\alpha_i}{\alpha_j}), iou_{ij}) \qquad (4)$$

$$h_i = max\{f_{i1}, \cdots, f_{ik}, \cdots, f_{im} | iou_{ij} > 0.2\} \qquad (5)$$

$$F_i = concat(f_i, h_i, s_i) \qquad (6)$$

In the above equations, $xmin$, $ymin$, $xmax$, $ymax$ indicate the coordinates of top-left and bottom-right corners, whereas the $w_i$, $h_i$, $W$, $H$ denote the width and height of detection boxes and images. In addition, we use $\Delta xmin_{ij}$ and $\Delta ymin_{ij}$ to signify the distance of two detection boxes $i, j$ in the directions x and y. To make this feature more discriminative, we also use an aspect ratio difference item $log(\frac{\alpha_i}{\alpha_j})$. In short, we present the three factors as we described above respectively in the Eqs. (3–5). The final feature of each detection box is made up of three subfeatures $f_i, h_i, s_i$.

## 3.2 Graph Convolution

To get the related features between each detection boxes, we try to use the convolutional neural network. However, the input encoded features are not sequence data like images and texts. Thus, it is impossible to apply the normal convolutional layer to these features. Inspired by [7], we treat them as graph-based data whose structure is determined by the iou matrix. For each node (detection box) in the graph, we select $K - 1$ neighbors to form the receptive field where $K$ means the kernel size of convolutional layer. We call this process Graph Normalization and present an example in Fig. 4(b). In this example, we choose a

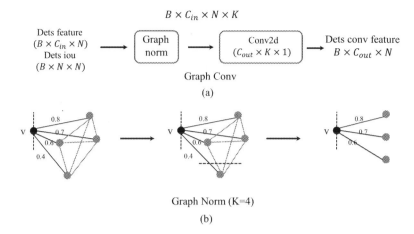

**Fig. 4.** The graph convolution.

convolutional kernel with size 4 and consider a node $v$ which is only adjacent to 4 boxes (nodes). To construct the receptive field, we rank the adjacent nodes according to their iou values and then choose the top 3 nodes (The receptive field includes the node itself). Following this rule, we discard all the extra nodes (the bottom node in this example). If the adjacent nodes are insufficient, we pad the zero node to this receptive field. After graph normalization, the shape of input tensor $(B \times C_{in} \times N)$ will be reorganized to $B \times C_{in} \times N \times K$ where $B, N, K$ denote batch size, number of nodes (detection boxes) and kernel size. Once we normalize the input tensor successfully, we apply a convolutional layer with kernel size $K \times 1$ to this reorganized tensor. Thus, the graph convolution in our NmsNet will not change the shape of input feature. In this paper, we use 4 graph convolutional layers with kernel size 5, 5, 7, 7 in each building block.

### 3.3 Self Attention

The attention function usually can be formulated as mapping a set of queries, keys and values to an output vector (In self attention, the queries, keys and values are derived from the same input). In our NmsNet, we adopt the multi-head attention mechanism which is introduced in [8,9]. For each node (box), the multi-head attention computes the weighted sum of all nodes (boxes) based on the similarity of inputs which is measured by scale dot product function. Equation (7) illustrates the scale dot product function. In this equation, $Q, K, V$ denote the queries, keys and values respectively while $d_k$ means the dimension of queries and keys.

$$Attention(Q, K, V) = softmax(\frac{QK^{\mathrm{T}}}{\sqrt{d_k}})V \tag{7}$$

It is worth noting that the multi-head attention does not perform the single attention function with the input vectors but first linearly projects the features

**Input**: $B = \{b_1,..,b_N\}, S = \{s_1,..,s_N\}, G = \{g_1,..,g_K\}$
  $B$:detection boxes,   $S$: detection scores,   G: gt boxes
**Begin**:

> $L \leftarrow \{l_1,..,l_N\}, l_i = 0$
> **while** $B \neq empty$ **do**
>> $m \leftarrow argmax\ S,\quad I \leftarrow 0.5$
>> $B \leftarrow B - b_m,\quad S \leftarrow S - s_m$
>> $M = \{m_1,..,m_K\}, m_i = 0,\quad match \leftarrow -1$
>> **for** $g_i\ in\ G$ **do**
>>> **if** $iou(b_m, g_i) < I$ **or** $m_i > 0$ **then**
>>>> continue
>>>
>>> **end**
>>> $I \leftarrow iou(b_m, g_i),\quad match \leftarrow i$
>>
>> **end**
>> **if** $match > -1$ **then**
>>> $m_{match} \leftarrow 1,\ l_i \leftarrow 1$
>>
>> **end**
>
> **end**
> **return** $L$

**end**

$g_i$ is already matched or $b_m$ is not a better one

**Fig. 5.** The pseudo code of matching strategy.

$h$ times and then applies the attention function to these features in parallel. The outputs are concatenated and projected again to get the final results. This process is interpreted in Eqs. (8–9). In our NmsNet, we first linearly projects the graph convolutional feature tree times and take these projected feature as the input of $Q, K, V$.

$$MultiHead(Q, K, V) = Concat(head_1, \cdots, head_h)W^O \tag{8}$$

$$head_i = Attention(QW_i{}^Q, KW_i{}^K, VW_i{}^V) \tag{9}$$

### 3.4   Detection Matching Module

The detection matching module takes the new scores and the ground truth boxes as input and generates the training label for each detection box. Revisiting the evaluation criterion of the benchmark, we found that the evaluation algorithm typically sorts the detection scores and matches each detection box to ground truth object. Note that each object can only be matched once. This strategy prefers the detection box with high iou and ensures that each object only has one detection result and all other boxes are false positive one. We use this evaluation strategy in the training phase of our NmsNet. Figure 5 shows the pseudo code of this algorithm. As we can see, the algorithm sets the box $b_m$ negative because it is not the better one (has higher iou) or the matched object $g_i$ is already selected.

## 4   Experimental Results

We evaluate our NmsNet on the VOC2007 Dataset. In the next few sections, we will compare our method with the traditional greedy NMS algorithm and the recently proposed Soft NMS. As for the evaluation of other detectors and datasets, we are still in practice.

### 4.1 Training Details

The VOC2007 dataset contains 9963 annotated images with 246640 labeled objects which belongs to 20 classes. In this dataset, about 5012 images are used to train the object detector. Before training our NmsNet, we first train a Faster R-CNN detector on the VOC2007 dataset and then generate detection results for each image in the training set. In this stage, we follow the settings in original Faster R-CNN paper [2] which means we set the learning rate 0.001 for the first 50 K mini-batches, and 0.0001 for the next 30 K iterations. After that, we train the NmsNet with detection results generated by Faster R-CNN detector, starting with a learning rate 0.001. Furthermore, We use the SGD (stochastic gradient descent) algorithm to train the Faster R-CNN whereas optimize the NmsNet with Adam.

### 4.2 Comparisons with Other Methods

We compare our NmsNet with the traditional greedy NMS and the recently proposed Soft NMS. The evaluation results are reported in Table 2. It is clear that our NmsNet improves the performance in most cases. For example, we get an improvement of 2.04% and 1.10% over classes bus and aeroplane. In terms of the metric MAP (mean average precision), our method achieves a significant promotion of 0.36% compared with Soft NMS, not to mention the greedy NMS. One may note that our method gets the worse result in some cases, such as bottle, bird, chair, pottleplant and so on. The reason for this phenomenon may be that the ground truth objects in these classes are small and thus the encoded features which are defined in Eqs. (3–6) are not discriminative like features of other classes. As for the efficiency, our NmsNet takes only 0.04 s to process detection results of one image whereas the Faster R-CNN needs 0.2 s to detect objects for each image. This means that the non-maximum suppression takes much less time than the detection phase (0.04 s vs 0.2 s). Last but not least, our NmsNet dose not receive any image features but only the information from detection results.

### 4.3 Visualize the NmsNet Feature

To figure out what the NmsNet learns, we randomly choose an example which contains four ground truth objects from the validation set and feed this sample to the network. After that, we transform the retrieved features with high dimension ($N \times 12$ for input feature, $N \times 32$ for output feature) to visible features by PCA (principal component analysis). The visualization result is presented in Fig. 6. We use the red dot to show the true positive samples while the false positives are illustrated by blue one. Visualizing the input encoded feature (Fig. 6(a)), we find that the detection samples are divided into four parts and each part is centered on a true positive example. In addition, all true positives are overwhelmed by their neighborhood or in other words, the redundant boxes. This is in line with our experimental results (In Table 1, the detector without NMS will have poor

**Table 2.** Comparisons with greedy nms and soft nms.

| Method | MAP | Aeroplane | Bicycle | Boat | Bottle | Bus | Car | Cat |
|---|---|---|---|---|---|---|---|---|
| NMS | 0.7516 | 0.7702 | 0.7909 | 0.6402 | 0.6040 | 0.8210 | 0.8555 | 0.8712 |
| Soft-NMS | 0.7623 | 0.7867 | 0.8144 | **0.6680** | **0.6050** | 0.8249 | 0.8619 | 0.8700 |
| NmsNet (ours) | **0.7659** | **0.7977** | **0.8273** | 0.6513 | 0.5911 | **0.8453** | **0.8677** | **0.8949** |

| Method | Diningtable | Cow | Dog | Pottedplant | Sofa | Tvmonitor | Person |
|---|---|---|---|---|---|---|---|
| NMS | 0.6749 | 0.8385 | 0.8651 | 0.5051 | 0.7427 | 0.7765 | 0.7810 |
| Soft-NMS | 0.6821 | **0.8454** | 0.8682 | **0.5214** | 0.7389 | **0.7791** | **0.8143** |
| NmsNet (ours) | **0.7184** | 0.8414 | **0.8880** | 0.5091 | **0.7492** | 0.7734 | 0.7887 |

| Method | Horse | Motorbike | Bird | Chair | Sheep | Train | - |
|---|---|---|---|---|---|---|---|
| NMS | 0.8392 | 0.7994 | 0.7754 | 0.5473 | 0.7601 | 0.7733 | - |
| Soft-NMS | 0.8473 | 0.8045 | **0.7825** | **0.5469** | **0.7727** | 0.8225 | - |
| NmsNet (ours) | **0.8546** | **0.8141** | 0.7635 | 0.5335 | 0.7673 | **0.8313** | - |

result on the metric of MAP). In contrast to input encoded feature, the output NmsNet feature seems to be more discriminative as shown in Fig. 6(b).

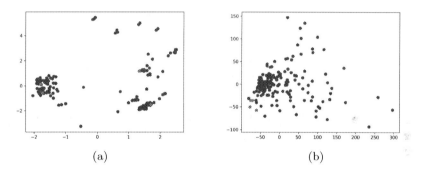

(a)                    (b)

**Fig. 6.** The visualization of input encoded feature and output NmsNet feature, (a) input encoded feature, (b) output NmsNet feature.

## 5   Conclusion

In this paper, we propose a novel network called NmsNet to deal with non-maximum suppression. Our method is equipped with graph convolution and self attention mechanism. By doing so, the proposed NmsNet predicts each detection box as true positive sample or redundant one. The experimental results on VOC2007 dataset demonstrate the effectiveness of our NmsNet. It is worth noting that our NmsNet takes no image features but only the information extracted from detection itself. In the future work, it may be interesting to add the ROI (region of interest) feature to the NmsNet. In conclusion, our proposed NmsNet is an effective way to perform non-maximum suppression with acceptable time cost.

**Acknowledgement.** This work was supported in part by National Natural Science Foundation of China under grants 61771145 and 61371148.

# References

1. Uijlings, J.R., Van De Sande, K.E., Gevers, T., Smeulders, A.W.: Selective search for object recognition. Int. J. Comput. Vis. **104**(2), 154–171 (2013)
2. Ren, S., He, K., Girshick, R., Sun, J.: Faster R-CNN: towards real-time object detection with region proposal networks. IEEE Trans. Pattern Anal. Mach. Intell. **39**(6), 1137–1149 (2017)
3. Liu, W., et al.: SSD: single shot MultiBox detector. In: Leibe, B., Matas, J., Sebe, N., Welling, M. (eds.) ECCV 2016. LNCS, vol. 9905, pp. 21–37. Springer, Cham (2016). https://doi.org/10.1007/978-3-319-46448-0_2
4. Bodla, N., Singh, B., Chellappa, R., Davis, L.S.: Soft-NMS—improving object detection with one line of code. In: Proceedings of the IEEE International Conference on Computer Vision, pp. 5562–5570. IEEE Press, Venice (2017)
5. Hosang, J., Benenson, R., Schiele, B.: learning non-maximum suppression. In: Proceedings of the IEEE Conference on Computer Vision and Pattern Recognition, pp. 6469–6477. IEEE Press, Honolulu (2017)
6. Ba, J.L., Kiros, J.R., Hinton, G.E.: Layer normalization. arXiv preprint arXiv:1607.06450 (2016)
7. Niepert, M., Ahmed, M., Kutzkov, K.: Learning convolutional neural networks for graphs. In: Proceedings of the 33rd International Conference on Machine Learning, pp. 2014–2023. ACM, New York (2016)
8. Vaswani, A., et al.: Attention is all you need. In: Advances in Neural Information Processing Systems, pp. 5998–6008 (2017)
9. Yu, A.W., et al.: QANet: combining local convolution with global self-attention for reading comprehension. In: International Conference on Learning Representations (2018)

# Dynamic Graph CNN with Attention Module for 3D Hand Pose Estimation

Xu Jiang and Xiaohong Ma$^{(\boxtimes)}$

School of Information and Communication Engineering,
Dalian University of Technology, Dalian, China
jiangxu@mail.dlut.edu.cn, maxh@dlut.edu.cn

**Abstract.** Recently, 3D hand pose estimation methods taking point cloud as input show the most advanced performance. We present a new 3D deep learning hand pose estimation network for an unordered point cloud. Our approach utilizes EdgeConv layer as the basic element, where an attention embedding version EdgeConv layer is proposed for feature extraction in hand pose estimation task. To improve the result, we design a hand pose improvement network that inputs points whose are in the neighbor of the estimated fingers and outputs a rectify hand pose. We evaluate our method on several famous datasets to prove that our method can get excellent result compared to some most advanced methods.

**Keywords:** 3D hand pose estimation · Point cloud ·
Attention embedding module

## 1 Introduction

Recently, 3D hand pose estimation has aroused great concerns because of its correlative to a broad range of human-computer interaction applications, or augmented reality applications. Nevertheless, there are still some difficulties in exploiting a highly accurate hand pose estimation method, due to the quality of depth images are bad, high flexibility of joints and great variations of hand posture.

Early studies on hand pose estimation mainly input depth image and regress hand pose with a simple convolutional neural networks (CNNs) [1,15,31,33], which cannot make use of spatial structure from depth image. Recently, 3D representation data have shown outperformance than depth image based methods. One method is to transfer 2D depth image to 3D volumetric representation, thus a 3D CNN can be used to regress hand pose [5,13]. But the voxelized data are too computationally in data format converting and network complexity. Another method is to use point cloud as input [2,3,6]. Recently a novel network structure for unordered point cloud named PointNet, has shown great process in many

---

X. Jiang—This project was partially supported by the National Natural Science Foundation of China (Grant No. U1708263).

© Springer Nature Switzerland AG 2019
H. Lu et al. (Eds.): ISNN 2019, LNCS 11554, pp. 87–96, 2019.
https://doi.org/10.1007/978-3-030-22796-8_10

tasks. PointNet utilizes symmetric function to extract global feature from individual points which is invariant to the different sequences of the point cloud. Ge *et al.* use PointNet [17, 18] to estimate hand pose with comparable performance. PointNet is a highly efficient and effective network while it lacks the ability to make fully use of local information on the point sets. To better capture the local information of the point sets, EdgeConv [29] is proposed to generate edge features which can describe the relationships between a point and its neighbors while maintaining permutation invariance.

The architecture is shown in Fig. 1, we present a dynamic graph CNN with attention module network for hand pose estimation task. Using EdgeConv [29] as the basic element, we present an attention embedding version of EdgeConv (Att-EdgeConv) to construct a deep network for feature extraction from the point cloud. Then a hand pose improvement network is used to further improve the result by taking points whose are in the neighbor of the estimated hand pose as input. To the best of our knowledge, it is the first time that EdgeConv [29] is used to regress 3D hand pose from point cloud, which can better utilize 3D spatial structure of the point cloud and regress hand joints in real time.

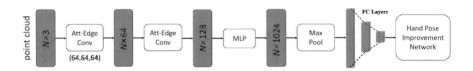

**Fig. 1.** The architecture of our proposed dynamic graph CNN with attention module network. Our network treats $N$ points as input, calculating edge features for each point by an Att-EdgeConv layer, and aggregates features within each set to compute Att-EdgeConv responses for corresponding points. The output features of the last Att-EdgeConv layer are aggregated globally to form an global descriptor, which is used to regress hand pose with several fully-connected layers. The hand pose improvement network can further increase the regression accuracy of hand joints locations.

The contributions of this paper are summarized as below:

- We present a new deep learning based hand pose estimation method for unordered point cloud. Using EdgeConv as the basic element to handle hand pose estimation task. Compared with PointNet based methods, our method can better capture local geometric features of point clouds while still maintaining permutation invariance.
- We design an attention embedding version of EdgeConv. From a great number of former layer features, our approach can generate weighted fusion features of point cloud models, which can capture value information across a large number of elements of a set of point cloud.

## 2   Related Work

**Depth-Based Methods.** There are normally three categories of approachs for hand pose estimation: generative approaches, discriminative approaches and hybrid approaches. Getting a predefined model, generative methods make the model suitable for observation, *i.e.* depth data including depth images, 3D point cloud and so on. Discriminative methods learn a predictive model from a great quantity of training data which maps input depth data to 3D hand gesture parameters such as joint coordinates [8,19,24,26]. Hybrid methods aim at combining the advantages of two methods mentioned above [11,20,23,32].

In the following, we concentrate on CNNs-based approachs. CNNs-based 3D hand pose estimation approach is first presented in [25]. They predict hand joints locations from 2D heat-maps by CNN. To withstand the negative effect that it cannot utilize 3D spatial structure in 2D heat-maps, Ge *et al.* [4] apply a structure of multi-view projection and fusion to estimate hand pose. Sun *et al.* [21] use hand-crafted viewpoint-invariant features to refine hand pose iteratively. Oberweger *et al.* [15] introduce a method that embedding constraint priori into CNNs in low dimensional space, then with some simple methods to increase accuracy [14]: adding ResNet mechanism, data augmentation, and better initial hand pose. Ye *et al.* [32] introduce an attention based structure with an iterative refinement strategy. Guo *et al.* [7,28] divide feature maps into several blocks and group them in a hierarchical way to regress the whole hand pose.

**3D Deep Learning.** 3D data can make fully use of 3D spatial structure information while it usually cannot directly utilize conventional CNN based method that worked on 2D images. Thus, some methods process the 3D data into 3D voxel representation and use 3D CNN to extract features [5,13,30]. Another way is to process the 3D data into 3D point cloud. Qi *et al.* [17] presented a novel network named PointNet, which can handle unordered point sets with symmetric operation. But PointNet [17] cannot make fully use of local information from the point sets. To solve the problem, some methods are provided. PointNet++ [18] develops PointNet by a hierarchical way to extract local features with different levels; Deep Kd-networks [9] directly adopts a Kd-tree structure to handle point cloud. Self-Organizing Net [10] produces a Self-Organizing Map to represent the point cloud. Dynamic Graph CNN [29] constructs a local graph structure and update the edge features of the adjacent points.

## 3   Methodology

The network we presented inputs a depth image and outputs the locations of hand joints. To make fully use of the 3D spatial structure on the hand pose, we firstly convert the depth image into point cloud. The architecture of our method is shown in Fig. 1. Our method takes $N$ points $P \in R^{N \times 3}$ from the 3D point cloud and normalizes them into an oriented bounding box. Then an Att-EdgeConv layer is used as fundamental element that inputs $N$ points from the point cloud and extract features of each point in regression task. In the end, a hand pose improvement network is proposed to improve the result of hand pose. The details of our proposed method will be described in the next sections. Section 3.1 introduces the EdgeConv block. Section 3.2 describes the Att-EdgeConv block. Section 3.3 presents the hand pose improvement network and implementation details are given in Sect. 3.4.

## 3.1  EdgeConv

The main limitation of PointNet based method is that it cannot make fully use of local information from point sets. To make better use of local information from point sets, we exploit EdgeConv [29] as the basic feature extraction layer recurrently applied in network for regressing hand pose. From Fig. 2 we can know, all features are chosen as centroids of local areas; then $K$-nearest features for each centroid feature are aggregated as a local area; a multi-layer perceptron layer and a max pooling operation are used to extract new local edge features of each centroid feature. After several EdgeConv layers, a global feature is abstracted from the whole features aggregated within each set of layers by applying a multi-layer perceptron.

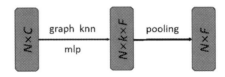

**Fig. 2.** EdgeConv block: the EdgeConv block inputs point features and computes edge features by applying a multi-layer perceptron and a max pooling.

## 3.2  Att-EdgeConv

To further increase the generalization performance of EdgeConv, we introduce an attention embedding version EdgeConv named Att-EdgeConv. The Att-EdgeConv encodes a wide range of contextual information into features which can well describe the relative relationships of different point sets and their contributions to the regress task. As illustrated in Fig. 3, given $C$-dim features of point cloud, an EdgeConv layer is employed to generate new local edge features. In the meanwhile, we also feed features into a multi-layer perceptron layer with a softmax layer to restrain the output of the attention scores to [0, 1], which represents the importance of corresponding features. Then the output of attention score branch and edge feature branch are merged in a residual connection way. We first combine the attention scores to the edge features by element-wise multiply operation to get refined features. Then we apply an element-wise sum operation with edge features and refined features to get the final output.

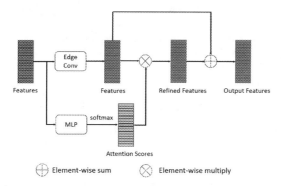

**Fig. 3.** Att-EdgeConv. It takes point cloud features as input, then with a multi-perception and a softmax to generate attention scores. Next, the scores are applied to the output features of EdgeConv in a residual way.

### 3.3   Hand Pose Improvement Network

It is apparently that the adjacent points of joint have a great influence on the regress result. Thus, we utilize a hand pose improvement network to polish up the result of hand pose. It inputs $M$ nearest points whose are in the neighbor of the estimated hand pose and outputs the improved locations of hand pose. As is illustrated in Fig. 4, the network is highly efficient and effective that we just use a simple PointNet module.

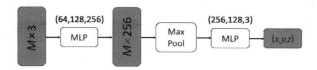

**Fig. 4.** Hand pose improvement network. $M$ nearest neighboring points of the estimated hand pose from the point cloud are fed into a simple PointNet and regress an improved hand pose.

### 3.4   Implementation Details

Our network takes point cloud as input. We first segment hand from depth image by using random decision forest (RDF) [25] and expand training data by cropping different lengths of arm, then transfer the cropped image to point clouds with the intrinsic parameters of depth camera. After that, we randomly select 1024 points from 3D point cloud and normalize them into an oriented bounding box (OBB). For EdgeConv layer, we use KNN to group features and set $K$ to 25. For hand pose improvement, we set $M$ to 64.

Our work is experimented on a computer with an NVIDIA GTX 2080Ti GPU. Our network is implemented in the Pytorch framework. In training phase, we use Adam optimizer with initial learning rate 0.001 and batch size 32. We finish the training phase after 60 epochs to avoid overfitting while we divide the learning rate by 10 after 30 epochs.

## 4    Experiments

In this section, we introduce three famous datasets and compare our approach with some advanced methods.

### 4.1    Datasets

We evaluate our method on three famous datasets: ICVL [22], NYU [25] and MSRA [21]. In ICVL dataset, there are 22 K training images and 1.6 K testing images, which each image has 16 labeled hand joints. In NYU dataset, there are 72 K training images and 8 K testing images, which each image has 14 labeled hand joints. In MSRA dataset, there are 76 K training images from nine different subjects, which each image has 21 labeled hand joints.

### 4.2    Self-comparisons

To explore the effect of our method in feature extraction, we experiment our method with two different feature extraction architectures by using a basic Point-Net and using a PointNet++ on ICVL [22] dataset. It is observed in Fig. 5 that the outcome of our method is better than the other methods. The average error of our method is 7.0 mm, which is 0.4 mm less than the basic PointNet and 0.3 mm less than the PointNet++. It shows the power of our method in hand pose feature extraction.

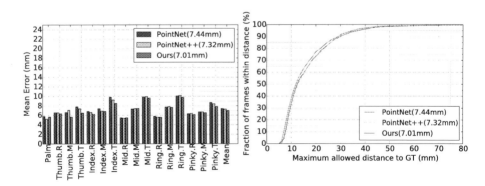

**Fig. 5.** Comparing average joint errors of our method with other structures on ICVL [22] dataset.

### 4.3 Comparisons with Advanced Methods

We evaluate our proposed approach on three famous datasets and compare with 12 advanced methods. The experiment results of the proportion of good frames over different error thresholds as well as mean error distances for per-joint with above approachs on three datasets can be seen in Figs. 6 and 7.

On ICVL [22] dataset, our approach performs the best with all these advanced approachs on the mean error distances. On NYU [25] dataset, our method outperforms some 3D deep learning methods as well as almost all 2D CNN based approachs over almost all of the error thresholds when the threshold is under 25 mm, which indicates that our method may not be robust enough with shelter occlusion. On MSRA [21] dataset, our method achieves comparable performance with other methods.

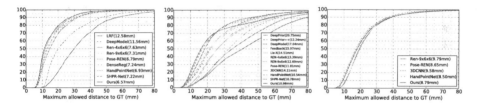

**Fig. 6.** The proportions of good frames on ICVL [22] (left), NYU [25] (middle) and MSRA [21] (right) datasets with some advanced methods.

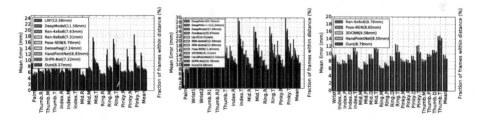

**Fig. 7.** The mean error distances on ICVL [22] (left), NYU [25] (middle) and MSRA [21] (right) datasets with some advanced methods.

Some qualitative results for ICVL [22], NYU [25] and MSRA [21] datasets can be seen in Fig. 8.

### 4.4 Time and Space Complexity

The testing time of our approach is 4.9 ms. The first step takes 1.3 ms to convert the image into point cloud and randomly select point cloud. Network forwarding

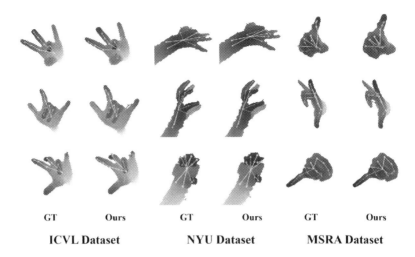

| GT | Ours | GT | Ours | GT | Ours |

**ICVL Dataset**          **NYU Dataset**          **MSRA Dataset**

**Fig. 8.** Qualitative results for ICVL [22] (left), NYU [25] (middle) and MSRA [21] (right) datasets.

is the next step, which takes 3.5 ms for hand pose estimation network and 0.1 ms for hand pose improvement network. So our approach is over 204 fps in real time. In the meanwhile, the size of our model is 10.3 MB, including 7.5 MB for hand pose estimation network and 2.8 MB for hand pose improvement network, while there is nearly 400 MB to 3D CNN [5].

## 5   Conclusion

Our work aims at exploring a generalized 3D deep learning network to regress hand joint coordinates. To better make use of the local structures from point cloud, we use EdgeConv layer as the basic element with an attention embedding module. Hand pose improvement network can further refine the result by a simple PointNet. The experimental results show that our approach makes superior performance on hand pose datasets. This approach makes a technical method for tracking in the interaction between environment and human.

## References

1. Chen, X., Wang, G., Guo, H., Zhang, C.: Pose guided structured region ensemble network for cascaded hand pose estimation. In: Neurocomputing (2018)
2. Chen, X., Wang, G., Zhang, C., Kim, T.K., Ji, X.: SHPR-Net: deep semantic hand pose regression from point clouds. IEEE Access **6**, 43425–43439 (2018)
3. Ge, L., Cai, Y., Weng, J., Yuan, J.: Hand PointNet: 3D hand pose estimation using point sets. In: Proceedings of the IEEE Conference on Computer Vision and Pattern Recognition, pp. 8417–8426 (2018)

4. Ge, L., Liang, H., Yuan, J., Thalmann, D.: Robust 3D hand pose estimation in single depth images: from single-view CNN to multi-view CNNs. In: Proceedings of the IEEE Conference on Computer Vision and Pattern Recognition, pp. 3593–3601 (2016)
5. Ge, L., Liang, H., Yuan, J., Thalmann, D.: 3D convolutional neural networks for efficient and robust hand pose estimation from single depth images. In: 2017 IEEE Conference on Computer Vision and Pattern Recognition (CVPR), pp. 5679–5688. IEEE (2017)
6. Ge, L., Ren, Z., Yuan, J.: Point-to-point regression PointNet for 3D hand pose estimation. In: Proceedings of the European Conference on Computer Vision (ECCV), pp. 475–491 (2018)
7. Guo, H., Wang, G., Chen, X., Zhang, C.: Towards good practices for deep 3D hand pose estimation. arXiv preprint arXiv:1707.07248 (2017)
8. Khamis, S., Taylor, J., Shotton, J., Keskin, C., Izadi, S., Fitzgibbon, A.: Learning an efficient model of hand shape variation from depth images. In: Proceedings of the IEEE Conference on Computer Vision and Pattern Recognition, pp. 2540–2548 (2015)
9. Klokov, R., Lempitsky, V.: Escape from cells: deep Kd-networks for the recognition of 3D point cloud models. In: Proceedings of the IEEE International Conference on Computer Vision, pp. 863–872 (2017)
10. Li, J., Chen, B.M., Hee Lee, G.: SO-Net: self-organizing network for point cloud analysis. In: Proceedings of the IEEE Conference on Computer Vision and Pattern Recognition, pp. 9397–9406 (2018)
11. Madadi, M., Escalera, S., Baró, X., Gonzalez, J.: End-to-end global to local CNN learning for hand pose recovery in depth data. arXiv preprint arXiv:1705.09606 (2017)
12. Maturana, D., Scherer, S.: VoxNet: a 3D convolutional neural network for real-time object recognition. In: 2015 IEEE/RSJ International Conference on Intelligent Robots and Systems (IROS), pp. 922–928. IEEE (2015)
13. Moon, G., Yong Chang, J., Mu Lee, K.: V2V-PoseNet: voxel-to-voxel prediction network for accurate 3D hand and human pose estimation from a single depth map. In: Proceedings of the IEEE Conference on Computer Vision and Pattern Recognition, pp. 5079–5088 (2018)
14. Oberweger, M., Lepetit, V.: Deepprior++: improving fast and accurate 3D hand pose estimation. In: 2017 IEEE International Conference on Computer Vision Workshop (ICCVW), pp. 585–594. IEEE (2017)
15. Oberweger, M., Wohlhart, P., Lepetit, V.: Hands deep in deep learning for hand pose estimation. arXiv preprint arXiv:1502.06807 (2015)
16. Oberweger, M., Wohlhart, P., Lepetit, V.: Training a feedback loop for hand pose estimation. In: Proceedings of the IEEE International Conference on Computer Vision, pp. 3316–3324 (2015)
17. Qi, C.R., Su, H., Mo, K., Guibas, L.J.: PointNet: deep learning on point sets for 3D classification and segmentation. In: Proceedings of the IEEE Conference on Computer Vision and Pattern Recognition, pp. 652–660 (2017)
18. Qi, C.R., Yi, L., Su, H., Guibas, L.J.: PointNet++: deep hierarchical feature learning on point sets in a metric space. In: Advances in Neural Information Processing Systems, pp. 5099–5108 (2017)
19. Romero, J., Tzionas, D., Black, M.J.: Embodied hands: modeling and capturing hands and bodies together. ACM Trans. Graph. (TOG) **36**(6), 245 (2017)

20. Sharp, T., et al.: Accurate, robust, and flexible real-time hand tracking. In: Proceedings of the 33rd Annual ACM Conference on Human Factors in Computing Systems, pp. 3633–3642. ACM (2015)
21. Sun, X., Wei, Y., Liang, S., Tang, X., Sun, J.: Cascaded hand pose regression. In: Proceedings of the IEEE Conference on Computer Vision and Pattern Recognition, pp. 824–832 (2015)
22. Tang, D., Jin Chang, H., Tejani, A., Kim, T.K.: Latent regression forest: structured estimation of 3D articulated hand posture. In: Proceedings of the IEEE Conference on Computer Vision and Pattern Recognition, pp. 3786–3793 (2014)
23. Tang, D., Taylor, J., Kohli, P., Keskin, C., Kim, T.K., Shotton, J.: Opening the black box: hierarchical sampling optimization for estimating human hand pose. In: Proceedings of the IEEE International Conference on Computer Vision, pp. 3325–3333 (2015)
24. Tkach, A., Tagliasacchi, A., Remelli, E., Pauly, M., Fitzgibbon, A.: Online generative model personalization for hand tracking. ACM Trans. Graph. (TOG) **36**(6), 243 (2017)
25. Tompson, J., Stein, M., Lecun, Y., Perlin, K.: Real-time continuous pose recovery of human hands using convolutional networks. ACM Trans. Graph. (TOG) **33**(5), 169 (2014)
26. Tzionas, D., Ballan, L., Srikantha, A., Aponte, P., Pollefeys, M., Gall, J.: Capturing hands in action using discriminative salient points and physics simulation. Int. J. Comput. Vis. **118**(2), 172–193 (2016)
27. Wan, C., Probst, T., Van Gool, L., Yao, A.: Dense 3D regression for hand pose estimation. In: Proceedings of the IEEE Conference on Computer Vision and Pattern Recognition, pp. 5147–5156 (2018)
28. Wang, G., Chen, X., Guo, H., Zhang, C.: Region ensemble network: towards good practices for deep 3D hand pose estimation. J. Vis. Commun. Image Represent. **55**, 404–414 (2018)
29. Wang, Y., Sun, Y., Liu, Z., Sarma, S.E., Bronstein, M.M., Solomon, J.M.: Dynamic graph CNN for learning on point clouds. arXiv preprint arXiv:1801.07829 (2018)
30. Wu, Z., et al.: 3D ShapeNets: a deep representation for volumetric shapes. In: Proceedings of the IEEE Conference on Computer Vision and Pattern Recognition, pp. 1912–1920 (2015)
31. Xu, C., Govindarajan, L.N., Zhang, Y., Cheng, L.: Lie-x: depth image based articulated object pose estimation, tracking, and action recognition on lie groups. Int. J. Comput. Vis. **123**(3), 454–478 (2017)
32. Ye, Q., Yuan, S., Kim, T.-K.: Spatial attention deep net with partial PSO for hierarchical hybrid hand pose estimation. In: Leibe, B., Matas, J., Sebe, N., Welling, M. (eds.) ECCV 2016. LNCS, vol. 9912, pp. 346–361. Springer, Cham (2016). https://doi.org/10.1007/978-3-319-46484-8_21
33. Zhou, X., Wan, Q., Zhang, W., Xue, X., Wei, Y.: Model-based deep hand pose estimation. arXiv preprint arXiv:1606.06854 (2016)

# Graph-FCN for Image Semantic Segmentation

Yi Lu[1,2], Yaran Chen[1,2(✉)], Dongbin Zhao[1,2], and Jianxin Chen[3]

[1] State Key Laboratory of Management and Control for Complex Systems
Institute of Automation, Chinese Academy of Sciences, Beijing 100190, China
[2] University of Chinese Academy of Sciences, Beijing 101408, China
{luyi2017,chenyaran2013,dongbin.zhao}@ia.ac.cn
[3] Beijing University of Chinese Medicine, Beijing 100029, China
cjx@bucm.edu.cn

**Abstract.** Semantic segmentation with deep learning has achieved great progress in classifying the pixels in the image. However, the local location information is usually ignored in the high-level feature extraction by the deep learning, which is important for image semantic segmentation. To avoid this problem, we propose a graph model initialized by a fully convolutional network (FCN) named Graph-FCN for image semantic segmentation. Firstly, the image grid data is extended to graph structure data by a convolutional network, which transforms the semantic segmentation problem into a graph node classification problem. Then we apply graph convolutional network to solve this graph node classification problem. As far as we know, it is the first time that we apply the graph convolutional network in image semantic segmentation. Our method achieves competitive performance in mean intersection over union (mIOU) on the VOC dataset (about 1.34% improvement), compared to the original FCN model.

**Keywords:** Graph neural network · Graph convolutional network · Semantic segmentation

## 1 Introduction

The semantic segmentation is an essential issue in the computer vision field, which is much more complex than the classification and detection task [11]. This is a dense prediction task which needs to predict the category of each pixel, namely it needs to learn the object outline, object position and object category from the high-level semantic information and local location information [16].

Deep learning-based semantic segmentation methods, particularly, the convolution neural networks have taken a series of significant progress to this domain.

This work is supported partly by National Key Research and Development Plan under Grant No. 2017YFC1700106, and No. GJHZ1849 International Partnership Program of Chinese Academy of Sciences.

The powerful generalization ability of obtaining the high-level features brings the outstanding performance of the image classification and detection task [5,19]. But the generalization accompanies the loss of local location information, which increases difficulties for dense prediction tasks. The high-level semantic information with a large receptive field corresponds to a small feature map in the convolution neural networks, which brings the loss of local location information at the pixel-level [6,7]. Many deep learning-based methods have made improvements on this problem, such as fully convolutional network (FCN) [16], Segent [1], Deeplab methods [2–4]. These works use the fully convolutional layer, dilated convolution, and pyramid structure to lessen the location information loss in extracting high-level features.

In order to solve this problem, firstly, we establish a graph-based model for the image semantic segmentation problem. The graph-based methods have been widely used in segmentation problems [10]. The methods regard the pixels as the nodes, and the dissimilarity between the nodes as the edges. The best segmentation is equivalent to the maximum cut in the graph. And combining the probability and graph theory, the probabilistic graphical model methods, such as Markov random field and conditional random field, are applied to refine the semantic segmentation result [13,20]. These methods model the detected object as the nodes of a graph in the image, and by extracting the relation between the objects to improve the detection accuracy [15]. Compared with the grid structure representation of input data in the deep convolution model, the graph model has a more flexible skip connection, so it can explore a variety of relationships among the nodes in the graph [9,17,18].

Limited by the amount of calculation, we initialize the graph model by the FCN. The graph model is established on a small size of the image with the nodes annotation initialized by FCN [16] and the weights of the edges initialized by the Gauss kernel function.

Then we use the graph convolutional network (GCN) to solve this graph model. GCN is one of the state-of-the-art method to deal with graph structure data [8,12,14]. The node-based GCN uses the message propagation to exchange information between neighbor nodes. This process can extracts the features in a large neighborhood of the graph acted the similar role of convolution and pooling layer in the convolutional network. Because there is no nodes disappear in this process, the node-based GCN expands the receptive field and avoids the loss of local location information.

In this paper, a novel model Graph-FCN is proposed to solve the semantic segmentation problem. We model a graph by the deep convolutional network, and firstly apply the GCN method to solve the image semantic segmentation task. The Graph-FCN can enlarge the receptive field and avoid the loss of local location information. In experiments, the Graph-FCN shows outstanding performance improvement compared to FCN.

## 2   Problem Formulation

Semantic segmentation is a pixels classification problem in the image. In 2015, Jonathan Long et al. used the convolution layer instead of the fully connected layer to establish the end-to-end FCN for pixels classification. The FCN adopts the convolutional layer to extract the local feature on the receptive field. Then it uses the upsampling to restore the feature map to the original size of the image. The model implements pixels-to-pixels mapping, and all the pixels in a single image are propagated forward and backward in parallel. The label image can be obtained by arranging the categories of pixels by pixel position. The input of the FCN is the image $\mathbf{X}, \mathbf{X} \in \mathbf{R}^{3 \times w \times h}$, the $w$ is the weight of the image and the $h$ is the height of the image. The output is the predict label image $\mathbf{Y}, \mathbf{Y} \in \mathbf{R}^{w \times h}, y_{i,j} \in \mathbf{R}^{w \times h}$. For semantic segmentation task, it is usually uses the cross-entropy loss function of all the pixels in the label image:

$$L_{FCN} = \sum_{i=1}^{w} \sum_{j=1}^{h} -p(y_{i,j}^{*}) \log \left( p(y_{i,j}) \right) \tag{1}$$

where the truth label of the pixel $(i, j)$ of the label image is denoted as $y_{i,j}^{*}$, and $p(y_{i,j}^{*})$ represents the probability of the $y_{i,j}^{*}$. FCN model can be trained end-to-end by minimizing the cross-entropy loss $L_{FCN}$ using the SGD algorithm.

For the deep learning methods, generalization facilitates identification of deformed objects in image classification and recognition tasks. The pool layer increases the receptive field and decreases the resolution which leads to the loss of pixel position information [16].

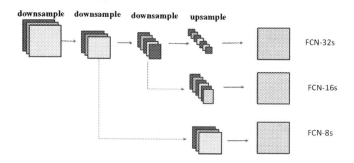

**Fig. 1.** The structure of FCN.

FCN introduces the skip connection to fuse feature layers of different scales, as shown in Fig. 1. Considering that the FCN-16s is just under FCN-8s 0.3 % mean intersection over union(mIOU) and has a more concise structure than the FCN-8s, we adopt the FCN-16s as the basic model to initialize the node annotation for the GCN nodes. More details of the nodes describes in Sect. 3.1.

# 3   Graph Model in Semantic Segmentation

The GCN is designed for solving the learning problem defined on the graph structure data set. The graph structure data can respect as a triple tuple $\mathbf{G}(N, E, U)$. $N$ respects the nodes set of the graph, it is a $|N| * S$ matrix, $|N|$ is the number of the graph nodes, $S$ is the dimension of the node annotation vector. $E$ is the graph edges set. $U$ respects the graph feature, and we omit the $U$ for it not involved in our task. Different from the data representation in Euclidean spatial, the matrix $N$ and edges $E$ are not unique in representation. Matrix $N$ corresponded to $E$, and they are according to the sequence of the nodes. We train the model by supervised learning. The node $n_j$ means the node set in graph $j$, $t_j$ is the label set to node set $n_j$. So the graph model in our task can be shown as the Eq. (2).

$$\min_{w} \; Loss(F_w(\mathbf{G}(N, E)), t)$$
$$s.t. \; \mathbf{G}_j(N, E) \longrightarrow t_j, j \; \epsilon \; T_r. \tag{2}$$

We use the cross entropy function as the loss function in our model. $T_r$ means the training set.

## 3.1   Node

In our model, the node annotations are initialized by the FCN-16s. By the end-to-end training, FCN-16s can get the feature map with a stride of 16 and 32, as shown in Fig. 2. The feature map with strides 32 can obtain the same size of the feature map with strides 16 by upsampling with the factor 2. The annotation $x_j$ (to node $j$) is initialized by the concatenation of the two feature vectors and the location of each node in the feature map. This annotation contains the extracted features on the local receptive field. In the training process, we obtain the label of the node by pooling the raw label image.

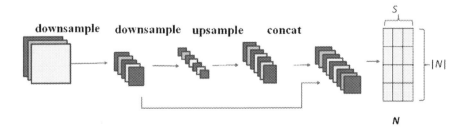

**Fig. 2.** The node annotation initialization process. The node annotation is the concatenation of two layers of the FCN-16s. $|N|$ is the number of the nodes and $S$ is the dimension of node annotations.

## 3.2   Edge

In the graph model, the edge is respected by the adjacent matrix. We assume that each node connects to its nearest $l$ nodes. The connection means that the nodes annotation can be transferred by the edges in the graph neural network. We give an instance to describe the receptive field in the graph neural network, as shown in Fig. 3. For example, $l$ is 4. In the view of the influence of distance on correlation, we adopt the weight adjacent matrix $A$ by the Gauss kernel function.

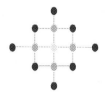

**Fig. 3.** The receptive field of a 2-layer GCN when $l$ is 4, which is different from the convolutional layer.

## 3.3   Training with Graph-FCN

We use GCN to classify the nodes of the graph model that we have established. The GCN is one of the deep learning methods to process graph structure [8,12]. For a graph the normalized Laplacian matrix $L$ has the form in Eq. (3).

$$L = I - D^{-1/2}AD^{-1/2}, \tag{3}$$

where matrix $D$ is the diagonal degree matrix, $D_{ii} = \sum_j A_{ij}$. For the Laplacian matrix $L$ has the orthogonal decomposition $L = U\Lambda U^T$, the matrix $U$ is orthogonal eigenvectors, the matrix $\Lambda$ is the diagonal matrix of eigenvalues. The graph fourier transform $g_\theta$ is defined as

$$g_\theta(L) * x = g_\theta(U\Lambda U^T)x = Ug_\theta(\Lambda)U^T x = U diag(\theta)U^T x, \tag{4}$$

where "*" represents the convolutional operator.
Use the Chebyshev polynomials as an approximation of $g_\theta$, we get

$$g_\theta * x \approx \theta_0 x - \theta_1 D^{-1/2}AD^{-1/2}x. \tag{5}$$

Due to that $\theta_0 = -\theta_1$ hold in the first order Chebyshev polynomials, we get the Eq. (6),

$$g_\theta * x = \theta_0(I + D^{-1/2}AD^{-1/2})x. \tag{6}$$

In order to ensure convergence, the equation Eq. (7) exactly is one layer operator in graph convolutional network. This operator takes the role of convolutional and pool layer in the convolutional newtwork and the features are propagated between nodes in this process.

$$X^{k+1} = \hat{A}X^k\Theta, \quad \hat{A} = \hat{D}^{-1/2}(I + A)\hat{D}^{-1/2}, \tag{7}$$

where $\hat{D}$ is the degree matrix of $I + A$.

The GCN is a form of Laplacian smoothing. When the messages propagate among the neighbor nodes, the neighbor nodes tend to have similar features [14]. This means that the GCN can not be very deep for the over-smoothing, so we adopt a 2-layers GCN network. The maximum range of node message the current node received can be regarded as the receptive field in the graph. For the instance described in Sect. 3.2, the size of receptive field is $5 \times 32 \times 32$, which is five times than that of FCN-16s. Moreover, there is no nodes disappeared in this progress which means that there is no loss of local location information.

In the Graph-FCN, the FCN-16s realize the nodes classification and initialization of the graph model in a small feature map. Meanwhile, the 2-layers GCN gets the classification of the nodes in the graph. We calculate the cross-entropy loss to the both outputs of these two parts. The same as the FCN-16s model, the Graph-FCN is also end-to-end training. The network structure shows in Fig. 4.

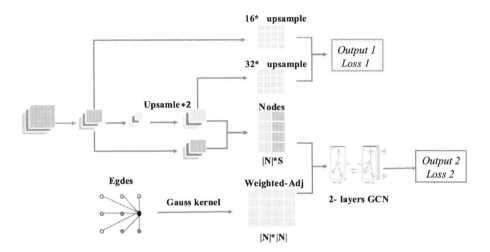

**Fig. 4.** The structure of the Graph-FCN. There are two outputs of the model, and two losses Loss 1 and Loss 2. They share the weights of the feature extracted by convolutional layer. Loss 1 is calculated by output1 and L2 calculated by the output2. Through minimizing Loss 1 and Loss 2, the FCN-16s can improve performance.

## 4   Experiments

In the experiments, we test our model on the VOC2012 dataset and get the performance improvement than the original FCN model.

## 4.1   Implementation

We take a $366 \times 500$ image in VOC dataset as an instance to describe the input and output in detail. In the FCN-16s, after the several pool layers we obtain 512 channels feature map $f1$ and 4096 channels feature map $f2$ of the image. By upsampling, the feature map $f_2$ achieves the same size as feature map $f_1$ ($4096 \times 23 \times 32$). As described in Sect. 3.1, we get the nodes annotation with the size of $4096 + 512 + 2$.

In experiments, the input images are the raw images of the VOC data with different size. In order to adapt to the different sizes of the images, we set the batch size 1. The weights of FCN-16s part are initialized by the pre-trained weights, the results of the FCN-16s are shown in the Table 1. The GCN part is initialized randomly. In the first 8,000 iterations, we only adjust the parameters of the GCN part with the learning rate 0.1. Then set the total learning rate 0.00001 to train the whole model with the weight decay 0.1. In the training, we adopt Adam as the optimizer.

## 4.2   Results

The GCN part in the Graph-FCN model can be regarded as a special loss function. After the model training, the forward output is still the FCN-16s model's output. In the test, the forward part of the Graph-FCN has the same structure as the FCN-16s. But by adding the GCN parts as an additional loss the model the semantic segmentation mIOU has improved 1.34%.

**Table 1.** Graph-FCN vs. FCN-16s

| Method (%) | mIOU | ACC | f.w.IU |
|---|---|---|---|
| FCN-16s | 64.57 | 90.67 | 84.19 |
| Graph-FCN | 65.91 | 91.98 | 85.68 |

Figure 5 shows some samples predicted by Graph-FCN and FCN. From the Fig. 5, we can see that the proposed Graph-FCN has much smoother results compared with FCN-16s. It may be due to that Graph-FCN applies the function of Laplacian smoothing to smooth the predictions. Moreover, the proposed method reduces classification error rate. For example, FCN-16s classfies a part of a sheep as a part of a dog, shown in the second line of Fig. 5. It reflects that the Graph-FCN can extract the messages from the neighbour nodes which help the current node classification.

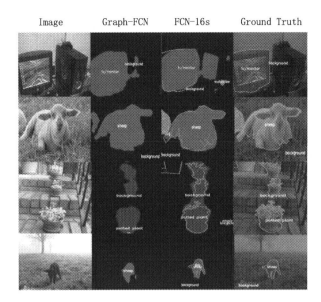

**Fig. 5.** The image semantic segmentation results. The second column is the Graph-FCN results. The third column is the FCN-16s results. The fourth column is the ground truth.

## 5    Conclusion

We model a graph network on the image by the FCN-16s, and propose a Graph-FCN model for semantic segmentation task. The Graph-FCN model can extract the feature on a larger receptive field than the FCN-16s. In the experiment, for the same forward structure, the Graph-FCN achieves a higher mIOU than the FCN-16s, which proves that the Graph-FCN enhance the feature extracting for the pixel classification.

## References

1. Badrinarayanan, V., Kendall, A., Cipolla, R.: SegNet: a deep convolutional encoder-decoder architecture for image segmentation. IEEE Trans. Pattern Anal. Mach. Intell. **39**, 2481–2495 (2017)
2. Chen, L.C., Papandreou, G., Kokkinos, I., Murphy, K., Yuille, A.L.: Semantic image segmentation with deep convolutional nets and fully connected CRFs. In: International Conference on Learning Representations, San Diego (2015)
3. Chen, L.C., Papandreou, G., Schroff, F., Adam, H.: Rethinking atrous convolution for semantic image segmentation. arXiv preprint arXiv:1706.05587 (2017)
4. Chen, L.C., Zhu, Y., Papandreou, G., Schroff, F., Adam, H.: Encoder-decoder with atrous separable convolution for semantic image segmentation. In: European Conference on Computer Vision, Munich, pp. 833–851 (2018)
5. Chen, Y., Zhao, D., Lv, L., Li, C.: A visual attention based convolutional neural network for image classification. In: 12th World Congress on Intelligent Control and Automation, Guilin, pp. 764–769 (2016)

6. Chen, Y., Zhao, D.: Multi-task learning with cartesian product-based multi-objective combination for dangerous object detection. In: Cong, F., Leung, A., Wei, Q. (eds.) ISNN 2017. LNCS, vol. 10261, pp. 28–35. Springer, Cham (2017). https://doi.org/10.1007/978-3-319-59072-1_4

7. Chen, Y., Zhao, D., Lv, L., Zhang, Q.: Multi-task learning for dangerous object detection in autonomous driving. Inf. Sci. **432**, 559–571 (2018)

8. Defferrard, M., Bresson, X., Vandergheynst, P.: Convolutional neural networks on graphs with fast localized spectral filtering. In: Neural Information Processing Systems, Barcelona, pp. 3844–3852 (2016)

9. Gilmer, J., Schoenholz, S.S., Riley, P.F., Vinyals, O., Dahl, G.E.: Neural message passing for quantum chemistry. In: International Conference on Machine Learning, Sydney, pp. 1263–1272 (2017)

10. Gori, M., Monfardini, G., Scarselli, F.: A new model for learning in graph domains. In: IEEE International Joint Conference on Neural Networks, pp. 729–734 (2005)

11. Huang, K.Q., Ren, W.Q., Tan, T.N.: A review on image object classification and detection. Chin. J. Comput. **37**, 1225–1240 (2014)

12. Kipf, T.N., Welling, M.: Semi-supervised classification with graph convolutional networks. In: International Conference on Learning Representations, Toulon (2017)

13. Krahenbuhl, P., Koltun, V.: Efficient inference in fully connected CRFs with Gaussian edge potentials. In: Neural Information Processing Systems, Granada, pp. 109–117 (2011)

14. Li, Q., Han, Z., Wu, X.M.: Deeper insights into graph convolutional networks for semi-supervised learning. In: Thirty-Second AAAI Conference on Artificial Intelligence, New Orleans, pp. 3538–3545 (2018)

15. Liu, Y., Wang, R., Shan, S., Chen, X.: Structure inference net: object detection using scene-level context and instance-level relationships. In: IEEE Conference on Computer Vision and Pattern Recognition, Salt Lake City, pp. 6985–6994 (2018)

16. Long, J., Shelhamer, E., Darrell, T.: Fully convolutional networks for semantic segmentation. In: IEEE Conference on Computer Vision and Pattern Recognition, Boston, pp. 3431–3440 (2015)

17. Scarselli, F., Gori, M., Tsoi, A.C., Hagenbuchner, M., Monfardini, G.: The graph neural network model. IEEE Trans. Neural Netw. **20**, 61–80 (2009)

18. Wang, X., Girshick, R.B., Gupta, A., He, K.: Non-local neural networks. In: IEEE Conference on Computer Vision and Pattern Recognition, Salt Lake City, pp. 7794–7803 (2018)

19. Zhao, D., Chen, Y., Lv, L.: Deep reinforcement learning with visual attention for vehicle classification. IEEE Trans. Cogn. Dev. Syst. **9**, 356–367 (2017)

20. Zheng, S., Jayasumana, S., Romera-Paredes, B., Vineet, V., Su, Z., Du, D., Huang, C., Torr, P.H.S.: Conditional random fields as recurrent neural networks. In: IEEE International Conference on Computer Vision, Santiago, pp. 1529–1537 (2015)

# Architecture Search for Image Inpainting

Yaoman Li[1,2(✉)] and Irwin King[1]

[1] The Chinese University of Hong Kong, Shatin, Hong Kong
{ymli,king}@cse.cuhk.edu.hk
[2] Lenovo Group Ltd., Beijing, China
yli6@lenovo.com

**Abstract.** Neural Architecture Search (NAS) shows the ability to automate the architecture engineering for specific tasks recently which is extremely promising. Many published works apply reinforcement learning or evolutionary algorithm to design the neural architecture for image classification and achieve state-of-the-art performance. However, using NAS to perform other challenging tasks, such as inpainting irregular regions in an image, has not been explored yet. The target of image inpainting is to generate plausible image regions to fill the missing regions in the original image. It has been widely used in many applications. In this paper, we are interested in applying neural architecture search methods to image inpainting tasks. We propose to use reinforcement learning to automatically design the network architecture. Our method can efficiently explore new network structure based on existing architecture. The experiment result demonstrates that the proposed method can design an efficient and high-quality architecture for image inpainting.

**Keywords:** Reinforcement learning · Neural architecture search ·
Image inpainting · Partial convolution · AutoML · U-Net

## 1 Introduction

Neural Architecture Search (NAS) is a subfield of Automating Machine Learning (AutoML). It has been an impressive and interesting research areas form decades ago. Most of the research works are based on evolutionary algorithm (EA) and reinforcement learning (RL) [7,14,16,27,28]. The EA-based methods can evolve the architecture by selecting better neural architecture generations with crossover and mutation [18,21]. These methods usually will need a huge number of computational resources and time. Deep reinforcement learning has obtained some great successes in many areas. Many RL-based neural architecture search methods have also been proposed for image classification [7,27,28]. These RL-based methods use the evaluation metrics of validation dataset as a reward and explore the new architecture by a sequence of actions. Both these methods have shown the ability to find an architecture which can outperform human designed architecture.

© Springer Nature Switzerland AG 2019
H. Lu et al. (Eds.): ISNN 2019, LNCS 11554, pp. 106–115, 2019.
https://doi.org/10.1007/978-3-030-22796-8_12

The existing NAS methods mainly focused on the image classification, which is a typical image task. There are many other challenging tasks that many researchers are working on. Researchers need to manually design the architectures for each specific task. The network architectures for different tasks may be different. In order to reach a better performance, researchers need to spend much time to try different architectures and fine-tune the hyper-parameters. The success of NAS on image classification shows a way to let the machine to automatically design architectures for these tasks.

Image inpainting is an automatic digital inpainting problem. It is more challenging and complicated than image classification. The architectures for image inpainting need to focus more on the pixel value in order to fill the missing regions of an image with the semantically meaningful pixel. Some neural network architectures have been proposed to solve this task, e.g., [3,5,9,13,23].

Two latest state-of-the-art methods, i.e., [24,25], present two new architectures to generate a visually realistic and semantically plausible image. Both of them apply a U-Net like architecture in their approaches [19]. It was originally used for the segmentation of neural structures. In U-Net, an upsampling operation has been proposed to replace the pooling operation in the usual contracting network. The upsampling operation allows the neural network to propagate the information from a lower resolution to a higher resolution.

In our paper, we will show how to apply reinforcement learning NAS methods to automate the exploration of new architectures for image inpainting. We propose to apply a network transformation architecture search method to search the network architecture, which can re-use the existing architectures and parameters. This method is time-saving and has high computational efficiency than other NAS methods. We also propose a new type of actor operation in reinforcement learning controller so that the NAS method can generate a U-Net like architecture. The experiment result shows that our NAS method can create a competitive architecture. To the best of our knowledge, we are the first one to apply NAS in image inpainting tasks. This result shows that the automatic neural architecture search methods can be easily extended and applied to more challenging tasks other than image classification. It can help us to save a lot of time to design new architecture manually.

In summary, our contributions are:

- The first one to apply the NAS method to image inpainting task.
- Propose a net reinforcement learning actor in NAS and extend the NAS method to generate more various architecture, e.g., U-Net like architecture.
- Show the ability of NAS to design architecture for complex tasks.

## 2   Related Work

### 2.1   Image Inpainting

Image inpainting [5] is the task to reconstitute the damaged or missing regions of an image, in order to restore its unity and make it well defined. If the missing regions, which are also known as holes, is not the regular rectangle shape, then they will be called irregular holes. There are many applications of image inpainting, e.g., fixing the broken image, removing unwanted object, improving image resolution. The methods that apply to image inpainting can be mainly divided into non-learning approaches and learning approaches.

The Non-learning approach [4,9,10] usually will propagate the pixel from near field. This method can obtain a good result when the holes are small and inpainting regions are not complex. However, if the impainting regions contain non-repetitive objects, it will create the over smooth and semantically meaningless results.

The learning approach is mainly based on deep learning. A number of Learning approaches have been proposed recently [12,13,23,25]. These methods apply deep learning to learn the semantic representation of the image. For example, [25] applied GAN to perform the image impainting task. In their search, they proposed to add a contextual attention layer, a feed-forward generative network. The contextual attention layer can learn how to generate the missing regions by copy or borrow feature image. The neural network in that work contains two stages. The first stage will rough out the missing content by using the dilated convolutional network. The second stage is to use the contextual attention layer to refine the generated patches. This feed-forward neural network can generate a great inpainting result on different datasets.

There are also a number of state-of-the-art image inpainting approaches are using U-Net [19] like architecture, e.g., Liu et al. [15] propose to use partial convolutional layers to replace the typical convolutional layers. There is an automatic mask update step in the partial convolutional layers. This partial convolutional layer is used to replace the original CNN in the U-Net. The method can handle the irregular holes efficiently. Yan et al. [24] propose to add a shift-connection layer, a new type of CNN, to the U-Net. These works have shown that network architecture for image inpainting has a great difference to the image classification tasks.

## 2.2   Architecture Search

Many techniques have been applied on neural architecture search, such as, evolutionary algorithms, reinforcement learning, Bayesian optimization, random search, etc. Many researchers have started working on NAS decades ago, e.g., [1,11,20]. NAS has shown its ability on the automating architecture design.

Recent works of [2,17,26,27] treat NAS as a reinforcement learning problem. The action of the agent will be considered to generate the structure of a neural network. The agent will get a reward from the evaluation result from validation dataset. Zoph and Le [27] use a recurrent neural network (RNN) as a controller to generate the new structure. The number of filters, filter height, filter width, etc., can be predicted by the RNN controller.

Most of these architecture search is started from scratch. It will take a lot of time and computer resources to get an outperformed human designed architecture. And it sometimes will also fail to get a good architecture especially when the computer resources is limit. So [6] propose to make use of the

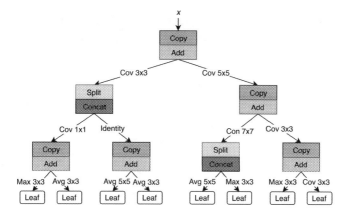

**Fig. 1.** Tree structure representation of neural network

existing human designed architecture, which can get a good architecture with much fewer resources and time. The search is started from an existing network, e.g., DenseNet, ResNet and Inception Net. The reinforcement learning controller will apply the Net2Net technique [8] to perform the Net2WiderNet and Net2DeeperNet transformation. Net2WiderNet transformation will replace the existing layer with a wider layer. Net2DeeperNet will replace the layer with and deeper layer. Both of these two transformations are function-preserving transformation.

In the work of [7], they proposed to use the path-level transformation to allow the bidirectional LSTM reinforcement learning meta-controller to explore a more flexible architecture. It can support multi-branch network architecture, e.g., ResNets and DenseNets. The path transform is based on the Net2Net technique [8]. These path transform operations are function-preserving. The training for new architecture can be based on previous architecture status, so that training is much faster. It can achieve competitive results in image classification tasks with 200 GPU-hours compared to 48,000 GPU-hours in [28]. However, it cannot generate U-Net like architecture. This search method can be problematic when applying to image inpainting tasks.

## 3    Approach

Similar to [15], we will use tree-structured architecture to represent the neural network. Each node (except leaf nodes) in the tree contains an assign operation and a merge operation. Each edge in the tree represents the primitive operation, e.g., pooling, convolution, up-sampling. Figure 1 shows an example of a tree-structured network architecture. The input will go down from the root node, by applying the assign operation on each node and edge. Then go up from a leaf, apply the merge operation to obtain the output.

Let $x$ be the input, output of a node is $T$, the output of $i^{th}$ child node is $T_i^c$, the edge operation of the $i^{th}$ child is $E_i$. The number of child node is $m$, so that the output of a node can be denoted as:

$$y_i = T_i^c(E_i(assign(x,i))), \quad 1 <= i <= m,$$
$$T(x) = merge(y_1, ..., y_m), \tag{1}$$

where $assign(x,i)$ is an function that assign the input $x$ to $i^{th}$ child based on the assign operation, $merge(\cdot)$ denotes merging the output of $m$ children based on the merge operation, $y_i$ is the output of $i^{th}$ child. If the child is a leaf node, then we do not do any operation. We just return the input $x$.

The available assign operations are: *copy*, *split* and *none*. The merge operations are: *add*, *concat* and *none*. When the operation is *none*, we will do nothing on the input. We just directly pass to the next operation.

The architecture of RL meta-controller is similar to the one proposed by [15]. We use tree-structured LSTM in our reinforcement learning meta controller [7] to encode the network architecture. Two kinds of tree-structured LSTM will be used. The Child-Sum Tree-LSTM will be applied for *add* merge operation node, N-ary Tree-LSTM will be applied for *concat* merge operation [22].

The controller will sample the transform operations and then apply to the existing architecture iteratively to generate new architectures. The new architecture will be trained and evaluated. The evaluation result will be used as a reward to update the controller by policy gradient algorithms. There is an encoder network in the controller, which is used to encode the network architecture and generate a low-dimensional representation. The transform operations are decided by softmax classifiers.

The RL meta-controler will automatic design the architecture for image inpaiting. Most of the image inpaiting methods can only handle rectangular regions. Liu et al. [15] introduce the partial convolutional layer, which can handle irregular regions and get a plausible inpainting result. The partial convolutional layer contains partial convolution operation and mask update operation. The partial convolution can be defined as:

$$x' = \begin{cases} W^T(X \odot M)\frac{1}{sum(M)} + b, & \text{if } sum(M) > 0 \\ 0, & \text{otherwise} \end{cases} \tag{2}$$

Here, $x'$ is the partial convolution output per location, $W$ is the convolution filter weights, $X$ is feature input of current convolution window, $M$ is the corresponding mask. $\odot$ denotes element-wise multiplication. The mask can be updated by

$$m' = \begin{cases} 1, & \text{if } sum(M) > 0 \\ 0, & \text{otherwise} \end{cases}, \tag{3}$$

where $m'$ is the updated mask per location. This partial convolutional layer would be used in our approach.

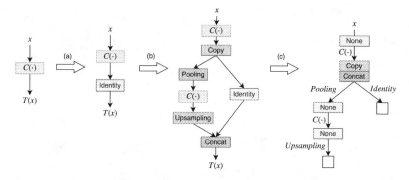

**Fig. 2.** Architecture transform. From left to right, (a) adds a child node with an identity edge, (b) applies an U-Shape transform to the identity edge, (c) use the tree architecture to represent the network architecture

We also noticed that the network architecture for image inpainting is different to image classification. Most of the architecture for image inpainting are U-Net like architecture. In order to generate a U-Net like architecture, except the basic layers, e.g. partial convolution, identity, etc., we also need a new type of layer, up-sampling layer. The available layers in the edge are:

- conv $1 \times 1$
- identity
- partial conv $3 \times 3$
- partial conv $5 \times 5$
- partial conv $7 \times 7$
- up-sampling $2 \times 2$

The transform operations are similar to [7], except the U-Shape transform. The U-Shape transform will add a new branch to the network. The new branch contains a *copy* operation, a pooling layer, a convolutional layer, an upsampling layer and a *concat* operation. The available actors in our reinforcement learning meta-controller are:

- Choose one child node, decide which assign operation (*copy, split, none*) and merge operation (*add, concat, nont*) will be applied, how many child nodes should be added
- Choose leaf node and decide whether expand it by an identity edge
- Select an identity edge to replace with one available layer type
- Select an identity edge and decide whether apply an U-Shape transform

Figure 2 shows an example of network architecture transform. The network starts from a simple convolutional architecture. By applying the add identity edge actor, U-Shape transformation actor, it will become a U-Net like architecture. This kind of network architecture can be easily represented by a tree structure.

(a) The inpainting result generated by Patch Match

(b) The inpainting result generated by the best model after 150 hours

**Fig. 3.** Inpainting comparison

**Fig. 4.** Mask data sample

When we start the neural network architecture search. We can also use an existing well know architecture to initialize the parent architecture for the reinforcement learning meta-controller. Half of the actors are function-preserving, so that after transforming to the new architecture, many parameters can be re-used. This will greatly speed-up our training and architecture search.

We also accelerate the architecture search by parallel training. The meta-controller will generate multiple new architectures and distributed to multiple machines to do the training. When the training for each architecture is finished, the evaluation result will be collected.

We use the loss functions introduced by [15]. These loss functions will evaluate both composition and per-pixel reconstruction accuracy.

## 4  Experiments

We evaluate our NAS model on one of the most challenging Places2. There are more than 10 million images with 400+ sense categories in this dataset. We use the irregular mask dataset which provided by [15] to generate the irregular hole. There are 55,116 masks in training set and 24,866 masks in the testing set. All the mask size for both training and testing set is $512 \times 512$. The mask will be randomly selected and combine with the image to generate a new image with an irregular hole (Fig. 4).

(a) The sample images before inpainting

(b) The ground truth images

**Fig. 5.** Sample experiment images

It will take a long time to train each architecture generated by the meta-controller. To speed up the training process, we distribute the generated multiple machines to do parallel training. We use four worker machines to do training. The meta-controller will put all new architecture to the network pool. Once the worker machine is available, it will select one untrained architecture to do training. When the training is finished, it will send the loss of the architecture to the meta-controller. Each worker machine contains four NVIDIA TITAN X GPUs (12 GB).

We train three epochs for each generated network architecture. We select the best network architecture after 150 h and fine-tuning it by training another epoch. Figure 5(a) is the several sample images before inpainting. Figure 3(a) shows the image that generated Patch Match [4]. Figure 3(b) shows the generated images by that model. We can see the model can fill the missing regions with plausible content. It is competitive to the state-of-the-art approaches.

We also do the quantitative comparisons of the methods. We compare Patch-Match and our method by total variation (TV) loss and peak signal-to-noise ratio (PSNR).

**Table 1.** Comparison of the PatchMatch and the best model reported by our method after 150 h

| Method | TV loss | PSNR |
|---|---|---|
| PatchMatch | 28.5% | 20.30 |
| Our method | **28.4%** | **25.28** |

From Table 1, we can see that our NAS generated architecture can get a better result than PatchMatch. These evaluation metrics are not perfect since there are many different solutions fill the missing regions to the original image, but it still reflects the reality of the generated image in some aspects.

## 5   Discussion

Some NAS works have shown that the auto-generated neural network can outperform the human design network. However, most of these works only focus on several specific tasks. The method or search space of these works can only work on the specific task. In this paper, we extend the method to another task, image inpainting. We found that we need to put a lot of effort to extend the existing methods to this new task. In order to let the NAS work on other more general tasks naturally, we may need to re-design the search method and extend the search space. The NAS process is time-consuming. We also need to pay attention to the efficiency. In our future work, we will consider modifying the tree-structured representation. Let it cover more network architectures. And we might add a meta-learning process, which can identify what task it is working on and suggest the candidate initial architectures, to our NAS method.

**Acknowledgments.** The work described in this paper was partially supported by the Research Grants Council of the Hong Kong Special Administrative Region, China (No. CUHK 14208815 of the General Research Fund).

## References

1. Angeline, P.J., Saunders, G.M., Pollack, J.B.: An evolutionary algorithm that constructs recurrent neural networks. IEEE Trans. Neural Netw. **5**(1), 54–65 (1994)
2. Baker, B., Gupta, O., Naik, N., Raskar, R.: Designing neural network architectures using reinforcement learning. CoRR abs/1611.02167 (2016)
3. Ballester, C., Bertalmío, M., Caselles, V., Sapiro, G., Verdera, J.: Filling-in by joint interpolation of vector fields and gray levels. IEEE Trans. Image Process. **10**(8), 1200–1211 (2001)
4. Barnes, C., Shechtman, E., Finkelstein, A., Goldman, D.B.: PatchMatch: a randomized correspondence algorithm for structural image editing. ACM Trans. Graph. **28**(3), 24:1–24:11 (2009)
5. Bertalmío, M., Sapiro, G., Caselles, V., Ballester, C.: Image in painting. In: SIGGRAPH, pp. 417–424. ACM (2000)
6. Cai, H., Chen, T., Zhang, W., Yu, Y., Wang, J.: Efficient architecture search by network transformation. In: AAAI, pp. 2787–2794. AAAI Press (2018)
7. Cai, H., Yang, J., Zhang, W., Han, S., Yu, Y.: Path-level network transformation for efficient architecture search. In: JMLR Workshop and Conference Proceedings, ICML, vol. 80, pp. 677–686. JMLR.org (2018)
8. Chen, T., Goodfellow, I.J., Shlens, J.: Net2Net: accelerating learning via knowledge transfer. In: ICLR (2016)
9. Efros, A.A., Freeman, W.T.: Image quilting for texture synthesis and transfer. In: SIGGRAPH, pp. 341–346. ACM (2001)
10. Efros, A.A., Leung, T.K.: Texture synthesis by non-parametric sampling. In: ICCV, pp. 1033–1038 (1999)

11. Floreano, D., Dürr, P., Mattiussi, C.: Neuroevolution: from architectures to learning. Evol. Intell. **1**(1), 47–62 (2008)
12. Iizuka, S., Simo-Serra, E., Ishikawa, H.: Globally and locally consistent image completion. ACM Trans. Graph. **36**(4), 107:1–107:14 (2017)
13. Köhler, R., Schuler, C., Schölkopf, B., Harmeling, S.: Mask-specific inpainting with deep neural networks. In: Jiang, X., Hornegger, J., Koch, R. (eds.) GCPR 2014. LNCS, vol. 8753, pp. 523–534. Springer, Cham (2014). https://doi.org/10.1007/978-3-319-11752-2_43
14. Liu, C., et al.: Progressive neural architecture search. In: Ferrari, V., Hebert, M., Sminchisescu, C., Weiss, Y. (eds.) ECCV 2018. LNCS, vol. 11205, pp. 19–35. Springer, Cham (2018). https://doi.org/10.1007/978-3-030-01246-5_2
15. Liu, G., Reda, F.A., Shih, K.J., Wang, T.-C., Tao, A., Catanzaro, B.: Image inpainting for irregular holes using partial convolutions. In: Ferrari, V., Hebert, M., Sminchisescu, C., Weiss, Y. (eds.) ECCV 2018. LNCS, vol. 11215, pp. 89–105. Springer, Cham (2018). https://doi.org/10.1007/978-3-030-01252-6_6
16. Liu, H., Simonyan, K., Vinyals, O., Fernando, C., Kavukcuoglu, K.: Hierarchical representations for efficient architecture search. In: ICLR (2018)
17. Pham, H., Guan, M.Y., Zoph, B., Le, Q.V., Dean, J.: Efficient neural architecture search via parameter sharing. In: JMLR Workshop and Conference Proceedings, ICML, vol. 80, pp. 4092–4101. JMLR.org (2018)
18. Real, E., et al.: Large-scale evolution of image classifiers. In: Proceedings of Machine Learning Research, ICML, vol. 70, pp. 2902–2911. PMLR (2017)
19. Ronneberger, O., Fischer, P., Brox, T.: U-Net: convolutional networks for biomedical image segmentation. In: Navab, N., Hornegger, J., Wells, W.M., Frangi, A.F. (eds.) MICCAI 2015. LNCS, vol. 9351, pp. 234–241. Springer, Cham (2015). https://doi.org/10.1007/978-3-319-24574-4_28
20. Stanley, K.O., Miikkulainen, R.: Evolving neural networks through augmenting topologies. Evol. Comput. **10**(2), 99–127 (2002)
21. Suganuma, M., Shirakawa, S., Nagao, T.: A genetic programming approach to designing convolutional neural network architectures. In: IJCAI, pp. 5369–5373. ijcai.org (2018)
22. Tai, K.S., Socher, R., Manning, C.D.: Improved semantic representations from tree-structured long short-term memory networks. In: ACL (1), pp. 1556–1566. The Association for Computer Linguistics (2015)
23. Xu, L., Ren, J.S.J., Liu, C., Jia, J.: Deep convolutional neural network for image deconvolution. In: NIPS, pp. 1790–1798 (2014)
24. Yan, Z., Li, X., Li, M., Zuo, W., Shan, S.: Shift-Net: image inpainting via deep feature rearrangement. In: Ferrari, V., Hebert, M., Sminchisescu, C., Weiss, Y. (eds.) Computer Vision – ECCV 2018. LNCS, vol. 11218, pp. 3–19. Springer, Cham (2018). https://doi.org/10.1007/978-3-030-01264-9_1
25. Yu, J., Lin, Z., Yang, J., Shen, X., Lu, X., Huang, T.S.: Generative image in painting with contextual attention. In: CVPR, pp. 5505–5514. IEEE Computer Society (2018)
26. Zhong, Z., Yan, J., Wu, W., Shao, J., Liu, C.: Practical block-wise neural network architecture generation. In: CVPR, pp. 2423–2432. IEEE Computer Society (2018)
27. Zoph, B., Le, Q.V.: Neural architecture search with reinforcement learning. CoRR abs/1611.01578 (2016)
28. Zoph, B., Vasudevan, V., Shlens, J., Le, Q.V.: Learning transferable architectures for scalable image recognition. In: CVPR, pp. 8697–8710. IEEE Computer Society (2018)

# An Improved Capsule Network Based on Newly Reconstructed Network and the Method of Sharing Parameters

Chunyan Lu[1,2,3], Shukai Duan[1,2,3(✉)], and Lidan Wang[1,2,3]

[1] The College of Electronic and Information Engineering, Southwest University, Chongqing 400715, China
duansk@swu.edu.cn
[2] National & Local Joint Engineering Laboratory of Intelligent Transmission and Control Technology, Chongqing 400715, China
[3] Brain-inspired Computing & Intelligent Control of Chongqing Key Lab, Chongqing 400715, China

**Abstract.** The capsule network is considered as the latest technology in the field of computer vision. However, it needs a large amount of storage space due to the large amount of parameters. In this paper, we have adopted two methods to solve this problem. First, a method of sharing the parameters of capsule layer is proposed to solve the problem of too many parameters in capsule layer, which can decrease by 18% parameters compared with the previous. Second, we redesigned the structure of the reconstructed network to replace the original, reducing the network's parameters by 16%. Moreover, we combine the two methods to further reduce the parameters, which can decrease by 34%. Finally, we use the improved capsule network for MNIST handwritten digit recognition, the result is almost the same as or even slightly higher than the original capsule network, and the reconstructed images also can smooth the noise. This article provides new ideas for the future optimization methods of various capsule networks.

**Keywords:** Capsule network · Shared parameters · Reconstructed network

## 1 Introduction

In 2012, the CNN structure-AlexNet won the championship of ILSVRC in image classification competition [1]. From then on, CNN began to receive widespread academic attention, resulting in the famous CNN structure of ZFNet [2], VGGNet [3], GoogleNet [4] and ResNet [5]. These structures are widely used in the fields of image classification [1–5], object detection and segmentation [6–10], text detection and recognition [11] and greatly promote the development of artificial intelligence.

H. Lu et al. (Eds.): ISNN 2019, LNCS 11554, pp. 116–123, 2019.
https://doi.org/10.1007/978-3-030-22796-8_13

Although CNN made rapid breakthroughs in recent years, there are also some shortcomings. It requires many training datas, and cannot handle the image ambiguously expressed. Because of many pooling layers in CNN, some information would be lose. Its scalar output also cannot well reflect the transformation of the spatial location. In response to these deficiencies of CNN, Hinton proposed a capsule network structure [12] whose output is a vector that can well represent the attitude information of an object, such as direction, position, thickness, etc., and the length of a vector is used to represent the probability of existence of an object. Moreover, the capsule network can well complete the network training when only needs much less data than CNN.

Hinton's capsule network has significantly reduced the amount of parameters compared to the baseline CNN, but its parameters in capsule and reconstruction layer are still quite large and the reconstruction layer uses a three-layer fully connected network that will cause gradient disappear and other issues in training process. Therefore, in this paper, a method of parameter sharing is proposed for the capsule layer of capsule network to further reduce the parameters of the capsule layer and reduce the training complexity. Moreover, we redesigned the structure of reconstruction layer of capsule network, using the form of deconvolution network, adding Batch Normalization optimization technology to further simplify the reconstruction layer structure, greatly reducing the amount of parameters.

## 2    Related Work

In 2011, Hinton proposed for the first time that neural network can output an overall instantiation vector to represent the learned features and validated that a change to the input vector on Transforming Auto-encoders can cause the output change accordingly [13]. This is called equivariance. In 2017, Hinton proposed a capsule network which expressed the attitude of the object in the direction of the vector, the probability of the existence of the entity in terms of the length of the vector, and proposed a routing-by-agreement algorithm which was successfully applied to the task of handwritten digit recognition [12].

In this paper, we try to reduce the parameters of the capsule network while keeping the recognition accuracy, and further reduce the network complexity. We mainly use the deconvolution network and Batch Normalization optimization techniques to redesign the structure of the reconstruction layer. In pixel-wise prediction such as image segmentation [14] and image generation [15], because of the need to do the original image size space prediction, and the convolution tend to reduce the size of the image, so often need to upsample to restore the original picture size, deconvolution acting as an upsampling role. The deconvolution network used in this article is similar to the usage of upsampling. Batch Normalization optimization technology was proposed in 2015 [16], and now it has been widely used in the field of deep learning. This article applies it to the design of the reconstruction layer to maintain the stability of the network.

## 3    Proposed Method

A three-layer capsule network with new reconstructed network structure is shown in Fig. 1. The first layer is an ordinary convolutional layer, using 256, 9 × 9 convolution kernel with stride of 1 and ReLU activation function. The second layer is the primary capsule layer, which consists of 6 × 6 × 32 capsules, each with 8 vectors. In this case, each capsule in this layer obtained from the feature map of the first layer using a 9 × 9 convolution kernel and taking a stride of two. The third layer is a digit capsule layer that outputs 10 capsules, each with 16 vectors. In this case, the output of each capsule in this layer connected with all the capsules in the second layer, i.e. the two capsule layers are fully connected. Calculate the length of the 10 capsule vectors to represent the output. Then connect the correct vector to reconstructed network.

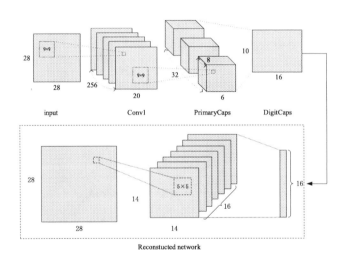

**Fig. 1.** Three-layer capsule network with new reconstructed network structure

### 3.1    Shared Parameters

The second layer of the primary capsule layer has 32 × 6 × 6 capsules. In the original Hinton's paper, each capsule between the second layer and the third layer have an 8 × 16 weight matrix connection, the number of weight parameters is 32 × 6 × 6 × 10 × 8 × 16 = 1,474,560, and the number is very large. Therefore, in the research of this paper, a method of sharing weight parameters is proposed. As shown in Fig. 2, the capsules that the primary capsule layer are arranged in a column and only the weights between the front n capsules and all the capsules in the latter layer are trained. The following capsules of (32 × 6 × 6/n − 1) replicate the previous weight. With this method of sharing weight, the number of parameters can be reduced to n × 10 × 8 × 16. We take different n for experiments and the specific experimental results are shown in the fourth part.

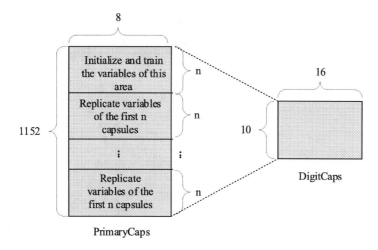

**Fig. 2.** Shared the weights from the primary capsule layer to the digital capsule layer

## 3.2   Reconstructed Network

At Hinton's work [12], he used three fully connected layers to reconstruct the image, in order to motivate the digit capsule layer's output vector to better represent the characteristics of the input image. The total number of neurons in the three fully connected layer is 512, 1024,784 and the input is 16, so the number of parameters in the reconstruction layer is 1,337,616. The parameters in these three fully connected layers are quite large, and excessive numbers of fully connected layers may cause problems such as the disappearance of the gradient in the training process. Therefore, this paper redesigned the reconstruction layer of capsule network, using deconvolution network instead of the original multi-layer fully connected network, can not only greatly reduce the number of parameters but also slightly improve the accuracy on the task of handwritten digit recognition.

Reconstructed network structure, detail chart as shown in Figs. 1 and 3. First, we masked out the incorrect output vectors of the digit capsule layer, leaving only the correct output of a capsule corresponding to the 16 vectors. Therefore, the input of reconstructed network is 16, and after a full-connected layer, and then using the optimization method of batch normalization, the $14 \times 14 \times 16$ feature maps are obtained through ReLU activation function. Then we used a 5 $\times$ 5 kernel for deconvolution, resulting in a reconstructed $28 \times 28 \times 1$ output picture.

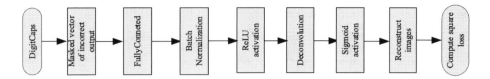

**Fig. 3.** Detail chart of reconstructed network

## 4    Experiments and Results

### 4.1    Training Process

The experiment uses MNIST handwritten digital data set [17]. The framework uses Mxnet [18], the optimizer uses Adam optimizer [19], the initial learning rate is 0.0005, and the decaying learning rate is 0.99. The routing between the two capsule layers is 3 and trained for 200 epochs. All other parameters are designed in accordance with Hinton's capsule network.

### 4.2    Experimental Results

First, we validate the method of sharing the weight parameters between capsule layers proposed in this paper. The experimental results are shown in Fig. 4(a). We take the value of n for 96, 48, and 36 respectively. As you can see, the accuracy of training the first epoch is higher than the original when taking different values of n. When n is taken as 96 or 48, the accuracy is higher than the original one after convergence of the training. When n is 36, the final convergence's accuracy is about the same as the original one, and the parameters between the two capsule layers decrease to 1/32 of the original network, the total number of network parameters decreased by 1.5M, as shown in Table 1.

Next, we verify the validity of the newly designed reconstruction network. The experimental results are shown in Fig. 4(b). It can be seen that the accuracy obtained when the reconstructed layer is converged is higher than the original one and the amount of parameters is reduced to 1/26 of the original reconstructed layer, which greatly reduces the complexity of the network. In this case, the total number of network parameters are decreased by 1.3M, as shown in Table 1.

Moreover, we combine the newly designed reconstruction network with the method of sharing the weight parameters between capsule layers. The experimental results are shown in Fig. 4(c). It can be seen that the accuracy of training the first epoch is higher than that of the original when n takes different values. The accuracy of convergence is slightly lower than that of the network without shared parameters, but when n takes 36, the parameters of the whole network are only 5.4M, compared with the original network's 8.2M parameters, it has greatly reduced, as shown in Table 1.

Finally, we present the reconstruction images that compared with the original images. As shown in Fig. 5, the first column is the original images, the second column is the reconstructed images of the previous capsule network, the middle

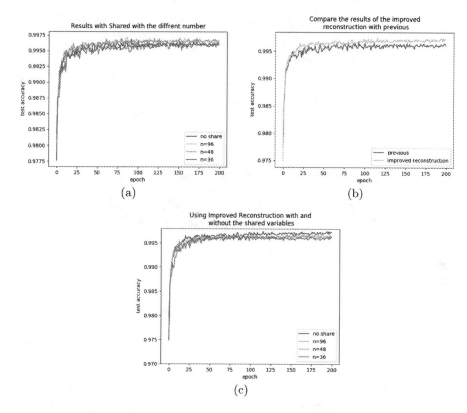

**Fig. 4.** The experimental results. (a) Recognition accuracy when takes different values of n (b) Results of improved reconstructed network compared with the original (c) Results of reconstruction network with and without the method of sharing the weight parameters

**Table 1.** Total number of parameters of improved network parameters with the original network

| | |
|---|---|
| Previous | 8.2M |
| Shared n = 36 | 6.7M |
| New reconstruction network | 6.9M |
| New reconstruction network and shared n = 36 | 5.4M |

column is the reconstructed images when only sharing parameters, the fourth column is the reconstructed images when using newly designed reconstructed network, the last column is the reconstructed images when using newly designed reconstructed network with the method of sharing parameters. Compared the three rightmost columns with the first column, we can see that our improved capsule network also preserves many of the details while smoothing the noise.

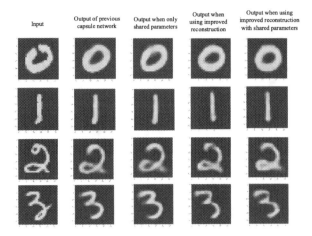

**Fig. 5.** Sample MNIST test reconstruction of various experiments

## 5   Conclusion

Based on Hinton's capsule network, in this paper, we proposed a method to share the parameters between capsule layers in order to reduce the parameters between capsule layers and reduce the complexity of network training. In addition, we also redesigned the structure of the reconstruction layer of the capsule network, using the deconvolution network combined with the optimization method of Batch normalization, compared to the previous reconstruction layer consisting of three-layer fully connected network, greatly simplify the network structure, and to a greater extent reduce the number of parameters. We use the improved network structure for MNIST handwritten digit recognition tasks, and the accuracy is not greatly reduced compared with the original one, and sometimes even higher than the original results. Is there any useless capsule exists? We will continue to delve deeper into these issues later, and will use the improved network for verification on some of the larger datasets.

**Acknowledgments.** This work was supported by the National Natural Science Foundation of China under Grant 61571372, 61672436 and 61601376, the Fundamental Research Funds for the Central Universities under Grant XDJK2016A001 and XDJK2017A005, and the Fundamental Science and Advanced Technology Research Foundation of Chongqing under Grant cstc2016jcyjA0547.

## References

1. Krizhevsky, A., Sutskever, I., Hinton, G. E.: ImageNet classification with deep convolutional neural networks. In: Advances in neural information processing systems, pp. 1097–1105 (2012)

2. Zeiler, M.D., Fergus, R.: Visualizing and understanding convolutional networks. In: Fleet, D., Pajdla, T., Schiele, B., Tuytelaars, T. (eds.) ECCV 2014. LNCS, vol. 8689, pp. 818–833. Springer, Cham (2014). https://doi.org/10.1007/978-3-319-10590-1_53

3. Simonyan, K., Zisserman, A.: Very deep convolutional networks for large-scale image recognition. arXiv preprint arXiv:1409.1556 (2014)

4. Szegedy, C., et al.: Going deeper with convolutions. In: CVPR (2015)

5. He, K., Zhang, X., Ren, S., Sun, J.: Deep residual learning for image recognition. In: Proceedings of the IEEE Conference on Computer Vision and Pattern Recognition, pp. 770–778 (2016)

6. Girshick, R., Donahue, J., Darrell, T., Malik, J.: Rich feature hierarchies for accurate object detection and semantic segmentation. In: Proceedings of the IEEE Conference on Computer Vision and Pattern Recognition, pp. 580–587 (2014)

7. He, K., Zhang, X., Ren, S., Sun, J.: Spatial pyramid pooling in deep convolutional networks for visual recognition. In: Fleet, D., Pajdla, T., Schiele, B., Tuytelaars, T. (eds.) ECCV 2014. LNCS, vol. 8691, pp. 346–361. Springer, Cham (2014). https://doi.org/10.1007/978-3-319-10578-9_23

8. Ren, S., Ren, S., He, K., Girshick, R., Sun, J.: Faster R-CNN: Towards real-time object detection with region proposal networks. In: Advances in Neural Information Processing Systems, pp. 91–99 (2015)

9. Redmon, J., Divvala, S., Girshick, R., Farhadi, A.: You only look once: unified, real-time object detection. In: Proceedings of the IEEE Conference on Computer Vision and Pattern Recognition, pp. 779–788 (2016)

10. He, K., Gkioxari, G., Dollár, P., Girshick, R.: Mask R-CZNN. In: European Conference on Computer Vision, pp. 2980–2988 (2017)

11. Kim, Y., Jernite, Y., Sontag, D., Rush, A.M.: Character-aware neural language models. In: AAAI, pp. 2741–2749 (2017)

12. Sabour, S., Frosst, N., Hinton, G.E.: Dynamic routing between capsules. In: Advances in Neural Information Processing Systems, pp. 3859–3869 (2017)

13. Hinton, G.E., Krizhevsky, A., Wang, S.D.: Transforming auto-encoders. In: Honkela, T., Duch, W., Girolami, M., Kaski, S. (eds.) ICANN 2011. LNCS, vol. 6791, pp. 44–51. Springer, Heidelberg (2011). https://doi.org/10.1007/978-3-642-21735-7_6

14. Long, J., Shelhamer, E., Darrell, T.: Fully convolutional networks for semantic segmentation. In: Proceedings of the IEEE Conference on Computer Vision and Pattern Recognition, pp. 3431–3440 (2015)

15. Radford, A., Metz, L., Chintala, S.: Unsupervised representation learning with deep convolutional generative adversarial networks. arXiv preprint arXiv:1511.06434 (2015)

16. Ioffe, S., Szegedy, C.: Batch normalization: accelerating deep network training by reducing internal covariate shift. In: International Conference on Machine Learning, pp. 448–456 (2015)

17. LeCun, Y.: The MNIST database of handwritten digits (1998). http://yann.lecun.com/exdb/mnist/

18. Chen, T., et al.: Mxnet: A flexible and efficient machine learning library for heterogeneous distributed systems. arXiv preprint arXiv:1512.01274 (2015)

19. Kingma, D.P., Ba, J.: Adam: A method for stochastic optimization. arXiv preprint arXiv:1412.6980 (2014)

# Bidirectional Gated Recurrent Unit Networks for Relation Classification with Multiple Attentions and Semantic Information

Bixiao Meng[1], Baomin XU[1(✉)], Erjing Zhou[1], Shuangyuan YU[1],
and Hongfeng Yin[2]

[1] School of Computer and Information Technology, Beijing Jiaotong University,
Beijing, China
{16120476, bmxu, 17127099, shyyu}@bjtu.edu.cn
[2] Department of Computer Science, Beijing Jiaotong University Haibin College,
Huanghua, China
06120560@bjtu.edu.cn

**Abstract.** Relation classification is an important part in natural language processing (NLP) field. The main task of relation classification is extracting the relations between target entities. In recent years, there are many methods for relation classification and some of them have achieved quite good results, but these methods have not given enough attention to the target words, and the semantic information of words is also lack of utilization. In order to make good use of the contextual information in the sentences as much as possible, we adopt the bidirectional gated recurrent unit networks (BGRU). On this basis, in order to focus on the computing process of target entities and target sentences, we add the multiple attention mechanism. Meanwhile, other semantic information such as the named entity and part of speech information of the word are also added as input data so as to make full use of the words' information in the corpus. We have conducted some experiments on the widely used datasets, and we got up to 3% improvement in the F1 value compared to previous optimal method.

**Keywords:** Relation extraction · Attention mechanism · Bidirectional GRU

## 1 Introduction

The essence of the relation extraction task is to find the semantic relation of the words in the sentences, which plays a big role in many natural language processing applications, such as "Question-answer System [1]", "Knowledge Graph [2]".

The relation extraction is actually a multi-category process. According to the methods, relation extraction method can be roughly divided into two types: traditional and deep learning based. Traditional methods include the kernel function-based approach, SVM, logistic regression and so on [3–5].

In recent years, the widely used deep learning models are CNN, RNN, LSTM, etc. On the basis of CNN, some researchers use methods such as attention mechanism to improve the experimental results [6–9]. Compared with CNN, the advantage of RNN lies in the extraction of the relationship between words that are far apart [10]. In 1997,

© Springer Nature Switzerland AG 2019
H. Lu et al. (Eds.): ISNN 2019, LNCS 11554, pp. 124–132, 2019.
https://doi.org/10.1007/978-3-030-22796-8_14

Hochreiter et al. proposed the long short-term memory network which is more suitable for the processing of long text [11]. Later, a variety of LSTM-based relationship extraction models emerged [12–14].

The main contributions of this paper are:

1. We added sentence-level and word-level attention mechanism on the basis of variant of long short-term memory network (LSTM). In order to use the corpus effectively and reduce the noise impact.
2. We added semantic information as input, enrich the characteristic information of the sentences. Combine with multiple attention mechanisms to reduce the negative impact of semantic information.
3. We have made multiple sets of comparative experiments on different datasets, verified the effects of multiple-level attention mechanism and semantic descriptions on extract results.

## 2   Gated Recurrent Unit

Our model is based on the variant of the long-short term memory network(LSTM). LSTM is similar to a variant of RNN, which is a chain structure with repeating units. LSTM can solve the vanishing gradient problem by determining whether information is useful and "forgetting" useless information.

There are many variants of LSTM since its creation. We use the gated recurrent unit (GRU) [15]. This variant is simpler and more efficient, with fewer parameters than LSTM.

As shown in Fig. 1, there are only two gates in the GRU model: "reset gate" and "update gate", and remove the cell status in LSTM. GRU controls whether to retain the information of the original hidden state through the reset gate, but no longer restricts the incoming of current information. The formulas are as follows:

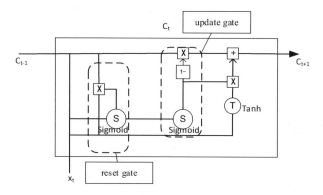

**Fig. 1.** GRU memory block.

$$z_t = \sigma(W_z[h_{t-1}, x_t]) \tag{1}$$

$$r_t = \sigma(W_r[h_{t-1}, x_t]) \tag{2}$$

$$\tilde{h}_t = \tanh(W[r_t * h_{t-1}, x_t]) \tag{3}$$

$$h_t = (1 - z_t) * h_{t-1} + z_i * \tilde{h}_t \tag{4}$$

$z_t$ represents the update gate, $r_t$ represents the reset gate, W is the weight matrix, h is the cell status information, x represents the input information.

The original GRU model is one-way, and information can only be transferred in one direction. We use a two-way GRU model, which can better combine context information and process sentence corpus more accurately through such forward and backward propagation. In this paper, we perform element summation on the forward and backward generated word feature vectors to obtain the final vector.

## 3   Methodology

The content of this chapter is mainly to introduce the specific solutions proposed in this paper, including the following parts: input layer, multiple attention mechanism and classification operation.

### 3.1   Input

We use the skip-gram method in word2vec proposed by Mikolov to vectorize the corpus [16], and introduce the position vector of the word as a supplement [8]. Each word has a two-dimensional position vector, indicating the distance of the current word from the target word. This distance is determined in a similar way to the coordinate system using the current word as the origin.

On this basis, we also introduce the semantic information to enrich the input data, such as part-of-speech (POS), named entity recognition (NER), dependency feature (Dep), hyponymy (HP), relative-dependency feature (RDF). We obtain the above semantic information through the existing open source NLP tool, convert it into one-hot vector and then perform embedding operation. Combine the generated vector into the input vector.

### 3.2   Multiple Attentions

In deep learning, the attention mechanism means providing more computing power and resources for the key data to obtain more accurate results.

In this paper, the application of attention mechanism is divided into two layers, which are the word-layer attention mechanism and the sentence-layer attention mechanism. The multiple attention model is shown in Fig. 2:

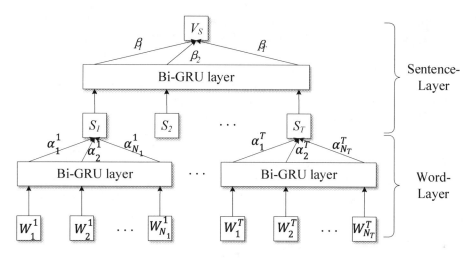

**Fig. 2.** Multiple attentions.

## Word-layer Attention

Each sentence has target words that need to be treated with emphasis. In the process of generating sentence vectors, we introduce a word-layer attention mechanism. As shown in Fig. 2, W represents the original word vector, S represents the sentence vector generated by word-layer GRU, and α are the word-layer weights.

Suppose there are N words in the sentence, we use h to represent the sentence vector generated by the GRU hidden layer. The formula for the attention mechanism of the word layer is as follows:

$$\overline{h_w} = \tanh(W_w h_w + \mathbf{b}) \tag{5}$$

$$\omega_w = \text{softmax}(\overline{h_w}) \tag{6}$$

$$v = S\omega_w^T \tag{7}$$

Where ω denotes a weight matrix with attention added, and $v$ represents the sentence feature vector matrix after the introduction of the attention mechanism. The weight measures the relation between each word and the target word. Words that have a greater effect on the extraction of target words' relation will be given higher weights.

## Sentence-layer Attention

In the case of insufficient labeling, we generally think that the sentence containing the target word can represent the relation of the target word. But there are exceptions, such as these two sentences:

- *Los Angeles is a city in California;*
- *In 1850, Los Angeles was officially established. In the same year, California became the 31st state in the United States.*

Both sentences contain these two place names: "Los Angeles" and "California", but only the first sentence clearly indicates the "contain" relation between them. We introduce an attention mechanism at the sentence layer, in order to increase the weight of effective sentences and reduce the influence of noise.

The method we use is similar to the attention mechanism of the word layer. We treat the sentences that contain the same target words as a collection, and input the sentences from the same collection to the GRU, get the weight of each sentence. Finally, each collection gets a weighted sentence vector for classification. As shown in Fig. 2, V represents the vector of the "Collection", and β represents the weight in sentence-layer.

The formula for the attention mechanism of the sentence layer is similar to the word layer attention mechanism:

$$\overline{h_s} = \tanh(W_s h_s + b) \tag{8}$$

$$\omega_s = \mathrm{softmax}(\overline{h_s}) \tag{9}$$

$$v = S\omega_s^T \tag{10}$$

Sentence-layer attention take the sentence vectors as input, the weights of all sentences are combined into the weight matrix of the collection: $\omega_s$.

### 3.3   Classifying

This paper used the softmax classifier to predict the relationship labels of corpus feature vectors from the set of labels. Taking the feature vector $o \in R^{n_o}$ containing the weight corpus as input, the conditional probability that the corpus represents the relationship $r_i$ is:

$$p(r_i|C, \theta) = \frac{\exp(o_i)}{\sum_{j=1}^{n_0} \exp(o_j)} \tag{11}$$

$\theta$ denotes all parameters and C represents a "Collection".

Then we used cross entropy to define the objective function as follows:

$$J(\theta) = \sum_{i=1}^{N} logp(r_i|C, \theta) \tag{12}$$

In the experiments, we used the dropout strategy [17] to alleviate the problem of overfitting.

## 4   Experiments

Our experiments mainly aimed to prove the improvement brought by semantic information of the words and multiple attention mechanisms. Meanwhile we compared our model with other neural network models.

## 4.1 Datasets and Evaluation Metrics

We used two widely used datasets, first is the SemiEval-2010 task 8 dataset. This dataset has nine relation types and one Other type, means sentences that do not represent any relation in the 9 types. This dataset contains 8000 training sentences and 2717 testing sentences. The second dataset we used is NYT10 dataset. It has 19 relation types and 522611 training sentences, 172448 testing sentences.

In order to facilitate comparison with experiments of other models, we used accuracy/recall rate and F1 value as evaluation criteria.

## 4.2 Experiments Setting

We used skip-gram model to vectorize the words, and obtained the position vector of the words. We used the existing NLP tools to obtain the semantic information of the words, and used the skip-gram to obtain the semantic information embeddings. Then, we combined the word embeddings and semantic information embeddings as the input vectors of the model.

Through experimental measurement, we've set the dimension of the word embedding is 50, the dimension of PF is 2*5, the dimension of each kind of semantic information embedding is 20, the mini-batch size is 10, and the dropout rate is 0.5.

We used GPU (GTX1080Ti) to run the experiments. The computer memory size is 32G.

## 4.3 Experimental Results and Analysis

Firstly, we compared the GRU model with the word-layer attention to the best-performing CNN model. And based on this, we did the comparative experiments on semantic information. These experiments used the SemEval-2010 task 8 dataset.

As can be seen from Table 1, the bidirectional GRU has at least 1% advantage over the traditional CNN in F1 value when only the word embeddings and the PF are used as inputs; The introduction of the word layer attention mechanism can increase the F1 value by about 1.7% compared to the bidirectional GRU network that does not have a single layer attention mechanism; Adding semantic information to bidirectional GRU +ATT model can increase the result by more than 2%. From the experimental results, it can be concluded that both word-layer attention and semantic can significantly improve the results of relation extraction.

Then we added the sentence-layer attention and chosen some classic methods to conducted comparative experiments without semantic information. These experiments used the NYT10 datasets.

From Fig. 3, we can see that after the curve is stable, the effect of the bidirectional GRU combined with the multiple attention is slightly better than the best model before: PCNN+ATT. However, in the case of high recall rate, the accuracy of BGRU+2ATT drops faster, even slight lower than PCNN+ATT. The performance of the model needs to be improved in the case of high retrieval rate, but the overall effect is better than other models.

**Table 1.** Comparison test of single-layer attention mechanism plus semantic description model.

| Model | Feature set | F1 |
|---|---|---|
| CNN [6] | Only word | 69.7 |
| | Add PF WordNet | 80.6 |
| BLSTM [18] | Only word | 81.6 |
| | Add PF POS NER etc. | 83.5 |
| Att-BLSTM[14] | Only word+PF | 84.0 |
| BGRU+ATT | Only word | 82.2 |
| | Add PF POS NER etc. | 84.5 |

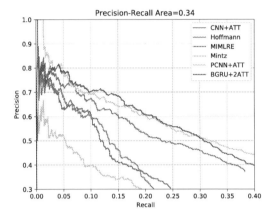

**Fig. 3.** PR curves for BGRU+2ATT and other classic models.

Finally, we added semantic information based on BGRU+2ATT and compared it with other models (Table 2).

**Table 2.** Comparison between BGRU+2ATT combining semantic information and previously published results.

| Model | Feature set | F1 |
|---|---|---|
| PCNN+ATT [9] | Only word | 72.2 |
| APCNNs [19] | Only word | 72.0 |
| | Add Entity Descriptions. | 74.0 |
| BGRU+ATT | Only word | 73.1 |
| | Add PF POS NER etc. | 74.8 |
| BGRU+2ATT | Only word | 78.6 |
| | Add PF POS NER etc. | 80.8 |

Since the NYT10 dataset has a lot of unlabeled corpus, the overall effect is not as good as the SemEval-2010 task 8 dataset. But by comparison, it can be seen that the GRU-based model still has a slight advantage over the CNN-based model. In the case where only the word embeddings are used as the input vector and the model has only a single layer attention mechanism, the effect of BGRU is about 1% higher than that of CNN. But the effect of BGRU+ATT model with semantic information is only 0.8% higher than the CNN model with entity descriptions. In fact, in the experimental results of TOP100, the BGRU+ATT model is even slightly lower than the CNN model. By analyzing the experimental data, we believe that the entity description used by Ji [19]. is different from the semantic description we use. They use the background information of the word, in simple terms, the model is trained by reducing the difference between the word vector and the background information vector. The problem with this approach is that the background information for most of the words is not easily accessible and is prone to generate noise data. But after introducing the sentence-layer attention, the effect of the BGRU+2ATT model will exceed the CNN model. The advantage can even reach nearly 5%. Compared with the experimental data on the SemEval-2010 task 8 dataset, the BGRU+2ATT model has a more significant improvement than the BGRU model with only word-layer attention. Therefore, we believe that the sentence-layer attention mechanism plays a larger role in the case of incomplete corpus annotation.

## 5  Conclusion

On the basis of the bidirectional GRU model, we combine the word-layer attention mechanism and the sentence-layer attention mechanism, and at the same time we add the semantic information of the word as a supplement to the input vector, and get better results than other models. Especially after the introduction of the sentence-layer attention mechanism, the effect of the model is obviously improved when the corpus is not fully marked. In the comparative experiments, the entity description used by Ji [19] caught our attention, we envision whether the introduction of semantic descriptions can be stripped from the model and calculated separately. Replacing the background information of words with semantic information can also avoid the problem that background information is difficult to obtain. We hope to make progress in this area in the future.

**Acknowledgments.** This work was supported by the Key Projects of Science and Technology Research of Hebei Province Higher Education [ZD2017304].

## References

1. Cui, H., Sun, R., Li, K., Kan, M.Y., Chua, T.S.: Question answering passage retrieval using dependency relations. In: 28th Annual International ACM SIGIR Conference on Research and Development in Information Retrieval, pp. 400–407. ACM Press, Salvador (2005)
2. Mottin, D., Lissandrini, M., Velegrakis, Y., Palpanas, T.: Exemplar queries: give me an example of what you need. Proc. VLDB Endow. **7**(5), 365–376 (2014)
3. Zelenko, D., Aone, C., Richardella, A.: Kernel methods for relation extraction. Mach. Learn. Res. **3**(Feb), 1083–1106 (2003)

4. Kambhatla, N.: Combining lexical, syntactic, and semantic features with maximum entropy models for extracting relations. In: ACL 2004 on Interactive Poster and Demonstration Sessions, p. 22. Association for Computational Linguistics. Barcelona (2004)

5. Qu, M., Ren, X., Zhang, Y., Han, J.: Weakly-supervised relation extraction by pattern-enhanced embedding learning. In: 2018 World Wide Web Conference on World Wide Web, pp. 1257–1266. Lyon (2018)

6. Zeng, D., Liu, K., Lai, S., Zhou, G., Zhao, J.: Relation classification via convolutional deep neural network. In: the 25th International Conference on Computational Linguistics: Technical Papers, pp. 2335–2344, Dublin (2014)

7. Santos, C.N.D., Xiang, B., Zhou, B.: Classifying relations by ranking with convolutional neural networks. In: International Joint Conference on Natural Language Processing, pp. 626–634 (2015)

8. Zeng, D., Liu, K., Chen, Y., Zhao, J.: Distant supervision for relation extraction via piecewise convolutional neural networks. In: Proceedings of the 2015 Conference on Empirical Methods in Natural Language Processing, pp. 1753–1762, Lisbon (2015)

9. Lin, Y., Shen, S., Liu, Z., Luan, H., Sun, M.: Neural relation extraction with selective attention over instances. In: Proceedings of the 54th Annual Meeting of the Association for Computational Linguistics, pp. 2124–2133. Berlin (2016)

10. Zhang, D., Wang, D.: Relation classification via recurrent neural network. J. Comput. Lang. arXiv:1508.01006v2 (2015)

11. Hochreiter, S., Schmidhuber, J.: Long short-term memory. Neural Comput. 9(8), 1735–1780 (1997)

12. Xu, Y., Mou, L., Li, G., Chen, Y., Peng, H., Jin, Z.: Classifying relations via long short term memory networks along shortest dependency paths. In: 2015 Conference on Empirical Methods in Natural Language Processing, pp. 1785–1794. Lisbon (2015)

13. Peng, N., Poon, H., Quirk, C., Toutanova, K., Yih, W.T.: Cross-sentence N-ary relation extraction with graph LSTMs. Transact. Assoc. Comput. Linguist. 5(1), 101–115 (2017)

14. Zhou, P., et al.: Attention-based bidirectional long short-term memory networks for relation classification. In: the 54th Annual Meeting of the Association for Computational Linguistics, pp. 207–212. Berlin (2016)

15. Cho, K., et al.: Learning phrase representations using RNN encoder–decoder for statistical machine translation. In: Empirical Methods in Natural Language Processing, pp. 1724–1734 (2014)

16. Mikolov, T., Sutskever, I., Chen, K., Corrado, G.S., Dean, J.: Distributed representations of words and phrases and their compositionality. In: Advances in Neural Information Processing Systems, pp. 3111–3119 (2013)

17. Hinton, G.E., Srivastava, N., Krizhevsky, A., Sutskever, I., Salakhutdinov, R.R.: Improving neural networks by preventing co-adaptation of feature detectors. Comput. Sci., arXiv preprint arXiv:1207.0580 (2012)

18. Zhang, S., Zheng, D., Hu, X., Yang, M.: Bidirectional long short-term memory networks for relation classification. In: the 29th Pacific Asia Conference on Language, Information and Computation, Shanghai, pp. 73–78 (2015)

19. Ji, G., Liu, K., He, S., Zhao, J.: Distant supervision for relation extraction with sentence-level attention and entity descriptions. In: the 31st AAAI Conference on Artificial Intelligence, San Francisco, pp. 3060–3066 (2017)

# Functional and Structural Features of Recurrent Neural Networks with Controlled Elements

Vasiliy Osipov[(✉)] and Viktor Nikiforov

St. Petersburg Institute for Informatics and Automation of Russian Academy
of Sciences, 39, 14 Liniya, St. Petersburg 199178, Russia
osipov_vasiliy@mail.ru, nik@iias.spb.su

**Abstract.** The features of recurrent neural networks with controlled elements are considered. The functions of these networks for controlled associative processing of signals are refined. A number of the space-time structures of such networks are analyzed. Among them, the neural networks with one-, two-, and three-level structures of layers are investigated. The results of studies on the stability of the associative processing of distorted signals by these networks are reflected. Based on the simulation results, recommendations are formulated to expand the possibilities of associative signal processing in recurrent neural networks with controlled elements.

**Keywords:** Recurrent neural network · Logical structure ·
Associative signal processing · Control

## 1 Introduction

In recent years, artificial recurrent neural networks have received significant development with regard to intelligent processing of various signal flows. Among the tasks of such processing are: the purification of signals from possible defects, recognition of speech and dynamic images, prediction of events, control of the behavior of robots and others. In the interests of this, a number of widely known recurrent neural networks (RNNs) are applicable [1, 2]. These include RNNs on the basis of the perceptron, associative neural network structures, networks with long-term and short-term memory.

Representatives of the RNNs based on the perceptron are the recurrent multi-layer perceptron, the Elman network, the real-time recurrent network [1]. These RNNs allow for quick, but not deep signal processing.

Among associative neural network structures Hopfield, Hamming networks, bidirectional associative memory, and others are used [1–6]. The capabilities of these associative structures in the depth of signal processing are significantly higher than the characteristics of the RNNs based on the perceptron, however, they in many respects do not meet the requirements of real time. In addition, these RNNs do not provide the control the signal associative processing.

The RNNs with controlled elements were proposed in [7, 8] and were developed further in [9–12]. Potentially, these RNNs can provide, not only operational, but also

© Springer Nature Switzerland AG 2019
H. Lu et al. (Eds.): ISNN 2019, LNCS 11554, pp. 133–140, 2019.
https://doi.org/10.1007/978-3-030-22796-8_15

deep associative processing of signal flows. Such networks can be endowed with different logical structures due to the realization of spatial shifts of signals during transmission from one layer to another.

The RNNs also provide the control of the associative interaction of the signals. They extend the possibilities for space-time signal binding and solving intellectual problems. In a sense, the RNNs with controlled elements can be called streaming recurrent neural networks with deep signal associative processing. The ability of the RNNs to solve creative problems largely depends on the perfection of their space-time structures and the stability of signal associative processing.

The latest results of the studies performed in this part are reflected in [12]. The linear and spiral structures of the RNNs with controlled elements have been studied. The aspects of building multilevel recurrent neural networks are partially affected. A number of local tasks to ensure the sustainability of the functioning of these RNNs have been solved. However, the results require further study. In particular, it is necessary to search for new space-time structures of the RNNs with enhanced capabilities and a deeper study of issues related to the stability of these networks.

The article clarifies the features of the RNNs with controlled elements, analyzes a number of the space-time structures of these networks. The results of studies on the stability of their associative processing of distorted signals are reflected.

## 2    Functional Features of RNNs with Controlled Elements

A generalized scheme of the RNNs is shown in Fig. 1. The essential difference of the RNNs from other neural networks is the presence of the control of network elements and new methods of processing information. At the input of the RNN, various signal streams (speech, music, video, and others) can be supplied only after preliminary direct conversion. In the general case, signals must be decomposed into spatial- frequency components before being fed into the RNN. Then each component is transformed into a sequence of single pulses with a frequency and a phase, as functions of the amplitude and phase of the component, respectively.

In discrete time the signals, which are supplied to the RNN, can be considered as sequences of sets of single pulses (SSP). After processing in the RNN, the sequences of the SSP at its output should be transformed by inverse transformation into the corresponding original signals. In this network, the matching of entry and exit is ensured by prioritizing short connections between interacting neurons. In the RNN, each pulsed neuron of one layer is associated in the general case with all the neurons of the other layer. Connections between neurons in the layers of the network are absent. Neurons can be in three states: waiting, excitation and refractory state. Each excited neuron generates a pulse and goes into a state of refractoriness. The generated pulse after a single delay enters the synapses. Passing through synapses, the impulse reads information about previous effects and leaves a "trace" through the change of weights (conductivities) of synapses. The delay time of each pulse in the formed bilayer circuits of this RNN does not exceed the neuron refractory time after excitation. With this in mind and the desire to minimize conflicts between neurons in the RNN, spatial shifts of signals are realized when they are transferred from layer to layer.

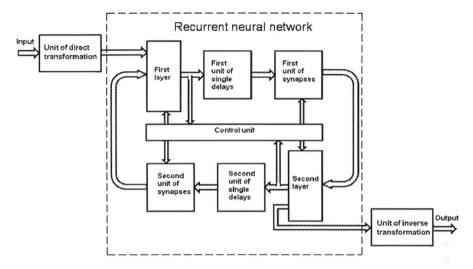

**Fig. 1.** Scheme of a recurrent neural network with controlled elements

The presence of the control unit in the RNN, in addition to ensuring spatial shifts, allows control the associative interaction of the processed signals through changing the parameters of synapses and neurons, to ensure the stable operation of the network.

The weights (conductivities) $w_{ij}(t)$ of artificial synapses in this RNN are defined as [8, 12],

$$w_{ij}(t) = k_{ij}(t) \cdot \beta_{ij}(t) \cdot \eta_{ij}(t),$$

where $k_{ij}(t)$ is the weighting coefficient of the synapse connecting the $i$th neuron of one layer with the $j$th neuron of the other layer; $\beta_{ij}(t)$ – function of attenuation of diverging signals from the $i$th neuron; $\eta_{ij}(t)$ – function of attenuation of converging signals to the $j$th neuron. As the basic functions $\beta_{ij}(t)$ и $\eta_{ij}(t)$ the known radial functions [1] can be taken. In the particular case when the distance between the interacting layers of a neural network tends to zero $\beta_{ij}(t)$ can be specified as:

$$\beta_{ij}(r_{ij}(t)) = 1/(1 + \alpha \cdot (r_{ij}(t))^{1/h}),$$

where $\alpha$ is the positive coefficient; $h$ – the degree of root. The value $r_{ij}(t)$ in this expression is measured in the units of neurons and depending on the realized spatial shifts of SSPs along the layers is defined as:

$$r_{ij}(t) = ((\Delta x_{ij}(t) + n_{ij}(t)d)^2 + (\Delta y_{ij}(t) + m_{ij}(t)q)^2)^{1/2},$$
$$n_{ij}(t) = \pm 0, 1, \ldots, D - 1; m_{ij}(t) = \pm 0, 1, \ldots, B - 1.$$

Here $\Delta x_{ij}(t)$, $\Delta y_{ij}(t)$ are the projections of the connection of the $j$th neuron with the $i$th neuron on the $X$, $Y$ axes without taking into account the spatial shifts; $d$, $q$ are the values of unit shifts in X, Y coordinates respectively; $D$ and $B$ are the numbers of columns and rows respectively each layer of the neural network is being divided by shifts.

Conductivities (weights) $w_{ij}(t)$ can be considered as the functions of the charges transferred through artificial synapses. In some cases they can be defined as:

$$w_{ij}(t) = 1/(R_{ON_{ij}}(t) + (R_{OFF_{ij}}(t) - R_{ON_{ij}}(t)) \cdot \exp(-A \cdot (q_{ij}(t - \Delta t) + q_{ij}(\Delta t)))),$$
$$R_{ON_{ij}}(t) = R_{ON}/\psi_{ij}(t), \ R_{OFF_{ij}}(t) = R_{OFF}/\psi_{ij}(t), \ \psi_{ij}(t) = \beta_{ij}(t) \cdot \eta_{ij}(t),$$

$R_{ON_{ij}}(t), R_{OFF_{ij}}(t)$ are the smallest and the largest resistances of the $ij$th synapse; $A$ – the coefficient with dimension $1/Q$; $q_{ij}(t-\Delta t)$ is the charge transferred through the $ij$th synapse up to the time moment $t-\Delta t$; $q_{ij}(\Delta t)$ – the charge carried over the time interval $\Delta t$. There are other models available for $w_{ij}(t)$ [1, 2, 12].

The amount of charge transferred through the synapse to the input of a single neuron also depends on the charges transferred from the other neurons. Note that in this case the values of $q_{ij}(\Delta t)$ can be either positive or negative. In the case of a negative charge transferred through the synapse the effect of a partial erasure of the previously memorized information is manifested.

All these circumstances can explain not only the spatial shifts of the signals along the layers, but also changes in the shapes and directionality of the cross sections of the pulses diverging and converging in space.

## 3    Logical Structures of RNNs

An example of endowing the RNN with a logical structure due to the spatial shifts of the processed signals is shown in Fig. 2, where 1 are lines of splitting the network layers into logical fields due to signal shifts during transmission from one layer to another; 2 is the direction of movement of SSPs along the layers of the network; 3 are neurons of the network; 4 - the directions of signal transmission from layer to layer.

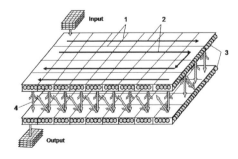

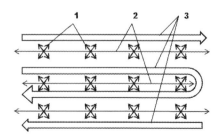

**Fig. 2.** Recurrent neural network with a three-level structure of layers

**Fig. 3.** The structure of the RNN at the level of neural network channels

According to Fig. 2 the signals as SSPs are fed to the input logic field of the first layer of the RNN, and are removed from the output field of the second layer. In Fig. 2 layers of the RNN are endowed with space-time interconnected semi-coil and linear structures. Each such structure has its own neural network channels of promoting and associative signal interaction in the network. With this in mind, the RNN (Fig. 2) at the level of such channels can be represented as Fig. 3, where 1 is the designation of the processes of associative interaction of the SSPs advancing through neural network channels; 2 – the control processes of associative signal interactions in a network; 3 – the neural network channels. In order to simplify the display of neural network channels in Fig. 3 lines breaking them into logical fields, not shown.

The peculiarity of this structure is that it is possible to control the associative interactions between neural network channels. When processing signals in the RNN, they are not blurred. This is ensured by prioritizing short connections between neurons of the network. Due to the inherent mechanisms, these RNNs can be endowed with various logical structures with different capabilities of associative information processing (Fig. 4).

(a)  Linear structure of the RNN with memory based on time delays

(b)  Linear structure of the RNN with associative memory and memory based on time delays

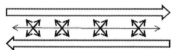

(c)  Structure of the RNN with counter-advancing signals

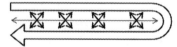

(d) Structure the RNN in the form of a semi-coil

(e)  Spiral Structure of the RNN

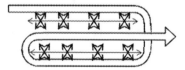

(f)  Structure of the RNN in the form of a loop

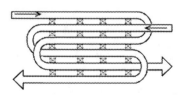

(g)  Structure of the RNN in form of a double spiral

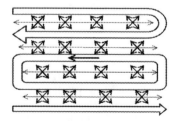

(h)  Multi-level RNN structure

**Fig. 4.** Possible variants of logical structures of RNNs at the level of neural network channels

The memorization of information in these structures is feasible both in the neural network channels themselves and in inter-channel connections. According to the information storage capabilities, neural network channels can be divided into two types. The first type is neural network channels with memory based on the time delays in changing the states of neurons in the network. The second type of channels is characterized additionally by associative memory at synapses. Inter-channel synaptic connections carry the main load of memorization signals in the studied RNNs.

The experiments conducted by the authors showed that by controlling the associative signal interactions in the RNNs with the structures in Figs. 4*b, c, d. e, f, g and h* can significantly expand the possibilities of associative intellectual processing.

By controlling associations in neural network channels (Fig. 4*b*), it is possible to make priority calls to previous or subsequent events. In the RNNs with the structures shown in Figs. 4*d, e, f, and g* in addition to the consistent associative interaction of signals, the organization of controlled cycles is possible. In the RNNs with multilevel structures (Figs. 3 and 4*h*), controlled interaction between different neural network channels is feasible. These RNNs are the most promising with regard to providing broad intellectual capabilities.

## 4  Parameters and Stability of RNNs

In the interests of determining the capabilities of the RNNs under study for stable associative processing of distorted signals, special modeling was performed. At first, the limits of successful elimination of possible defects in the signal flows were investigated. These defects are false and missed signal elements. Experiments were implemented on the RNNs with the structures shown in Figs. 4*b and d*. The number of neurons in each layer did not exceed 960 units. Each layer was divided into logical fields of $5 \times 6$ neurons. The RNNs were processing sequences of images. Each image represented a distortion of one of the four patterns.

The possibilities for eliminating seven different defects were estimated for different values of $L$ - learning cycles (epochs) of RNNs, at different values of *Thr* - excitation threshold of neurons and scaling factor $\alpha$ of distances between neurons. The number of threshold values was $J$, and the number of distance scale values was $I$. The evaluation was carried out for the number $Q(L)$ of successful defect elimination options as

$$Q(L) = \sum_{i=1}^{I} \sum_{j=1}^{J} q_{ij}(\alpha_i, Thr_j, L),$$

where $q_{ij}(\alpha_i, Thr_j, L)$ is the function of eliminating defects in the *ij*th version of the study after $L$ cycles (epochs) of learning. This function for the case of complete elimination of defects is equal to 1, in other cases - 0.

Results are shown in Fig. 5. According to these dependencies, the RNN with the semi-coil structure is more stable than the linear RNN.

It should be noted that the obtained borders for the elimination of the defects of the first and second types are not wide. This is due to the fact that there is an objective contradiction. In the RNNs, false elements in signals are easily eliminated at elevated thresholds of neuron excitation. The missing elements, on the contrary, are successfully restored at lower thresholds of neuron excitation.

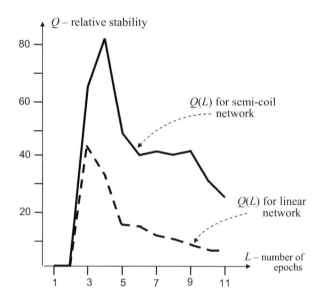

**Fig. 5.** Dependence of relative stability from the number of epochs

With this in mind, it was proposed to train the RNNs under study and eliminate false elements in the processed signals at high thresholds. After elimination of defects of this type, it is recommended to reduce the thresholds of neuron excitation and strengthen the associative call of signals from the network memory. The experiments on such processing on the RNNs with structures of the type of Figs. 4d, e and f, by 2016 neurons in each layer showed that the studied borders can be significantly extended.

## 5   Conclusion

The research develops the ideas outlined in [12], refines the functions for controlled associative processing of signals and reveals new features of the RNNs with controlled elements. The considered features of the RNNs with controlled elements indicate a significant potential of these networks for intelligent signal processing. In a logical sense, such networks are transparent. Endowing the RNNs with various single-channel and multi-channel interconnected structures allows to significantly diversify the working and long-term memory of networks, and to expand the set of operations on the processed signals.

RNNs realize controlled binding of signals and their transmission channels. This makes it possible to organize cycles, signal swaps, to form other signals from the support structures, to increase the depth of signal associative processing while maintaining the possibility of operational responses to input influences. Controlling signal associations in the RNNs is possible through changing the parameters of synapses and neurons.

As shown by the results of the performed simulation, due to such control, the boundaries of stable associative processing of distorted signals are significantly extended. The capabilities of the RNNs with controlled elements for intelligent signal processing depend directly on the sizes of the networks. The more neurons and synaptic connections in the RNN, the higher the possibilities of endowing network with intelligent logical structures and functions. For the implementation of large RNNs with the considered properties, it is advisable to use modern memristive technologies.

Recurrent neural networks with controlled elements can be used to create promising autonomous cognitive self-learning robots.

**Acknowledgement.**  This research is partially supported by the RFBR foundation grant No 16-29-09482.

# References

1. Haykin, S.: Neural Networks and Learning Machines, 3rd edn. Prentice Hall, New-York (2008)
2. Galushkin, A.I.: Neural Networks Theory. Springer, Berlin Heidelberg (2007). https://doi.org/10.1007/978-3-540-48125-6
3. Hopfield, J.J.: Neural networks and physical systems with emergent collective computational abilities. Proc. Nat. Acad. Sci. USA **79**(8), 2554–2558 (1982)
4. Kosko, B.: Bidirectional associative memories. IEEE Trans. Syst. Man Cybern. **18**(1), 49–60 (1988)
5. Hawkins, J., Blakeslee, S.: On Intelligence. Brown Walker (2006)
6. Palm, G.: Neural associative memories and sparse coding. Neural Networks **37**, 165–171 (2013)
7. Osipov, V.Y.: Associative intellectual machine. Inf. Technol. Comput. Mach. **2**, 59–67 (2010)
8. Osipov, V., Osipova, M.: Method and device of intellectual processing of information in neural network. RU Patent No 2413304 (2011)
9. Osipov, V.: Space-time structures of recurrent neural networks with controlled synapses. In: Cheng, L., Liu, Q., Ronzhin, A. (eds.) ISNN 2016. LNCS, vol. 9719, pp. 177–184. Springer, Cham (2016). https://doi.org/10.1007/978-3-319-40663-3_21
10. Osipov, V.: Structure and basic functions of cognitive neural network machine. In: 12th International Scientific-Technical Conference on Electromechanics and Robotics, Zavalishin's Readings, pp. 1–5 (2017)
11. Osipov, V., Nikiforov, V.: Formal aspects of streaming recurrent neural networks. In: Huang, T., Lv, J., Sun, C., Tuzikov, Alexander V. (eds.) ISNN 2018. LNCS, vol. 10878, pp. 29–36. Springer, Cham (2018). https://doi.org/10.1007/978-3-319-92537-0_4
12. Osipov, V., Osipova, M.: Space-time signal binding in recurrent neural networks with controlled elements. Neurocomputing **308**, 194–204 (2018)

# A GAN-Based Data Augmentation Method for Multimodal Emotion Recognition

Yun Luo[1], Li-Zhen Zhu[1], and Bao-Liang Lu[1,2,3(✉)]

[1] Center for Brain-Like Computing and Machine Intelligence,
Department of Computer Science and Engineering, Shanghai Jiao Tong University,
800 Dong Chuan Road, Shanghai 200240, China
[2] Key Laboratory of Shanghai Education Commission for Intelligent Interaction
and Cognition Engineering, Shanghai Jiao Tong University, 800 Dong Chuan Road,
Shanghai 200240, China
[3] Brain Science and Technology Research Center, Shanghai Jiao Tong University,
800 Dong Chuan Road, Shanghai 200240, China
{angeleader,zhulz98,bllu}@sjtu.edu.cn

**Abstract.** The lack of training data is an obstacle to build satisfactory multimodal emotion recognition models. Generative adversarial network (GAN) has recently shown great successes in generating realistic-like data. In this paper, we propose a GAN-based data augmentation method for enhancing the performance of multimodal emotion recognition models. We adopt conditional Boundary Equilibrium GAN (cBEGAN) to generate artificial differential entropy features of electroencephalography signal, eye movement data and their direct concatenations. The main advantage of cBEGAN is that it can overcome the instability of conventional GAN and has very quick converge speed. We evaluate our proposed method on two multimodal emotion datasets. The experimental results demonstrate that our proposed method achieves 4.6% and 8.9% improvements of mean accuracies on classifying three and five emotions, respectively.

**Keywords:** EEG · Eye movement · Emotion recognition ·
Generative adversarial network · Data augmentation

## 1 Introduction

Affective computing [12], which aims to equip machines with the ability to recognize, interpret, process, and simulate human affects, has drawn increasing attention in recent years. In the framework of affective computing, emotion recognition is the first critical phase since machines can never process human moods without precise emotion recognition. Researchers have made great progress in recognizing emotions from different signals, such as facial expressions, speeches, and some physiologoical signals including EEG and eye movement signals.

© Springer Nature Switzerland AG 2019
H. Lu et al. (Eds.): ISNN 2019, LNCS 11554, pp. 141–150, 2019.
https://doi.org/10.1007/978-3-030-22796-8_16

In recent years, researchers focused on studying multimodal emotion recognition methods to leverage the complementary property among different kinds of signals. Lu *et al.* introduced a multimodal emotion recognition framework for three emotions by combining EEG and eye movement signals [9]. By taking advantages of the deep neural networks, Liu *et al.* further improved the performance of multimodal framework [7]. Zhao *et al.* also adopted multimodal framework and extended it for recognizing five emotions [15]. Although these studies have developed various promising approaches for multimodal emotion recognition, the performance of emotion recognition models is unsatisfactory due to the lack of training data.

The popular multimodal emotion datasets contain physiological signals such as EEG and eye movement signals, which are difficult to collect. The high prices of EEG and eye movement acquisition devices and the high cost of performing multimodal emotion experiments limit the scale of the datasets. As a result, the training set is very small in size in comparison with image dataset such as ImageNet.

Data augmentation is a promising approach to dealing with the problem of lack of training data mentioned above. It enlarges the dataset by applying a transformation to the real data and generating realistic-like data [3]. Lotte generated artificial EEG data by relevant combinations and distortions of the original trials [8], and this approach increased the recognition accuracy when the training set is small. Krell *et al.* proposed to generate EEG data by rotational distortions [6]. Wang *et al.* improved the performance of the emotion recognition models by adding Gaussian noise to EEG features to generate artificial data [14]. However, the basic idea behind these methods is to generate more data by using geometric transformation and it is difficult to capture the deep information inside data.

By taking advantages of deep neural networks and adversarial training, GAN could learn information about data probability distribution and generate artificial data from real data distribution. In the field of computer vision, GAN has demonstrated its ability of generating realistic-like images by playing a zero-sum non-cooperative game [4,13]. Inspired by GAN, Hartmann *et al.* proposed EEG-GAN to generate EEG signals [5]. However, they did not use the generated data for classification. Luo and Lu generated EEG data in DE feature form by adopting cWGAN and enlarged the training dataset [10]. Their experimental results indicated that the accuracies of EEG-based emotion recognition models could be improved by adding training data generated by cWGAN.

In this paper, we propose a GAN-based data augmentation method for enhancing the performance of multimodel emotion recognition models. Since the original GAN suffers from instability and non-convergence problems, we implement cBEGAN to generate training data [1]. The main advantage of cBEGAN is that it has quick convergence speed and has an indicator for the training process. Meanwhile, we can control the category of the generated data by adding auxiliary conditional label information [11]. In this paper, we generate EEG signals and eye movement data in DE (differential entropy) feature form instead of raw data, because our previous studies have shown that the DE features of EEG and eye movement data are more effective for emotion recognition [2,9]. The multimodal data, which is the direct concatenation of EEG and eye movement data, are also generated in DE feature form with cBEGAN and cWGAN.

We evaluate our method on three-category and five-category multimodal emotion datasets. To the best of our knowledge, this is the first research work regarding GAN-based data augmentation for multimodal emotion recognition. Our experimental results demonstrate that cBEGAN has a better performance than cWGAN and significantly improves the accuracies of multimodal emotion recognition models.

## 2   Methods

### 2.1   GAN

A standard GAN consists of two competing parts which are both parameterized as deep neural networks. A generator $G$ produces synthetic data given a noise variable input while a discriminator $D$ tries to identify whether a sample comes from the real data distribution $X_r$ or the generated data distribution $X_g$. In other words, the discriminator is trained to estimate the probability of a given sample coming from the real data distribution. And the generator is optimized to trick the discriminator to offer a high probability for the generated data. The two parts are optimized simultaneously to find a Nash equilibrium. More formally, the procedure can be expressed as a minimax function:

$$
\begin{aligned}
\min_{\theta_G} \max_{\theta_D} L(X_r, X_g) &= \mathbb{E}_{x_r \sim X_r}[\log(D(x_r))] + \mathbb{E}_{z \sim Z}[\log(1 - D(G(z)))] \\
&= \mathbb{E}_{x_r \sim X_r}[\log(D(x_r))] + \mathbb{E}_{x_g \sim X_g}[\log(1 - D(x_g))]
\end{aligned}
\tag{1}
$$

where $\theta_g$ and $\theta_d$ represent the parameters of generator and discriminator, respectively, and $Z$ can be a Uniform noise distribution or a Gaussian noise distribution.

This function is optimized in two steps; (i) maximize it by fixing $G$ and $X_g$, and get the optimum of $D$; and (ii) minimize the function by the optional $D$, and then it equals to minimizing the Jensen-Shannon divergence between $X_r$ and $X_g$. The game will achieve equilibrium if and only if $X_r = X_g$.

Although GAN has shown great successes in realistic data generation, it suffers from some major problems such as non-convergence, mode collapse and diminished gradient. Researchers believed that the Jensen-Shannon divergence could lead to vanishing gradients, which was the main reason of the GAN's instability. In real world tasks such as image generation, the distribution of real images always lies in low dimensional manifolds, and the distribution of generated images also rests in low dimensional manifolds. The two distributions are almost certainly disjoint and have no overlaps. In this situation, Jensen-Shannon divergence between the two distributions is a fixed number, which can not provide useful gradients for the GAN's training.

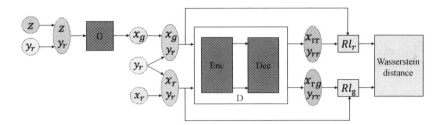

**Fig. 1.** Illustration of cBEGAN. cBEGAN adopts the auto-encoder to handle the difference between two reconstruction losses distributions. Here, G, D, Enc, Dec, $Rl_r$ and $Rl_g$ represent generator, discriminator, encoder, decoder and the two reconstruction losses, respectively.

## 2.2  cBEGAN

The discriminator in BEGAN adopts an auto-encoder which uses an encoder to extract the latent features from the input data and applies a decoder to reconstruct the data from the latent representations as shown in Fig. 1. And now the discriminator aims to matching the reconstruction loss distribution of real data and generated data. Berthelot *et al.* believe that matching auto-encoder loss could lead to the matching of the data distribution of real data and generated data directly [1], which is adopted in typical GANs. In other words, the generated data will have the similar data distribution when their reconstruction loss distributions are similar. In this way, BEGAN avoids the instability problem of conventional GAN.

BEGAN chooses Wasserstein distance to measure the difference between the two reconstruction loss distributions. The Wasserstein distance is also called Earth Mover's distance (EM distance). The distance formula for continuous probability domain is:

$$W(X_r, X_g) = \inf_{\gamma \sim \Pi(X_r, X_g)} \mathbb{E}_{(x_r, x_g) \sim \gamma}[||x_r - x_g||] \tag{2}$$

where $\Pi(X_r, X_g)$ is the set of all possible joint probability distributions between $X_r$ and $X_g$. The reconstruction loss is defined as the pixel-wise $L_1$ or $L_2$ distance between input data and reconstructed data, which can be formulated as:

$$L_r(x) = |x - D(x)|^\eta \tag{3}$$

where $D$ is the discriminator (auto-encoder) function, and $\eta \in 1, 2$, and $x$ can be a sample of real data distribution or generated data distribution.

Let $\mu_{rr}$ and $\mu_{rg}$ be the real and generated reconstruction loss distributions, respectively, and let $m_{rr}, m_{rg} \in \mathbb{R}$ be their respective means, and $\Pi(\mu_{rr}, \mu_{rg})$ is the set of all possible joint probability distributions between two distributions. By using Jensens inequality, the formula can be expressed as:

$$
\begin{aligned}
W(\mu_{rr}, \mu_{rg}) &= \inf_{\gamma \sim \Pi(\mu_{rr}, \mu_{rg})} \mathbb{E}_{(x_{rr}, x_{rg}) \sim \gamma}[||x_{rr} - x_{rg}||] \\
&\geq \inf |\mathbb{E}[x_{rr} - x_{rg}]| = |m_{rr} - m_{rg}|
\end{aligned} \tag{4}
$$

so now we are aiming to optimize a lower bound of the Wasserstein distance between the two reconstruction losses. Then the loss of BEGAN is:

$$\min_{\theta_G} \max_{\theta_D} L(X_r, X_g) = -\mathbb{E}_{x_r \sim X_r}[(L_r(x_r)] + \mathbb{E}_{z \sim Z}[L_r(G(z))]$$
$$= -\mathbb{E}_{x_r \sim X_r}[(L_r(x_r)] + \mathbb{E}_{x_g \sim X_g}[L_r(x_g)] \tag{5}$$

where $\theta_G$ and $\theta_D$ represent the respective parameters of the generator and the discriminator in BEGAN.

In BEGAN, the discriminator has two goals: auto-encode real data and discriminate real data from generated ones. In order to maintain the balance between the generator and discriminator losses, we can apply a hyper-parameter $\gamma \in [0,1]$ defined as:

$$\gamma = \frac{\mathbb{E}[L_r(G(z))]}{\mathbb{E}[L_r(x_r)]} \tag{6}$$

To maintain the equilibrium $\mathbb{E}[L_r(G(z))] = \gamma\mathbb{E}[L_r(x_r)]$, we use Proportional Control Theory by adopting an extra variable $k_t \in [0,1]$ to control the proportion of $L_r(G(z))$ during gradient descent. Similar with cWGAN, we add an extra label information to control the generated categories. The cBEGAN can be formulated as:

$$\max_{\theta_D} L(X_r, X_g, Y_r) =$$
$$- \mathbb{E}_{x_r \sim X_r, y_r \sim Y_r}[(L_r(x_r|y_r)] + k_t\mathbb{E}_{x_g \sim X_g, y_r \sim Y_r}[L_r(x_g|y_r)] \tag{7}$$

$$\min_{\theta_G} L(X_g, Y_r) = \mathbb{E}_{x_g \sim X_g, y_r \sim Y_r}[L_r(x_g|y_r)] \tag{8}$$

$$k_{t+1} = k_t + \lambda_k(\gamma L_r(x_r) - L(G(z))) \tag{9}$$

where $Y_r$ is the label distribution. We initialize $k_0 = 0$ and set $\lambda_k = 0.001, \gamma = 0.75$ in this paper. Now we can define a convergence measure as:

$$Mglobal = L_r(x_r) + |\gamma L_r(x_r) - L_r(G(z)| \tag{10}$$

Mglobal can be used as an indicator for the convergence of the network.

In this paper, we also extend cWGAN, used in our previous work [10], to multimodal emotion recognition. For cWGAN and cBEGAN, the losses of generator and discriminator are optimized in an alternating procedure. The distribution of the generated data is similar with the real data when the networks converge.

## 3  Experiment Settings

### 3.1  EEG Datasets

We evaluate our proposed method on two multimodal emotion datasets SEED [1] [16] and SEED-V.

---

[1]  http://bcmi.sjtu.edu.cn/~seed/index.html.

SEED dataset contains 62-channel EEG signals and eye movement signals of three different emotions (happy, sad, and neutral). The original EEG signals were recorded at a sampling rate of 1000 Hz with ESI NeuroScan System and the eye movement signals were collected with SMI ETG eye tracking glass, which contained information about blink, saccade, fixation and so on. In this dataset, 15 participants watched 15 emotional film clips for 3 times. In this work, 9 subjects' data (27 experiments) are used because they have completed multimodal data.

SEED-V dataset were also formed with 62-channel EEG signals and eye movement signals. 16 participants watched 15 emotional film clips to elicit five emotions: happy, sad, neutral, fear, and disgust. They took part in the experiments for three times, so there were totally 48 experiments. The EEG and eye movement signals were collected by the same device used in SEED.

## 3.2    Feature Extraction

We use a band pass filter (1–50 Hz) to eliminate low-frequency noise and high-frequency noise in the EEG signals. Then we extract DE features by adopting a 4s-length non-overlapping Hanning window for five frequency bands: $\delta$: 1–3 Hz, $\theta$: 4–7 Hz, $\alpha$: 8–13 Hz, $\beta$: 14–30 Hz, and $\gamma$: 31–50 Hz. In order to eliminate the rapid changes of the DE features, we also adopt a linear dynamic system. Each EEG sample has 310 dimensions since there are 62 channels for each band.

As for eye movement signals, we extracted the same features as in [9, 15]. The features include blink, saccade, fixation and so on. Notably, each eye movement sample has 41 dimensions in SEED dataset and it has 33 dimensions in SEED-V dataset since we simplify the eye movement features in SEED-V.

## 3.3    Evaluation Details

In order to demonstrate the effectiveness of the proposed method, we conduct cross validation on both datasets. Since, each experiment of the two datasets has 15 trials, so there are 5 trials for each emotion category in SEED dataset and 3 trails for each emotion category in SEED-V dataset. As for SEED dataset, we use 5-fold cross validation for each experiment. And as for SEED-V dataset, we adopt 3-fold cross validation for each experiment to make sure each fold has 5 emotion categories. We normalize both DE and eye movement features by min-max normalization before feeding them to the networks.

We perform grid search on the number of network layers and hidden nodes to optimize the network structure of cWGAN and cBEGAN. The numbers of layers are searched from 2 to 4 for both generator and discriminator. The input dimension is decided by the dimension of the corresponding input feature and the dimension of auxiliary label is 3 for SEED and 5 for SEED-V. The output dimension of cWGAN's discriminator is 1 while the output dimension is the same with its input for cBEGAN's discriminator.

The numbers of hidden nodes for each layer are randomly searched from 50 to 600. For cBEGAN, the hidden nodes of encoder and decoder are the same. The outputs of the two generators have the same dimension as the input data.

And ReLU activation function is used for all hidden layers. The batch size is set to 128. Adam with initial learning rate 0.0001 is used as the optimizer. The noises are sampled from a uniform distribution $U[-1,1]$.

As for the classifier, we apply an SVM with linear kernel. The parameter $c$ is searched from $2^{-10} \sim 2^{10}$ to find the optimal value.

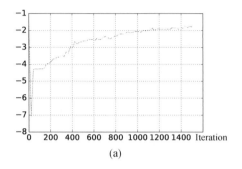
(a)

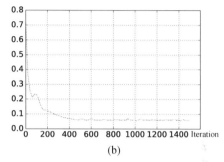
(b)

**Fig. 2.** Dloss for cWGAN (a), and Mglobal for cBEGAN (b) tendency along with training steps of SEED dataset.

## 4    Experimental Results

To evaluate the performance of the proposed method, we generate different features for both datasets. We generate DE features of EEG signals and eye movement data. As for multimodal data augmentation, we directly concatenate DE features of EEG signals and DE features of eye movement data, and generate realistic-like multimodal feature from the concatenated features. The number of the generated features for each emotion category is the same. In this section, we will first compare the convergence speed between cBEGAN and cWGAN, then discuss the performance of data augmentation for the two datasets.

### 4.1    Convergence Performance

As mentioned above, cWGAN and cBEGAN can overcome the instability problem of conventional GANs and both of them have an indicator for training procedure. Figure 2(a) shows the convergence curve of cWGAN. Dloss rises to $-2$ after 1000 iterations, which indicates the network have a good convergence performance. Besides, as the Wasserstein distance between real data distribution and generated data distribution, Dloss converging to a small value means the two data distributions are similar. As shown in Fig. 2(b), Mglobal decreases to about 0.6 and also has a stable convergence trend. cBEGAN has a better convergence performance since it converges after 500 iterations.

**Table 1.** Mean accuracies and standard deviations of the models trained on SEED dataset and appending datasets generated by cBEGAN and cWGAN

| No. of appended training data | EEG | | Eye movement | | Multimodal | |
|---|---|---|---|---|---|---|
| | cWGAN | cBEGAN | cWGAN | cBEGAN | cWGAN | cBEGAN |
| 0 | 0.8190 | 0.8190 | 0.7715 | 0.7715 | 0.8573 | 0.8573 |
| | 0.1074 | 0.1074 | 0.1327 | 0.1327 | 0.0879 | 0.0879 |
| 50 | 0.8331 | 0.8423 | 0.7881 | 0.7938 | 0.8606 | 0.8814 |
| | 0.1014 | 0.1020 | 0.1241 | 0.1202 | 0.0864 | 0.0906 |
| 200 | 0.8392 | 0.8557 | 0.7924 | 0.8043 | 0.8621 | 0.8878 |
| | 0.1028 | 0.0941 | 0.1249 | 0.1262 | 0.0877 | 0.0888 |
| 600 | 0.8372 | 0.8601 | 0.7956 | 0.8100 | 0.8539 | 0.9021 |
| | 0.1045 | 0.0876 | 0.1225 | 0.1228 | 0.0913 | 0.0858 |
| 700 | 0.8373 | 0.8641 | 0.7907 | 0.8063 | 0.8589 | **0.9033** |
| | 0.1086 | 0.0894 | 0.1262 | 0.1241 | 0.0883 | 0.0837 |
| 800 | 0.8377 | 0.8651 | 0.7929 | 0.8093 | 0.8558 | 0.9033 |
| | 0.1084 | 0.0914 | 0.1213 | 0.1139 | 0.0887 | 0.0837 |
| 2000 | 0.8338 | **0.8756** | 0.7958 | **0.8160** | 0.8586 | 0.9000 |
| | 0.1030 | **0.0852** | 0.1276 | **0.1042** | 0.0874 | **0.0776** |

### 4.2　SEED Results

For SEED dataset, the number of samples for each experiment is 842. And we generate 50, 200, 600, 700, 800, and 2000 artificial samples of the three features and add them to their respective original training datasets. Table 1 illustrates the performance at different number of augmented training data. 0 means the model is trained by original training dataset. As for single modality, cBEGAN achieves the best mean accuracies of 87.56% and 81.60% when we add 2000 samples of generated EEG and eye movement data, respectively. For multimodal data augmentation, cBEGAN reaches the best mean accuracy of 90.33% when adding 700 generated multimodal data.

### 4.3　SEED-V Results

For each subject, the number of sample for each experiment is 681, 541 and 601 since they watched different movie clips for each time. Considering these numbers are approximate, we neglect the difference and generate 50, 200, 400, 500, 1000 and 2000 samples of the three data and enlarge their respective original dataset for each experiment. As shown in Table 2, cBEGAN achieves the best mean accuracies of 62.87%, 60.19%, and 68.32% when we add 2000, 2000 and 1000 samples to the training datasets of EEG, eye movement and multimodal data.

**Table 2.** Mean accuracies and standard deviations of the models trained on SEED-V dataset and appending datasets generated by cBEGAN and cWGAN

| No. of appended training data | EEG | | Eye movement | | Multimodal | |
|---|---|---|---|---|---|---|
| | cWGAN | cBEGAN | cWGAN | cBEGAN | cWGAN | cBEGAN |
| 0 | 0.5434 | 0.5434 | 0.4862 | 0.4862 | 0.5946 | 0.5946 |
| | 0.1525 | 0.1525 | 0.1432 | 0.1432 | 0.1603 | 0.1603 |
| 50 | 0.5793 | 0.6064 | 0.5207 | 0.5533 | 0.6260 | 0.6485 |
| | 0.1534 | 0.1655 | 0.1381 | 0.1388 | 0.1599 | 0.1595 |
| 100 | 0.5846 | 0.6124 | 0.5336 | 0.5555 | 0.6279 | 0.6568 |
| | 0.1546 | 0.1616 | 0.1345 | 0.1382 | 0.1626 | 0.1559 |
| 200 | 0.5901 | 0.6181 | 0.5457 | 0.5609 | 0.6294 | 0.6674 |
| | 0.1536 | 0.1592 | 0.1437 | **0.1369** | 0.1594 | 0.1598 |
| 400 | 0.5946 | 0.6207 | 0.5446 | 0.5816 | 0.6366 | 0.6775 |
| | 0.1580 | 0.1558 | 0.1417 | 0.1434 | 0.1582 | 0.1584 |
| 500 | 0.5954 | 0.6225 | 0.5349 | 0.5815 | 0.6330 | 0.6810 |
| | 0.1571 | 0.1544 | 0.1400 | 0.1430 | 0.1606 | 0.1548 |
| 1000 | 0.5965 | **0.6287** | 0.5486 | 0.5892 | 0.6326 | **0.6832** |
| | 0.1594 | 0.1526 | 0.1456 | 0.1388 | 0.1590 | 0.1549 |
| 2000 | 0.5912 | 0.6278 | 0.5518 | **0.6019** | 0.6325 | 0.6831 |
| | 0.1593 | **0.1442** | 0.1470 | 0.1399 | 0.1620 | **0.1504** |

Compared with cWGAN, cBEGAN has higher accuracies for single and multimodal data augmentation on the two datasets. Besides, cBEGAN also has a better convergence performance. By measuring the difference between the two reconstruction loss distributions instead of two data distributions, cBEGAN can capture deeper information of the real data distribution than cWGAN, and generate artificial samples with rich information and diverse distribution, which leads better margins for the recognition models. Although cWGAN-based data augmentation has a poorer performance in terms of accuracy than cBEGAN-based data augmentation, the mean accuracies also has improvements on the two datasets, which demonstrates the multimodal emotion recognition models are more robust when adopting the proposed GAN-based data augmentation method.

## 5   Conclusion and Future Work

In this paper, we have proposed a GAN-based data augmentation method for improving the accuracy of multimodal emotion recognition models. We have generated realistic-like EEG, eye movement and their direct concentration data with cBEGAN and cWGAN. Our experimental results on two multimodal emotion datasets indicate the effectiveness of the proposed method and cBEGAN achieves the biggest improvements of mean accuracies on classifying three and five emotions with a better convergence speed. In the future, we will evaluate our method on more multimodal emotion recognition tasks and employ recurrent neural networks to consider temporal information of EEG and eye movement signals.

**Acknowledgments.** This work was supported in part by grants from the National Key Research and Development Program of China (Grant No. 2017YFB1002501), the National Natural Science Foundation of China (Grant No. 61673266), and the Fundamental Research Funds for the Central Universities.

# References

1. Berthelot, D., Schumm, T., Metz, L.: BEGAN: Boundary equilibrium generative adversarial networks. arXiv preprint arXiv:1703.10717 (2017)
2. Duan, R.N., Zhu, J.Y., Lu, B.L.: Differential entropy feature for EEG-based emotion classification. In: IEEE EMBS NER 2013, pp. 81–84. IEEE (2013)
3. Dyk, D.A.V., Meng, X.L.: The art of data augmentation. J. Comput. Graph. Stat. **10**(1), 1–50 (2001)
4. Goodfellow, I., et al.: Generative adversarial nets. In: NIPS 2014, pp. 2672–2680 (2014)
5. Hartmann, K.G., Schirrmeister, R.T., Ball, T.: EEG-GAN: generative adversarial networks for electroencephalograhic (EEG) brain signals (2018)
6. Krell, M.M., Su, K.K.: Rotational data augmentation for electroencephalographic data. In: Engineering in Medicine and Biology Society, pp. 471–474 (2017)
7. Liu, W., Zheng, W.-L., Lu, B.-L.: Emotion recognition using multimodal deep learning. In: Hirose, A., Ozawa, S., Doya, K., Ikeda, K., Lee, M., Liu, D. (eds.) ICONIP 2016. LNCS, vol. 9948, pp. 521–529. Springer, Cham (2016). https://doi.org/10.1007/978-3-319-46672-9_58
8. Lotte, F.: Signal processing approaches to minimize or suppress calibration time in oscillatory activity-based brain-computer interfaces. Proc. IEEE **103**(6), 871–890 (2015)
9. Lu, Y., Zheng, W.L., Li, B., Lu, B.L.: Combining eye movements and EEG to enhance emotion recognition. In: IJCAI 2015, pp. 1170–1176 (2015)
10. Luo, Y., Lu, B.L.: EEG data augmentation for emotion recognition using a conditional Wasserstein GAN. In: IEEE EMBS 2018, pp. 2535–2538. IEEE (2018)
11. Mirza, M., Osindero, S.: Conditional generative adversarial nets. arXiv preprint arXiv:1411.1784 (2014)
12. Picard, R.W.: Affective Computing. MIT Press, Cambridge (2000)
13. Radford, A., Metz, L., Chintala, S.: Unsupervised representation learning with deep convolutional generative adversarial networks. In: ICLR 2016 (2016)
14. Wang, F., Zhong, S., Peng, J., Jiang, J., Liu, Y.: Data augmentation for EEG-based emotion recognition with deep convolutional neural networks. In: Schoeffmann, K., et al. (eds.) MMM 2018. LNCS, vol. 10705, pp. 82–93. Springer, Cham (2018). https://doi.org/10.1007/978-3-319-73600-6_8
15. Zhao, L.M., Li, R., Zheng, W.L., Lu, B.L.: Classification of five emotions from EEG and eye movement signals: complementary representation properties. In: IEEE NER 2019. IEEE (2019)
16. Zheng, W.L., Lu, B.L.: Investigating critical frequency bands and channels for EEG-based emotion recognition with deep neural networks. IEEE Transact. Auton. Ment. Dev. **7**(3), 162–175 (2015)

# FG-SRGAN: A Feature-Guided Super-Resolution Generative Adversarial Network for Unpaired Image Super-Resolution

Shuailong Lian, Hejian Zhou, and Yi Sun$^{(\boxtimes)}$

School of Information and Communication Engineering,
Dalian University of Technology, Dalian, China
{note,zhouhj}@mail.dlut.edu.cn, lslwf@dlut.edu.cn

**Abstract.** Recently, the performance of single image super-resolution has been significantly improved by convolution neural networks (CNN). However, most of these networks are trained with paired images and take the bicubic-downsampled images as inputs. It's impractical if we want to super-resolve low-resolution images in the real world, since there is no ground truth high-resolution images corresponding to the low-resolution images. To tackle this challenge, a Feature-Guided Super-Resolution Generative Adversarial Network (FG-SRGAN) for unpaired image super-resolution is proposed in this paper. A guidance module is introduced in FG-SRGAN, which is utilized to reduce the space of possible mapping functions and help to learn the correct mapping function from low-resolution domain to high-resolution domain. Furthermore, we treat the outputs of guidance module as fake examples, which can be leveraged using another adversarial loss. This is beneficial for the main task as it forces FG-SRGAN to learn valid representations for super-resolution. When applied to super-resolve low-resolution face images in the real world, FG-SRGAN is able to achieve satisfactory performance both qualitatively and quantitatively.

**Keywords:** Image super-resolution · Unsupervised learning · GAN

## 1 Introduction

Single image super-resolution (SISR) is a fundamental low-level vision task aiming to estimate a high-resolution image from its low-resolution counterpart. With the success of deep convolutional neural networks (CNN) in computer vision, many super-resolution models were proposed, such as the pioneer work of SRCNN [3] and various other networks [4,6,8,10–15,17,18,20,21,23,26], which continuously improved the SR performance.

---

S. Lian and H. Zhou—Equal contributions.

© Springer Nature Switzerland AG 2019
H. Lu et al. (Eds.): ISNN 2019, LNCS 11554, pp. 151–161, 2019.
https://doi.org/10.1007/978-3-030-22796-8_17

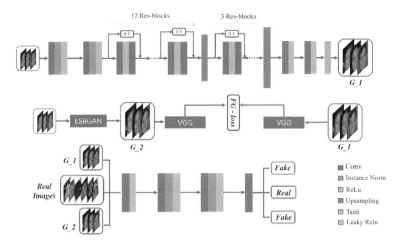

**Fig. 1.** Overall architecture of FG-SRGAN.

However, most of these networks are trained with paired images and take bicubic-downsampled images as inputs. It's impractical when we want to super-resolve low-resolution images in the real world, since there is no ground truth high-resolution images corresponding to the low-resolution images. Although methods mentioned above have achieved good performance in terms of the Peak Signal-to-Noise Ratio (PSNR), their ability to super-resolve low-resolution images is limited since degradation model in the real world doesn't fit their assumption. To tackle these challenges, we propose a Feature-Guided Super-Resolution Generative Adversarial Network (FG-SRGAN), which aims at solving SISR problem of unpaired images. Since GAN is hard to train and tends to suffer from mode dropping, it is necessary to exploit extra information to help learn the correct mapping function from low-resolution domain to high-resolution domain. Although general SISR models are not suitable for SISR problem of unpaired images, they can provide us with basic guidance to generate correct high-resolution images. Therefore, we choose ESRGAN [23] to help us accomplish this goal since it achieves state-of-the-art performance in SISR problem of paired images. More concretely, ESRGAN takes the same inputs as the generator in FG-SRGAN, and the results of ESRGAN are encouraged to share specific content features with the outputs of generator in FG-SRGAN. The specific content features of the two super-resolution networks are both extracted by VGG [19], and we check the consistence of them by $l_1$ loss. Since we use ESRGAN's results as guidance of our method, there is a risk of generating the same images as ESRGAN's results. To avoid this situation, we treat the outputs of ESRGAN as fake high-resolution images when we train adversarial network in FG-SRGAN. Our overall network architecture is shown in Fig. 1 and our contributions can be summarized as follows:

- We propose a Feature-Guided Super-Resolution Generative Adversarial Network (FG-SRGAN) that effectively learns the mapping function from low-resolution to high-resolution image manifold in the real world. To the best of our knowledge, this is the first approach that tackles SISR problem of unpaired images through training a single GAN. And we propose a guidance module in FG-SRGAN, which is utilized to reduce the space of possible mapping functions.
- We propose a new loss function in FG-SRGAN. In the generator network, we restrain the generator by conditioning it with intermediate supervision of ESRGAN. In the discriminator network, we treat outputs of ESRGAN as fake samples, which can be leveraged using another adversarial loss. The new formulation of loss function is beneficial for learning valid representations for super-resolution.
- Experiments on real-world face images show that our approach has achieved satisfactory super-resolution results visually. Furthermore, quantitative comparisons with other super-resolution models illustrate improvements obtained by the proposed method.

## 2   Related Work

Due to the superior performance of CNN on SISR problem, we mainly introduce the progress of CNN-based super-resolution approaches, and we refer readers to [16] for a complete survey of super-resolution. As a pioneer work, Dong *et al.* proposed SRCNN to learn the mapping from low-resolution to high-resolution images and achieved superior performance against previous works. Later on, various networks [10,11,20,21,26] designed for SISR were proposed and they were all trained end-to-end. Specifically, Lim *et al.* [14] proposed EDSR model by removing unnecessary BN layers in its architecture and expanding the model size, which achieved significant improvement on SISR. Zhang *et al.* [26] proposed to use effective residual dense block in its model, and they further explored a deeper network with channel attention [25] which achieved a satisfactory performance on SISR. However, all these approaches tend to output over-smoothed results without sufficient high-frequency details because loss functions of their models are based on pixel space. To improve the visual quality of super-resolution results, perceptual loss [9] was proposed to optimize super-resolution models and it was defined in a feature space instead of pixel space. To make super-resolution results realistic, it is beneficial coupling perceptual loss with adversarial loss defined in GAN framework [5], since adversarial loss forces models to learn valid mapping for super-resolution by adversarial training. To accomplish this goal, SRGAN [13] applied GAN to SISR for the first time and significantly improved the visual quality of super-resolution results. To further reduce the differences between SRGAN results and the ground-truth images, ESRGAN [23] redesigned the architecture of SRGAN and achieved more visually pleasing performance.

All methods mentioned above take bicubic-downsampled images as inputs and belong to supervised super-resolution approach. However, it's impractical when we apply these methods to super-resolve low-resolution images in the real world, because there are no corresponding high-resolution images for training these networks. Moreover, the super-resolution results are poor since the degradation model in the real world doesn't fit their assumption. As for unsupervised approaches, Cycle-GAN [27] has achieved good performance using the concept of cycle consistence. However, parameters of Cycle-GAN are twice the number of single GAN since there are two GANs in it. To make a tradeoff between model size and super-resolution performance on unpaired images, we propose a GAN-based approach named Feature-Guided Super-Resolution Generative Adversarial Network (FG-SRGAN). Moreover, we train FG-SRGAN using unpaired low-resolution and high-resolution images.

## 3    Network Architectures

GAN-based networks using paired images for training have achieved impressive results in image super-solution, such as ESRGAN. However, it's impractical to obtain corresponding high-resolution images when processing low-resolution images in the real world. What's more, using two sets of unlabeled and unpaired images directly to learn a super-resolution mapping function is also difficult, since GAN tends to suffer from mode dropping. Aiming to tackle this challenge, FG-SRGAN is proposed in this paper, see Fig. 1 for an overview. ESRGAN is utilized to produce roughly high-resolution images, and these images are encouraged to share specific content features with the outputs of generator. The guided features of the ESRGAN's outputs are extracted by VGG network. By conditioning the generator on this intermediate supervision, it is possible to guide image super-resolution and map low-resolution images to corresponding high-resolution images. While in adversarial training, we treat the results of ESRGAN as fake high-resolution images which contributes to the main task as it forces FG-SRGAN to learn valid representations for super-resolution. We describe the detail of our network architecture in Sect. 3.1. Two loss functions proposed in this paper are presented in Sect. 3.2.

### 3.1    FG-SRGAN

There are three main modules in FG-SRGAN, including generator $G$, guidance module and discriminator $D$ as illustrated in Fig. 1.

The generator network $G$ is used to map low-resolution images into corresponding high-resolution images. $G$ begins with two convolution layers each followed by one instance normalization layer, and useful low-level features are extracted in this stage. In the following stage, we employ twelve residual blocks, each with identical layout, and one upsampling block to enlarge resolution two times. Afterwards, three similar residual blocks are used to enlarge resolution another two times. Finally, images of high resolution are obtained by two convolution blocks. It is worth noting that we employ the residual block proposed in [14], where the balance between calculation speed and the performance is guaranteed by this design.

As for guidance module of FG-SRGAN, we employ it to help achieve one-to-one mapping from the input to the output. There are two parts in guidance module, one is a pre-trained model ESRGAN which is trained with paired images and the other is pre-trained VGG network. First, ESRGAN takes low-resolution images as inputs and generates rough high-resolution images. And these images are encouraged to share specific content features with the outputs of FG-SRGAN even though there may be some slight differences. The specific content features of the two super-resolution networks are both extracted by VGG and checked by $l_1$ loss. By conditioning the generator on this intermediate supervision, we can train FG-SRGAN more easily because it further reduces the space of possible mapping functions.

Complementary to the generator and guidance module of FG-SRGAN, the discriminator network $D$ is used to judge whether the input is a real high-resolution image. There are two main differences from common discriminators where one is the discriminator follows shallow fully convolutional structure as used in [27], and the other is the discriminator takes more fake samples as inputs. More concretely, despite the outputs of $G$, the discriminator also takes super-resolution results from ESRGAN as fake samples. This is beneficial for the main task as it forces the generator to learn correct representations for super-resolution.

### 3.2 Loss Function

In this section, we describe the proposed loss function. Let $L$ be the loss function, $G^*$ and $D^*$ be the optimum generator and discriminator respectively. Our objective is to solve the min-max problem:

$$(G^*, D^*) = arg \min_G \max_D L(G, D) . \tag{1}$$

The loss function $L(G, D)$ consists of two parts, including feature-guidance loss $L_{fg}(G, D)$ and adversarial loss $L_{adv}(G, D)$. We use a simple additive form for the loss function:

$$L(G, D) = L_{fg}(G, D) + \omega L_{adv}(G, D) . \tag{2}$$

Where $\omega$ shows the emphasis paid on adversarial loss. Larger $\omega$ leads to produce more realistic high-resolution images and can prevent generating images the same as the outputs of ESRGAN. We describe $L_{fg}(G, D)$ in detail and present the detail of $L_{adv}(G, D)$ in the following parts of this section.

In addition to transform images from low-resolution domain to high-resolution domain, the first thing we should consider is to ensure the outputs of generator retain the same semantic content with the inputs. We accomplish this goal with the help of a pre-trained model ESRGAN and VGG. ESRGAN has the same inputs as generator in FG-SRGAN, afterwards we employ VGG to extract high level feature maps of both ESRGAN's outputs and generator's outputs. Finally, $l_1$ loss is used to measure the similarity between their feature maps. Accordingly, we define the feature-guidance loss as:

$$L_{fg}(G, D) = \mathbb{E}_{p_i \sim s_{data}(lr)} ||VGG_l(G(p_i)) - VGG_l(ESRGAN(p_i))||_1 . \quad (3)$$

Where $l$ refers to the feature maps of a specific VGG layer. In our experiment, we use the feature maps in the layer 'Conv3-2' of VGG to achieve a good performance.

The adversarial loss is applied to both generator and discriminator, which helps to produce high-resolution images of high quality. In common GAN frameworks, the task of discriminator is to figure out whether the input image is real or synthesized from the generator. However, in FG-SRGAN, it is not sufficient for generating high quality images. The generator $G$ tends to produce images similar to the ESRGAN's outputs since we use ESRGAN's results as guidance of our method. To circumvent this problem, we propose a new formulation of adversarial loss. We treat the results of ESRGAN as fake high-resolution samples while in adversarial training. For each image $l$ in the low-resolution domain, the generator $G$ outputs a fake high-resolution image $G(l)$ and the pre-trained model ESRGAN outputs another fake high-resolution image $E(l)$. In FG-SRGAN, the discriminator is used to assign correct labels to the input images, including $G(l) \sim s_{data}(g_1)$, $E(l) \sim s_{data}(g_2)$ and real high-resolution images R$\sim s_{data}(r)$, so that the generator $G$ can be guided correctly and transform the low-resolution images to the corresponding high-resolution images. Therefore, we define the adversarial loss in FG-SRGAN as:

$$\begin{aligned} L_{adv}(G, D) =& \mathbb{E}_{r_i \sim s_{data}(r)} \left[ (D(r_i) - 1)^2 \right] \\ &+ \mathbb{E}_{g_j \sim s_{data}(g_1)} \left[ (D(g_j))^2 \right] \\ &+ \mathbb{E}_{g_k \sim s_{data}(g_2)} \left[ (D(g_k))^2 \right] . \end{aligned} \quad (4)$$

## 4   Experiments

### 4.1   Experimental Settings

**Methods Compared.** We compare our method against four related works both numerically and qualitatively: one image deblurring work named SRN [22], and three image super-resolution works named SRGAN [13], ESRGAN [23] and RCAN [25]. All of these methods are tested on a computer with Intel Core i7 CPU, 32 GB of RAM and an NVIDIA GTX1080 GPU.

**Datasets.** Following [2], we select 50,000 high-resolution face images and 50,000 low-resolution face images as training set. For testing, we randomly select 3,000 low-resolution images from the Widerface dataset [24] as evaluation set.

**Evaluation Metrics.** Since there are no corresponding ground-truth high-resolution images of the evaluation set, we numerically assess the super-resolution results using the Fréchet Inception Distance (FID) [7]. Furthermore, we provide PSNR results on 1,000 test images from the LS3D-W [1] dataset. The same as the standard super-resolution experimental setting, all methods take bicubic-downsampled images as inputs when calculating PSNR.

## 4.2   Super-Resolution Results

To explore the effectiveness of guidance module proposed in this paper, we evaluate the performance of our method with or without guidance module. As shown in Fig. 2, super-resolution results are of poor quality and have no correlation with the inputs when we remove guidance module. However, FG-SRGAN yields satisfactory high-resolution images. The results in Fig. 2 demonstrate the effectiveness of our proposed guidance module.

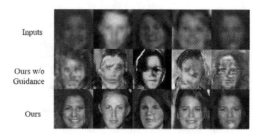

**Fig. 2.** Experiments on evaluating the effectiveness of guidance module.

**Table 1.** FID results on evaluation set and PSNR results on test set from LS3D-W.

| Method | FID | PSNR |
|---|---|---|
|  | LR test set | LS3D-W |
| SRN [22] | 231.59 | 23.11 |
| RCAN [25] | 166.35 | 25.52 |
| SRGAN [13] | 143.17 | 23.19 |
| ESRGAN [23] | 167.98 | 25.70 |
| Ours | 25.81 | 22.74 |

We compare our method with the other four models quantitatively and qualitatively. Table 1 shows FID and PSNR results of methods mentioned above. We provide visual results for several images in Fig. 3. It can be observed that our method is able to achieve satisfactory performance in terms of FID while out-performs all the other four methods, and visual quality of our proposed method is better than the other four methods. For instance, FG-SRGAN can produce sharper and more natural faces than the other four methods, which tend to generate blurry results. Furthermore, FG-SRGAN is capable of generating clear and distinct facial organs while other methods tend to yield face images of bad quality.

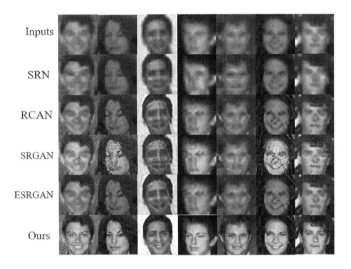

**Fig. 3.** Qualitative results on evaluation set from Widerface. The methods compared are described in Sect. 4.1.

In addition, with regard to our experiment on LS3D-W, our method achieves competitive results compared to four related state-of-the-art methods including ESRGAN, SRGAN, SRN and RCAN. It is worth noting that we trained FG-SRGAN with unpaired images, however, the other four methods were trained on pair of bicubic-downsampled and original high-resolution images.

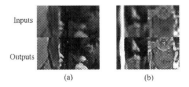

**Fig. 4.** Examples of failure cases. (a) Failures caused by large pose. (b) Failures caused by heavily degraded inputs.

### 4.3 Failure Cases

Although our method can achieve satisfactory results in solving unpaired super-resolution problems in many cases, it has unsatisfactory performance in some samples as illustrated in Fig. 4. For most of these failure cases, we note that the input face images are either degraded significantly or of large pose, and both factors have negative impacts on the performance of our method.

## 5 Conclusion

In this paper, we propose a Feature-Guided Super-Resolution Generative Adversarial Network for unpaired image super-resolution and it is worth noting that our method does not assume the low-resolution images as bicubic-downsampled images. We introduce a guidance module in the proposed method, which is utilized to reduce the space of possible mapping functions and help learn the correct mapping funtion from low-resolution domain to high-resolution domain. Furthermore, we treat the outputs of guidance module as fake examples, and this is beneficial for the main task as it forces the generator to learn valid representations for super-resolution. When applied to super-resolve low-resolution face images in the real world, our method is able to achieve satisfactory performance in terms of FID and visual quality.

**Acknowledgement.** This project was partially supported by the National Natural Science Foundation of China (Grant No.61671104), and the National Major Scientific Instruments Project (Grant No. 2014YQ24044501)

## References

1. Bulat, A., Tzimiropoulos, G.: How far are we from solving the 2D & 3D face alignment problem? (and a dataset of 230,000 3D facial landmarks). In: International Conference on Computer Vision, pp. 1021–1030 (2017)
2. Bulat, A., Yang, J., Tzimiropoulos, G.: To learn image super-resolution, use a GAN to learn how to do image degradation first. In: Ferrari, V., Hebert, M., Sminchisescu, C., Weiss, Y. (eds.) ECCV 2018. LNCS, vol. 11210, pp. 187–202. Springer, Cham (2018). https://doi.org/10.1007/978-3-030-01231-1_12
3. Dong, C., Loy, C.C., He, K., Tang, X.: Learning a deep convolutional network for image super-resolution. In: Fleet, D., Pajdla, T., Schiele, B., Tuytelaars, T. (eds.) ECCV 2014. LNCS, vol. 8692, pp. 184–199. Springer, Cham (2014). https://doi.org/10.1007/978-3-319-10593-2_13
4. Dong, C., Loy, C.C., Tang, X.: Accelerating the super-resolution convolutional neural network. In: Leibe, B., Matas, J., Sebe, N., Welling, M. (eds.) ECCV 2016. LNCS, vol. 9906, pp. 391–407. Springer, Cham (2016). https://doi.org/10.1007/978-3-319-46475-6_25
5. Goodfellow, I., et al.: Generative adversarial nets. In: Advances in Neural Information Processing Systems, pp. 2672–2680 (2014)
6. Han, W., Chang, S., Liu, D., Yu, M., Witbrock, M., Huang, T.S.: Image super-resolution via dual-state recurrent networks. In: Computer Vision and Pattern Recognition, pp. 1654–1663 (2018)

7. Heusel, M., Ramsauer, H., Unterthiner, T., Nessler, B., Hochreiter, S.: GANs trained by a two time-scale update rule converge to a local nash equilibrium. In: Advances in Neural Information Processing Systems, pp. 6626–6637 (2017)
8. Hui, Z., Wang, X., Gao, X.: Fast and accurate single image super-resolution via information distillation network. In: Computer Vision and Pattern Recognition, pp. 723–731 (2018)
9. Johnson, J., Alahi, A., Fei-Fei, L.: Perceptual losses for real-time style transfer and super-resolution. In: Leibe, B., Matas, J., Sebe, N., Welling, M. (eds.) ECCV 2016. LNCS, vol. 9906, pp. 694–711. Springer, Cham (2016). https://doi.org/10.1007/978-3-319-46475-6_43
10. Kim, J., Kwon Lee, J., Mu Lee, K.: Accurate image super-resolution using very deep convolutional networks. In: Computer Vision and Pattern Recognition, pp. 1646–1654 (2016)
11. Kim, J., Kwon Lee, J., Mu Lee, K.: Deeply-recursive convolutional network for image super-resolution. In: Computer Vision and Pattern Recognition, pp. 1637–1645 (2016)
12. Lai, W.S., Huang, J.B., Ahuja, N., Yang, M.H.: Deep Laplacian pyramid networks for fast and accurate super-resolution. In: Computer Vision and Pattern Recognition, pp. 5835–5843 (2017)
13. Ledig, C., et al.: Photo-realistic single image super-resolution using a generative adversarial network. In: Computer Vision and Pattern Recognition, pp. 105–114 (2017)
14. Lim, B., Son, S., Kim, H., Nah, S., Lee, K.M.: Enhanced deep residual networks for single image super-resolution. In: Computer Vision and Pattern Recognition Workshops, pp. 1132–1140 (2017)
15. Mao, X.J., Shen, C., Yang, Y.B.: Image restoration using convolutional auto-encoders with symmetric skip connections. arXiv preprint arXiv:1606.08921 (2016)
16. Nasrollahi, K., Moeslund, T.B.: Super-resolution: a comprehensive survey. Mach. Vis. Appl. **25**(6), 1423–1468 (2014)
17. Shi, W., et al.: Real-time single image and video super-resolution using an efficient sub-pixel convolutional neural network. In: Computer Vision and Pattern Recognition, pp. 1874–1883 (2016)
18. Shocher, A., Cohen, N., Irani, M.: "Zero-shot" super-resolution using deep internal learning. In: Computer Vision and Pattern Recognition, pp. 3118–3126 (2018)
19. Simonyan, K., Zisserman, A.: Very deep convolutional networks for large-scale image recognition. arXiv preprint arXiv:1409.1556 (2014)
20. Tai, Y., Yang, J., Liu, X.: Image super-resolution via deep recursive residual network. In: Computer Vision and Pattern Recognition, pp. 2790–2798 (2017)
21. Tai, Y., Yang, J., Liu, X., Xu, C.: MemNet: A persistent memory network for image restoration. In: Computer Vision and Pattern Recognition, pp. 4539–4547 (2017)
22. Tao, X., Gao, H., Shen, X., Wang, J., Jia, J.: Scale-recurrent network for deep image deblurring. In: Computer Vision and Pattern Recognition, pp. 8174–8182 (2018)
23. Wang, X., et al.: ESRGAN: Enhanced super-resolution generative adversarial networks. arXiv preprint arXiv:1809.00219 (2018)
24. Yang, S., Luo, P., Loy, C.C., Tang, X.: Wider face: A face detection benchmark. In: Computer Vision and Pattern Recognition, pp. 5525–5533 (2016)
25. Zhang, Y., Li, K., Li, K., Wang, L., Zhong, B., Fu, Y.: Image super-resolution using very deep residual channel attention networks, In: European Conference on Computer Vision. pp. 286–301 (2018)

26. Zhang, Y., Tian, Y., Kong, Y., Zhong, B., Fu, Y.: Residual dense network for image super-resolution. In: Computer Vision and Pattern Recognition, pp. 2472–2481 (2018)
27. Zhu, J.Y., Park, T., Isola, P., Efros, A.A.: Unpaired image-to-image translation using cycle-consistent adversarial networks. In: International Conference on Computer Vision, pp. 2242–2251 (2017)

# Time Series Analysis, Dynamic
# Prediction, and Uncertain Estimation

# Artificial Neural Networks for Realized Volatility Prediction in Cryptocurrency Time Series

Ryotaro Miura, Lukáš Pichl$^{(\boxtimes)}$, and Taisei Kaizoji

International Christian University, Osawa 3-10-2, Mitaka, Tokyo 181-8585, Japan
lukas@icu.ac.jp
http://www.icu.ac.jp/

**Abstract.** Realized volatility (RV) is defined as the sum of the squares of logarithmic returns on high-frequency sampling grid and aggregated over a certain time interval, typically a trading day in finance. It is not a priori clear what the aggregation period should be in case of continuously traded cryptocurrencies at online exchanges. In this work, we aggregate RV values using minute-sampled Bitcoin returns over 3-h intervals. Next, using the RV time series, we predict the future values based on the past samples using a plethora of machine learning methods, ANN (MLP, GRU, LSTM), SVM, and Ridge Regression, which are compared to the Heterogeneous Auto-Regressive Realized Volatility (HARRV) model with optimized lag parameters. It is shown that Ridge Regression performs the best, which supports the auto-regressive dynamics postulated by HARRV model. Mean Squared Error values by the neural-network based methods closely follow, whereas the SVM shows the worst performance. The present benchmarks can be used for dynamic risk hedging in algorithmic trading at cryptocurrency markets.

**Keywords:** ANN · MLP · LSTM · GRU · CNN · SVM · HARRV · Ridge regression · Realized volatility

## 1 Introduction

Uncertainty modeling in financial markets in econometrics has traditionally been based on the notion of volatility inferred indirectly from daily time series data, an approach which has extended to high frequency data since the pioneering work on realized volatility (RV) by Andersen et al. [1]. The quantity is defined on aggregation period between $t$ and $t + T$ sampled $N$ times as

$$RV(t) = \sum_{i=1}^{N} R_i^2, \quad R_i = \log\left(P_i/P_{i-1}\right) \quad P_i \equiv P(t + iT/N). \tag{1}$$

There has been a recent surge in the the number of work related to deep learning (DL) algorithm for market prediction [3,4] including lately cryptocurrency markets [5–7], however, the applications of machine learning (ML) to RV

© Springer Nature Switzerland AG 2019
H. Lu et al. (Eds.): ISNN 2019, LNCS 11554, pp. 165–172, 2019.
https://doi.org/10.1007/978-3-030-22796-8_18

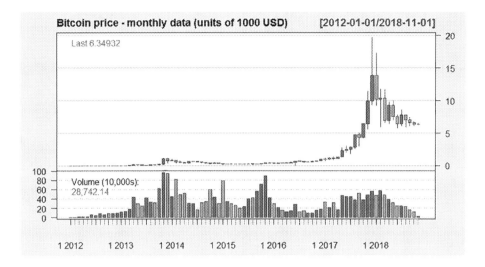

**Fig. 1.** Candlestick diagram for monthly Bitcoin prices in USD (aggregated from minute data at the Bitstamp exchange). The body of each box ranges between open and close prices, with low and high values indicated by vertical lines. Green color codes price increase and red color price decrease. The volume series in the bottom are counted in units of 10 thousand Bitcoin. (Color figure online)

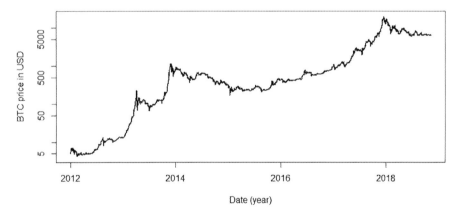

**Fig. 2.** BTCUSD daily-aggregated close price in logarithmic scale. Time period ranges from 2012-01-01 to 2018-11-11 contains 2507 days. Source data from [2].

of digital currencies have not appeared yet. It is the purpose of this work to explore Bitcoin time series data depicted in Figs. 1 and 2 (source: [2]) and provide a benchmark study of modern ML algorithms to RV prediction. The paper is organized as follows. The next section sums up empirical properties of data and lists the ML/DL models applied. Results are given and discussed in Sect. 3, followed by a brief conclusion.

## 2    Data and Models

Figures 3 and 4 show the distribution of the log returns, from which volatility clustering and a fat-tail distribution may be observed.

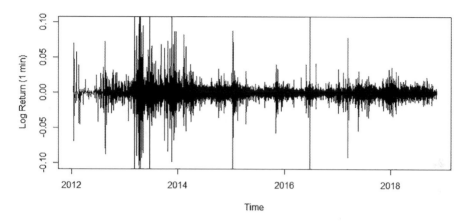

**Fig. 3.** BTCUSD minute log returns. After removal of missing prices and non-adjacent returns, there are 1,804,479 values. Source data fom [2].

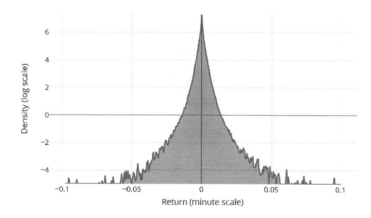

**Fig. 4.** Distribution of log returns of BTCUSD price on minute-sampling scale.

The distribution in Fig. 4 is practically symmetric, and the absolute values of the log return can be fitted with power law as shown in Fig. 5. One-lag correlations are shown in Fig. 6. Realized volatility distribution is shown in Fig. 7(a), along with a log-value differencing transform in Fig. 7(b). Figure 8 shows the autocorrelation function (ACF) values for the first 20 lags. Correlations between RV and

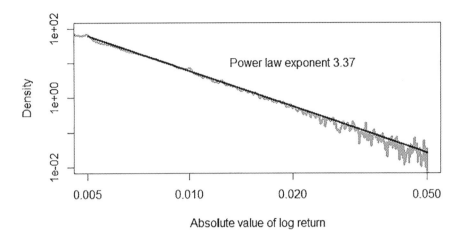

**Fig. 5.** Power-law distirbution of the fat-tail of BTCUSD minute log returns and the fitted exponent.

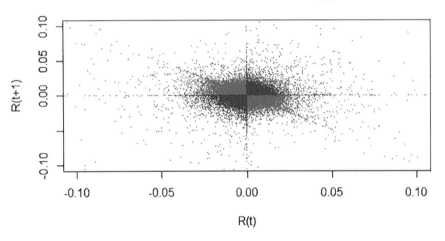

**Fig. 6.** Distirbution patterns of adjacent minute returns. Note the line structure on the reverse diagonal of II and IV quadrant, which corresponds to negative autocorrelation for lag 1.

intraday variance, defined as $2(H-L)/(H+L)$, where $H$ stands for the highest and $L$ for the lowest price within each 3-h bin interval, are shown in Fig. 9(a). Approximate relation between RV and aggregated trading volume is depicted in Fig. 9(b). The methods of analysis are the heterogenous auto-regressive model of realized volatility (HARRV) [1] with 3 lags, multi-layer perceptron (MLP, dense layer neural network) [8], convolutional neural network (CNN) [9], long short-term memory (LSTM) [10] (cf. Fig. 10), gated recurrent unit (GRU) [11],

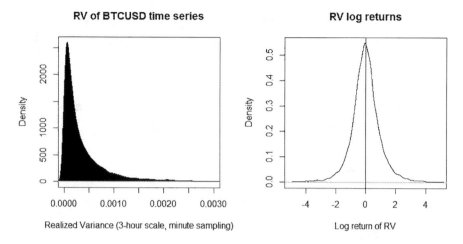

**Fig. 7.** Realized Volatility distributions: (a) Sum of the squared log returns on minute scale aggregated over 180 min, (b) log returns taken from adjacent values of RV in (a).

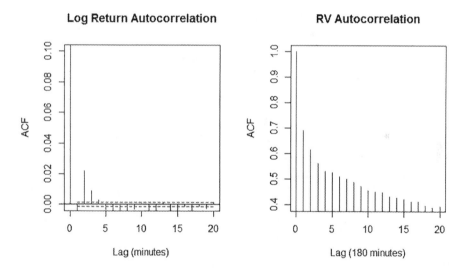

**Fig. 8.** Autocorrelation function (a) Left: correlation of log returns: notice the negative value for lag 1, and the significant yet small values for lags 2 and 3; (b) Right: persistent correlations of RV time series.

support vector machine (SVM) [12] and ridge regression [13]. In addition, we implemented Dropout mechanism to LSTM, GRU, and Batch Normalisation (BN) to MLP and CNN.

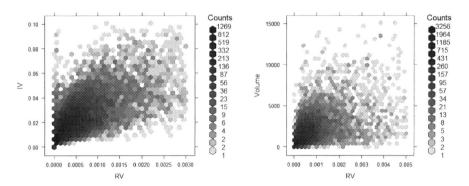

**Fig. 9.** 2D Histograms in the form of heatmap diagram: (a) Correlation of RV and intraday variance (see text for definition) (b) Right: correlation of RV and trading volume (time scale of 3 h).

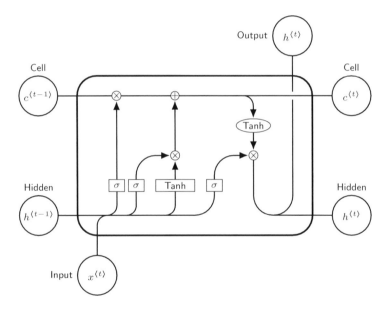

**Fig. 10.** Long short-term memory unit operation following notation in Ref. [10].

## 3    Results

Table 1 shows the results for several models referenced above, with model configuration listed in the first column. The models were validated with 10-fold cross validation (CV) for time series using 100 runs, and benchmark on the test set, using 30 random weight initializations. Optimized parameters (length of sequence used for prediction) are shown in column 2, and the statistical characteristics of the mean squared error (MSE) and the rooted MSE (RMSE) are listed. Result variances are also displayed where applicable. The best performing method on the validation data set is the GRU 1 layer + 2 dense layer result; however, it does not generalize well on the test data set, where the ridge regression method benchmarks as the most superior method with penalty parameter 0.6951.

**Table 1.** Resuts of machine learning models trained with cross validation and benchmarked on the test set.

| Model | Sequence length | CV MSE | CV RMSE | Test MSE | Test RMSE |
|---|---|---|---|---|---|
| HARRV | 1, 6, 16 | 3.1617e-04 | 1.7781e-02 | 4.8442e-06 | 2.2010e-03 |
| MLP 4 layers with dropout | 10 | 2.9613e-04 | 1.7209e-02 | 4.8704e-06 (var 1.3311e-14) | 2.2067e-03 (var 6.8202e-10) |
| MLP 4 layers with BN | 10 | 3.0151e-04 | 1.7364e-02 | 4.8212e-06 (var 4.9997e-14) | 2.1952e-03 (var 2.5325e-09) |
| LSTM 2 layers + 1 Dense | 12 | 2.9845e-04 | 1.7276e-02 | 4.8233e-06 (var 2.0082e-14) | 2.1960e-03 (var 1.0293e-09) |
| GRU 1 layer + 2 Dense | 5 | **2.9607e-04** | **1.7207e-02** | 4.7433e-06 (var 8.8148e-15) | 2.1778e-03 (var 4.5781e-10) |
| CNN 2 layers + 1 Dense | 6 | 3.0608e-04 | 1.7495e-02 | 4.7605e-06 (var 1.2904e-14) | 2.1817e-03 (var 6.7603e-10) |
| SVM | 7 | 3.2120e-04 | 1.7922e-02 | 4.3463e-05 | 6.5293e-03 |
| Ridge Regression | 6 | 3.0615e-04 | 1.7497e-02 | **4.6667e-06** | **2.1603e-03** |

# 4  Conclusion

We have analyzed high-frequency Bitcoin time series sampled on minute scale using statistical method and machine learning algorithms. First, the auto-covariance function of the minute-based log return values not only shows a negative significant value at lag 1, but also small positive values distinct from zero at lags 2, 3 and 4. Since cryptocurrency exchanges continue trading with customers all over the globe, 24 h a day, there is no a priori reason to sample the realized volatility at daily scale. Given the length of the data set (about 6 years) and the data-savvy machine learning algorithms, we decided to aggregate the RV values using 3-h long intervals. The RV values show a weak correlation with relative values of the 3-h interval based high-low price extent. This work has focused solely on the heterogeneous autoregressive dynamics. We have found that albeit at the validation data set level neural network algorithms provide good fits of the RV dynamics, this does not carry over to the test set benchmarks. In particular, the best performing method is the ridge regression, in which past RV values are used as predictors. The optimized lags were 1, 6, and 16. We remark, nevertheless, that the assumptions of heterogenous time scales for the auto-regressive process is not necessary valid at crypto-currency exchanges, which have their specific dynamics. For instance, it may be possible to use differenced log values of realized volatility to model the increment process. Future work along these lines will include predicting the RV values from a broader set of indicators, in additiion to the past RV data, especially the minute-scale time series of logarithmic returns, and transaction volume series.

# References

1. Andersen, T.G., Bollerslev, T., Diebold, F., Labys, P.: Modeling and forecasting realized volatility. Econometrica **71**, 579–625 (2003)
2. Kaggle, Bitcoin historical data. https://www.kaggle.com/mczielinski/bitcoin-historical-data. Accessed 1 Dec 2018. Released under CC BY-SA 4.0 license
3. Moews, B., Herrmann, J.M., Ibikunle, G.: Lagged correlation-based deep learning for directional trend change prediction in financial time series. Expert Syst. Appl. **120**, 197–206 (2019)
4. Cao, J., Li, Z., Li, J.: Financial time series forecasting model based on CEEMDAN and LSTM. Phy. A: Stat. Mech. Appl. **519**, 127–139 (2019)
5. Mallqui, D.C.A., Fernandes, R.A.S.: Predicting the direction, maximum, minimum and closing prices of daily Bitcoin exchange rate using machine learning techniques. Appl. Soft Comput. **75**, 596–606 (2019)
6. Lahmiri, S., Bekiros, S.: Cryptocurrency forecasting with deep learning chaotic neural networks. Chaos, Solitons Fractals **118**, 35–40 (2019)
7. Nakano, M., Takahashi, A., Takahashi, S.: Bitcoin technical trading with artificial neural network. Phys. A: Stat. Mech. Appl. **510**, 587–609 (2018)
8. Rosenblatt, F.: Principles of Neurodynamics Perceptrons and the Theory of Brain Mechanisms. Spartan Books, Washington (1961)
9. LeCun, Y., et al.: Back-propagation applied to handwritten zip code recognition. Neural Compu. **1**(4), 541–551 (1989)
10. Hochreiter, S., Schmidhuber, J.: Long short-term memory. Neural Comput. **9**(8), 1735–1780 (1997)
11. Cho, K., van Merrienboer, B., Bahdanau, D., Bengio, Y.: On the properties of neural machine translation: encoder-decoder approaches. In: 8th Workshop on Syntax. Semantics and Structure in Statistical Translation, pp. 102–111. Association for Computational Linguistics, Doha (2014)
12. Vapnik, V.N.: The Nature of Statistical Learning Theory. Springer, Heidelberg (1995). https://doi.org/10.1007/978-1-4757-3264-1
13. Hoerl, A.E., Kennard, R.W.: Ridge regression: biased estimation for nonorthogonal problems. Technometrics **12**, 55–67 (1970)

# Pattern Matching in Sequential Data Using Reservoir Projections

Sebastián Basterrech[(✉)]

Department of Computer Science, Faculty of Electrical Engineering and Informatics,
VSB-Technical University of Ostrava, Ostrava-Poruba, Czech Republic
Sebastian.Basterrech@vsb.cz

**Abstract.** A relevant problem on data science is to define an efficient and reliable algorithm for finding specific patterns in a given signal. This type of problems often appears in medical applications, biophysical systems, complex systems, financial analysis, and several other domains. Here, we introduce a new model based in the ability of Recurrent Neural Networks (RNNs) for modelling time series. The technique encodes temporal information of the reference signal and the given query in a feature space. This encoding is done using a RNN. In the feature space, we apply similarity techniques for analysing differences among the projected points. The proposed method presents advantages with respect of state of art, it can produce good results using less computational costs. We discuss the proposal over three benchmark datasets.

**Keywords:** Recurrent Neural Networks · Reservoir Computing ·
Time series matching · Similarity · Echo State Property

## 1 Introduction

During the last decades, Neural Networks (NNs) have been successfully applied for solving machine learning problems, time series forecasting and modelling data [22]. In particular, a NN with a recurrent topology is a very powerful computational model for analyzing and predicting sequential data [15]. In this work, we analyze an approach based in the power of recurrent NN for memorizing sequential data. Given a specific pattern (query), which can be a noisy segment of a reference time series, the goal is to identify if the given pattern belongs to the reference time series. The problem has provoked a lot of attention, and there are already several employed techniques [5,13]. A common approach is to apply a sequential scanning of the query and the reference segments. Other approaches are based in comparing features from the query and the reference signal [19].

In this work, we discuss a new approach based in the literature introduced in the area of Reservoir Computing [15]. The proposed method encodes the reference signal in a feature space, the encoding is done using a Recurrent Neural Network (RNN). The same encoding algorithm is performed to the query. Then, we analyze two variations. In one case, we study the simularity of the

© Springer Nature Switzerland AG 2019
H. Lu et al. (Eds.): ISNN 2019, LNCS 11554, pp. 173–183, 2019.
https://doi.org/10.1007/978-3-030-22796-8_19

projected reference signal and the projected query on the projected space (lattice of points projected by a RNN). Another variation analyzed in this study consists of applying a dimensionality reduction technique of the points in the feature map. Once the redundant information is reduced, we apply similarity techniques among points in a lower dimensional space. Our presented technique is an attempt of analyzing similarities between the dynamics generator of the reference time series and the dynamics generator of the query.

In this preliminary study, we describe the proposed technique and we present results over three benchmark datasets. The rest of the article contains an overview of RNNs that includes a description of Reservoir Computing models. Section 3 introduces the specification of the problem and the proposed pattern matching approach. Section 4 presents the experiments and discuss the results. Finally, the article ends with a brief conclusion.

## 2   Overview on Recurrent Neural Network

Conceptually a standard NN is a parallel distributed processing system, which is composed by interconnected simple processing units known as neurons [21]. Both processing units and the interconnection among them are mathematical abstractions of the biological nervous system [22]. The model design depends on two independent factors: (i) defining the schedule of time-dependent interactions among the elements, that can be single neurons or group of neurons; and (ii) in its architecture. The system architecture is characterized by several factors such as: the selection of the activation functions, the pattern of connectivity of the network, the weight connections, as well as the protocol of communication among the neurons with the environment (how to provide information to the networks and how to extract the results) [21]. Often the NN field is categorized into two major types: Feedforward Neural Network (FNN) and Recurrent Neural Network (RNN). A FNN is a parametric function where the signals are traveling through the network from the input to the output neurons [11]. Last ten years, deep multi-layer FNNs has became especially popular on the area of ML. Among the most significant results are the ones produced by the research groups of LeCun, Bengio [3], and Hinton [27]. By contrary, a RNN has at least one cyclic synaptic connection that enables neurons to feed their output signals back to the system. Using the graph terminology, a RNN has at least a circuit in the network topology. The recurrent topology of the graph makes a significant difference between FNN and RNN, while a FNN is a parametric function, the RNN model is a dynamical system [10]. The recurrent network topology ensures that a transformation of the input history can be stored in internal states. Furthermore, an important property of the RNN model is its computational power, at the early 90s, it was shown that a RNN with a finite number of units and sigmoidal activation functions are universal Turing machines [6,23].

## 2.1 Formal Specification

We formulate the targeting problem in terms of a discrete time system with an input $\mathbf{u}(t)$, an output $\mathbf{y}(t)$ and a hidden state $\mathbf{x}(t)$. A standard RNN is a dynamical defined system as follows:

$$\mathbf{x}(t) = \phi_{\mathrm{h}}(\boldsymbol{\theta}^{\mathrm{h}}, \mathbf{u}(t), \mathbf{x}(t-1)), \tag{1}$$

and

$$\mathbf{y}(t) = \phi_{\mathrm{out}}(\boldsymbol{\theta}^{\mathrm{out}}, \mathbf{x}(t)), \tag{2}$$

where $\phi_{\mathrm{h}}(\cdot)$ and $\phi_{\mathrm{out}}(\cdot)$ are two pre-defined coordinate-wise functions, and $\boldsymbol{\theta}^{\mathrm{h}}$ and $\boldsymbol{\theta}^{\mathrm{out}}$ are adjustable control variables. A numerical optimization problem appears when a NN is applied for solving supervised Machine Learning problems, which consists in finding the optimal set of parameters $\boldsymbol{\theta}^{\mathrm{h}}$ and $\boldsymbol{\theta}^{\mathrm{out}}$ such that a distance between the matching target and the network output is minimized [18].

## 2.2 Reservoir Computing Methods

In spite of the computational power of the RNN model, in many real-world problems with real-time constraints and computational restrictions is a hard challenge to find the optimal network topology [26]. At the beginning of 2000s, a new computational concept for designing and training NNs was introduced with the names of Liquid State Machine (LSM) [16] and Echo State Network (ESN) [10]. Since 2007 the approach has become popular under the name of Reservoir Computing (RC) [24]. The RC model performs a convolution of at least two operations. In the first operation is performed a dynamical system (so-called *reservoir*), which has the double role of memorizing the sequence of input data and to enhance the linear separability of the data. The reservoir function acts as a temporal kernel method projecting the data in a feature space [15]. The second operation is a simple supervised learning mapping between the feature map (projected data by the reservoir) and the output space (so-called *readout*). Most often this mapping is a simple linear regression. A distinguishing characteristic of a RC technique is that the model has two types of parameters, the ones in the reservoir are randomly assigned and the parameters in the readout structure are adjusted according to the training data [26]. As a consequence, the training algorithm is robust and fast.

In the following we define the canonical ESN model [10]. Let $N_{\mathrm{u}}$ be the dimension of the input space (number of input neurons), $N_{\mathrm{x}}$ denotes the number of neurons in the reservoir and $N_{\mathrm{y}}$ is the number of output neurons. The network weights are collected in three matrices. The input-reservoir weight matrix $\boldsymbol{\theta}^{\mathrm{in}}$ with dimensions $N_{\mathrm{x}} \times N_{\mathrm{u}}$, the $N_{\mathrm{x}}$ squared matrix $\boldsymbol{\theta}^{\mathrm{h}}$ with hidden-hidden weights, and the readout matrix $\boldsymbol{\theta}^{\mathrm{out}}$ collecting weights from the projected space to the output space of dimensions $N_{\mathrm{y}} \times N_{\mathrm{x}}$. For the sake of notation simplicity, we omit the bias term in these matrices. The reservoir state is characterised by the following recurrence:

$$\mathbf{x}(t) = \phi_{\mathrm{h}}(\boldsymbol{\theta}^{\mathrm{in}}\mathbf{u}(t) + \boldsymbol{\theta}^{\mathrm{h}}\mathbf{x}(t-1)), \tag{3}$$

where $\phi_{\mathrm{h}}(\cdot)$ is a predefined Lipschitz function. The model prediction is computed by a linear regression:

$$\hat{\mathbf{y}}(t) = \phi_{\mathrm{out}}(\boldsymbol{\theta}^{\mathrm{out}}\mathbf{x}(t)), \tag{4}$$

where $\phi_{\mathrm{out}}(\cdot)$ is a predefined coordinate-wise functions.

The model performance depends on the initialization of several global parameters [15]:

- Dimension of the projected space: the dimension of the feature space is given by the reservoir size. To increase the number of neurons in the reservoir may improve the linear separability of the data.
- Input scaling factor: note that expression (3) contains the input pattern, then how to scale and pre-process the input sequence can impact in expression (3).
- Density of the reservoir matrix: it is suggested to use around a 20% of non-zero values on the reservoir matrix [15].
- Network controllability: there are several studied about the stability of the recurrent expression (3). In the next section, we discuss with more details the technical issues of the reservoir dynamics.

The family of RC models is very large. Other variations of the original ESN and LSM methods have been introduced during the last years. We can classify the proposed variations in two types: the family of methods that modifies the reservoir projection and the family that modifies the readout model. The projection can change according to the network topology and according to the reservoir activation function. A type of cascade of reservoir projections was presented in [8], particular topologies composed by regular cycles was analyzed in [20]. Another variation of the reservoir projection consists in using different neural activation functions. For example: hyperbolic tangent and linear neurons were studied in [4,15], leaky neurons were presented in [12], neurons inspired from self-organization were studied in [14], and random spiking neurons in stationary state were developed in [2]. Another variation of the RC models developed in the community is related to the supervised learning structure (readout) [7]. Actually, any type of supervised learning can be used, such as Support Vector Machine (SVM), FNN, CART, and so on.

## 3   Proposed Approach

A time-series data is a sequence of events obtained over repeated measurements of time. Most commonly, the arrival event at time $t$ may impact on the arrival event at time $t + k$ ($k > 0$), i.e. the sequence of events are dependent of each of other. The problem specification is as follows. Given a reference time series $\mathcal{U} = \{u(-T), \ldots, u(t), u(t + 1), \ldots\}$, then we are interested in matching a given sequential pattern $\mathcal{Q}$ (often called query) in $\mathcal{U}$ [13]. We consider that $\mathcal{Q}$ matches a segment $\mathcal{S}$ ($\mathcal{S} \subset \mathcal{U}$) iff $dist(\mathcal{Q}, \mathcal{S}) < \varepsilon$, where $\varepsilon$ is an arbitrary threshold and $dist(\cdot)$ is a selected distance function.

Our proposal consists in projecting the reference time series in a feature space. The projection is done using a classic reservoir. On other words, the reservoir captures the data dependency and encodes the temporal information of the input data in a spatial space (feature map). Instead of using time windows and to compare the query with sliding segments directly in $\mathcal{U}$, we realise the similarity analysis in the feature space. On other words, we compare the dynamic generator of the query and the dynamic generator of the reference segment. Note that the reservoir projections should be stable and able of memorizing the input time series. For this reason, the projection should be done with a reservoir that satisfies the Echo State Property (ESP) [1,15,17,25,28].

Roughly speaking, the ESP defines conditions of reservoir matrix in order to guarantee stability in the dynamical system of expression 3. In addition, if the system satisfies the ESP, then the initial conditions of the dynamical system do not impact in the forecasting of the time series. This aspect is very important in our proposal because the query projection and segments can't be sensitive to the reservoir initial state. More formal, the ESP was defined as follows [10]: given a recurrent network without feedback connections, input sequences belong to the input space $\mathcal{A}$, and the network states $\mathbf{x}(t)$ are in a compact set $\mathcal{S}$; the dynamical system (3) satisfies the ESP if $\mathbf{x}(t)$ is uniquely determined by any left-infinite input sequence $\{\mathbf{u}(t - k) : k \in \mathbb{N}\}$ [25,29]. Let $\rho(\boldsymbol{\theta}^{\mathrm{h}})$ be the spectral radius of the hidden-hidden weight matrix $\boldsymbol{\theta}^{\mathrm{h}}$, and let $\eta(\boldsymbol{\theta}^{\mathrm{h}})$ be the largest singular values of $\boldsymbol{\theta}^{\mathrm{h}}$. A necessary condition of the ESP is that $\rho(\boldsymbol{\theta}^{\mathrm{h}}) \leq 1$, and a sufficient condition of the ESP is that $\eta(\boldsymbol{\theta}^{\mathrm{h}}) < 1$. However, the ESP can be preserved in the situation of $\rho(\boldsymbol{\theta}^{\mathrm{h}}) > 1$ with the additional condition that $\mathcal{A}$ doesn't contain the zero input sequence [17,29].

We apply the same input and reservoir matrices for projecting the query and the reference data. The reservoir matrix is scaled for satisfying the ESP [17]. Once the projection is done, we analyze the results using two variations (Proposal A and Proposal B). In one variation, we apply dimensionality reduction, in another one we directly compare the reservoir states of the projected query and projected time series.

**Proposal A:** In this first presentation of our proposal, we reduce the feature map dimensionality using Principal Component Analysis (PCA). However, there are several techniques in the literature for dimensionality reduction that may also be applied. Figure 1 illustrates the proposal A. Let $\mathbf{y}^{\mathrm{u}}(t)$ be the projection of the reference data, and let $\mathbf{y}^{\mathrm{q}}(t)$ be the projected vector of the query. Note that, the distance between $\mathbf{y}^{\mathrm{u}}(t)$ and $\mathbf{y}^{\mathrm{q}}(t)$ is computed for all $t$ in the query, and the cumulated distance is compared with a threshold. The goal is to define a rule for deciding if the subset $\{\mathbf{y}^{\mathrm{u}}(t), \mathbf{y}^{\mathrm{u}}(t + 1), \ldots, \mathbf{y}^{\mathrm{u}}(t + k)\}$ is close enough to $\{\mathbf{y}^{\mathrm{q}}(t), \mathbf{y}^{\mathrm{q}}(t + 1), \ldots, \mathbf{y}^{\mathrm{q}}(t + k)\}$. Note that, $k$ can be lower than the query size. This is due to the fact that the reservoir is able to memorize the temporal information, then $k$ (number of projected points to be compared) can be smaller than the query size. This is an advantage with respect of the sliding windows using the data in the original space.

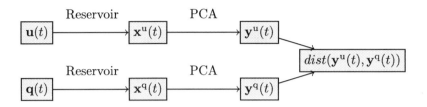

**Fig. 1.** Proposed method using Reservoir projection and PCA.

**Proposal B:** In this variation, we avoid the application of the dimensionality reduction technique. The analysis of similarity is done directly in the feature map (lattice of points created by the reservoir). Therefore, we analyze if the distance between two sequences $\{\mathbf{x}^u(t), \mathbf{x}^u(t+1), \ldots, \mathbf{x}^u(t+k)\}$ is close enough to $\{\mathbf{x}^q(t), \mathbf{x}^q(t+1), \ldots, \mathbf{x}^q(t+k)\}$. However, the main characteristic of a RNN is its ability to encode the temporal dependences, therefore the state $\mathbf{x}(t)$ already contains information of the previous states $\mathbf{x}(t-1), \mathbf{x}(t-2), \ldots$. As a consequence, in order of reducing the computational costs is possible to compare the last few points of the projected query. We also analyze the results when is compared only the last projected time-step of the query with the projection of the reference segment.

## 4    Experimental Results

### 4.1    Data Description

We evaluate the proposal A and B over three datasets. The Noisy Multiple Superimposed Oscillator (MSO), Rossler time series, and a financial data example. The noisy MSO is a sequential dataset generated for two sine waves and a gaussian noise. The dynamics are created following the system

$$u(t) = \sin(0.2t) + \sin(0.311t) + z,$$

where $t = 1, 2, \ldots$, and $z$ is a Gaussian random variable with distribution $\mathcal{N}(0, 0.01)$. Rossler attractor is a well-analyzed time series generated for the following dynamics

$$\frac{\partial x}{\partial t} = -z - y, \qquad \frac{\partial y}{\partial t} = x + ry, \qquad \frac{\partial z}{\partial t} = b + z(x - c),$$

where the parameters values are $r = 0.15$, $b = 0.20$, $c = 10.0$. The last benchmark dataset is a financial time series available in the repository [9].

### 4.2    Results

For each benchmark problem, we study three types of queries. The query can perfectly match a segment in the reference data. The query has a white noise, then the segment and query exactly match each other. Finally, also we discuss examples where the query doesn't belong to the reference time series. The last example it was made just for analyzing eventual mistakes produced by the reservoir projections. We present results using a reservoir of 100 neurons, hyperbolic tangent and the spectral radius of the reservoir matrix equal to 0.5. In this preliminary work, we don't analyze the impact of the reservoir global parameters in the proposed approach. Figure 2 illustrates the results when the reference data is the MSO time series. The figure has two graphics, in the left side is presented the problem when the target query fits perfectly with the reference time series. This problem can be easily solved using other methods based on sliding windows. However, here we wanted to visualize the impact of the reservoir projections that were randomly generated. The figure shows the reservoir state projected in one dimension using the PCA (proposal A), the red curve is the projection of the query. The graphic in the right side shows the errors computed using sliding windows. In this example the query has a size of 50 time steps. Then, in the time 800 is shown the cumulated distance point by point between $\mathbf{y}^u(t)$ and $\mathbf{y}^q(t)$ for $t \in [800, 850]$. The minimum error is presented at the time 800. In Figure 3 is shown the original signal and two errors according to the different methods.

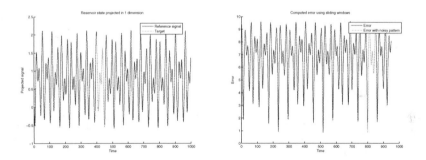

**Fig. 2.** Pattern matching example in the MSO benchmark problem.

Figure 4 shows an example computed on the Rossler time series. The goal was to match a noisy query, the figure contains three graphics. The first one presents the projected signals, the last two presents the analysis of similarities. Figure 4 shows another example where the query doesn't fit with any segment in the reference time series. The figure in the graphic at the top shows in blue colour the reference signal and in red colour the query signal. In the graphics on the bottom, it is shown the projected reservoir state in one dimension. The figure illustrates that the distance among the curves is large enough for discarding a similarity. Figure 5 presents two graphics regarding the financial time series problem. In the left side, the figure presents results when we analyze a noisy query and the analysis of similarity was done using sliding windows of the projected reservoir states using the PCA. In the right side, we shows results when the error was computed only with the difference between $\{\mathbf{x}^u(t), \mathbf{x}^u(t+1), \ldots, \mathbf{x}^u(t+k)\}$ and

$\{\mathbf{x}^q(t), \mathbf{x}^q(t+1), \ldots, \mathbf{x}^q(t+k)\}$, with $k = 50$ (last 50 reservoir states). It is possible to see that the minimum errors are produced between the segment 1000–1100, also depending of the selected threshold the query matches the reference signal between the values 2300–2500.

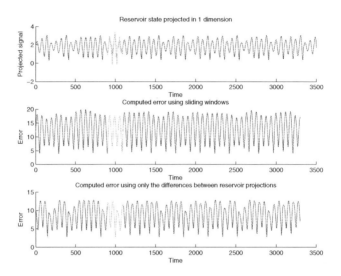

**Fig. 3.** Pattern matching example in the Rossler benchmark problem with a noisy query.

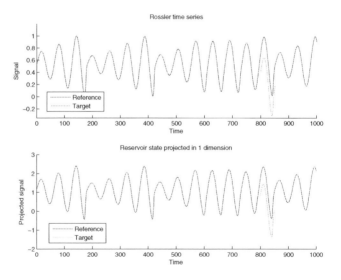

**Fig. 4.** Pattern matching example in the Rossler benchmark problem.

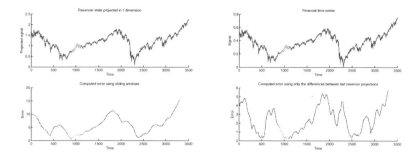

**Fig. 5.** Pattern matching example in the Financial benchmark problem.

# 5 Conclusions and Future Work

In this work, we introduce a new technique for finding specific patterns in a time series. The approach is based in the ability of a RNN for memorizing the input data. Instead of comparing the reference signal with the given pattern, we use a recurrent network for encoding temporal information and for projecting the data in a feature space. Once the data is transformed in a new space, we apply analysis of similarity among the projected points. We presented two variations of our approach. In one variation we use PCA in the analysis of similarities. In another one, we applied distance comparisons among large vectors in the feature map. The technique presents an important advantage with respect of state of art based in sliding widows of segments and comparisons with the query. The sequence of projected points by the RNN encodes temporal information from the past, therefore it is possible to compare the last time-step of the projected query with the projected reference signal. This reduces the computational costs and it is possible to store less data.

This is a preliminary study that can open several research avenues. In the near future we expect to analyze different temporal encoding techniques. Besides, it is necessary to study the minimum required size of the query in order of vanishing the impact of the initial reservoir state. Furthermore, we would like to apply the method on video frames.

**Acknowledgements.** This work was supported by the projects SP2019/135 and SP2019/141 of the Student Grant System, VSB-Technical University of Ostrava, Czech Republic, and by the Ministry of Education, Youth and Sports from the Specific Research Projects (SP2019/135 and SP2019/141) and by The Technology Agency of the Czech Republic in the frame of the project TN01000024 National Competence Center-Cybernetics and Artificial Intelligence.

# References

1. Basterrech, S.: Empirical analysis of the necessary and sufficient conditions of the echo state property. In: 2017 International Joint Conference on Neural Networks, IJCNN 2017, Anchorage, AK, USA, 14–19 May 2017, pp. 888–896 (2017). https://doi.org/10.1109/IJCNN.2017.7965946
2. Basterrech, S., Rubino, G.: Echo state queueing networks: a combination of reservoir computing and random neural networks. Probab. Eng. Inf. Sci. **31**, 457–476 (2017). https://doi.org/10.1017/S0269964817000110
3. Bengio, Y.: Learning deep architectures for AI. Found. Trends Mach. Learn. **2**(1), 1–127 (2009). https://doi.org/10.1561/2200000006
4. Butcher, J.B., Verstraeten, D., Schrauwen, B., Day, C.R., Haycock, P.W.: Reservoir computing and extreme learning machines for non-linear time-series data analysis. Neural Netw. **38**, 76–89 (2013)
5. Christos Faloutsos, M. Ranganathan, Y.M.: Fast subsequence matching in time-series databases. In: SIGMOD Conference, pp. 419–429 (1994)
6. Funahashi, K., Nakamura, Y.: Approximation of dynamical systems by continuous time recurrent neural networks. Neural Netw. **6**, 801–806 (1993)
7. Gallicchio, C., Micheli, A.: Architectural and Markovian factors of echo state networks. Neural Netw. **24**(5), 440–456 (2011). https://doi.org/10.1016/j.neunet.2011.02.002
8. Gallicchio, C., Micheli, A., Pedrelli, L.: Deep reservoir computing: a critical experimental analysis. Neurocomputing **268**(Supplement C), 87–99 (2017). https://doi.org/10.1016/j.neucom.2016.12.089, http://www.sciencedirect.com/science/article/pii/S0925231217307567. advances in artificial neural networks, machine learning and computational intelligence
9. Hyndman, R.: Time series data library. http://robjhyndman.com/TSDL
10. Jaeger, H.: The "echo state" approach to analysing and training recurrent neural networks. Technical Report 148, German National Research Center for Information Technology (2001)
11. Jaeger, H.: Tutorial on training recurrent neural networks, covering BPPT, RTRL, EKF and the "echo state network" approach, Technical Report 148, German National Research Center for Information Technology (2002)
12. Jaeger, H., Lukoševičius, M., Popovici, D., Siewert, U.: Optimization and applications of echo state networks with leaky-integrator neurons. Neural Netw. **20**(3), 335–352 (2007)
13. Keogh, E., Smyth, P.: A probabilistic approach to fast pattern matching in time series databases. AAAI Technical Report WS-98-07, pp. 52–57 (1998)
14. Lukoševičius, M.: On self-organizing reservoirs and their hierarchies. Technical Report 25, Jacobs University, Bremen (2010)
15. Lukoševičius, M., Jaeger, H.: Reservoir computing approaches to recurrent neural network training. Comput. Sci. Rev. **3**, 127–149 (2009). https://doi.org/10.1016/j.cosrev2009.03.005
16. Maass, W.: Noisy spiking neurons with temporal coding have more computational power than sigmoidal neurons. Technical Report TR-1999-037, Institute for Theorical Computer Science. Technische Universitaet Graz. Graz, Austria (1999). http://www.igi.tugraz.at/psfiles/90.pdf
17. Manjunath, G., Jaeger, H.: Echo state property linked to an input: exploring a fundamental characteristic of recurrent neural networks. Neural Comput. **25**(3), 671–696 (2013). https://doi.org/10.1162/NECO_a_00411

18. Martens, J., Sutskever, I.: Training deep and recurrent networks with hessian-free optimization. In: Montavon, G., Orr, G.B., Müller, K.-R. (eds.) Neural Networks: Tricks of the Trade. LNCS, vol. 7700, pp. 479–535. Springer, Heidelberg (2012). https://doi.org/10.1007/978-3-642-35289-8_27

19. Mueen, A., et al.: The fastest similarity search algorithm for time series subsequences under euclidean distance, August 2017. http://www.cs.unm.edu/~mueen/FastestSimilaritySearch.html

20. Rodan, A., Tino, P.: Simple deterministically constructed cycle reservoirs with regular jumps. Neural Comput. **24**, 1822–1852 (2012). https://doi.org/10.1162/NECO_a_00297

21. Rumelhart, D.E., Hinton, G.E., McClelland, J.L.: A general framework for parallel distributed processing. In: Parallel Distributed Processing: Explorations in the Microstructure of Cognition, Computational Models of Cognition and Perception, vol. 1, chap. 2, pp. 45–76. MIT Press, Cambridge (1986)

22. Schmidhuber, J., Wierstra, D., Gagliolo, M., Gomez, F.: Training recurrent networks by Evolino. Neural Netw. **19**, 757–779 (2007)

23. Siegelmann, H.T., Sontag, E.D.: Turing computability with neural nets. Appl. Math. Lett. **4**(6), 77–80 (1991). https://doi.org/10.1016/0893-9659(91)90080-F

24. Verstraeten, D., Schrauwen, B., D'Haene, M., Stroobandt, D.: An experimental unification of reservoir computing methods. Neural Netw. **20**(3), 287–289 (2007)

25. Wainrib, G., Galtier, M.N.: A local Echo State Property through the largest Lyapunov exponent. Neural Netw. **76**, 39–45 (2016)

26. Wang, D., Li, M.: Stochastic configuration networks: fundamentals and algorithms. IEEE Trans. Cybern. **47**(10), 3466–3479 (2017). https://doi.org/10.1109/TCYB.2017.2734043

27. LeCun, Y., Bengio, Y., Hinton, G.: Deep learning. Nature **521**, 436–444 (2015)

28. Yildiza, I.B., Jaeger, H., Kiebela, S.J.: Re-visiting the echo state property. Neural Netw. **35**, 1–9 (2012). https://doi.org/10.1016/j.neunet.2012.07.005

29. Zhang, B., Miller, D.J., Wang, Y.: Nonlinear system modeling with random matrices: echo state networks revisited. Neural Netw. **76**, 39–45 (2016)

# Multi Step Prediction of Landslide Displacement Time Series Based on Extended Kalman Filter and Back Propagation Trough Time

Ping Jiang[1], Jiejie Chen[2], and Zhigang Zeng[3(✉)]

[1] Computer School, Hubei Polytechnic University, Huangshi 435000, China
[2] College of Computer Science and Technology, Hubei Normal University,
Huangshi 435002, China
[3] School of Artificial Intelligence and Automation,
Huazhong University of Science and Technology, Wuhan 430074, China
zgzeng@mail.hust.edu.cn

**Abstract.** Landslide is a complex geological natural disaster that brings harm or damage to human beings and their living environment. By strengthening landslide monitoring and forecasting technology, people can avoid or reduce the impact of disasters more reasonably. At present, the single step prediction of landslide displacement time series mainly uses t time to predict the data of t+1 moment, which obviously makes it difficult for people to take appropriate measures to deal with landslide changes. In this paper, a time reverse recursive algorithm based on extended Kalman filter (EKF)and Back propagation trough time (BPTT) method, is used to predict landslide displacement in order to extend the time width of landslide prediction. The EKF is firstly used to optimize the BPTT weights, and then the network parameters are adjusted in real time to improve the reliability of the prediction. Finally, the landslide displacement data of Liangshuijing (LSJ) in the three Gorges Reservoir area is used as experimental samples to verify the feasibility and practicability of EKF-BPTT.

**Keywords:** Landslide · Time series · Prediction · EKF-BPTT

## 1 Introduction

China is one of the countries with the most landslide disasters and serious losses, because of its large land area and complicated geological conditions. According to the national geological hazard notification data published annually by the State Geological Survey (see [1]), geological disasters such as landslides in the past five years (2012–2016) have caused great human and economic losses in China (See Fig. 1).

Landslide is a geological phenomenon in which a soft shear surface appears in the interior of a natural slope or a man-made slope under the combined action

© Springer Nature Switzerland AG 2019
H. Lu et al. (Eds.): ISNN 2019, LNCS 11554, pp. 184–193, 2019.
https://doi.org/10.1007/978-3-030-22796-8_20

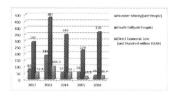

**Fig. 1.** Statistics on losses caused by geological hazards (2012–2016).

of various factors (see [1–4]), and the slope rock and soil gradually slip along this surface. The three Gorges Reservoir area in Hubei Province is a region with extremely complicated geology and steep topography. Since the storage of water in 2003, landslide disasters and landslides have occurred frequently in the reservoir area, which greatly intensifies ecological destruction, resource damage, waste and depletion, and also has a destructive impact on economic and social development (see [5]). The development of a series of new technologies based on landslide prediction and prevention to improve the ability of landslide emergency treatment and ensure the safety of resources has always been a hot spot in the engineering field (see [6–8]).

Some scholars have begun to study the induced factors of landslide and understand the process of landslide deformation and evolution (see [9,10]). Generally, rainfall is regarded as the main factor to induce landslide (see [11]). Lin Xiaosong and Huang Runqiu have studied the relationship between rainfall and landslide, and established the relative stability evaluation system (see [12,13]). By establishing a regression model of exponential function, Jiancong et al. revealed the law of correlation between landslide displacement and rainfall (see [14]). At the same time, many other scholars all over the world, such as the angle of landslide monitoring, the displacement response and deformation principle between rainfall intensity and landslide (see [15,16]). Yao wei and Liancheng established dynamic models of landslide displacement prediction using improved ESN and ultimate learning machine, respectively (see [17,18]).

Based on the above researchs, ANNs have the disadvantage of slow convergence rate and local minimum trap in landslide prediction. In this paper, a new algorithm EKF-BPTT is used, which mainly uses BPTT recurrent neural network to predict landslide displacement in multi-step, and EKF is used to optimize the weight of BPTT. The data of LSJ landslide displacement in the Three Gorges reservoir area are used as experimental samples to verify the feasibility and practicability of EKF-BPTT.

## 2  Preliminary Data

### 2.1  EKF

Kalman Filter (KF) is an algorithm used to find the optimal state estimation by means of the state equation of the linear system and the external representation of the input and output data. Apollo's navigation computer has used this filter. The application of KF requires several assumptions: the probability distribution of the current state must be a linear function of the previous state and the control quantity to be performed, and then a Gaussian noise is superimposed. The EKF is the KF for nonlinear systems.

Considering a simple and special example of a discrete time system with added value without observing noise:

$$x(n + 1) = f(x(n)) + q(n) \tag{1}$$

$$y(n) = h(x(n)) + v(n) \tag{2}$$

where $x(n)$ is state matrix, $f(*)$ is state update function, $q(n)$ is external input for system (uncorrelated Gaussian noise process, can also be regarded as process noise). $y(n)$ is output for system, $h(*)$ is a time-varying observation function (based on the original linear Kalman filter). At time $n = 0$, the state of the system $x(0)$ is obtained by the multidimensional normal distribution of the mean and covariance matrices. Until the moment $n$, the systematic observations are obtained by $y(0), y(1), \cdots, y(n)$.

## 2.2  BPTT

BPTT is an improved algorithm for the well-known Back-propagation (BP) algorithm of feedforward networks. It uses stacked recurrent neural network (RNN) with the same replicas and reconnects the connections between subsequent replicas in the network to "expand" the recursive network in time, as shown in Fig. 2, where $A$ represents the original RNN and $B$ represents the feedforward network acquired through $A$.

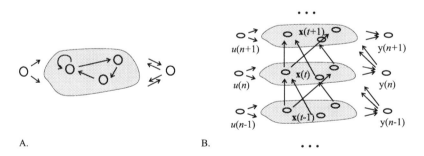

**Fig. 2.** Basic schematic diagram of BPTT.

The weights between cell layers are equal in or between replicas. Sample set data now includes a single input-output time series:

$$\mathbf{u}(n) = (u_1(n), \cdots, u_K(n))', \mathbf{d}(n) = (d_1(n), \cdots, d_L(n))' \tag{3}$$

where $n = 1, \cdots, T$.

# 3    Application of EKF-BPTT in Multi-step Prediction of Landslide Displacement Time Series

## 3.1    Thought of Model Design

EKF is used to update the weights of BPTT neural network and avoid local minimal problems caused by gradient descent algorithm, and new effective information is used reasonably to improve the convergence speed of learning algorithm. The learning network can not only learn autonomously, but also have the optimal estimation performance of fading filtering.

## 3.2    Model Implementation Process

According to the design idea of the model, the design and implementation process is as follows:

Firstly, a training set containing input and output is generated by BPTT:

$$u(n) = (u_1(n), ..., u_K(n))^t, n = 1, ..., T \tag{4}$$

$$d(n) = (d_1(n), ..., d_K(n))^t, n = 1, ..., T \tag{5}$$

where the weights of input layer, middle layer, output layer and back projection connection layer are related as follows:

$$W^{in} = (\omega_{ij}^{in}), W^{in} = (\omega_{ij}), W^{out} = (\omega_{ij}^{out}), W^{back} = (\omega_{ij}^{back}) \tag{6}$$

Secondly, instead of subdividing the weights of each level, we use a weight matrix $w$ to describe them and EKF to optimize the weights of BPTT. The output variable $d(n)$ of BPTT is an equation about weight h and input variable n:

$$d(n) = h(w, u(0), ..., u(n)) \tag{7}$$

The transient state of the initial network state has disappeared, and the input and output variables make it change with time. Assuming that the network needs to update some noise including process, we add some Gaussian noise $q(n)$. The dynamic behavior of BPTT is described as follows:

$$w(n+1) = w(n) + q(n), d(n) = h_n(w(n)) \tag{8}$$

EKF is a sub-optimal solution algorithm. For non-linear Gauss, it expands the non-linear part by Taylor expansion. Considering the single-input and single-output neural networks, there are no intermediate elements.

## 4    Case Study

### 4.1    LSJ Landslide

The mechanism of landslide formation is very complicated, and there are many factors affecting landslide. We choose LSJ landslide as a case study, which is the center of the Three Gorges Dam. It is located in Shuirang Village on the South Bank of the Yangtze River, Guling Town, Yunyang County, Chongqing City, northeastern China as shown **(a)** in Fig. 3. There are 24 GPS monitoring points on the surface of the landslide. We select the measured data on the monitoring point ZJG24 as the input data of the prediction model in this paper.

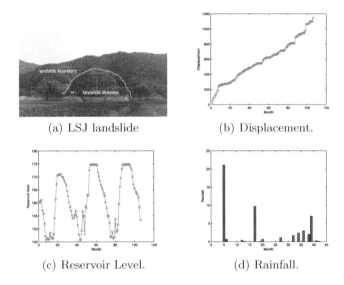

(a) LSJ landslide                    (b) Displacement.

(c) Reservoir Level.                  (d) Rainfall.

**Fig. 3.** LSJ landslide and its Displacement, Reservoir Level and Rainfall Time Series of ZJG24.

**(b)** and **(c)** in Fig. 3 depict the time series of displacement and reservoir water level from April 6, 2009 to May 25, 2011. For the acquired data, we use the data of the sixth, sixteenth and twenty-six days of each month. The total length of the data is 106, which is divided into two parts. The first part takes the first 71 data as training set to build the prediction model, and the remaining 35 data as test set. **(d)** in Fig. 3 depicts the rainfall time series curve from April 6, 2009 to June 16, 2010. The total non-zero data length is 43, while the rest of the non-rainfall data is 0.

### 4.2    Correlation and Stability Analysis of Time Series

MI and PCC can be used to identify the linear and nonlinear statistical dependence between a set of candidate input and output variables (see [19]). For calculating PCC and MI between displacement, reservoir level and rainfall of LSJ landslide, we first need to normalize all data by using the following formula, that is, to transform all data into $[-1, 1]$:

$$x_i^{(new)} = \frac{x_i^{(old)} - min(x_i)}{max(x_i) - min(x_i)}, \tag{9}$$

$$x_i^{(old)} = x_i^{(new)}(max(x_i) - min(x_i)) + min(x_i). \tag{10}$$

where $x_i^{(old)}$ and $x_i^{(new)}$ are input variables and output variables, $min(x_i)$ and $max(x_i)$ represent the maximum and minimum input variables in this interval, respectively.

Table 1. PCC and MI.

| Parameter | $Dis^a$ vs $ResL^b$ | Dis vs Rainfall | ResL vs Rainfall |
|---|---|---|---|
| PCC | 0.9338 | 0.1100 | 0.0091 |
| MI | 0.7236 | −0.0063 | 0.1367 |

[a] Displacement, [b] Reservoir Level

It can be seen from the Table 1 that the PCC (MI) between the displacement and reservoir level is larger, which reflects the close relationship between them. Then, the variables of displacement and reservoir level are choosed inputs for our forecast model.

## 4.3   Analysis and Results

The total length of LSJ landslide displacement and reservoir water level is 106 in Fig. 3. We use 70% data as the test set and 30% data as the prediction set. But here we use multi-step algorithm. If the step size is different each time, the corresponding data length will change correspondingly. Here we mainly use two-step, four-step and six-step prediction in advance. At the same time, we use two kinds of recurrent neural network RTRL and BPTT methods to do comparative experiments, using four common evaluation and prediction Mean Absolute Error (MAE), Root Mean Square Error (RMSE), Relative Error (RE) and PCC (R) to evaluate.

Figures 4, 5 and 6 show multi-step predicted displacement values, which are two-step, four-step and six-step predicted displacement values, respectively. Among them, blue represents the original data and red represents the prediction data.

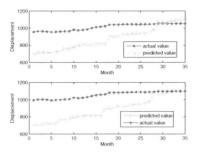

**Fig. 4.** EKF-BPTT two-step prediction of displacement.

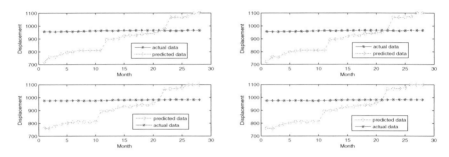

**Fig. 5.** EKF-BPTT four-step prediction of displacement.

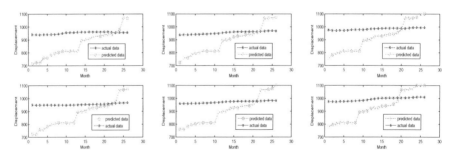

**Fig. 6.** EKF-BPTT six-step prediction of displacement.

Table 2 shows the characteristics of calculating the RE value of EKF-BPTT and extracting the RE value. It is also mainly considered from three aspects: maximum, minimum and mean. From the calculated data, it can be seen that the minimum RE value is 0.0006 and the maximum RE value is 0.3709.

**Table 2.** RE statistics.

| VARIABLE | MIN | MAX | MEAN |
|---|---|---|---|
| Two-step variable1 | 0.0006 | 0.3709 | 0.2126 |
| Two-step variable2 | 0.0240 | 0.3212 | 0.2066 |
| Four-step variable1 | 0.0091 | 0.3192 | 0.1296 |
| Four-step variable2 | 0.0053 | 0.3324 | 0.2366 |
| Four-step variable3 | 0.0035 | 0.3265 | 0.1258 |
| Four-step variable4 | 0.0011 | 0.2655 | 0.1176 |
| Six-step variable1 | 0.0058 | 0.3449 | 0.1482 |
| Six-step variable2 | 0.0003 | 0.3450 | 0.1430 |
| Six-step variable3 | 0.0087 | 0.2956 | 0.1127 |
| Six-step variable4 | 0.0016 | 0.2637 | 0.1186 |
| Six-step variable5 | 0.0105 | 0.2837 | 0.1197 |
| Six-step variable6 | 0.1197 | 0.2470 | 0.1145 |

**Table 3.** Two-step prediction comparison between RTRL, BPTT and EKF-BPTT.

| VARIABLE | MAE | RMSE | R |
|---|---|---|---|
| RTRL two-step variable1 | 151.5080 | 174.5301 | **_0.9319_** |
| RTRL two-step variable2 | **173.8630** | **202.8133** | **0.8122** |
| BPTT two-step variable1 | 155.2981 | 177.4132 | 0.9103 |
| BPTT two-step variable2 | 166.3599 | 189.3157 | 0.9176 |
| EKF-BPTT two-step variable1 | **_146.1874_** | **_165.3521_** | 0.9152 |
| EKF-BPTT two-step variable2 | 166.7754 | 190.7905 | 0.9203 |

Tables 3 and 4 represent the MAE, RMSE and R values calculated by three algorithms in multi-step prediction. Among them, the smaller the values of MAE and RMSE, the better, while R is the opposite. Due to space limitation, the table represent the MAE, RMSE and R values calculated by three algorithms in six-step is omitted, and experimental results show that the maximum MAE, RMSE and minimum R between RTRL, BPTT and EKF-BPTT are 116.9876, 136.2621 and 0.7162 respectively, all for BPTT six-step variable 1. And the minimum MAE, RMSE and maximum R are 78.2898 for EKF-BPTT six-step variable 3, 82.5877 for EKF-BPTT six-step variable 6 and 0.9647 for EKF-BPTT six-step variable 3 respectively. In summary, the experimental results in this set of landslide data show that the EKF-BPTT algorithm has the best prediction effect in two, four and six steps. This result shows EKF-BPTT algorithm is more reasonable in weight adjustment.

**Table 4.** Four-step prediction comparison between RTRL, BPTT and EKF-BPTT.

| VARIABLE | MAE | RMSE | R |
|---|---|---|---|
| RTRL four-step variable1 | 113.0197 | 126.1280 | 0.8901 |
| RTRL four-step variable2 | 112.8539 | 124.6291 | 0.8938 |
| RTRL four-step variable3 | 100.3440 | 110.5544 | 0.9171 |
| RTRL four-step variable4 | 97.3896 | 107.4008 | 0.9261 |
| BPTT four-step variable1 | 113.4523 | 132.5690 | 0.8156 |
| BPTT four-step variable2 | **117.8957** | **132.8315** | 0.8615 |
| BPTT four-step variable3 | 104.5090 | 121.6359 | 0.8906 |
| BPTT four-step variable4 | 109.7865 | 125.6540 | **0.5892** |
| EKF-BPTT four-step variable1 | 96.6189 | 97.6799 | *0.9817* |
| EKF-BPTT four-step variable2 | 102.6799 | 98.7399 | 0.9712 |
| EKF-BPTT four-step variable3 | 95.4461 | *94.4792* | 0.9592 |
| EKF-BPTT four-step variable4 | *93.2015* | 98.0521 | 0.9366 |

## 5   Concluding Remarks

A landslide displacement forecast approach has been proposed in this paper. And the causes of landslides are very complex and the harm is very great. In order to avoid these crises as far as possible, we need to try different methods to predict and deal with the changes of landslides. There are many methods to deal with the time series prediction of landslide displacement from different perspectives, such as one-step prediction, interval prediction and probability prediction. In this paper, RNNs and its improved EKF-BPTT learning algorithm are used for multi-step prediction. Using EKF to optimize RNs can not only accurately predict displacement, but also have some advantages compared with BPTT and RTRL. How to reasonably link control with prediction can be regarded as an important and difficult problem for future research.

**Acknowledgements.** The work was supported by the Natural Science Foundation of China under Grants 61841301, 61603129 and 61673188, the Research Project of Hubei Provincial Department of Education under Grant Q20184504, the Scientific Research Project of Hubei PolyTechnic University under Grant 18xjz02C.

## References

1. Qin, S.Q., Jiao, J.J., Wang, S.J.: The predictable time scale of landslides. Bull. Eng. Geol. Environ. **59**, 307–312 (2001)
2. Qin, S.Q., Jiao, J.J., Wang, S.J.: A nonlinear dynamical model of landslide evolution. Geomorphology **43**, 77–85 (2002)
3. Chen, C.T., Lin, M.L., Wang, K.L.: Landslide seismic signal recognition and mobility for an earthquake-induced rockslide in Tsaoling, Taiwan. Eng. Geol. **171**, 31–44 (2014)

4. Sorbino, G., Sica, C., Cascini, L.: Susceptibility analysis of shallow landslides source areas using physically based models. Nat. Hazards **53**, 313–332 (2010)
5. Miao, H.B., Wang, G.H., Yin, K.L., Kamai, T., Lin, Y.Y.: Mechanism of the slow-moving landslides in Jurassic red-strata in the Three Gorges Reservoir, China. Eng. Geol. **171**, 59–69 (2014)
6. Zhang, Y.B., Chen, G.Q., Zheng, L., Li, Y., Wu, J.: Effects of near-fault seismic loadings on run-out of large-scale landslide: a case study. Eng. Geol. **166**, 216–236 (2013)
7. Inoussa, G., Peng, H., Wu, J.: Nonlinear time series modeling and prediction using functional weights wavelet neural network-based state-dependent AR model. Neurocomputing **86**, 59–74 (2012)
8. Li, X.Z., Kong, J.M., Wang, Z.Y.: Landslide displacement prediction based on combining method with optimal weight. Nat. Hazards **61**, 635–646 (2012)
9. Jakob, M.: The impacts of logging on landslide activity at Clayoquot Sound. Br. Columbia Catena **38**(4), 279–300 (2000)
10. Melchiorre, C., Matteucci, M., Azzoni, A., Zanchi, A.: Artificial neural networks and cluster analysis in landslide susceptibility zonation. Geomorphology **94**, 379–400 (2008)
11. Lajtai, E.Z., Schmidtke, R.H., Bielus, L.P.: The effect of water on the time-dependent deformation and fracture of a granite. Int. J. Rock Mech. Min. Sci. Geomech. Abs. **24**(4), 247–255 (1987)
12. Lin, X.S., Guo, Y.: A study on coupling relationship between landslide and rainfall. J. Catastrophol. **16**(2), 87–92 (2001)
13. Huang, R.Q., Zhao, S.J., Song, X.B.: The formation and mechanism analysis of Tiantai landslide, Xuanhan County Sichuan Province. Hydrogeol. Eng. Geol. **32**(1), 13–15 (2005)
14. Xu, J.C., Shang, Y.Q., Wang, J.L.: Study on relationship between slope-mass slide displacement and precipitation of loose soil landslide. Chin. J. Rock Mech. Eng. **1**, 2854–2860 (2006)
15. Herrera, G., Fcmaudez-Merodo, J.A., Mulas, J., et al.: A landslide forecasting model using ground based SAR data: the Portalet case study. Eng. Geol. **105**(3–4), 220–230 (2009)
16. Bui, D.T., Pradhan, B., Lofman, O., et al.: Regional prediction of landslide hazard using probability analysis of intense rainfall in the Hoa Binh province, Vietnam. Nat. Hazards **66**(2), 1–24 (2012)
17. Yao, W., Zeng, Z.G., Lian, C., et al.: Ensembles of echo state networks for time series prediction. In: Proceedings of the 6th International Conference on Advanced Computational Intelligence Hangzhou, China, pp. 299–304 (2013)
18. Lian, C., Zeng, Z.G., Yao, W., Tang, H.M.: Displacement prediction model of landslide based on a modified ensemble empirical mode decomposition and extreme learning machine. Nat. Hazards **66**, 759–771 (2013)
19. Frenzel, S., Pompe, B.: Partial mutual information for coupling analysis of multivariate time series. Phys. Rev. Lett. **99**, 1–4 (2007)

# Noise Filtering in Cellular Neural Networks

Mikhail S. Tarkov[(✉)]

Rzhanov Institute of Semiconductor Physics SB RAS, Novosibirsk, Russia
tarkov@isp.nsc.ru

**Abstract.** A cellular neural network (CNN) with a bipolar stepwise activation function is considered. A comparative analysis of CNN learning algorithms on a given set of binary reference images for various degrees of noise (inversion of randomly selected pixels) and various cell neighborhood sizes is performed. For CNN training a local projection method, which provides much higher noisy images quality filtering than the classical local perceptron learning algorithm, is proposed.

**Keywords:** Cellular neural network · Noise filtering ·
Perceptron training algorithm · Local projection method · Noisy images ·
Cell neighborhood

## 1 Introduction

The main drawback of Hopfield networks [1–4] is a huge interneuron connections (complete graph) number which hampers their hardware implementation. As shown in [5], the connections number can be reduced by an order of magnitude due to the weights matrix reduction when training the Hopfield network according to Hebb, but the number of connections is still large. A possible way out of this difficult situation is the use of cellular neural networks [6–11], which makes this work relevant.

Cellular neural networks (CNN) were introduced in [6]. A CNN consists of a large number of simple processing elements (cells) usually located in nodes of an orthogonal or hexagonal grid, where each cell is connected to many nearest neighbors. The connections between the cells are weighed and each cell calculates a nonlinear function of its internal state which is modified depending on the sum of cell neighbor weighed outputs.

All cells calculate their next states in parallel and synchronously. Calculations begin when all cells are established in the initial state and are completed in a steady state when all the cells no longer change their states. For learning CNN, the perceptron algorithm is usually used. A classical version of this algorithm was proposed in [12]. In [7] this algorithm is adapted to CNN.

An orthogonal CNN example is presented in Fig. 1. A cell $C_{ij}$ corresponds to the states pattern $X_{ij}$ of its neighborhood and the pattern of weights $W_{ij}$ of the cell connections with its neighbors.

© Springer Nature Switzerland AG 2019
H. Lu et al. (Eds.): ISNN 2019, LNCS 11554, pp. 194–201, 2019.
https://doi.org/10.1007/978-3-030-22796-8_21

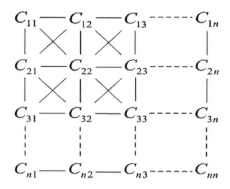

**Fig. 1.** Example of cellular neural network ($N_\varepsilon = 3 \times 3$)

For the neighborhood with the cell number $N_\varepsilon = (2r+1) \times (2r+1)$, $r = 1$ we have

$$X_{ij} = \begin{bmatrix} x_{i-1,j-1} & x_{i-1,j} & x_{i-1,j+1} \\ x_{i,j-1} & x_{ij} & x_{i,j+1} \\ x_{i+1,j-1} & x_{i+1,j} & x_{i+1,j+1} \end{bmatrix} \text{ and } W_{ij} = \begin{bmatrix} w_{i-1,j-1} & w_{i-1,j} & w_{i-1,j+1} \\ w_{i,j-1} & w_{ij} & w_{i,j+1} \\ w_{i+1,j-1} & w_{i+1,j} & w_{i+1,j+1} \end{bmatrix}.$$

We introduce the notation $X_{ij} \otimes W_{ij} = \sum\limits_{k=i-r}^{i+r} \sum\limits_{l=j-r}^{j+r} x_{kl} w_{kl}$, $r \geq 1$.

Then the cell $C_{ij}$ functioning is described by the expression:

$$x_{ij} = \text{sgn}(X_{ij} \otimes W_{ij})$$

where

$$\text{sgn}(a) = \begin{cases} 1, & a > 0 \\ -1, & a \leq 0 \end{cases}$$

is a bipolar stepwise activation function.

## 2  Methods of Learning Cellular Neural Networks

### 2.1  Hebb Method

Hopfield networks [1–4] can be considered as a CNN in which any cell is adjacent to all the others. When entering training vectors $x^t$, $t = 1, \ldots, p$, weights $w_{ij}$ are calculated according to the generalized Hebb rule

$$w_{ij} = \frac{1}{N} \sum_{k=1}^{p} x_i^k x_j^k,$$

where $N$ is the number of cells (neurons).

## 2.2 Projection Method

The Hopfield network projection learning method [3, 4] has the iterative weight matrix $W$ dependence on the learning vector sequence $x^t$, $t = 1, \ldots, p$:

$$y^t = (W^{t-1} - E) \cdot x^t,$$
$$W^t = W^{t-1} + \frac{y^t \cdot y^{tT}}{y^{tT} \cdot y^t}$$

under the initial conditions $W^0 = 0$ ($E$ is the identity matrix). As the vectors presentation result, the network weight matrix takes on a value $W = W^p$.

## 2.3 Cellular Version of the Perceptron Learning Algorithm

A cellular version of the perceptron learning algorithm can be represented by the formula [7]

$$W_{ij}^{t+1} = \begin{cases} W_{ij}^t, & \text{if } x_{ij}^t \cdot (X_{ij}^t \otimes W_{ij}^t) > 0, \\ W_{ij}^{t+1}, & \text{otherwise.} \end{cases}$$

Here, index $t$ indicates the states of cells and their interconnection weights at time $t$.

## 2.4 Cellular Version of the Hebb Method

As an alternative to the perceptron learning algorithm, this paper considers the cellular version of the Hebb algorithm, which is described by the formula

$$W_{ij} = \frac{1}{N_\varepsilon} \sum_{t=1}^{p} x_{ij}^t X_{ij}^t,$$

where $N_\varepsilon$ is the number of cells in neighborhood $\varepsilon$.

## 2.5 Cellular Version of the Projection Method

In this paper, we propose to use the following local version of the projection method for learning CNN:

$$y_{ij}^t = \sum_{j=i-r}^{i+r} \sum_{k=i-r}^{i+r} (w_{jk} - 1) \cdot x_{jk}^t, \quad r \geq 1.$$

$$W_{ij}^t = W_{ij}^{t-1} + \frac{y_{ij}^t \cdot Y_{ij}^t}{N_\varepsilon}, \quad t = 1, 2, \ldots, p, \quad W_{ij}^0 = 0.$$

## 3  Noise Filtering

As reference vectors (matrices) $x^t$, $t = 1, 2, \ldots, p$, $p = 10$, the binary images of digits (Fig. 2) sized $N = 16 \times 16$ and $N = 32 \times 32$ are used.

**Fig. 2.**  Reference patterns

To the input of the CNN trained by the above algorithms, the noisy versions of the reference vectors are fed. The noise is introduced by inverting randomly selected pixels. The noise level is set as a fraction of the pixels image total number. The filtration result is compared to the reference vector and the number of the corresponding mismatched components (Hamming distance) is recorded. This result is averaged over all reference patterns.

Experiments show that the global learning methods (Hebb method and projection method), traditionally used in Hopfield networks, generate a high-level noise (tens of percent) at the CNN output even if the input noise is absent.

In Figs. 3, 4, 5, 6, 7, 8 are the Hamming distance (residual noise) dependences on the percentage of the component inversions in the reference images for different sizes $N_\varepsilon = (2r + 1) \times (2r + 1)$, $r = 1, 2, 3$ of the cell neighborhood, for the local Hebb method, the perceptron algorithm, and the local projection method. These dependences show that:

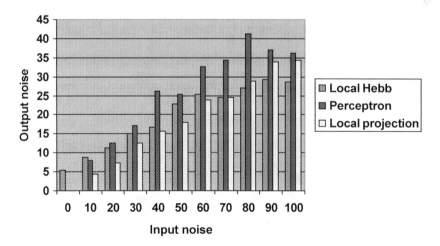

**Fig. 3.**  CNN output noise level dependence on the input noise level, $N = 16 \times 16$, $N_\varepsilon = 3 \times 3$.

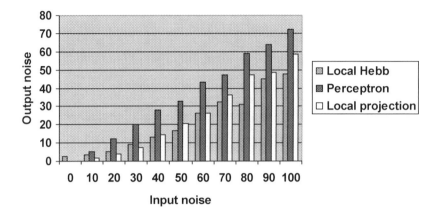

**Fig. 4.** CNN output noise level dependence on the input noise level, $N = 32 \times 32$, $N_\varepsilon = 3 \times 3$.

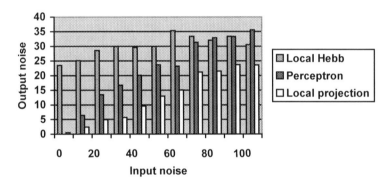

**Fig. 5.** CNN output noise level dependence on the input noise level, $N = 16 \times 16$, $N_\varepsilon = 5 \times 5$.

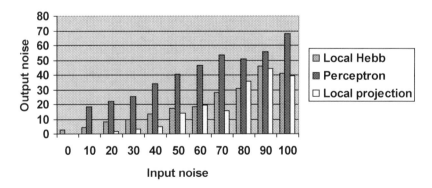

**Fig. 6.** CNN output noise level dependence on the input noise level, $N = 32 \times 32$, $N_\varepsilon = 5 \times 5$.

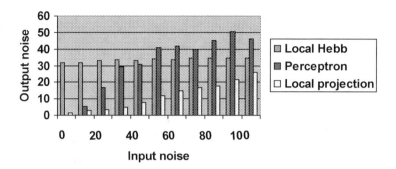

**Fig. 7.** CNN output noise level dependence on the input noise level, $N = 16 \times 16$, $N_\varepsilon = 7 \times 7$.

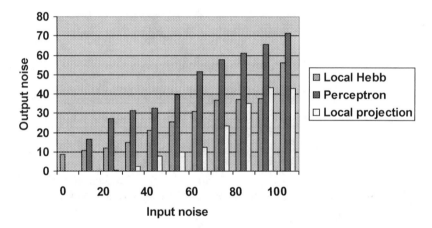

**Fig. 8.** CNN output noise level dependence on the input noise level, $N = 32 \times 32$, $N_\varepsilon = 7 \times 7$.

1. The perceptron method ideally reproduces reference images only in the absence of distortion. In other cases, it loses, almost everywhere, to the local projection method in terms of residual noise.
2. The local Hebb method works better than the other two methods only with minimum neighborhood size $N_\varepsilon = 3 \times 3$ and high noise levels (at least 70%) in the input signal.
3. The local projection method almost perfectly reproduces reference images in the absence of input noise (the output noise level does not exceed 1.56% with $N = 16 \times 16$ and 0.14% with $N = 32 \times 32$) and works better than the other two methods with small input noise levels (up to 60%).

For the local projection method, the residual noise level dependences on the input noise level are presented in Fig. 9 for the cell neighborhood sizes $N_\varepsilon = (2r+1) \times (2r+1)$, $r = 1, 2, 3$, and the cells number $N = 16 \times 16$ (Fig. 9a) and $N = 32 \times 32$ (Fig. 9b). It follows from Fig. 9 that the residual noise increase slows down when the neighborhood size $N_\varepsilon$ increases.

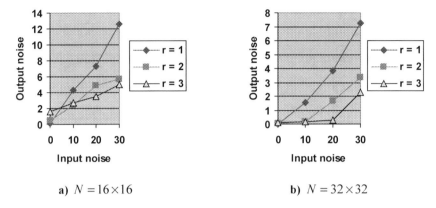

**a)** $N = 16 \times 16$                              **b)** $N = 32 \times 32$

**Fig. 9.** CNN output noise level dependence on the input noise level for the local projection method

## 4    Conclusion

A cellular neural network (CNN) with a bipolar stepwise activation function is considered. A comparative analysis of learning algorithms for a CNN on a given set of binary reference images is performed.

A local projection method, which provides much higher quality filtering of noisy images than the classical local perceptron learning algorithm, is proposed.

The trained CNN versions are tested with different noise (inversion of randomly selected pixels) degrees in the input reference images and various cell neighborhood sizes. Determined that:

1. Global learning methods (Hebb method and projection method), traditionally used in Hopfield networks, in cellular networks generate a high level noise (tens of percent) at the CNN output even in the absence of input noise.
2. The perceptron training method ideally reproduces reference patterns only in the absence of distortion. In other cases, it loses, almost everywhere, to the local projection method in terms of residual noise.
3. The local Hebb method works better than the other two methods only with minimum neighborhood $3 \times 3$ and high noise levels (at least 70%) of the input signal.
4. The local projection method almost perfectly reproduces the reference patterns in the absence of input noise (the output noise level does not exceed 1.56% for images sized $16 \times 16$ and 0.14% for images sized $32 \times 32$) and works better than other methods at a low level input noise (up to 60%).

# References

1. Osowski, S.: Neironnye seti dlya obrabotki informatsii (neural networks for information processing). Finansy i statistika, Moskwa (2002). (in Russian)
2. Hopfield, J.: Neural networks and physical systems with emergent collective computational abilities. Proc. Nat. Acad. Sci. USA **79**, 2554–2558 (1982)
3. Personnaz, L., Guyon, I., Dreyfus, G.: Collective computational properties of neural networks: new learning mechanisms. Phys. Rev. A **34**(5), 4217–4228 (1986)
4. Michel, A.N., Liu, D.: Qualitative analysis and synthesis of recurrent neural networks. Marcel Dekker Inc., New York (2002)
5. Tarkov, M.S.: Synapses reduction in autoassociative hopfield network. In: Proceedings of the 2017 International Multi-Conference on Engineering, Computer and Information Sciences (SIBIRCON), Novosibirsk, pp. 158–160 (2017). https://ieeexplore.ieee.org/document/8109860
6. Chua, L.O., Yang, L.: Cellular neural networks: theory and application. IEEE Trans. Circ. Syst. **35**(10), 1257–1290 (1988). **CAS**
7. Pudov, S.G.: Learning of cellular neural networks. Future Gener. Comput. Syst. **17**, 689–697 (2001)
8. Park, J., Kim, H.-Y., Park, Y., Lee, S.-W.: A synthesis procedure for associative memories based on space-varying cellular neural networks. Neural Netw. **14**(1), 107–113 (2001)
9. Bise, R., Takahashi, N., Nishi, T.: An improvement of the design method of cellular neural networks based on generalized eigenvalue minimization. IEEE Trans. Circuits Syst.-I **50**(12), 1569–1574 (2003)
10. Li, H., Liao, X., Li, C., Huang, H., Li, C.: Edge detection of noisy images based on cellular neural networks. Commun. Nonlinear Sci. Numer. Simul. **16**, 3746–3759 (2011)
11. Duan, S., Hu, X., Dong, Z., Wang, L., Mazumder, P.: Memristor-based cellular nonlinear/neural network: design, analysis, and applications. IEEE Trans. Neural Netw. Learn. Syst. **26**(6), 1202–1213 (2015)
12. Rosenblatt, F.: Principles of Neurodynamics. Spartan, Washington (1959)

# Learning Agents with Prioritization and Parameter Noise in Continuous State and Action Space

Rajesh Mangannavar$^{(\boxtimes)}$ and Gopalakrishnan Srinivasaraghavan

International Institute of Information Technology, Bangalore, Bangalore, India
{rajesh,gsr}@iiitb.ac.in

**Abstract.** Among the many variants of RL, an important class of problems is where the state and action spaces are continuous—autonomous robots, autonomous vehicles, optimal control are all examples of such problems that can lend themselves naturally to reinforcement based algorithms, and have continuous state and action spaces. In this paper, we introduce a prioritized form of a combination of state-of-the-art approaches such as Deep Q-learning (DQN) and Deep Deterministic Policy Gradient (DDPG) to outperform the earlier results for continuous state and action space problems. Our experiments also involve the use of parameter noise during training resulting in more robust deep RL models outperforming the earlier results significantly. We believe these results are a valuable addition for continuous state and action space problems.

**Keywords:** Reinforcement learning · Policy search ·
Prioritized learning · Parameter noise · RL · Deep learning · Mujoco ·
Policy gradient · DDPG

## 1 Introduction

Reinforcement learning (RL) is about an agent learning an optimal way to control and/or navigate through an environment that requires sequential decision making by the agent. The agent does this by trying to maximize a numerical performance measure that expresses a long-term objective, by trial-and-error. RL arises naturally in a wide range of domains including autonomous control, gaming, natural language processing & dialogue management, etc. [1].

### 1.1 RL in Continuous State and Action Space

Many interesting real-world control tasks, such as driving a car or riding a snowboard, require smooth continuous actions taken in response to high-dimensional, real-valued sensory inputs. In applying RL to continuous problems, the predominant approach in the past involved discretizing the state and action spaces and then applying an RL algorithm for a discrete stochastic system [2]. However

© Springer Nature Switzerland AG 2019
H. Lu et al. (Eds.): ISNN 2019, LNCS 11554, pp. 202–212, 2019.
https://doi.org/10.1007/978-3-030-22796-8_22

the drawbacks of discretization are, they do not scale, do not allow fine-grained smooth control characteristic of continuous space-action systems, and are too sensitive to the (arbitrary) choice of the granularity with which the discretization is carried out.

Hence, the formulation of the reinforcement learning problem with continuous state and action space holds great value in solving more real-world problems.

## 1.2   Deep Reinforcement Learning

The advent of deep learning has had a significant impact in many areas in machine learning, dramatically improving the state-of-the-art tasks such as object detection, speech recognition and language translation [3]. Deep Neural Networks are very powerful function approximators and can be trained to automatically find low-dimensional representations of high-dimensional data. This enables deep learning to scale to problems which were previously intractable including reinforcement learning problems with high dimensional, continuous state and action spaces [4].

Few of the current state of the art methods in the area of Deep RL are:

- Deep Q-learning Networks (DQN) - introduced novel concepts which helped in using neural networks as function approximators for reinforcement learning algorithms (for continuous state space) [5].
- Prioritized Experience Replay (PER) - builds upon DQN with some newer approaches to outperform DQN (for continuous state space) [6].
- Deep Deterministic Policy Gradients (DDPG) - follows a different paradigm as compared to the above methods. It uses DQN as the function approximator while building on the seminal work of [7] (for both continuous state and action space) on deterministic policy gradients.

## 1.3   Prioritized Experience Replay in DDPG ( PDDPG)

We propose a new algorithm, Prioritized DDPG using the ideas proposed in DQN, prioritized experience replay and DDPG such that it outperforms the DDPG in the continuous state and action space. Prioritized DDPG uses the concept of prioritized sampling in the function approximator of DDPG. Our results show that prioritized DDPG outperforms DDPG in a majority of the continuous action space environments. We then use the concept of parameter space noise for exploration and show that this further improves the rewards achieved.

## 2   Previous Work

### 2.1   Critic Methods of RL

Critic methods rely exclusively on a value or Q-function approximation and aim at learning a "good" approximation of the value/Q-function [8]. We survey a few of the recent best-known critic methods in RL.

**Deep Q-Learning Networks.** As in any value function-based approach, DQN method tries to find the optimal value function for any given state/state-action-pair. The novelty of their approach is that they efficiently use a non-linear function approximator to learn the function. Prior to their work, though the potential for using non-linear function approximators was recognized, there was no tangible demonstration of their use in practice in an efficient manner.

The challenge in using function approximators was that the Monte-Carlo sampling typically used for collecting experience in the form of state-action-rewards by generating episodes did not guarantee i.i.d data that most machine learning algorithms assume for training the models from. To overcome this drawback, the novel ideas that were proposed which made it possible to use non-linear function approximators for reinforcement learning are the following:

- Experience Replay Buffer: In this technique, the neural network is trained from random samples from a large buffer of stored observations.
- Periodic Target Network Updates: Two sets of parameters are maintained for the same neural network — one for generating behaviour from (possibly using an $\epsilon$-greedy strategy) and the other (the target network) for the value function estimation. The target network parameters are used in the loss computation and are updated by the behaviour network parameters periodically.

**Prioritized Experience Replay.** The prioritized experience replay algorithm is a further improvement on the deep Q-learning methods and can be applied to both DQN and Double DQN. The idea proposed by the authors is as follows: instead of selecting the observations at random from the replay buffer, they can be chosen based on some criteria which will help in making the learning faster. Intuitively the selection criterion from the replay buffer is biased towards the 'more useful observations' and less on the 'stale' ones. To select these more useful observations, the criteria they use is the error of that particular observation.

This criterion helps select those observations which provide the highest learning opportunity for the agent. The problem with this approach is that greedy prioritization focuses on a small subset of the experience and this lack of diversity may lead to over-fitting. Hence, the authors introduce a stochastic sampling method that interpolates between pure greedy and uniform random prioritization. Hence, the new probability of sampling a transition $i$ is

$$P(j) = \frac{p_j^\alpha}{\sum_k p_k^\alpha} \tag{1}$$

where $p_i$ is the priority of transition $i$ and $\alpha$ determines how much prioritization is used. The approach, while it improves the results has a problem of changing the distribution of the expectation. This is resolved by the authors by using Importance Sampling (IS) weights

$$w_i = \left( \frac{1}{N} \cdot \frac{1}{P(i)} \right)^\beta \tag{2}$$

where if $\beta = 1$, the non-uniform probabilities $P(i)$ are fully compensated ([6]).

## 2.2   Actor Methods in RL

Actor methods work with a parameterized family of policies. The gradient of the performance, with respect to the actor parameters, is directly estimated by simulation, and the parameters are updated in the direction of improvement ([8]).

**Deterministic Policy Gradients (DPG).** The most popular policy gradient method is the deterministic policy gradient (DPG) method and in this approach, instead of having a stochastic policy, the authors make the policy deterministic and then determine the policy gradient.

The deterministic policy gradient is the expected gradient of the action-value function, which integrates over the state space, whereas in the stochastic case, the policy gradient integrates over both state and action spaces. What this leads to is that the deterministic policy gradient can be estimated more efficiently than the stochastic policy gradient.

The DPG algorithm, presented by [7] maintains a parameterized actor function $\mu(s|\theta^\mu)$ which is the current deterministic policy that maps a state to a particular action. The authors use the Bellman equation to update the critic $Q(s, a)$. They then go on to prove that the derivative expected return with respect to actor parameters is the gradient of the policy's performance.

## 2.3   Actor-Critic Methods

Actor-critic models (ACM) are a class of RL models that separate the policy from the value approximation process by parameterizing the policy separately. The parameterization of the value function is called the critic and the parameterization of the policy is called the actor. The actor is updated based on the critic which can be done in many ways, while the critic is update based on the current policy provided by the actor ([8,9]).

**Deep Deterministic Policy Gradients (DDPG).** The DDPG algorithm tries to solve the reinforcement learning problem in continuous action and state space setting. The authors of this approach extend the idea of deterministic policy gradients. What they add to the DPG approach is the use of a non-linear function approximator ([10]).

While using a deterministic policy, the action value function reduces from

$$Q^\pi(s_t, a_t) = \mathbb{E}_{r_t, s_{t+1} \sim E}[r(s_t, a_t) + \gamma \mathbb{E}_{a_{t+1} \sim \pi}[Q^\pi(s_{t+1}, a_{t+1})]] \qquad (3)$$

to

$$Q^\mu(s_t, a_t) = \mathbb{E}_{r_t, s_{t+1} \sim E}[r(s_t, a_t) + \gamma Q^\mu(s_{t+1}, \mu(s_{t+1}))] \tag{4}$$

as the inner expectation is no longer required. Here, $\gamma \in [0, 1]$ is the discounting factor. What this also tells us is that the expectation depends only on the environment and nothing else. Hence, we can learn off-policy, that is, we can train our reinforcement learning agent by using the observations made by some other agent. The authors use this as well as the novel concepts used in DQN to construct their function approximator. These concepts cannot be applied directly to continuous action space, as there is an optimization over the action space at every step which is in-feasible when there is a continuous action space [10].

Once we have both the actor and the critic networks with their respective gradients, we can then use the DQN concepts - replay buffer and target networks to train these two networks. They apply the replay buffer directly without any modifications but make small changes in the way target networks are used. Instead of directly copying the values from the temporary network to the target network, they use soft updates to the target networks.

**Parameter Space Noise for Exploration.** There is no best exploration strategy in RL. For some algorithms, random exploration works better and for some greedy exploration. But whichever strategy is used, the important requirement is that the agent has explored enough about the environment and learns the best policy. Plappert et al. [11] in their paper explore the idea of adding noise to the agent's parameters instead of adding noise in the action space. In their paper Parameter Space Noise For Exploration, they explore and compare the effects of four different kinds of noises

- Uncorrelated additive action space noise
- Correlated additive Gaussian action space noise
- Adaptive-param space noise
- No noise

They show that adding parameter noise vastly outperforms existing algorithms or at least is just as good on a majority of the environments for DDPG as well as other popular algorithms such as Trust Region Policy Optimization ([12]).

## 3   Prioritized Deep Deterministic Policy Gradients

The proposed algorithm is primarily an adaptation of DQN and DDPG with ideas from the work of Schaul et al. [6] on continuous control with deep reinforcement learning to design an RL scheme that improves on DDPG significantly. The intuition behind the idea is as follows: The DDPG algorithm uses the DQN method as a sub-algorithm and any improvement over the DQN algorithm should ideally result in the improvement of the DDPG algorithm. But from the above-described methods, not all algorithms which improve DQN can be used to improve DDPG. That is because some of them need the environment to have discrete action spaces. So, for our work, we will consider only the prioritized experience replay method which does not have this constraint.

## 3.1   Prioritized DDPG Algorithm

Now, the improvement to the DQN algorithm, the prioritized action replay method can be integrated into the DDPG algorithm in a very simple way. Instead of using just DQN as the function approximator, we can use DQN with prioritized action replay. That is, in the DDPG algorithm, instead of selecting observations randomly, we select the observations using the stochastic sampling method as defined in Eq. 1. The pseudo-code for the prioritized action replay is given in Algorithm 1. The algorithm runs in a for loop $M$ times where $M$ is the number of episodes we want to train the agent for.

---

**Algorithm 1** PDDPG algorithm

---

Randomly initialize critic network $Q(s, a|\theta^Q)$ and actor $\mu(s|\theta^\mu)$ with weights $\theta^Q$ and $\theta^\mu$.

Initialize target network $Q'$ and $\mu'$ with weights $\theta^{Q'} \leftarrow \theta^Q, \theta^{\mu'} \leftarrow \theta^\mu$

Initialize replay buffer $R$

**for** episode = 1, M **do**

    Initialize a random process $N$ for action exploration

    Receive initial observation state $s_1$

    **for** t = 1, T **do**

        Select action $a_t = \mu(s_t|\theta^\mu) + N_t$ using to the current policy

        Execute action $a_t$ and observe reward $r_t$ and new state $s_{t+1}$

        Store transition $(s_t, a_t, r_t, s_{t+1})$ in $R$ ▷ Storing to the replay buffer

        Sample a mini-batch of $N$ transitions $(s_i, a_i, r_i, s_{i+1})$ from $R$ from $R$ each such that - $iP(i) = p_i^\alpha / \Sigma_i p_i^\alpha$ ▷ Stochastic sampling

        Set $y_i = r_i + \gamma Q'(s_{i+1}, \mu'(s_{i+1}|\theta^{\mu'})|\theta^{Q'})$

        Update critic by minimizing the loss: $L = \frac{1}{N} \Sigma_i (y_i - Q(s_i, a_i|\theta^Q))^2$

        Update the actor policy using the sampled policy gradient

$$\nabla_{\theta^\mu} J \approx \frac{1}{N} \Sigma_i \nabla_a Q(s, a|\theta^Q)|_{s=s_i, a=\mu(s_i)} \nabla_{\theta^\mu} \mu(s|\theta^\mu)|_{s_i}$$

        Update the target networks:

$$\theta^{Q'} \leftarrow \tau \theta^Q + (1 - \tau)\theta^{Q'}$$
$$\theta^{\mu'} \leftarrow \tau \theta^\mu + (1 - \tau)\theta^{\mu'}$$

        Update the transition priorities for the entire batch based on the error

    **end for**

**end for**

---

This algorithm is quite similar to the original DDPG algorithm with the only changes being the way the observations are selected for training and the transition probabilities are being updated. The first change ensures we are selecting the better set of observations which help in learning faster and the second change helps in avoiding over-fitting as it ensures all the observations have a non-zero probability of being selected to train the network and only a few high error observations are not used multiple times to train the network.

## 4    PDDPG with Parameter Noise

As introducing parameter noise improved the results obtained by DDPG, we introduce the parameter noise with PDDPG in the same way. The noise is added such we can achieve structured exploration by applying the noises to the parameter of the current policy. Also, the policy on which we have applied our noise is sampled at the beginning of each episode.

The noise we add to the parameter are the ones discussed before -

- Uncorrelated additive action space noise
- Correlated additive Gaussian action space noise
- Adaptive-param space noise

We also have a result without adding any noise (original PDDPG) for comparison.

## 5    Results

The proposed, prioritized DDPG algorithm was tested on many of the standard RL simulation environments that have been used in the past for benchmarking the earlier algorithms. The environments are available as part of the Mujoco platform ([13]).

### 5.1    Mujoco Platform

Mujoco is a physics environment which was created to facilitate research and development in robotics and similar areas, where fast simulation is an important component.

This set of environments provide a varied set of challenges for the agent as environments have continuous action as well as state space. All the environments contain stick figures with some joints trying to perform some basic task by performing actions like moving a joint in a particular direction or applying some force using one of the joints.

### 5.2    Empirical Evaluation

The implementation used for making the comparison was the implementation of DDPG in baselines ([14]). The prioritized DDPG algorithm was implemented by extending the existing code in baselines.

**Results for PDDPG.** The following are the results of the prioritized DDPG agent as compared DDPG agent ([10]). The overall reward - that is the average of the reward across all epochs until that point and reward history - the average of the last 100 epochs on four environments are plotted. The y-axis represents the reward the agent has received from the environment and the x-axis is the number of epochs with each epoch corresponding to 2000 time steps.

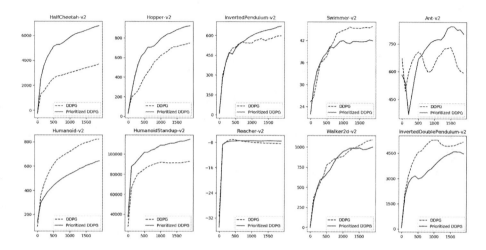

**Fig. 1.** Prioritized DDPG vs DDPG

As seen in Fig. 1, the Prioritized DDPG algorithm reaches the reward of the DDPG algorithm in less than 300 epochs for the HalfCheetah environment. This shows that the prioritized DDPG algorithm is much faster in learning.

The same trend can be observed in Fig. 1 for HumanoidStandup, Hopper and Ant environments. That is, the prioritized DDPG agent learns and gets the same reward as DDPG much faster. This helps is in reducing overall training time. Prioritized DDPG algorithm can also help in achieving results which might not be achieved by DDPG even after a large number of epochs. This can be seen in the case of the Ant environment. Figure 1 shows that DDPG rewards are actually declining with more training. On the other hand, Prioritized DDPG has already achieved a reward much higher and is more stable.

One a few environments such as the Reacher, InvertedDoublePendulum and Walker2d, it can be seen from Fig. 1 the prioritized DDPG only, marginally outperforms the DDPG algorithm.

**Results for PDDPG with Parameter Noise.** The PDDPG algorithm with parameter noise was run on the same set of environments as the PDDPG algorithm - the Mujoco environments. The empirical results are as follows.

We can see from Fig. 2 that noise improves the overall reward obtained only some of the environment. This is because the idea behind adding noise is for

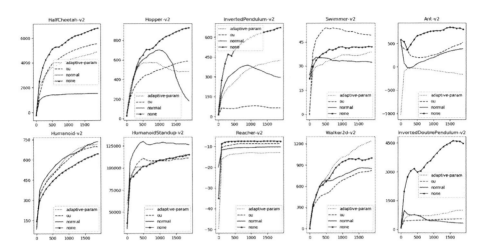

**Fig. 2.** Prioritized DDPG across all noi ses - adaptive-param, uncorrelated, co related and with no noise

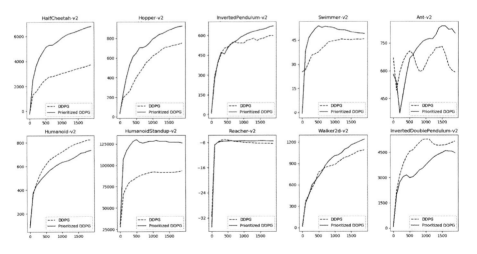

**Fig. 3.** Prioritized DDPG with noise vs DDPG

better exploration and PDDPG explores and learns much faster as seen in 1. In environments such as Inverted Pendulum or inverted double pendulum where lesser exploration is required, the addition of noise does not improve the rewards achieved, but in environments such as Humanoid-Stand up or Walker-2d, where a lot of exploration is required, the noise does improve the overall reward achieved.

Overall, with the proposed changes, prioritization with the addition of noise, the proposed algorithm outperforms DDPG on eight of the ten environments as seen in Fig. 3 and does reasonably well in the others.

## 6     Conclusions

To summarize, this paper discusses the state of the art methods in reinforcement learning with our improvements that have led to RL algorithms in continuous state and action spaces that outperform the existing ones.

The proposed algorithm combines the concept of prioritized action replay with deep deterministic policy gradients. As it has been shown, on a majority of the mujoco environments this algorithm vastly outperforms the DDPG algorithm both in terms of overall reward achieved and the average reward for any hundred epochs over the thousand epochs over which both were run.

The proposed algorithm seems to learn much faster than the DDPG algorithm. Also after 2000 iterations, as the graphs above show, the proposed algorithm had accumulated significantly more rewards than DDPG while still retaining a positive slope compared to the flattened curve for DDPG. This indicates that it is unlikely for the DDPG algorithm to surpass the results of the proposed algorithm on a majority of the environments. Also, certain kinds of noises further improve PDDPG to help attain higher rewards. One other important conclusion is that different kinds of noises work better for different environments which is evident in how drastically the results changed based on the parameter noise.

The presented algorithm can also be extended and improved further by finding more concepts in value based methods, which can be used in policy-based methods. The overall improvements in the area of continuous space and action state space can help in making reinforcement learning more applicable in real-world scenarios. These methods can potentially be extended to safety-critical systems, by incorporating the notion of safety during the training of an RL algorithm. This is currently a big challenge because of the necessary unrestricted exploration process of a typical RL algorithm.

## References

1. Li, Y.: Deep reinforcement learning: an overview. CoRR. abs/1701.07274 (2017)
2. Doya, K.: Reinforcement learning in continuous time and space. Neural Comput. **12**, 219–245 (2000)
3. LeCun, Y., Bengio, Y., Hinton, G.: Deep learning. Nature **521**(7553), 436–444 (2015)
4. Arulkumaran, K., Deisenroth, M.P., Brundage, M., Bharath, A.A.: A brief survey of deep reinforcement learning. CoRR. abs/1708.05866 (2017)

5. Mnih, V., et al.: Playing atari with deep reinforcement learning. CoRR. abs/1312.5602 (2013)
6. Schaul, T., Quan, J., Antonoglou, I., Silver, D.: Prioritized experience replay. CoRR. abs/1511.05952 (2015)
7. Silver, D., Lever, G., Heess, N., Degris, T., Wierstra, D., Riedmiller, M.: Deterministic policy gradient algorithms. In: Proceedings of the 31st International Conference on International Conference on Machine Learning, Beijing (2014)
8. Konda, V.R., Tsitsiklis, J.N.: Actor-critic algorithms. In: Advances in Neural Information Processing Systems (2000)
9. Feinberg, E.A., Shwartz, A.: Handbook of Markov Decision Processes: Methods and Applications. Springer, Heidelberg (2012). https://doi.org/10.1007/978-1-4615-0805-2
10. Lillicrap, T.P., et al.: Continuous control with deep reinforcement learning. CoRR. abs/1509.02971 (2015)
11. Plappert, M., et al.: Parameter space noise for exploration. CoRR. abs/1706.01905 (2017)
12. Schulman, J., Levine, S., Moritz, P., Jordan, M.I., Abbeel, P.: Trust region policy optimization. CoRR. abs/1502.05477 (2015)
13. Todorov, E., Erez, T., Tassa, Y.: MuJoCo: a physics engine for model-based control. In: IEEE/RSJ International Conference on Intelligent Robots and Systems, pp. 5026–5033 (2012)
14. OpenAI Baselines Implementation. https://github.com/openai/baselines

# Demand Forecasting Techniques for Build-to-Order Lean Manufacturing Supply Chains

Rodrigo Rivera-Castro[1(✉)], Ivan Nazarov[1], Yuke Xiang[2], Alexander Pletneev[1], Ivan Maksimov[1], and Evgeny Burnaev[1]

[1] Skolkovo Institute of Science and Technology, Moscow, Russia
{rodrigo.riveracastro,alexander.pletneev,ivan.maksimov,
e.burnaev}@skoltech.ru,
ivan.nazarov@skolkovotech.ru
[2] Huawei Noah's Ark Lab, Hong Kong, China
yuke.xiang@huawei.com

**Abstract.** Build-to-order (BTO) supply chains have become common-place in industries such as electronics, automotive and fashion. They enable building products based on individual requirements with a short lead time and minimum inventory and production costs. Due to their nature, they differ significantly from traditional supply chains. However, there have not been studies dedicated to demand forecasting methods for this type of setting. This work makes two contributions. First, it presents a new and unique data set from a manufacturer in the BTO sector. Second, it proposes a novel data transformation technique for demand forecasting of BTO products. Results from thirteen forecasting methods show that the approach compares well to the state-of-the-art while being easy to implement and to explain to decision-makers.

**Keywords:** Demand forecasting · Supply chain modelling · Kernels · Neural networks

## 1 Introduction

Supply chain management (SCM) represents the managerial backbone of the logistics and production sector. Due to its relevance, new methodologies appear regularly in the literature. One of them is Build-To-Order Supply Chain Management (BTO-SCM). This technique has seen significant adoption. Motivated by the lack of work in the demand forecasting literature on BTO-SCM problems, this research develops and presents *Diagonal Feeding*, a data transformation technique, specially tailored for this setting together with a relevant and novel data set from an electronics manufacturer. Data set and implementations of all methods are available for download[1].

---

[1] https://github.com/rodrigorivera/isnn19.

© Springer Nature Switzerland AG 2019
H. Lu et al. (Eds.): ISNN 2019, LNCS 11554, pp. 213–222, 2019.
https://doi.org/10.1007/978-3-030-22796-8_23

An accurate demand forecasting is essential for the global economy with stockpiled or in-transit inventories representing 17% of the world's Gross Domestic Product (GDP), [8]. Yet, imprecise demand planning is still pervasive with forecast errors of up to 44-53% for new products, [17,18]. As a result, retailers experience out-of-stock (OOS) events with rates amounting to 8.3% worldwide, [14]. BTO-SCM helps address the uncertainty that arises from variability in demand.

The aims and backgrounds of this research is to present a technique for data pre-processing accessible to non-technical business experts. Traditional forecasting methods are still being used primarily by over 40% of demand planners in the industry, [10], and the use of novel machine learning methods is a promising area with little academic research and insufficient efforts to expose practitioners to them, [24,25].

The significance of developing demand forecasting methods that can be easily adopted by demand planners is evident. For discrepancies as low as 2%, it is worth improving the accuracy of a forecast, [12], and a 10% reduction in OOS can increase retailers' revenue by up to 0.5%, [19]. Yet, companies struggle hiring the adequate personnel to address these tasks. For example by 2020, Vietnam is expected to face a shortage of over 500,000 employees with data science and analytic skills and over 80% of the local workforce is unsuited to fill this gap, [23]. In Europe, over 70% of surveyed businesses struggle hiring data science personnel and over 60% are resorting to internal training to upgrade the skills of existing business analysts, [1]. This work seeks to alleviate this situation by presenting a feature engineering technique well-suited for demand forecasting in manufacturing that is both accurate as well as easy to communicate to decision-makers.

The research goal is to propose a method for time series forecast of BTO products that can be adopted by practitioners. For this purpose, the study poses the questions: (1) What is the state of the art in academic research in demand forecasting for BTO products? (2) How does *Diagonal Feeding* support the BTO supply chain? To achieve the research goal, two objectives have been assigned: (a) To review the existing theory on time series prediction, specially on demand forecasting; (b) To make a performance comparison of the proposed technique. The object of research is the balance between accessibility and precision of methods for time series in an industry setting. The subject of the research is forecasting product demand for BTO-SCM.

## 2 Literature Review

BTO-SCM can be defined according to [15] as "the system that produces goods and services based on individual customer requirements in a timely and cost competitive manner by leveraging global outsourcing, the application of information technology and through the standardization of components and delayed product differentiation strategies". Proponents of BTO supply chains such as [27] argue that they promote sales, reduce costs and increase customer satisfaction. For example, [11] mentions that 74% of US car buyers would purchase a customized vehicle if the delivery time is less than three weeks. As an example of costs reduction, it is claimed that Nissan, a car manufacturer, could save up to 3600 USD per vehicle, if they were to transition their supply chain completely to BTO. In the technology industry, Dell, a computer maker, has generated up to a 160% return on its invested capital by implementing a BTO supply chain for its e-commerce website, [28].

Demand forecasting has commanded attention from different communities due to its importance in the supply chain management, [7]. A comprehensive treatment can be found in the works of [10] and [13]. Formally, it can be stated as predicting future values $x_{t+h}$, given a time series $x_{t-w+1}, \ldots, x_{t-1}, x_t$, where $w$ is the length of the window on the time series and $h$ is called the prediction horizon, [22]. To obtain these predictions, quantitative methods such as ARIMA, exponential smoothing models and alike are often used. Yet, [2] argues that there have been few large scale comparison studies of machine learning models for regression or time series aimed at forecasting problems. For the electronics manufacturing industry, [30] introduced SVM regression to the supply chain of various producers. They concluded that it yields good results compared to other forecasting methods. Although SVM regression is a popular method for forecasting, not everyone has identified it as the most effective method. For example, [21] used multivariate adaptive regression splines (MARS) to construct sales forecasting models for computer wholesalers. Similarly, [31] proposed a Bayesian model for demand forecasting of computer parts and compared it against exponential smoothing and a judgment-based method.

## 3   Dataset

The data consists of three data sets of observations from an electronics manufacturer representing a subset of their total inventory. Each of them contains the demand of 854 different items, totaling 2562 items, with up to 45 periods of data. The items can be requested for delivery either for the same month or up to three months later. In addition, two hierarchical dependencies are provided in the form

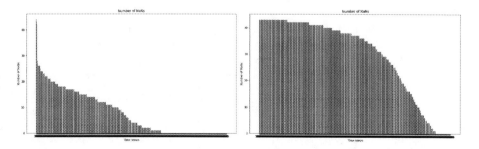

**Fig. 1.** Number of NaNs (zero orders) per item for each of the three data sets. First and second data set (left), third data set (right). X-axis: Item's ID, Y-axis: number of zeroes

of categories and parent-child relationships. These items have varying amounts
of required quantities, with many of them being requested only sporadically, as
seen in Fig. 1, and few of them being requested in large quantities. Noticeable,
a significant percentage of products lacks a continuous demand. There are no
periods in all three data sets where all 854 products are requested. The demand
for items both in variety as well as quantity is significantly higher for the delivery
date 0 (same month). From Fig. 2, it can be appreciated that quantities for all
periods are consistently higher on delivery date 0 versus others. In the case of
delivery date 3, the requested quantities for all periods are significantly smaller.
Similarly, the amounts required in the third data set are much higher and they
peak out, whereas the first and second data sets have an increasing demand
pattern.

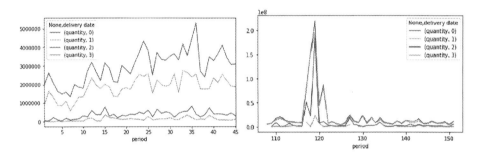

**Fig. 2.** Aggregated quantity by delivery date. First and second data set (left), third
data set (right). X-axis: period, Y-axis: quantity. Blue line: same month delivery. (Color
figure online)

## 4   Diagonal Feeding

This work introduces the practitioner to *Diagonal Feeding*, a data transformation technique well-suited for multi-step structured forecasting from anticipatory
data. As its main benefit, it levers the anticipatory nature of pre-orders' time
series data and makes forecasting the pre-order structure more streamlined. As
a result, the accuracy of a regression model is improved. This is made possible due to the data set containing information not only about the current
demand, but also on the volumes of pre-orders made in advance. Advance pre-
orders are expectation-driven, naturally forward-looking and known beforehand,
as they reflect planning and some anticipation of the market at the end of the
period, when the order is to be fulfilled. This information is leveraged for predicting the pre-order structure in the future by also taking into account the
cross-correlations between the pre-orders.

*Representation.* Let $q_t^h$ be the total quantity of some item *requested by the end of period t* through $h$ period advance pre-orders. Those made from the end of $t - h - 1$ until the end of $t - h$. The key property of the data set is that for every item the value $q_t^h$ is effectively known and available for use at period $t - h$, i.e. well before $t$. For example, $q_{t-1}^1$ is known at the end of $t - 2$ and corresponds to the quantity requested at the end of $t - 1$ accumulated via pre-orders made by the end of $t - 2$. In the definition of the data set structure the index $t$ corresponds to the "period" field, whereas $h$ is the "delivery date" and $q_t^h$ is the value in the "quantity" field. The "item_code" key is intentionally omitted in order to simplify the explanation of the key aspects of the *Diagonal Feeding* representation. Prior discussion implies that the $h$-period advance pre-order to be fulfilled at the end of period $t$ seems to reflect certain beliefs about the market environment in period $t$. Therefore, the time series $q_t^{h+1}$ of the volume of period-$t$ pre-orders made $h+1$ periods in advance can be considered to be *anticipatory* in relation to the time series $q_t^h$. For example, $q_{t+1}^2$ is known with certainty at $t - 1$, represents the total demanded quantity by the end of $t + 1$ accumulated between the end of $t - 2$ and the end of $t - 1$, and effectively reflects information 2 periods ahead forward-looking on $q_{t+1}^0$. In order to be able to utilize the anticipatory nature of time series with different "delivery dates" and at the same time to be able to forecast the pre-order structure, the data set in the *Diagonal Feeding* ($t$-frontier) representation is represented as following:

$$
\begin{pmatrix} q_{t+0}^0 & q_{t+0}^1 & q_{t+0}^2 & q_{t+0}^3 \\ q_{t+1}^0 & q_{t+1}^1 & q_{t+1}^2 & q_{t+1}^3 \\ q_{t+2}^0 & q_{t+2}^1 & q_{t+2}^2 & q_{t+2}^3 \\ q_{t+3}^0 & q_{t+3}^1 & q_{t+3}^2 & q_{t+3}^3 \\ q_{t+4}^0 & q_{t+4}^1 & q_{t+4}^2 & q_{t+4}^3 \end{pmatrix} \rightarrow \begin{pmatrix} x_{t0} & x_{t1} & x_{t2} & x_{t3} \\ y_{t0} & x_{t4} & x_{t5} & x_{t6} \\ y_{t1} & y_{t2} & x_{t7} & x_{t8} \\ y_{t3} & y_{t4} & y_{t5} & x_{t9} \\ y_{t6} & y_{t7} & y_{t8} & y_{t9} \end{pmatrix} \rightarrow \begin{pmatrix} z_{00} & z_{01} & z_{02} & z_{03} \\ z_{10} & z_{11} & z_{12} & z_{13} \\ z_{20} & z_{21} & z_{22} & z_{23} \\ z_{30} & z_{31} & z_{32} & z_{33} \\ z_{40} & z_{41} & z_{42} & z_{43} \end{pmatrix} \quad (1)
$$

In Eq. 1, the target $y_t$ is the output. Its final representation is the lower diagonal $z$. Thus, $y_t$ represents the pre-order structure for the next 3 periods beginning with $t + 1$. The objective is to predict the lower diagonal of the matrix by triangularly reshaping the multivariate time series for each "item" and introducing some redundancy that localizes relevant anticipatory features. Since the quantity $q_t^h$ is known at time $t - h$, each diagonal $(q_{t+s+h}^h)_{h \geq 0}$ in the scheme above is *known* at $t + s$, $s \in \mathbb{Z}$. This representation allows the value $q_s^f$ (the volume of period-$s$ pre-orders made $f$ periods in advance) to be used for forecasting the value $q_t^h$, whenever $t - h > s - f$, i.e. the moment $s - f$, when $q_s^f$ is revealed, is strictly earlier than the moment $t - h$, when $q_t^h$ becomes known. Therefore, the aim is to predict $y_t = \mathrm{cat}\left(q_{t+h}^{:h}\right)_{h \geq 0}$, the pre-order structure for the next 4 periods from $t + 1$, based on the preorders $x_t = \mathrm{cat}\left(q_{t+h}^{h:}\right)_{h \geq 0}$ and their history, known by $t$, where *cat* stands for concatenation of vectors and $q_s^{:f} = (q_s^j)_{j=0}^{f-1}$ ; further, $q_t^{:0}$ is empty. The values in the target $y_t$ that seem to be most relevant for practical demand forecasting are on its main diagonal $z_{pj}$ with $p = j + 1$, since they are the earliest future volumes.

*Correlation Analysis.* The demand volume data set contains time series for each items with many of them having sparse volume records due *absent periods*. They represent zero-volume orders. Thus, the multivariate time series for each item and "delivery date" were rebuilt from the data set by explicitly setting the volume to zero at absent periods. To test the viability of the proposed representation, only items with at least 60% periods with actual non-zero recorded volume were kept. After filling the missing periods on the time series of each item $i$ respectively with zeroes, they were transformed into the diagonal representation $(x_t{}^i, y_t{}^i)_{t \in period_i}$ and then pooled into one sample $(x_s, y_s)$. The pre-processed data set was split into development and holdout set, with the latter containing the last 8 periods available for each item (134–148 depending on the particular item).

*Results.* The Spearman Rank correlation[2] was computed on the development sub-sample to estimate the "informativeness"; it is invariant under monotonic transformations of the data and thus insensitive to scale. It was applied to the total demand data set $q_t^h$ for the selected subset of the items. A table containing the triangular grouping of the values is obtained, which the *Diagonal feeding* is built upon. Each value $(pr, ij)$ (left-right, top-bottom) in the table is the rank correlation between $q_{t+p}^i$ and $q_{t+r}^j$ computed across all periods $t$ and items. For instance, it shows that the most significant correlation ($\geq 0.9$) is between $x_{11} = q_{t+1}^1$ with $y_{10} = q_{t+1}^0$ computed over pooled sample of different *items* and *periods*. Since the values along the "delivery date" axis of the data set $q_t^h$ are in fact cumulative (gross), the correlation analysis is applied to the first differences $\delta_t^h = q_t^h - q_t^{h+1}$. Making this difference allows for measuring the pure anticipatory information content of $x_t$ in the target $y_t$ in the *Diagonal Feeding* representation of $\delta_t^h$. The correlations of this *differenced* data (net monthly orders) confirms that much of the observed correlation was due to the accumulated (integrated) nature of the $q_t^h$ data. The dramatic drop in the correlation between $x_{11}$ and $z_{10}$ shows that the orders $\delta_t^0$ of the *net current* periods are poorly predictable by the 1-period ahead net pre-orders $\delta_t^1$ meaning that they are likely to have different drivers. Nevertheless, the relatively small change in values within the sample Spearman rank correlation on the $q_t^h$ data and its counterpart on the $\delta_t^h$ data demonstrates that the 1 period ahead net pre-orders $\delta_t^1$ and lagged net current period orders $\delta_{t-1}^0$ have generally a little more predictive content for the future pre-order structure than the pre-orders made more than 1 period in advance.

## 4.1  Experimental Results

The *Diagonal Feeding* transforms the demand forecasting problem from a time series analysis setting, where the goal is to model $q_t^h \sim g_h\left((q_s^f)_{s-f<t-h}\right)$ simultaneously for every $h = 0, \ldots, 3$, to a supervised multi-output regression problem, where it seeks to learn $y_t \sim g(x_t, x_{t-1}, \ldots)$. Extensive numerical experiments were conducted using various common regression models involving

---

[2] http://bit.ly/2zLvNNf.

cross-validated grid search over the hyper-parameters. Besides the described pre-process, no further data-engineering was carried out. The best estimator in each model class was picked via 10-fold cross-validation on the development set (the first 38 observed periods of each selected item). The final test scores were computed on the held out data set using by averaging SMAPE across periods and items:

$$s_j = \frac{1}{|items|} \sum_{i \in items} s_{ji}, \quad s_{ji} = \frac{1}{|test_i|} \sum_{t \in test_i} s_{jit}, \quad s_{jit} = \frac{2|y_{tj}^i - \hat{y}_{tj}^i|}{|y_{tj}^i| + |\hat{y}_{tj}^i|},$$

where $\hat{y}_{tj}^i$ is the forecast integer volume of the $j$-th element in the advance preorder structure of item $i$ in the forecast starting with the future period $t$. An assessment of the SMAPE scores of the best models across all the targets in the diagonal representation $z_{pj}$ for $p > j$ reveals that the best model is a collection of 10 independent large ensembles of the gradient boosted regression trees with 500 estimators in each ensemble. The next best model with generally lower scores is the 2-layer dense ReLu network with $(80, 20)$ hidden units each. A histograms of the computed $s_{ji}$ for each $j$ in the structure of the gradient boosted ensemble will depict that most accurate predictions are for the next periods's gross order volume $(q_{t+1}^0)$. However, the best prediction accuracy drops dramatically when the full grid search experiment repeated for the *differenced* $q_t^h$ data $(\delta_t^h)$. An evaluation of the holdout SMAPE scores for the best models on the $\delta_t^h$ reveals that even the best model, $k$-nearest neighbor regression, completely fails to predict the net volume for the next period.

## 5    Experiments

To validate *Diagonal Feeding*, this study carried out an assessment of various methods for demand forecasting. In total, thirteen different methods were assessed. For conciseness, this work focuses on the third data set seen in Fig. 1 and on its delivery date 0 (same period). The evaluated methods are (1) Adaboost, (2) ARIMAX, (3) ARIMA, (4) Bayesian Structural Time Series (BSTS), (5) Bayesian Structural Time Series with a Bayesian Classifier (BSTS Classifier), (6) Ensemble of Gradient Boosting (Ensemble), (7) Ridge regression (Ridge), (8) Kernel regression (Kernel), (9) Lasso, (10) Neural Network (NN), (11) Poisson regression (Poisson), (12) Random Forest (RF), (13) Support Vector Regression (SVR).

Each of them had as a target value three different options: (a) Quantity (non-transformed), (b) Log-transformed quantity, (c) Min-Max transformed quantity. Additionally, *Diagonal Feeding*, presented in Sect. 4, was evaluated for regression methods. Thus, three settings were considered: (a) No *Diagonal Feeding*, (b) *Diagonal Feeding* with an item by item training (One by One). In this case, a vector containing the input of a specific item is fed individually to a model, (c) *Diagonal Feeding* fitting the model on the full data set (All Items). Here, a matrix with the input from all items is used. In all three cases, an individual vector corresponding to a given item is obtained as an output. For (a), extensive feature engineering was conducted and 360 features were generated. The specific features are documented in the code base provided. The training set consisted of 37 periods and the test set of 8. To evaluate the performance of the models, the Symmetric Mean Absolute Percent Error (SMAPE) is used. It is defined as $\text{SMAPE} = \frac{200\%}{n} \sum_{t=1}^{n} \frac{|F_t - A_t|}{|A_t| + |F_t|}$ with $F_t$ being the forecasted value and $A_t$ the actual value at time $t$ respectively. The results reveal that the median SMAPE for all methods is 0,42 and for methods with *Diagonal Feeding* is 0,43. The best Top 10 models had a median SMAPE of 0,31 and the Top 10 for those using Diagonal Feeding exclusively was 0,37.

## 6    Discussion

*Diagonal Feeding.* The key insight from the analysis of the data set is that the next period's gross total demand volume $q_{t+1}^0$ is mostly determined by the currently known one-period ahead pre-orders for the period $(q_{t+1}^1)$. The correlation analysis and the results of the grid search experiment confirm the observation that the net next period's volume $\delta_{t+1}^0$ is the difference between $q_{t+1}^0$ and $q_{t+1}^1$. Viewed through *Diagonal Feeding*, it is mostly independent of the history of net pre-orders for the period $t+1$ and is thus less predictable from advance pre-order data, as indicated by a correlation analysis and results from a grid search experiment. The success of forecasting the $q_{t+1}^0$, especially in contrast to the other next period's pre-order volumes $q_{t+1+j}^j$ for $j \in \{1, 2, 3\}$, might be attributed to the observed high correlation of the one-period ahead pre-order volume $q_{t+1}^1$. Further, *Diagonal Feeding* delivers results comparable to those obtained doing extensive feature engineering; in this case, more than 300 features were generated. Along these lines, exploring different transformations of the target value is essential. For some models using log-transform or min-max delivered good results. For example, a Neural Network without a transformed quantity fitted on the full data set had a SMAPE of 1,13, with a log transformation, it was 0,38.

*Experiments.* The best model was Adaboost with an SMAPE of 0,17. It was followed by the Ensemble of Random Forests with 0,18. Yet, it is important to remember that these models had extensive feature engineering with over 300 features generated. For *Diagonal Feeding*, the best method was a random forest with log transform and fitted on the full data set. It obtained 0,34. This was significantly better than the average of methods trained on the data set with feature engineering, which obtained a significantly higher SMAPE of 0,42.

## 7    Conclusion

This work introduced *Diagonal Feeding*, a technique specially useful for forecasting Build-To-Order products. It helps improving accuracy whenever future delivery dates are known. This approach does not require domain knowledge, extensive feature engineering or advanced technical skills. The results from multiple experiments show that there is no go-to technique for time series prediction. In addition, this research made available a highly-relevant and novel data set. A challenge in developing methods for demand forecasting of BTO products is the lack of public data sets. As a further line of work, it is worth exploring the impact on *Diagonal Feeding* of transforming the target variable, as it was shown, certain transformations perform better than others. From an algorithmic point of view the approach can be strengthened with non-parametric pre-processing techniques to filter out anomalies such as in [3, 16, 26, 29], including multichannel anomaly detection, [9], performing online aggregation of different forecasting models via long-term aggregation strategies, [20], along with approaches to model quasi-periodic data, [6], and extraction of trends in the presence of non-stationary noise with long tails, [4, 5]. Given the potential of BTO supply chains, it is expected that an increasing number of researchers will direct their attention to this area and add additional entries to the literature.

**Acknowledgements.** The research in Sect. 2 was partially supported by the Russian Foundation for Basic Research grant 16-29-09649 ofi m. The research presented in other sections was supported by the Mexican National Council for Science and Technology (CONACYT), 2018-000009-01EXTF-00154.

# References

1. Agell, N., Carricano, M.: Adopcion e impacto del Big Data y Advanced Analytics en Espana, ESADE, Institute for Data-Driven Decisions (2018)
2. Ahmed, N.K., et al.: An empirical comparison of machine learning models for time series forecasting. Economet. Rev. **29**(5–6), 594–621 (2010)
3. Artemov, A., Burnaev, E.: Ensembles of detectors for online detection of transient changes. In: Gammerman, A., et al. (eds.) Eighth International Conference on Machine Vision (2015)
4. Artemov, A., Burnaev, E.: Detecting performance degradation of software-intensive systems in the presence of trends and long-range dependence. In: 2016 IEEE 16th ICDMW, pp. 29–36 (2016)
5. Artemov, A., Burnaev, E.: Optimal estimation of a signal perturbed by a fractional Brownian noise. Theor. Probab. Appl. **60**(1), 126–134 (2016)
6. Artemov, A., et al.: Nonparametric decomposition of quasi-periodic time series for change-point detection. In: Proceedings of the SPIE, vol. 9875 (2015)
7. Attar, K.S.: Proposed methodology of intelligent system using support vector regression for demand forecasting. Int. J. ETTCS (IJETTCS) **5**(1), 122–123 (2016)
8. Bogataj, D., Bogataj, M.: NPV approach to material requirements planning theory - a 50-year review of these research achievements. Int. J. Prod. Res. 1–17 (2018)
9. Burnaev, E.V., Golubev, G.K.: On one problem in multichannel signal detection. Probl. Inf. Transm. **53**(4), 368–380 (2017)
10. Chase Jr., C.W.: Demand-Driven Forecasting: A Structured Approach to Forecasting. Wiley, Hoboken (2013)

11. Christensen, W.J., et al.: Build-to-order and just-in-time as predictors of applied supply chain knowledge and market performance. J. Oper. Manag. **23**(5), 470–481 (2005)
12. Fleisch, E., Tellkamp, C.: Inventory inaccuracy and supply chain performance: a simulation study of a retail supply chain. Int. J. Prod. Econ. **95**(3), 373–385 (2005)
13. Gilliand, M.: Business Forecasting. Wiley, Hoboken (2015)
14. Corsten, D., Gruen, T.: On shelf availability: an examination of the extent, the causes, and the efforts to address retail out-of-stocks. In: Consumer Driven Electronic Transformation, pp. 131–149 (2005)
15. Gunasekaran, A., Ngai, E.W.T.: Build-to-order supply chain management: a literature review and framework for development. J. Oper. Manag. **23**(5), 423–451 (2005)
16. Ishimtsev, V., et al.: Conformal k-NN Anomaly Detector for Univariate Data Streams. http://arxiv.org/abs/1706.03412 (2017)
17. Jain, C.L.: Benchmarking new product forecasting. J. Bus. Forecast. Methods Syst. **25**(4), 22 (2006)
18. Kahn, K.B.: An exploratory investigation of new product forecasting practices. J. Prod. Innov. Manag. **19**(2), 133–143 (2002)
19. Kaul, R.: Retail out-of-stock management: an outcome-based approach, WiPro (2013)
20. Korotin, A., et al.: Aggregating strategies for long-term forecasting. In: Gammerman, A., et al. (eds.) Proceedings of the Seventh Workshop on Conformal and Probabilistic Prediction and Applications, pp. 63–82 (2018)
21. Lu, C.-J., et al.: Sales forecasting for computer wholesalers: a comparison of multivariate adaptive regression splines and artificial neural networks. Decis. Support Syst. **54**(1), 584–596 (2012)
22. Gahirwal, M.: Inter time series sales forecasting. http://arxiv.org/abs/1303.0117 (2013)
23. Pompa, C., Burke, T.: Data science and analytics skills shortage: equipping the APEC workforce with the competencies demanded by employers. APEC Human Resource Development Working Group (2017)
24. Rivera, R., Burnaev, E.: Forecasting of commercial sales with large scale Gaussian Processes. In: Proceedings of the IEEE 17th ICDMW, pp. 625–634 (2017)
25. Rivera, R., et al.: Towards forecast techniques for business analysts of large commercial data sets using matrix factorization methods. J. Phys. Conf. Ser. **1117**(1), 012010 (2018)
26. Safin, A.M., Burnaev, E.: Conformal kernel expected similarity for anomaly detection in time-series data. Adv. Syst. Sci. Appl. **17**(3), 22–33 (2017)
27. Sharma, A., LaPlaca, P.: Marketing in the emerging era of build-to-order manufacturing. Ind. Mark. Manag. **34**(5), 476–486 (2005)
28. Swaminathan, J.M., Tayur, S.R.: Models for supply chains in E-Business. Manag. Sci. **49**(10), 1387–1406 (2003)
29. Volkhonskiy, D., et al.: Inductive conformal martingales for change-point detection. In: Gammerman, A., et al. (eds.) Proceedings of the Sixth Workshop on Conformal and Probabilistic Prediction and Applications, pp. 132–153 (2017)
30. Wan, X.-L., et al.: Exploring an interactive value-adding data-driven model of consumer electronics supply chain based on least squares support vector machine. Sci. Program. **2016**, 1–13 (2016)
31. Yelland, P.M., et al.: A Bayesian model for sales forecasting at sun microsystems. Inf. J. Appl. Anal. **40**(2), 118–129 (2010)

# Imputation Using a Correlation-Enhanced Auto-Associative Neural Network with Dynamic Processing of Missing Values

Xiaochen Lai[1,4], Xia Wu[1], Liyong Zhang[2(✉)], and Genglin Zhang[3]

[1] School of Software, Dalian University of Technology, Dalian 116620, China
laixiaochen@dlut.edu.cn, wxwuxiaa@gmail.com
[2] School of Control Science and Engineering, Dalian University of Technology,
Dalian 116624, China
zhly@dlut.edu.cn
[3] Dalian Chinacreative Technology Co., Ltd., Dalian 116021, China
zhanggl@china-creative.net
[4] Key Laboratory for Ubiquitous Network and Service Software of Liaoning Province,
Dalian 116620, China

**Abstract.** The missing value is a common phenomenon in real-world datasets, which makes the analysis of incomplete data become an active research area. In this paper, a correlation-enhanced auto-associative neural network (CE-AANN) is proposed for imputations of missing values. We design correlation-enhanced hidden neurons and combine them with traditional hidden neurons organically, thereby constructing CE-AANN. Compared with the traditional auto-associative neural network (AANN), the improved architecture can mine cross-correlations among attributes more effectively. The introduction of correlation-enhanced hidden neurons keeps the network from learning a meaningless identity mapping. Moreover, a training scheme named MVPT is used for network training. Missing values are regarded as variables of the loss function and adjusted dynamically based on optimization algorithms. The dynamic processing mechanism takes account of the incompleteness of data during training, which makes the imputation accuracy increase as the training goes further. Experiments validate the effectiveness of the proposed method.

**Keywords:** Incomplete data · Missing value imputation ·
Auto-associative neural network · Dynamic processing mechanism

## 1 Introduction

Real-world datasets in the industrial, medical, and financial fields are susceptible to missing values. The presence of missing values reduces the quality of data and further affects the accuracy of analysis results. As an active research area, neural networks provide effective solutions for missing value imputation [1,2].

© Springer Nature Switzerland AG 2019
H. Lu et al. (Eds.): ISNN 2019, LNCS 11554, pp. 223–231, 2019.
https://doi.org/10.1007/978-3-030-22796-8_24

The auto-associative neural network (AANN) learns relationships among attributes by a multi-layer perceptron whose numbers of nodes in the input and output layers equal to the dimension of attributes, then imputes missing values based on these relationships. The imputation method based on AANN can be regarded as a two-stage method composed of training stage and imputation stage [3]. It trains the network by only complete records, then inputs the pre-processed incomplete records into the trained network and uses the corresponding network outputs to impute missing values. In the imputation stage, mean imputation and K-means imputation are usually adopted to pre-fill missing values since incomplete records cannot be taken as the network inputs directly. Considering that there may exist a certain estimation error between pre-filled values and real values, [4] inputs the incomplete records into the trained network and treats missing values as variables of the loss function, then works out the optimal results through the genetic algorithm. Subsequently, several researchers made improvements based on the above idea [5,6].

Nowadays, the family of AANN architectures has obtained great achievements in the field of missing value imputation. Researchers construct several variations of AANN based on radial basis function neural network [4], general regression neural network [7], and counter-propagation neural network [8], etc., and they have achieved ideal performance of imputation.

Since in AANN, the network output has a certain dependence on its corresponding input. In order to avoid learning a meaningless identity mapping, the number of hidden neurons and the capacity of the network are always limited. In this paper, we design the correlation-enhanced hidden neuron and combine improved neurons and traditional neurons organically in the hidden layer, thereby building a correlation-enhanced auto-associative neural network (CE-AANN). The architecture uses the correlation among attributes to constrain the excessive dependence of network output on its corresponding input, which reduces the constraint of network architecture effectively. Moreover, a training scheme MVPT is used for maximizing the use of present values, in which the whole incomplete dataset is allowed to participate in network training, and missing values are regarded as variables adjusted dynamically by optimization algorithms.

The rest of this paper is organized as follows. Section 2 introduces the traditional AANN. Section 3 presents the architecture CE-AANN and the training scheme MVPT in detail. Section 4 discusses the imputation performance of the proposed method. Finally, conclusions are presented in Sect. 5.

## 2    The Auto-Associative Neural Network

AANN is a specific type of feed-forward neural networks where the output attempts to reconstruct its corresponding input, and the structure is shown in Fig. 1(a). The dimension of the input space $s$ is 5; $x = [x_1, x_2, ..., x_5]^T$ is the network input; $y = [y_1, y_2, ..., y_5]^T$ is the network output. The number of hidden neurons is usually smaller than those of the input and output neurons, which makes the network a bottleneck structure. The loss function of AANN is defined by

$$L(W) = \frac{1}{2}\|x - f(x, W)\|^2, \tag{1}$$

where $W$ is the set of network parameters; $x \in \Re^s$, representing the $s$-dimensional input vector; $f(\cdot) : \Re^s \to \Re^s$, representing the mapping from input to output.

Considering the case in Fig. 1(b) where the dimension of the hidden layer is larger than that of the input layer, the input node and the corresponding output node are connected directly if the weights of red lines tend to 1 and those of black lines tend to 0. As a result, AANN can achieve the reconstruction simply through an identity mapping, which may not mine the real association within data. And the above situation is more frequent when the network is over-fitted [9].

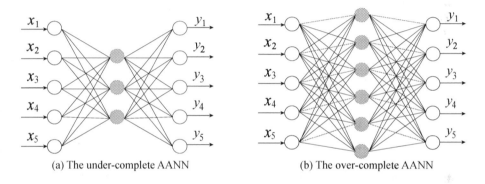

(a) The under-complete AANN      (b) The over-complete AANN

**Fig. 1.** The auto-associative neural network.

The training process of AANN involves a compromise between two forces. First, the network should reconstruct the inputs approximately. Second, architectural constraints such as limiting the dimension of hidden layer should be satisfied, which avoids learning an identity mapping [10]. Considering these two forces, we design the correlation-enhanced hidden neuron which can constrain the copy ability of the model output to its corresponding input by enhancing cross-correlations among attributes. Then, we combine the traditional hidden neurons and the improved neurons together, thus to improve the reconstitution capacity of the model while weakening the restriction on hidden layers.

## 3 CE-AANN-Based Imputation Scheme

### 3.1 Correlation-Enhanced Hidden Neuron

The hidden neurons are presented in Fig. 2, in which $0 < i < n$, with $n$ being the number of records in a dataset; $0 < j < s$, with $s$ being the dimension of records;

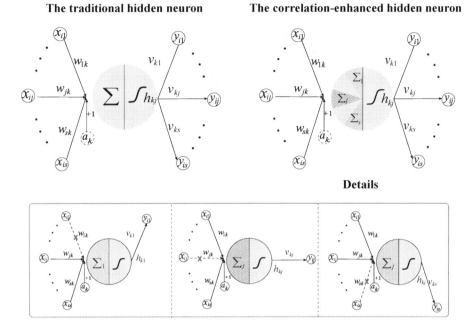

**Fig. 2.** Two types of hidden neurons.

$x_i=[x_{i1}, x_{i2}, ..., x_{is}]^T$ is the $i$th $s$-dimensional record; $y_i=[y_{i1}, y_{i2}, ..., y_{is}]^T$ is the network output; $w_{jk}$ is the weight of connection from the $j$th input neuron to the $k$th hidden neuron; $a_k$ is the threshold of the $k$th hidden neuron; $v_{kj}$ is the weight of connection from the $k$th hidden neuron to the $j$th output neuron; $h_{kj}$ is the output of the $k$th hidden neuron with respect to the $j$th output neuron. In the traditional hidden neuron, $h_{kj}$ is defined by

$$h_{kj} = \sigma(\sum_{l=1}^{s} w_{lk}x_{il}+a_k), \tag{2}$$

where $\sigma(\cdot)$ is the activation function of hidden neurons.

The red cross in Fig. 2 indicates that the connection from the input neuron to the hidden neuron is disabled when calculating the output $h_{kj}$ of the neuron. It means that the improved neuron performs different summation operations for different output neurons, and then passes these values into the activation function for calculating $h_{kj}$. Hence, $h_{kj}$ in the improved neuron is defined by

$$h_{kj} = \sigma(\sum_{l=1,l\neq j}^{s} w_{lk}x_{il}+a_k). \tag{3}$$

It can be seen from (3), we remove the dependence of $y_{ij}$ on $x_{ij}$, which avoids the direct connection from the output neuron to its corresponding input

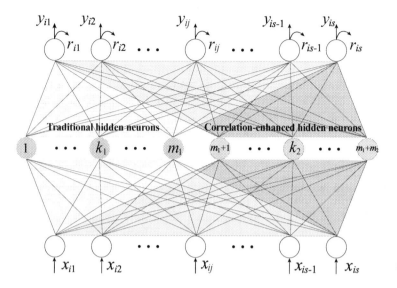

**Fig. 3.** The correlation-enhanced auto-associative neural network.

neuron when the dimension of hidden layer is larger than that of input layer. The improved neuron can be regarded as a multi-task learning structure which learns the correlation between each attribute and the other attributes in parallel, thereby avoiding the over-fitting of a single task.

### 3.2 The Architecture of CE-AANN

The architecture of CE-AANN is shown in Fig. 3. It can be seen that the network has three layers, and the hidden layer contains two types of neurons: $m_1$ traditional neurons and $m_2$ improved neurons. $y_i = [y_{i1}, y_{i2}, ..., y_{is}]^T$ is the network output calculated by only traditional neurons; $r_i = [r_{i1}, r_{i2}, ..., r_{is}]^T$ is the reference output calculated by only improved neurons.

The network output $y_{ij}$ is

$$y_{ij} = \theta(\sum_{k=1}^{m_1} h_{kj}v_{kj} + b_j), \qquad (4)$$

where $\theta(\cdot)$ is the activation function of output neurons; $b_j$ is the threshold of the $j$th output neuron. Besides, the corresponding reference output $r_{ij}$ is

$$r_{ij} = \theta(\sum_{k=m_1+1}^{m_1+m_2} h_{kj}v_{kj} + b_j). \qquad (5)$$

## 3.3    The Training Scheme MVPT

In this paper, we take the entire incomplete dataset as the training set and treat missing values as variables of the loss function. Let $E_i$ be the error between the input $x_i$ and the output $y_i$, which is defined as

$$E_i = \sum_{x_{ij} \in X_P} \frac{1}{2} \left[ (y_{ij} - x_{ij})^2 + (y_{ij} - r_{ij})^2 \right] + \sum_{x_{ij} \in X_M} \frac{1}{2} \left[ (y_{ij} - \widehat{x}_{ij})^2 + (y_{ij} - r_{ij})^2 \right],$$

$$(6)$$

in which $X_P$ is the set of present values; $X_M$ is the set of missing values; $\widehat{x}_{ij}$ is the dynamic estimate of a missing value $x_{ij}$. When adopting the gradient descent algorithm, we need to calculate derivatives of model parameters and missing values with respect to the loss function, as shown in (7)–(11).

$$\frac{\partial E_i}{\partial b_j} = (2y_{ij} - r_{ij} - I_{ij} \cdot x_{ij} - (1 - I_{ij}) \cdot \widehat{x}_{ij}) \frac{\partial y_{ij}}{\partial b_j} + (r_{ij} - y_{ij}) \frac{\partial r_{ij}}{\partial b_j},$$

$$(7)$$

where $I_{ij} = \begin{cases} 1, x_{ij} \in X_P \\ 0, x_{ij} \in X_M \end{cases}$. $\frac{\partial y_{ij}}{\partial b_j}, \frac{\partial r_{ij}}{\partial b_j}$ are determined by the activation function $\theta(\cdot)$ in (4). Since it is set as the linear function in this paper, $\frac{\partial y_{ij}}{\partial b_j} = 1$, and $\frac{\partial r_{ij}}{\partial b_j} = 1$.

$$\frac{\partial E_i}{\partial v_{kj}} = \begin{cases} (2y_{ij} - r_{ij} - I_{ij} \cdot x_{ij} - (1 - I_{ij}) \cdot \widehat{x}_{ij}) \cdot \frac{\partial y_{ij}}{\partial b_j} \cdot h_{kj}, & 0 < k \leq m_1 \\ (r_{ij} - y_{ij}) \cdot \frac{\partial r_{ij}}{\partial b_j} \cdot h_{kj}, & m_1 < k \leq m_1 + m_2 \end{cases}.$$

$$(8)$$

$$\frac{\partial E_i}{\partial a_k} = \sum_{l=1}^{s} \left( \frac{\partial E_i}{\partial v_{kl}} \cdot v_{kl} \cdot \frac{\partial h_{kl}}{\partial a_k} / h_{kl} \right),$$

$$(9)$$

where $\frac{\partial h_{kl}}{\partial a_k}$ is determined by the activation function $\sigma(\cdot)$ shown in (2). Since it is set as the sigmoid function, thus $\frac{\partial h_{kl}}{\partial a_k} = h_{kl} \cdot (1 - h_{kl})$.

$$\frac{\partial E_i}{\partial w_{jk}} = \begin{cases} \frac{\partial E_i}{\partial a_k} (I_{ij} \cdot x_{ij} + (1 - I_{ij}) \cdot \widehat{x}_{ij}), & 0 < k \leq m_1 \\ \sum_{l=1, l \neq j}^{s} \left( \frac{\partial E_i}{\partial v_{kl}} \cdot v_{kl} \cdot \frac{\partial h_{kl}}{\partial a_k} / h_{kl} \right), & m_1 < k \leq m_1 + m_2 \end{cases}.$$

$$(10)$$

$$\frac{\partial E_i}{\partial \widehat{x}_{ij}} = \sum_{k=1}^{m_1 + m_2} \left( \frac{\partial E_i}{\partial a_k} \cdot w_{jk} \right) - (r_{ij} - y_{ij}) \cdot \frac{\partial r_{ij}}{\partial b_j} \cdot \sum_{k=m_1+1}^{m_1+m_2} \left( v_{kj} \frac{\partial h_{kj}}{\partial a_k} w_{jk} \right) - (y_{ij} - \widehat{x}_{ij}).$$

$$(11)$$

# 4   Results and Analyses

## 4.1   The Design of Experiments

The compared methods include (1) Mean imputation (Mean), (2) Mean pre-filling based AANN imputation (Mean+AANN), (3) Genetic algorithm-based AANN imputation (AANN+GA) [4], (4) MVPT-based AANN imputation (AANN+MVPT), and (5) MVPT-based CE-AANN imputation (CE-AANN+MVPT). We take four datasets from the UCI machine learningrepository, as shown in Table 1. Incomplete datasets are generated manually based on complete datasets, and missing rates are set as 5%, 10%, 20%, 30%, 40%, and 50%.

Table 1. The description of the datasets.

| Dataset | Record | Attribute | Dataset | Record | Attribute |
|---|---|---|---|---|---|
| Glass identification (glass) | 214 | 9 | Iris | 150 | 4 |
| Concrete slump test (slump) | 103 | 10 | Seeds | 210 | 7 |

The imputation performance is measured by Mean Absolute Error (MAE):

$$MAE = \frac{1}{n_{miss}} \sum_{x_{ij} \in X_M} \left| \hat{x}_{ij} - t_{ij} \right|, \tag{12}$$

where $n_{miss}$ is the number of missing values; $\hat{x}_{ij}$ is the estimate of missing value; $t_{ij}$ is the corresponding true value. We set ranges of hyperparameters, as listed in Table 2, and select the optimal hyperparameters by 10-fold cross validation.

Table 2. The ranges of the hyperparameters.

| Parameters | Ranges |
|---|---|
| Learning rate | for 0.01 to 1.0, in 0.01 steps |
| Momentum factor | for 0.1 to 1.0, in 0.1 steps |
| The dimension of hidden nodes $m_1$, $m_2$ | 5, 10, 15, 20, 25, 30 |
| The number of training epochs $e$ | 30000 |

## 4.2   The Analyses of Results

The results are shown in Fig. 4. Compared with MAE values obtained from CE-AANN+MVPT and the other methods, we can find that CE-AANN+MVPT has the best imputation performance under different missing rates.

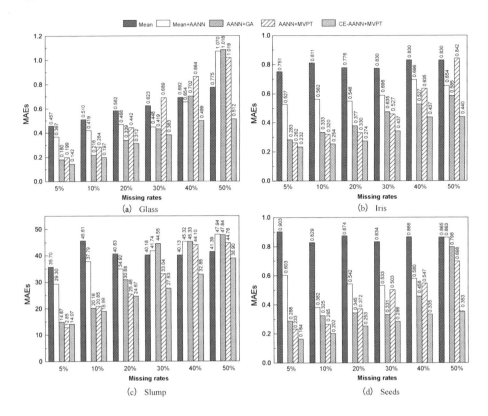

**Fig. 4.** The experimental results obtained from five imputation methods.

It can be seen that when the missing rate is lower than 40%, AANN+MVPT is obviously superior to the other three methods except CE-AANN+MVPT. It indicates that the dynamic processing mechanism of missing values takes account of the incompleteness of data, which improves the imputation accuracy effectively. However, when the missing rate gets large, AANN is prone to overfitting, thus further affecting the imputation performance.

According to results obtained from CE-AANN+MVPT and AANN+MVPT, the imputation accuracy of CE-AANN+MVPT is better than AANN+MVPT when adopting the same training scheme. CE-AANN uses the correlations among attributes to constraint theexcessive dependence of the network output to its corresponding input, which helps to improve the imputation performance.

## 5    Conclusions

In this paper, an improved architecture CE-AANN trained by the scheme MVPT is proposed for imputation of missing values. Experiments show that the proposed method has an ideal performance of imputation. The main reasons are as follows. First, the introduction of correlation-enhanced hidden neurons into CE-AANN reduces the constraint on the dimension of hidden layers. Meanwhile, it improves the capacity of the network to mine the cross-correlation among attributes. Second, the scheme MVPT where missing values are treated as variables makes full consideration of the incompleteness of data during training. The imputation accuracy increases gradually as the training goes further.

**Acknowledgement.** This work was supported by National Key R&D Program of China (2018YFB1700200).

# References

1. Garcíalaencina, P.J., Sanchogómez, J., Figueirasvidal, A.R.: Pattern classification with missing data: a review. Neural Comput. Appl. **19**(2), 263–282 (2010)
2. Masoud, S.A., Negin, D.: Missing value imputation using a novel grey based fuzzy c-means, mutual information based feature selection, and regression model. Expert Syst. Appl. **115**, 68–94 (2019)
3. Azim, S., Aggarwal, S.: Using fuzzy c means and multi layer perceptron for data imputation: simple v/s complex dataset. In: 2016 3rd International Conference on Recent Advances in Information Technology (RAIT), pp. 197–202. IEEE (2016)
4. Abdella, M., Marwala, T.: The use of genetic algorithms and neural networks to approximate missing data in database. In: 2015 3rd IEEE International Conference on Computational Cybernetics (ICCC), pp. 207–212. IEEE (2005)
5. Nelwamondo, F.V., Golding, D.: A dynamic programming approach to missing data estimation using neural networks. Inf. Sci. **237**, 49–58 (2013)
6. Aydilek, I.B., Arslan, A.: A novel hybrid approach to estimating missing values in databases using k-nearest neighbors and neural networks. Int. J. Innovative Comput. Inf. Control **7**(8), 4705–4717 (2012)
7. Ravi, V., Krishna, M.: A new online data imputation method based on general regression auto associative neural network. Neurocomputing **138**, 106–113 (2014)
8. Gautam, C., Ravi, V.: Counter propagation auto-associative neural network based data imputation. Inf. Sci. **325**, 288–299 (2015)
9. Mistry, F.J., Nelwamondo, F.V., Marwala, T.: Missing data estimation using principle component analysis and autoassociative neural networks. J. Syst. Cybern. Inf. **7**(3), 72–79 (2009)
10. Goodfellow, I., Bengio, Y., Courville, A.: Deep Learning. MIT press, Cambridge (2016)

# Broad Learning for Optimal Short-Term Traffic Flow Prediction

Di Liu[1], Wenwu Yu[1,2(✉)], and Simone Baldi[2,3]

[1] School of Cyber Science and Engineering, Southeast University,
Nanjing 210096, China
liud923@126.com, wwyu@seu.edu.cn
[2] School of Mathematics, Southeast University, Nanjing 210096, China
s.baldi@tudelft.nl
[3] Delft Center for Systems and Control, Delft University of Technology,
2628 CD Delft, The Netherlands

**Abstract.** In this work, we explore the use of a Broad Learning System (BLS) as a way to replace deep learning architectures for traffic flow prediction. BLS is shown to not only outperforms standard learning algorithms (Least absolute shrinkage and selection operator (LASSO), shallow and deep neural networks, stacked autoencoders) in terms of training time, but also in terms of testing accuracy.

**Keywords:** Broad Learning System · Traffic flow prediction ·
Flat network · Fast least-square methods

## 1 Introduction

Traffic congestion is becoming a serious problem due to the constantly increasing urban traffic volumes and incomplete transportation management system. It has been shown that precise real-time (15–40 min) traffic flow prediction (also called short-term traffic flow prediction) can help the travelers plan the better path in advance [21].

Traffic flow prediction is a complex process since it is affected by many different factors, such as traffic volumes, weather conditions, seasonal and weekly variability. In recent years, machine learning methods have been widely adopted to address such complexity, such as support vector machine (SVM) [7], LASSO [22], neural networks (NN) [2], etc. Among these methods, neural networks are considered to be the most popular. However, neural networks with shallow structures show deficiencies in capturing complex rules hidden traffic data [13].

This work was supported by the National Natural Science Foundation of China under Grant No. 61673107, the National Ten Thousand Talent Program for Young Top-notch Talents under Grant No. W2070082, the General joint fund of the equipment advance research program of Ministry of Education under Grant No. 6141A020223, and the Jiangsu Provincial Key Laboratory of Networked Collective Intelligence under Grant No. BM2017002.

© Springer Nature Switzerland AG 2019
H. Lu et al. (Eds.): ISNN 2019, LNCS 11554, pp. 232–239, 2019.
https://doi.org/10.1007/978-3-030-22796-8_25

In order to overcome the problems associated with shallow architectures, many researchers focused on the study of deep architectures (also called deep learning architectures or deep neural networks) [6,10,12,13,15,20]. Lv et al. presented a stacked autoencoder considering the spatial and temporal correlations to learn traffic flow features [13]. Polson and Sokolov developed a deep learning architecture able to capture special features of traffic flows under special events [15]. Koesdwiady et al. combined deep belief networks and data fusion decision-level to improve the traffic flow prediction accuracy [12]. Huang et al. combined deep belief network to obtain the effective traffic flow features and multitask regression for supervised prediction [10]. Yang et al. proposed a stacked autoencoder Levenberg-Marquardt model with optimized structure [19].

Although deep architectures have been shown to be effective, it is widely recognized that deep learning requires a time-consuming training process because of a large number of connecting parameters in filters and layers. In order to solve these time-consuming issues, a wide variety of methods have been proposed, such as exploiting sparsity and data structure [3,8,16]. However, with the growth of data size such methods may not sustain the original good performance any more.

Broad learning system (BLS) [4] is an emerging approach for effectively and efficiently modelling of complex systems. As compared to deep learning, where learning is improved by deepening the architecture of the neural network, in broad learning the learning is improved by extending the structure in the wide direction (or in the broad direction), i.e. by increasing the number of features and enhancement nodes [5].

In this paper, we explore the use of broad learning systems for short-term spatial and temporal traffic flow prediction. As compared to the deep learning structures used for traffic flow prediction, we are able to exploit fast least-square methods. To the best of the authors knowledge, this is the first time that BLS is applied to predict spatial and temporal traffic flow.

The rest of the paper is organized as follows. Section 2 recalls the broad learning model and gives the problem formulation, Sect. 3 discusses the traffic flow prediction test case and simulations, and Sect. 4 concludes the work.

## 2    One Shot Broad Learning System

In this section, first, the details of one shot structure of BLS (without incremental learning) are given [4]. BLS is constructed on the random vector functional-link neural networks (RVFLNN) [11,14]. Figure 1 illustrates the main idea: the network comprises features nodes (nonlinear transformations of the input data) and enhancement nodes (nonlinear transformations of the feature node data). Both feature and enhancement nodes are used to approximate the output data.

Let $X \in \mathrm{R}^{S_{num} \times M}$ represents the input training data, $Y \in \mathrm{R}^{S_{num} \times N}$ represents the target training data, where $S_{num}$ is the number of samples, $M, N$ are the dimensions of the input and output respectively. At this point we can define $\varphi_i(XW_{e_i} + b_{e_i})$ as the $i$th mapped features $Z_i$,

$$Z_i = \varphi_i(XW_{e_i} + b_{e_i}), i = 1, ..., n \tag{1}$$

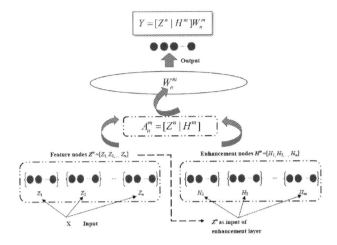

**Fig. 1.** Illustration of a Broad Learning System (BLS).

where $\varphi_i$ denotes the activation function. It is important to note that the weights $W_{e_i}$ and bias $b_{e_i}$ of the activation functions are randomly chosen (as typical in functional networks). After putting each $Z_i$ in row, we obtain $Z^n = [Z_1, Z_2, \cdots, Z_n]$, which denotes the whole set of feature nodes. At this point, $Z^n$ becomes the input to the enhancement nodes. Define the $j$th group enhancement nodes as

$$H_j = \zeta_j(Z^n W_{h_j} + b_{h_j}), j = 1, 2 \cdots, m \tag{2}$$

where $\zeta_j$ is the activation function, whose weights $W_{h_j}$ and bias $b_{h_j}$ are also randomly generated with proper dimensions. Similarly to what done before, connect all the $j$th group enhancement nodes in one row, so as to obtain $H^m = [H_1, H_2, \cdots, H_m]$. Then the one-shot broad learning system can be described as follows

$$
\begin{aligned}
Y &= [Z_1, \cdots, Z_n | \zeta(Z^n W_{h_1} + b_{h_1}), \cdots, \zeta(Z^n W_{h_m} + b_{h_m})] W^m \\
&= [Z_1, \cdots, Z_n | H_1, \cdots, H_m] W_n^m \\
&= [Z^n | H^m] W_n^m
\end{aligned} \tag{3}
$$

After defining $A_n^m = [Z^n | H^m]$, $W_n^m$ can be obtained through pseudoinverse of a partitioned matrix [1,4,5,9,18]

$$W_n^m = [(A_n^m)^T A_n^m + \lambda I]^{-1} (A_n^m)^T Y \tag{4}$$

where $\lambda$ is a positive constant representing the weight on the sum of squares. It is important to notice that the use of pseudoinverse or ridge regression represents a much faster way of training the network, as compared to gradient based algorithms. This is because the flat structure of the network allows for the use least-square based methods, which is not allowed in multilayer deep NN. The following algorithm summarized the training of a BLS.

---

**Algorithm 1.** Broad learning: one shot structure

---

1: input training data $X(k)$ and target matrix $Y(k)$
2: input testing data $X_t(k)$ and target matrix $Y_t(k)$
3: **for** $i = 1, i \leq n$
        Randomly initialize $W_{e_i}, b_{e_i}$
        Compute $Z_i = \varphi_i(XW_{e_i} + b_{e_i}), i = 1, ..., n$
        Obtain the feature mapping group

$$Z^n = [Z_1, Z_2, \cdots, Z_n]$$

4: **for** $j = 1, j \leq m$
        Randomly initialize $W_{h_j}, b_{h_j}$
        Compute $H_j = \zeta_j(Z^n W_{h_j} + b_{h_j}), j = 1, 2 \cdots, m$
        Obtain the enhancement nodes group

$$H^m = [H_1, H_2, \cdots, H_m]$$

5: Take $A_n^m = [Z^n | H^m]$, compute $(A_n^m)^+$
6: Output

$$W_n^m = [(A_n^m)^T A_n^m + \lambda I]^{-1}(A_n^m)^T Y$$

7: Testing
8: **for** $i = 1, i \leq n$
        Compute $Z_{t,i} = \varphi_i(X_t W_{e_i} + b_{e_i}), i = 1, ..., n$
        Obtain the feature mapping group

$$Z_t^n = [Z_{t,1}, Z_{t,2}, \cdots, Z_{t,n}]$$

9: **for** $j = 1, j \leq m$
        Compute $H_{t,j} = \zeta_j(Z_t^n W_{h_j} + b_{h_j}), j = 1, 2 \cdots, m$
        Obtain the enhancement nodes group

$$H_t^m = [H_{t,1}, H_{t,2}, \cdots, H_{t,m}]$$

10: Take $A_{t,n}^m = [Z_t^n | H_t^m]$, compute $\hat{Y}_t = A_{t,n}^m W_n^m$

---

## 3 Traffic Flow Prediction with BLS

### 3.1 Test Case

The Peformance Measurement System (PeMS) Data Source is one of the most popular datasets in the traffic field. The dataset is managed and updated by the California Department of Transportation (Caltrans). Data obtained from the Caltrans PeMS are collected in real-time from nearly $40,000$ individual detectors spanning the freeway system across all major metropolitan areas of the State of California.

    The dataset under consideration in this work is consistent with the one in [17]. It contains the flow, speed and occupancy rate at 33 different locations in $I405$ freeway (North-bound Interstate 405). The data are aggregated every

5 min, resulting in 30000 data samples for training and 10000 data samples for testing. Three different short term predictions are considered:

- Predicting the traffic flow 15 min ahead in time
- Predicting the traffic flow 30 min ahead in time
- Predicting the traffic flow 45 min ahead in time

The term 'Predicting the traffic flow' encompass predicting flow, speed and occupancy in all 33 locations: therefore it is not only a temporal prediction, but also a spatial one. As a comparison measure, we compare BLS against the 4 following standard prediction methods

(1) Backpropagation Neural Network (BPNN): train a shallow neural network (feedforward network with one hidden layer with 10 neurons), using the default Matlab options[1].
(2) Deep Neural Network (DNN): similar to the previous algorithm, but increasing the number of hidden layers (three hidden layers, with $5, 3, 2$ neurons respectively).
(3) Regularized least-squares regression (LASSO): using cross-validated fits (10-fold cross-validation with labeled predictor variables) and the elastic net method with $Alpha = 0.75$[2].
(4) Stacked Autoencoders (SAE): Training a stacked autoencoder (training the first autoencoder with hidden size of 10, training the second autoencoder with hidden size of 5, training the final softmax layer, forming the stacked neural network, perform backpropagation on the whole multilayer network)[3].

### 3.2  Comparison Results

The comparisons are provided in the following Tables 1, 2 and 3. As far as training time is concerned, it can be seen that the SAE method is the slowest (because it requires several training step), followed by the LASSO method and the BPNN. The DNN is faster than BPNN due to the relatively small structure. The BL-one shot is more than one order of magnitude faster.

The accuracy is measured according to the mean absolute error (MAE), calculated as

$$\text{MAE} = 1 - \frac{1}{S_{num}} \sum_{i=1}^{S_{num}} \frac{|y_i - x_i|}{\max\{|y_i|, 0.01\}} \tag{5}$$

---

[1] According to the documentation in https://www.mathworks.com/help/deeplearning/ref/train.html.

[2] Implemented in Matlab according to the documentation in https://www.mathworks.com/help/stats/lasso.html.

[3] The documentation for training a stacked autoencoder can be found in https://www.mathworks.com/help/deeplearning/examples/train-stacked-autoencoders-for-image-classification.html.

**Table 1.** Prediction accuracy comparisons for future 15 min prediction

| Method | Training time (s) | Training accuracy (%) | Test accuracy (%) |
|---|---|---|---|
| BPNN | 1852.3 | 91.11 | 86.05 |
| DNN | 615.17 | 91.24 | 86.78 |
| LASSO | 6188.7 | 89.86 | 84.28 |
| SAE | 7967.6 | 90.12 | 87.01 |
| BL-one shot | 29.37 | 91.24 | 87.43 |

**Table 2.** Prediction accuracy comparisons for future 30 min prediction

| Method | Training time (s) | Training accuracy (%) | Test accuracy (%) |
|---|---|---|---|
| BPNN | 1600.00 | 89.55 | 83.00 |
| DNN | 671.51 | 89.32 | 83.61 |
| LASSO | 6577.14 | 86.10 | 78.25 |
| SAE | 7766.74 | 87.53 | 83.66 |
| BL-one shot | 29.37 | 88.64 | 84.35 |

**Table 3.** Prediction accuracy comparisons for future 45 min prediction

| Method | Training time (s) | Training accuracy (%) | Test accuracy (%) |
|---|---|---|---|
| BPNN | 1779.70 | 87.70 | 79.46 |
| DNN | 734.36 | 87.54 | 80.84 |
| LASSO | 7660.45 | 82.00 | 72.34 |
| SAE | 7591.28 | 84.91 | 80.79 |
| BL-one shot | 30.77 | 86.01 | 81.70 |

where the max operator is used to avoid division by zero. Such performance measure can be calculated for both the training and the testing data.

As far as the training accuracy is concerned, the results are mixed: for the 15 min prediction, BLS and DNN have the same performance, whereas for 30 and 45 min BPNN provides the best performance. However, what clearly matters is the performance for the testing data. From here it can be clearly seen that BL-one shot outperforms all other methods, often by more than 1%. The other best performing algorithms are SAE and DNN.

## 4   Conclusions

In this work we have explored a Broad Learning System (BLS) as a way to replace deep learning architectures for traffic flow prediction. BLS not only outperforms standard learning algorithms (LASSO, shallow and deep neural networks, stacked autoencoders) in terms of training time, but also in terms of testing accuracy. Future work will focus on further improving accuracy by including weather factors or spatial/temporal features (such as seasonal and weekly variability).

# References

1. Ben-Israel, A., Greville, T.N.E.: Generalized Inverses: Theory and Applications. Wiley, New York (1974)
2. Chan, K.Y., Dillon, T.S., Singh, J., Chang, E.: Neural-network-based models for short-term traffic flow forecasting using a hybrid exponential smoothing and levenberg–marquardt algorithm. IEEE Trans. Intell. Transp. Syst. **13**(2), 644–654 (2012)
3. Chen, C.L.P., Zhang, C.Y., Chen, L., Gan, M.: Fuzzy restricted boltzmann machine for the enhancement of deep learning. IEEE Trans. Fuzzy Syst. **23**(6), 2163–2173 (2015)
4. Chen, C.P., Liu, Z.L.: Broad learning system: an effective and efficient incremental learning system without the need for deep architecture. IEEE Trans. Neural Netw. Learn. Syst. **29**(1), 10–24 (2018)
5. Chen, C.P., Wan, J.Z.: A rapid learning and dynamic stepwise updating algorithm for flat neural networks and the application to time-series prediction. IEEE Trans. Syst. Man Cybern. Part B Cybern. **29**(1), 62–72 (1999)
6. Chen, M., Yu, X., Liu, Y.: PCNN: deep convolutional networks for short-term traffic congestion prediction. IEEE Trans. Intell. Transp. Syst. **19**(11), 1–10 (2018)
7. Ghofrani, F., Jamshidi, A., Keshavarz-Haddad, A.: Internet traffic classification using hidden naive Bayes model. In: Electrical Engineering, Tehran, pp. 235–240. IEEE Press (2015)
8. Gong, M., Liu, J., Li, H., Cai, Q., Su, L.: A multiobjective sparse feature learning model for deep neural networks. IEEE Trans. Neural Netw. Learn. Syst. **26**(12), 3263–3277 (2017)
9. Hoerl, A., Kennard, R.: Ridge regression: biased estimation for nonorthogonal problems. Technometrics **12**(1), 55–67 (2000)
10. Huang, W., Song, G., Hong, H., Xie, K.: Deep architecture for traffic flow prediction: deep belief networks with multitask learning. IEEE Trans. Intell. Transp. Syst. **15**(5), 2191–2201 (2014)
11. Igelnik, B., Pao, Y.H.: Stochastic choice of basis functions in adaptive function approximation and the functional-link net. IEEE Trans. Neural Netw. **6**(6), 1320–1329 (1995)
12. Koesdwiady, A., Soua, R., Karray, F.: Improving traffic flow prediction with weather information in connected cars: a deep learning approach. IEEE Trans. Veh. Technol. **65**(12), 9508–9517 (2016)
13. Lv, Y., Duan, Y., Kang, W., Li, Z., Wang, F.Y.: Traffic flow prediction with big data: a deep learning approach. IEEE Trans. Intell. Transp. Syst. **16**(2), 865–873 (2015)
14. Pao, Y.H., Takefuji, Y.: Functional-link net computing: theory, system architecture, and functionalities. Computer **25**(5), 76–79 (1992)
15. Polson, N.G., Sokolov, V.O.: Deep learning for short-term traffic flow prediction. Transp. Res. Part C Emerg. Technol. **79**, 1–17 (2017)
16. Tang, J., Deng, C., Huang, G.B.: Extreme learning machine for multilayer perceptron. IEEE Trans. Neural Netw. Learn. Syst. **27**(4), 809–821 (2017)

17. Wu, Y., Tan, H., Qin, L., Ran, B., Jiang, Z.: A hybrid deep learning based traffic flow prediction method and its understanding. Transp. Res. Part C **90**(1), 166–180 (2018)
18. Xu, M., Han, M., Chen, C.L.P., Qiu, T.: Recurrent broad learning systems for time series prediction. IEEE Trans. Cybern. (2018). https://doi.org/10.1109/TCYB.2018.2863020
19. Yang, H.F., Dillon, T.S., Chen, Y.P.: Optimized structure of the traffic flow forecasting model with a deep learning approach. IEEE Trans. Neural Netw. Learn. Syst. **28**(10), 2371–2381 (2016)
20. Yi, H., Jung, H.J., Bae, S.: Deep neural networks for traffic flow prediction. In: IEEE International Conference on Big Data & Smart Computing, Jeju, pp. 328–331. IEEE Press (2017)
21. Yuan, J., Zheng, Y., Xie, X., Sun, G.: Driving with knowledge from the physical world. In: ACM SIGKDD International Conference on Knowledge Discovery & Data Mining, pp. 316–324 (2011)
22. Zhan, H., Gomes, G., Li, X.S., Madduri, K., Wu, K.: Consensus ensemble system for traffic flow prediction. IEEE Trans. Intell. Transp. Syst. **19**(12), 3903–3914 (2018)

# A Novel QGA-UKF Algorithm
# for Dynamic State Estimation
# of Power System

Lihua Zhou[1], Minrui Fei[1(✉)], Dajun Du[1], Wenting Li[1], Huosheng Hu[2],
and Aleksandar Rakić[3]

[1] Shanghai Key Laboratory of Power Station Automation Technology,
School of Mechatronical Engineering and Automation,
Shanghai University, Shanghai 200444, China
lhzhou561@163.com, mrfei@staff.shu.edu.cn, {ddj,lwting}@shu.edu.cn
[2] School of Computer Science and Electronic Engineering, University of Essex,
Colchester CO4 3SQ, UK
hhu@essex.ac.uk
[3] School of Electrical Engineering, University of Belgrade,
Bulevar kralja Aleksandra 73, 11000 Belgrade, Serbia
rakic@etf.rs

**Abstract.** To ensure the safe operation of a power system, it is necessary to conduct its state estimation continuously. In this paper, a novel quantum genetic algorithm (QGA) is combined with unscented Kalman filter (UKF) for dynamic state estimation of power systems. Firstly, an innovation matrix is used to improve the estimation accuracy by constructing an adaptive correction factor for correcting the prediction covariance matrix in real time. The prediction error of constant Holt's two-parameter model is then analysed for adaptive optimization, and QGA is employed to adjust the parameters dynamically. Finally, simulation tests are carried out on IEEE 30 bus system and the results indicate that the proposed approach, namely QGA-UKF, has good estimation accuracy and stability that are higher than GA-UKF and UKF.

**Keywords:** Power system · Dynamic state estimation ·
Unscented Kalman filter · Quantum genetic algorithm

## 1 Introduction

State estimation can ensure the safe operation of a power system, which is classified as static state estimation and dynamic ones [1]. More specifically, static state estimation mainly uses the currently measured information to estimate the operating state of the grid. In contrast, dynamic state estimation (DSE) uses multi-stage measured information to estimate the grid state [2]. Extended Kalman Filter (EKF) was widely used in dynamic state estimation, which however has the truncation error caused by its non-linear function approximation.

© Springer Nature Switzerland AG 2019
H. Lu et al. (Eds.): ISNN 2019, LNCS 11554, pp. 240–250, 2019.
https://doi.org/10.1007/978-3-030-22796-8_26

On the other hand, UKF used in dynamic state estimation does not require the non-linear function approximation and Jacobian matrix calculation, achieving accurate state estimation [3].

Zhao and Mili developed a new UKF by deriving a batch-mode regression form and minimizing a convex Huber cost function [4], which can suppress observation and innovation outliers and filter out non-Gaussian process and measurement noise, achieving satisfactory results. Ahmad, et al. proposed an improved UKF based dynamic state estimation algorithm for an electric power distribution systems, which can deal with the presence of load variations and noisy data [5]. Sun, et al. used an adaptive factor in UKF for the state estimation of distribution networks [6]. Zhong and Hong deployed genetic algorithm (GA) to optimize the parameters of self-adaptive Kalman filter for the dynamic state estimation on power System [7]. The approaches mentioned above require the parameters to be selected by experience and cannot adjust them according to the real-time measurement information.

To address this issue, this paper develops a novel QGA-UKF algorithm to conduct the dynamic state estimation of power system. The contribution is focused on constructing adaptive correction factor and optimizing Holt's two-parameter model. More specifically, an innovation matrix is created to construct the adaptive correction factor used for the online correction of the prediction covariance matrix. The QGA is adopted to adjust the parameter of Holt's two-parameter model adaptively. Simulation is conducted on IEEE 30 bus system to verify the feasibility of the adaptive correction factor. The results show that the proposed method can achieve good estimation accuracy and adaptive optimization.

The rest of the paper is organized as follows. UKF Principle is briefly introduced in Sect. 2. Section 3 develops the proposed QGA-UKF algorithm. Simulation tests are conducted in Sect. 4 to verify the feasibility and performance of the proposed approach. Finally, a brief conclusion and future work are given in Sect. 5.

## 2   UKF Principle

The state distribution in EKF is propagated analytically through the first-order linearization of the nonlinear system. In contrast, the UKF is a derivative-free alternative to EKF by using a deterministic sampling approach [8]. Like EKF, UKF consists of the same two steps: model prediction and data assimilation, except for another preceded step for the selection of sigma points.

### 2.1   UT

The unscented transformation (UT) calculates the statistical value of random variables based on the sigma point sampling strategy through non-linear transformation means [9]. In this paper, the proportional symmetric sampling strategy is selected. For the $n$-dimensional system, the initialization of the state variable $x_0$ and the error covariance matrix $p_0$ are as follow

$$\hat{x}_0 = E(x_0).  \tag{1}$$

$$p_0 = E\left[(x_o - \hat{x}_0)\right]\left[(x_0 - \hat{x}_0)^T\right].  \tag{2}$$

Then, the position and corresponding weight of $2n + 1$ Sigma points are calculated at time $k - 1$,

$$\chi_{k-1}^i = \hat{x}_{k-1}, i = 0.  \tag{3}$$

$$\chi_{k-1}^i = \hat{x}_{k-1} + (\sqrt{(n+k)\,P_{k-1}})_i.  \tag{4}$$

$$\chi_{k-1}^i = \hat{x}_{k-1} - (\sqrt{(n+k)\,P_{k-1}})_i.  \tag{5}$$

$$W_i^{(m)} = \frac{\lambda}{n+\lambda}, i = 0.  \tag{6}$$

$$W_i^{(c)} = \frac{\lambda}{n+\lambda} + \left(1 - \alpha^2 + \beta\right), i = 0.  \tag{7}$$

$$W_i^{(m)} = W_i^{(c)} = \frac{1}{2(n+\lambda)}, i = 1, 2, ..., 2n.  \tag{8}$$

where $\hat{x}_{k-1} = \left[\hat{x}_{k-1}^1, \hat{x}_{k-1}^2, ..., \hat{x}_{k-1}^n\right]^T$, and $\lambda = \alpha^2(n+k) - n$ is the scale factor, which determines the distribution state of the sampling points; $\beta$ is the parameter related to the high-order term, which is taken $\beta = 2$ under the Gaussian distribution; $W_i^{(m)}$ is the mean weighted value, $W_i^{(c)}$ is the covariance weighted value, $\left(\sqrt{(n+\lambda)\,P_{k-1}}\right)_i$ indicates the $i$ column of the square root of the matrix.

## 2.2 UKF Implementation Process

The mathematical model of general state equation and measurement equation are as follow

$$x_{k+1} = f(x_k) + w_k.  \tag{9}$$

$$y_k = h(x_k) + v_k.  \tag{10}$$

where $x_k$ and $y_k$ denote as the system state variable and measurement vector at time $k$ respectively, the state transition function $f$ and the measurement function $h$ are both nonlinear. Moreover, process excitation noise $w_k$, and observation noise $v_k$ , which are independent of each other and white noise with normal distribution, here, $E\left[w_k w_k^T\right] = Q$, and $E\left[v_k v_k^T\right] = R$.

For general dynamic system model, the UKF consists of prediction step and data assimilation step. The prediction step includes state quantity prediction value and observation quantity prediction value, wherein the state quantity prediction value is obtained by infinite transformation and Holt's two-parameter exponential smoothing method. Then the observation measurement prediction value is obtained according to the measurement equation. After the filter gain is obtained during the data assimilation step, the measured value is corrected to update the state variable. The specific implementation steps are as follow.

(1) Prediction step

$$\chi_{k|k-1}^i = f\left(\chi_{k-1}^i\right).  \tag{11}$$

where $\chi_{k|k-1}^i$ is a $n \times (2n+1)$ dimensional matrix, and the sigma point is weighted to obtain the predicted value $\hat{x}_{k|k-1}^-$ and the prediction error covariance matrix $P_{x,k|k-1}^-$.

$$\hat{x}_{k|k-1}^- = \sum_{i=0}^{2n} W_i^{(m)} \chi_{k|k-1|}^i  \tag{12}$$

$$P_{x,k|k-1}^- = \sum_{i=0}^{2n} W_i^{(c)} \left[\chi_{k|k-1|}^i - \hat{x}_k^-\right]\left[\chi_{k|k-1|}^i - \hat{x}_k^-\right]^T + Q_{k-1}.  \tag{13}$$

The state transition function $f$ is replaced by Holt's two-parameter exponential smoothing method. The sigma point of the measured prediction value $\gamma_{k|k-1}^i$ and the predicted value of the measurement quantity $\hat{y}_{k|k-1}^-$ are obtained according to the nonlinear transformation.

$$\gamma_{k|k-1}^i = h\left(\chi_{k|k-1}^i\right).  \tag{14}$$

$$\hat{y}_{k|k-1}^- = \sum_{i=0}^{2n} W_i^{(m)} \gamma_{k|k-1}^i.  \tag{15}$$

(2) Data assimilation step

Then, the weighted summation is used to calculate the covariance matrix $P_{y,k}$ and $P_{xy,k}$ of the measurement quantity prediction value, as well as the state quantity prediction value.

$$P_{y,k} = \sum_{i=0}^{2n} W_i^{(c)} \left[\gamma_{k|k-1}^i - \hat{y}_k^-\right]\left[\gamma_{k|k-1}^i - \hat{y}_k^-\right]^T + R_k.  \tag{16}$$

$$P_{xy,k} = \sum_{i=0}^{2n} W_i^{(c)} \left[\chi_{k|k-1}^i - \hat{x}_k^-\right]\left[\gamma_{k|k-1}^i - \hat{y}_k^-\right]^T.  \tag{17}$$

Thus, the filter gain matrix $K$, the state estimate $\hat{x}_k$, and the state estimate covariance matrix $P_k$ are updated as follow

$$K = P_{xy,k} P_{y,k}^{-1}.  \tag{18}$$

$$\hat{x}_k = \hat{x}_{k|k-1}^- + K\left(y_{k|k-1} - \hat{y}_{k|k-1}^-\right).  \tag{19}$$

$$P_k = P_{k|k-1} - K P_{y,k} K^T.  \tag{20}$$

The UKF does not require the calculation of the Jacobian matrix in the prediction and the data assimilation step, and only involve the simple function calculation. However, Holt's two-parameter is constant in its prediction step, which may increase prediction error and affect the data assimilation effect. As the filter gain matrix K affects estimation accuracy of UKF, we adopt an adaptive correction factor.

# 3    A Novel QGA-UKF for DSE

## 3.1    Adaptive Factor Construction

Define the innovation vector $z_k = y_k - \hat{y}_{k|k-1}^-$ and the innovation matrix $z_k z_k^T$. Through measuring the trace of prediction variance matrix $P_{y,k}$ and innovation matrix, we construct an adaptive correction factor $\omega_k$ [10]

$$\omega_k = \begin{cases} 1 & tr\left(z_k z_k^T\right) \le tr\left(P_{y,k}\right) \\ \frac{tr(P_{y,k})}{tr\left(z_k z_k^T\right)} & tr\left(z_k z_k^T\right) > tr\left(P_{y,k}\right) \end{cases} \tag{21}$$

where $tr\left(z_k z_k^T\right)$ and $tr\left(P_{y,k}\right)$ are indicated the trace of the innovation matrix and the measurement quantity prediction variance matrix respectively.

When the trace of the innovation matrix is less than or equal to the trace of the measurement quantity prediction variance matrix, the adaptive correction factor takes value of 1, which means the values of $P_{y,k}$ and $P_{xy,k}$ are not changed; Conversely, They are changed, which indicates that the operating state of the system changes. At this time, the prediction error is large, and the values of $P_{y,k}$ and $P_{xy,k}$ need to be corrected online to obtain a new one. The new equations are as follows

$$\grave{P}_{y,k} = \frac{1}{\omega_k} \sum_{i=0}^{2n} W_i^{(c)} \left[\gamma_{k|k-1}^i - \hat{y}_k^-\right] \left[\gamma_{k|k-1}^i - \hat{y}_k^-\right]^T + R_k. \tag{22}$$

$$\grave{P}_{xy,k} = \frac{1}{\omega_k} \sum_{i=0}^{2n} W_i^{(c)} \left[\chi_{k|k-1}^i - \hat{x}_k^-\right] \left[\gamma_{k|k-1}^i - \hat{y}_k^-\right]^T. \tag{23}$$

## 3.2    Holt's Two-Parameter Method Optimized by QGA

In the traditional UKF, Holt's two-parameter exponential smoothing method [11] is used instead of the state transfer function. The basic idea is to use historical data for prediction and estimation. The parameters $\alpha, \beta$ are constant and will reduce the prediction accuracy of the algorithm. Therefore, this paper uses quantum genetic algorithm to optimize its parameters.

Genetic Algorithm (GA) is an adaptive search algorithm that simulates biological evolution. It is based on the principle of survival of the fittest in nature. The GA optimization GA starts from the initial population. New individuals are generated by using three operators: selection, intersection and mutation, which will be more adaptable to the environment. Finally, the best individual is selected, and the corresponding solution is the optimal solution of the objective function.

In contrast, Quantum Genetic Algorithm (QGA) is a combination of GA and quantum computing. QGA introduces the quantum state vector expression into the genetic coding, realizes the superposition of multiple states of a chromosome, and uses the quantum revolving gate to perform chromosome update. Due to the superposition of quantum information, the gene expressed contains all possible information. So that the objective function optimization solution is more accurate.

The operation of quantum revolving door is defined as

$$U\left(\theta_i\right) = \begin{bmatrix} \cos\left(\theta_i\right) & -\sin\left(\theta_i\right) \\ \sin\left(\theta_i\right) & \cos\left(\theta_i\right) \end{bmatrix}. \tag{24}$$

where $\theta_i$ is the rotation angle. The update process by

$$\begin{bmatrix} \dot{\varphi}_i \\ \dot{\phi}_i \end{bmatrix} = U\left(\theta_i\right) \begin{bmatrix} \varphi_i \\ \varphi_i \end{bmatrix} = \begin{bmatrix} \cos\left(\theta_i\right) & -\sin\left(\theta_i\right) \\ \sin\left(\theta_i\right) & \cos\left(\theta_i\right) \end{bmatrix} \begin{bmatrix} \varphi_i \\ \varphi_i \end{bmatrix}. \tag{25}$$

where $|\phi|^2 + |\varphi|^2 = 1$, $\left(\phi_i, \varphi_i\right)^T$ and $\left(\dot{\phi}_i, \dot{\varphi}_i\right)$ are expressed as the probability amplitude before and after the update of the $i$ qubit revolving gate of the chromosome respectively.

QGA is used to optimize the parameters of Holt's two-parameter method and submitted to the following settings. The values of the two independent variables are both in $[0, 1]$, the solution is accurate to 4 decimal places, $2^{13} < 10^4 < 2^{14}$, so the quantum bit code length is 14, and the population size is 30. The fixed rotation angle strategy is adopted, and all population genes are initialized as $\left(\frac{1}{\sqrt{2}}, \frac{1}{\sqrt{2}}\right)$. The minimum error between state filter value and state prediction value is selected as $FitnessValue = \min\left\{\hat{x}_{k+1} - \hat{x}_{k+1}^-\right\}$ [12]. Then, the new optimized prediction model is given by

$$x_k = F_{k-1}x_{k-1} + u_{k-1}. \tag{26}$$

$$a_{k-1} = \alpha_k x_{k-1} + \left(1 - \alpha_k\right)x_{k-1|k-2}. \tag{27}$$

$$b_{k-1} = \beta_k\left(a_{k-1} - a_{k-2}\right) + \left(1 - \beta_k\right)b_{k-2}. \tag{28}$$

$$F_{k-1} = \alpha_k\left(1 + \beta_k\right)I. \tag{29}$$

$$u_{k-1} = \left(1 + \beta_k\right)\left(1 - \alpha_k\right)x_{k-1|k-2} - \beta_k a_{k-2} + \left(1 - \beta_k\right)b_{k-2}. \tag{30}$$

where $\alpha, \beta \in [0, 1]$.

According to the optimization result, the QGA-UKF state filter is performed, and the measurement quantity $y_{k+1}$ is performed to obtain the state value $\hat{x}_{k+1}$ of filter.

## 4   Case Analysis

### 4.1   Simulation Overview

For the IEEE 30 bus system, the UKF, GA-UKF and QGA-UKF, are simulated. In order to analyze their dynamic estimation effect, the following indicators are used to evaluate the prediction state quantity and the estimated state quantity in the algorithm.

(1) average relative error of predicted state quantity

$$\zeta_k = \frac{1}{n}\sum_{j=1}^{n}\left|\frac{\hat{x}_k^-(j) - x_k(j)}{x_k(j)}\right|. \tag{31}$$

(2) average relative error of estimated state quantity

$$\zeta_s = \frac{1}{n}\sum_{j=1}^{n}\left|\frac{\hat{x}_k(j) - x_k(j)}{x_k(j)}\right|. \tag{32}$$

(3) algorithm filter indicator

$$\eta_k = \frac{\sum_{i=1}^{m}|\hat{y}_k(i) - y_k(i)|}{\sum_{i=1}^{m}|\hat{y}_k^-(i) - y_k(i)|}. \tag{33}$$

where $\hat{x}_k^-(j)$, $x_k(j)$ and $\hat{x}_k(j)$ are expressed as the predicted value, the true value and the estimated value of the $j$ state quantity at time $k$ respectively; $\hat{y}_k(i)$, $y_k(i)$ and $\hat{y}_k^-(i)$ are expressed as the estimated value, the true value and the measured value of the $i$ quantity measurement at time $k$ respectively.

### 4.2   Analysis of Simulation Results

The parameter settings of IEEE 30 bus system are mentioned in [13]. The quantity measurement is obtained by adding a normal distribution random noise with a mean of zero and a standard deviation of 2% on the basis of the power flow calculation.

Figure 1 shows true value, estimate value of three filters of voltage amplitude and phase angle with bus 2 under IEEE 30 bus system. It can be seen that the estimate value of three algorithms fluctuate around true value, which indicate that the proposed method is effective in the dynamic state estimation of the power system.

Figure 2 shows the estimated value error $\zeta_k$ and predicted value error $\zeta_s$ of UKF, GA-UKF and QGA-UKF respectively under IEEE 30 bus system. It can be seen from the figure that the maximum value of estimated error and prediction error of three algorithms is all smaller than $4 \times 10^{-3}$. The fluctuation range of

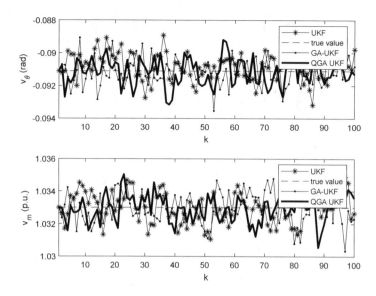

**Fig. 1.** Estimation of voltage amplitude and phase angle of node 2.

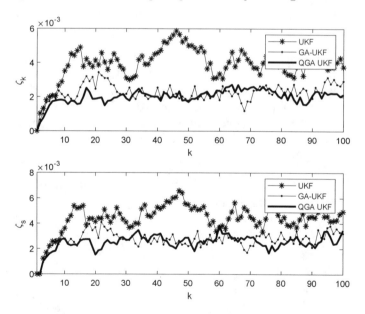

**Fig. 2.** UKF, GA-UKF, QGA-UKF estimation error and prediction error.

estimated value error and predicted value error of QGA-UKF are smaller than UKF and GA-UKF, which indicates QGA-UKF is with good stability.

Figure 3 shows the data assimilation performance index of three filters under IEEE 30 bus system, and the data assimilation performance index of all

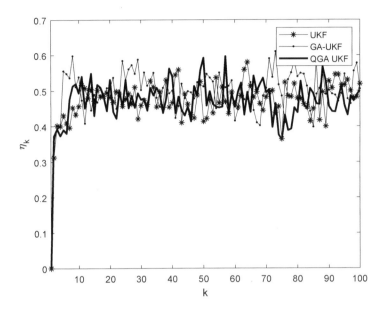

**Fig. 3.** UKF, GA-UKF, QGA-UKF data assimilation performance indicator.

fluctuates around 0.45. Especially the QGA-UKF fluctuates more slowly, which further indicates that it is feasible and effective of QGA-UKF in DSE of power system.

Table 1 shows the three performance indicators of three filters. In comparison to UKF and GA-UKF, the average value of estimation error, prediction error and data assimilation index of QGA-UKF are lower. The minimum and maximum data assimilation index value of QGA-UKF are higher than both of them, which indicates that there is a deviation between the predicted values and estimated values of QGA-UKF at a few sampling moments. The data assimilation indexes of three algorithms are less than 1, especially the indicators of QGA-UKF are best among of them, which indicates that the proposed QGA-UKF can effectively improve the data assimilation precision of UKF and reduce prediction error.

**Table 1.** Performance indicators of three methods in IEEE 30 bus system

| Methods | $\zeta_k$ | | | $\zeta_s$ | | | $\eta_k$ | | |
|---|---|---|---|---|---|---|---|---|---|
| | Min | Avg | Max | Min | Avg | Max | Min | Avg | Max |
| $UKF$ | 0.0010 | 0.0040 | 0.0059 | 1.2511e−11 | 0.0045 | 0.0066 | 0.3123 | 0.4767 | 0.5850 |
| $GA-UKF$ | 6.43634e−04 | 0.0023 | 0.0034 | 1.2206e−11 | 0.0028 | 0.0041 | 0.3542 | 0.5030 | 0.6103 |
| $QGA-UKF$ | 4.6764e−04 | 0.0020 | 0.0027 | 1.2206e−11 | 0.0025 | 0.0039 | 0.3622 | 0.4736 | 0.5968 |

# 5   Conclusion

Unlike constant parameters of Holt's two-parameter model used in the UKF prediction step, this paper proposed a novel QGA-UKF approach to optimize the parameters dynamically. The adaptive correction factor was created to reduce the prediction error and achieve the adaptive optimization of dynamic state estimation of a power system. Simulation tests were conducted by using IEEE 30 bus system and the results indicated that the proposed QGA-UKF has good estimation accuracy and system stability.

Our future work will be focused on how to improve its performance without affecting the convergence speed of dynamic state estimation of power systems.

**Acknowledgments.** This work was supported by the Natural Science Foundation of China under Grant 61633016 and Project 111 under Grant D18003.

# References

1. Uzunoğlu, B., Ülker, M.A.: Maximum likelihood ensemble filter state estimation for power systems. IEEE Trans. Instrum. Meas. **67**(9), 2097–2106 (2018)
2. Sun, Y.B., Fu, M.Y., Zhang, H.S.: Performance comparison of distributed state estimation algorithms for power systems. J. Syst. Sci. Complex. **30**(3), 595–615 (2017)
3. Rayyam, M., Zazi, M., Barradi, Y.: A new metaheuristic unscented Kalman filter for state vector estimation of the induction motor based on ant lion optimizer. COMPEL **37**(3), 1054–1068 (2018)
4. Zhao, J.B., Mili, L.: A robust generalized-maximum likelihood unscented Kalman filter for power system dynamic state estimation. IEEE J-STSP **12**(4), 578–592 (2018)
5. Ahmad, F., Muhammad Abdul Rashid, K., Rasool, A., Ozsoy, E.E., Sabanoviç, A., Elitas, M.: An improved unscented Kalman filter based dynamic state estimation algorithm for electric distribution systems. COMPEL **36**(4), 1220–1236 (2017)
6. Sun, J.S., Liu, M., Deng, L., Ying, L.Y., Li, Z.T.: State estimation of distribution network based on AUKF. Power Syst. Prot. Control **46**(11), 1–7 (2018)
7. Zhong, Z.J., Hong, B.Z.: Self-adaptive Kalman filter dynamic state estimation on power system based on genetic algorithm. Guangdong Electric Power **27**(7), 78–82 (2014)
8. Yu, S.L., Emami, K., Fernando, T., Lu, H.H.C., Wong, K.P.: State estimation of doubly fed induction generator wind turbine in complex power systems. IEEE Trans. Power Syst. **31**(6), 4935–4944 (2016)
9. Wang, L., Cheng, X.H.: Algorithm of Gaussian sum filter based on high-order UKF for dynamic state estimation. Int. J. Control. Autom. Syst. **13**(3), 652–661 (2015)
10. Jiao, W.B., Su, W.X., Gao, Q., Wang, M.: Application of adaptive UKF in initial alignment of MINS/GPS integrated navigation system. In: 3rd International Conference on Advanced Computer Theory and Engineering, Chengdu, pp. 224–228. IEEE Press (2010)
11. Da Silva, A.L., Do Coutto Filho, M.B., De Queiroz, J.F.: State forecasting in electric power systems. IEE P-Gener. Transm. Distrib. **130**(5), 237–244 (1983)

12. Kumar, S., Chaturvedi, D.K.: Optimal power flow solution using GA-fuzzy and PSO-fuzzy. J. Inst. Eng. India Ser. B **95**(4), 363–368 (2014)
13. Li, X., Li, W.T., Du, D.J., Sun, Q., Fei, M.R.: Dynamic state estimation of smart grid based on UKF under denial of service attack. ACTA Automatica Sinica (2018). https://doi.org/10.16383/j.aas.2018.c180431

# An Improved Result on $H_\infty$ Performance State Estimation of Delayed Static Neural Networks

Guoqiang Tan[1], Jidong Wang[1,2], Zhanshan Wang[1(✉)], and Xiaolong Qian[1]

[1] School of Information Science and Engineering,
Northeastern University, Shenyang 110819, China
guoqiangtan163@126.com, kfwjd@ncwu.edu.cn,
{wangzhanshan,qianxiaolong}@ise.neu.edu.cn
[2] School of Electrical Engineering,
North China University of Water Resources and Electric Power,
Zhengzhou 450011, China

**Abstract.** In this paper, a new state estimator with integral term is proposed for studying the $H_\infty$ performance of static neural networks with time-varying delay. Firstly, some integral inequalities are given to handle the derivative of Lyapunov functional. Secondly, a delay dependent criterion is derived for the estimation error system. Thirdly, in order to guarantee the $H_\infty$ performance, the gain matrices can be obtained by the linear matrix inequalities. Finally, an example is used to verify the effectiveness of our proposed method.

**Keywords:** $H_\infty$ performance · Time-varying delay ·
Static neural networks · Linear matrix inequalities.

## 1 Introduction

Since more and more applications of neural networks have been found in a large number of fields, neural networks have begun to be taken seriously in recent years. As a result, many papers about neural networks have been reported [1–3]. In fact, neural networks are classified into two kinds. One is static neural networks and the other is local field neural networks. Generally speaking, the two kinds of neural networks are different. But the two kinds of neural networks are equivalent only when they satisfy some critical conditions [4,5]. The recurrent results mainly focus on local field neural networks [6–8]. It is obvious that the static neural networks under some assumption can be taken to solve some control problems such as $H_\infty$ performance.

In the past decades, the state estimation problem of static time-varying delayed neural networks has been found attracting many scholars and making great achievements, such as classification, pattern recognition, static imagine processing, as well as combinatorial optimization. In fact, due to expensive cost

© Springer Nature Switzerland AG 2019
H. Lu et al. (Eds.): ISNN 2019, LNCS 11554, pp. 251–260, 2019.
https://doi.org/10.1007/978-3-030-22796-8_27

and little information obtained from neuron state [9], especially external disturbance is imposed to neural networks, it's very important to carry out the study of neuron state estimation based on available information. As a result, some practical performance can be achieved based on the successful application of neural networks. It is noted the proposed new state estimator is similar to the state estimator in [10] such that it can make a good performance. Thus, this kind of estimator can be applied in the practical state estimation.

In the study of recurrent neural networks, time delay will be inevitably encountered. Time delay is one of the main reasons leading to the neural system instability, so stability analysis of time delay was extensively studied. In recent years, the study of state estimation has achieved a good result [11–15]. In addition, some sufficient conditions of state estimator are presented. Therefore, less conservative criteria [16,17], time-varying state estimation issues begin to attract a large number of scholars. As discussed before, the static neural networks and local field neural networks are different, the criteria in many papers or books may connot be directly applied to the delayed static neural networks. Therefore, it should be considered that the practical significance of criteria is of importance. Many scholars made their efforts in designing state estimation. The traditional estimation is Luenberger state estimator [14] and Arcak's observer [18]. Recently, in [10], another type of observer was presented, in this estimator, the integral term is considered. However, in this paper, $e^{-t}$ is added to the integral term such that our state estimator accelerates the convergence rate. Therefore, our state estimator can estimate the system state more accurately.

This paper is concerned with the problem of $H_\infty$ state estimation for static neural networks with time-varying delay. A new state estimator, which is constructed to deal with the $H_\infty$ performance problem together with some integral inequalities. By choosing a suitable Lyapunov functional, and employing some integral inequalities such that the error system is globally asymptotically stable. Finally, an example is utilized to illustrate the advantage of our method.

## 2    Problem Description

The static neural network with time-varying delay is depicted by

$$\begin{cases} \dot{x}(t) = -Ax(t) + f(Wx(t - \tau(t)) + J) + B_1\omega(t) \\ y(t) = Cx(t) + Dx(t - \tau(t)) + B_2\omega(t) \\ z(t) = H_1x(t) + H_2x(t - \tau(t)) \\ x(t) = \phi(t), \quad t \in [-\tau, 0] \end{cases} \tag{1}$$

where $x(t) = [x_1(t), x_2(t), \ldots, x_n(t)] \in \mathbb{R}^n$ is the neuron state vector with $n$ neurons, $y \in \mathbb{R}^m$ is the network measurement vector, $z(t) \in \mathbb{R}^p$ is the linear combination of the states to be estimated, $\omega(t) \in \mathbb{R}^q$ is the noise disturbance vector belonging to $L_2[0, \infty]$. $A$ is a diagonal and positive definite matrix, $W$ is a delayed connection weight matrix, $B_1$, $B_2$, $C$, $D$, $H_1$, and $H_2$ are real matrices with appropriate dimensions. $f(x) = [f_1(x_1), f_2(x_2), \ldots, f_n(x_n)]^T$ is the continuous neuron activation function, $J = [J_1, J_2, \ldots, J_n]^T$ is an external input vector, $\tau(t)$ and $\phi(t)$ are the time-varying delay and the initial condition defined on $[-\tau, 0]$.

**Assumption 1.** The neural activation function $f_i(\cdot)$ satisfies

$$0 \leq \frac{f_i(a) - f_i(b)}{a - b} \leq l_i, \quad a \neq b \in \mathbb{R} \tag{2}$$

**Assumption 2.** There exist scalars $\tau > 0$ and $\mu \in \mathbb{R}$ such that

$$0 \leq \tau(t) \leq \tau, \quad 0 \leq \dot{\tau}(t) \leq \mu \tag{3}$$

where $\tau$, $\mu$ are real constants. In this paper, a new state estimator is designed as follows:

$$\begin{cases} \dot{\hat{x}}(t) = -A\hat{x}(t) + f(W\hat{x}(t - \tau(t)) + K_1(y(t) - \hat{y}(t)) + J) \\ \qquad + K_2(y(t) - \hat{y}(t)) + K_3 e^{-t}\xi \\ \dot{\xi}(t) = y(t) - \hat{y}(t) \\ \hat{y}(t) = C\hat{x}(t) + D\hat{x}(t - \tau(t)) \\ \hat{z}(t) = H_1\hat{x}(t) + H_2\hat{x}(t - \tau(t)) \\ \hat{x}(t) = 0, \quad t \in [-\tau, 0] \end{cases} \tag{4}$$

where $\hat{x}(t) \in \mathbb{R}^n$, $\hat{z}(t) \in \mathbb{R}^p$, $\xi(t) \in \mathbb{R}^m$, $K_1$, $K_2$, and $K_3$ are the gain matrices to be determined.

**Remark 1.** In this paper, a new state estimator with $e^{-t}$ is presented. The Luenberger state estimator is a special case of the proposed new state estimator when $K_3 = 0$. Due to our adding integral term, the proposed new state estimator will estimate neuron state more accurately.

Define $e(t) = x(t) - \hat{x}(t)$, $\tilde{z}(t) = z(t) - \hat{z}(t)$, and $\bar{\xi}(t) = e^{-t}\xi(t)$, then the error system is obtained with (1) and (4) as follows:

$$\begin{cases} \dot{e}(t) = -(A + K_2 C)e(t) + g(t) - K_2 D e(t - \tau(t)) \\ \qquad + (B_1 - K_2 B_2)w(t) - K_3\bar{\xi}(t) \\ \dot{\xi}(t) = Ce(t) + De(t - \tau(t)) + B_2 w(t) \\ \tilde{z}(t) = H_1 e(t) + H_2 e(t - \tau(t)) \end{cases} \tag{5}$$

where $g(t) = f(Wx(t - \tau(t)) + J) - f(W\hat{x}(t - \tau(t)) + K_1(y(t) - \hat{y}(t)) + J)$. From (2), it's easy to get that the following inequality holds

$$0 \leq -2g^T(t)\Lambda g(t) + 2g^T(t)\Lambda L[-K_1 Ce(t) \\ + (W - K_1 D)e(t - \tau(t)) - K_1 B_2 w(t)] \tag{6}$$

where $\Lambda = diag(\lambda_1, \lambda_2, \ldots, \lambda_n) > 0$, $L = diag(l_1, l_2, \ldots, l_n)$. $l_i$ is defined in (2). The problem of an $H_\infty$ state estimation is stated as follows:

For a prescribed $\gamma > 0$, it's better to find a suitable state estimator (4) such that the estimation error system (5) with $\omega(t) = 0$ is globally asymptotically stable under the condition that

$$\|\tilde{z}(t)\|_2 < \gamma \|\omega(t)\|_2. \tag{7}$$

The following Lemmas will be used in this paper.

**Lemma 1** [13]. *For real matrices $T > 0$ and $Z$ with appropriate dimensions, and any vector function $\zeta(t)$, the following inequality holds*

$$-\int_{-\tau}^{0}\int_{t+\theta}^{t} \dot{e}^T(s)T\dot{e}(s)dsd\theta \le \frac{1}{2}\tau^2\zeta^T(t)Z^TT^{-1}Z\zeta(t)$$

$$+2\zeta^T(t)Z^T[\tau e(t) - \int_{t-\tau}^{t} e(s)ds] \tag{8}$$

**Lemma 2** [14]. *Let $\pi(t) = [e^T(t - \tau(t)) - e^T(t - \tau), e^T(t) - e^T(t - \tau(t))]^T$. If the matrix $\begin{bmatrix} S & V \\ * & S \end{bmatrix} \ge 0$, then*

$$-\tau\int_{t-\tau}^{t} \dot{e}^T(s)S\dot{e}(s)ds \le -\pi^T(t)\begin{bmatrix} S & V \\ * & S \end{bmatrix}\pi(t) \tag{9}$$

## 3    Main Result

In this section, we have the following theorem.

**Theorem 1.** *For given scalars $\tau > 0, \mu > 0, \gamma > 0$, the problem of the $H_\infty$ performance state estimation is solvable if there exist real matrices $P_1 > 0, P_2 > 0, Q_1 > 0, Q_2 > 0, R > 0, S > 0, T > 0, \zeta_i(i = 1, 2, \ldots, 7), V, G_i(i = 1, 2, 3)$ and $\Lambda = diag(\lambda_1, \lambda_2, \ldots, \lambda_n) > 0$ such that the LMIs (10) and (11) are satisfied:*

$$\begin{bmatrix} S & V \\ * & S \end{bmatrix} \ge 0 \tag{10}$$

$$\begin{bmatrix} \Phi_1 & \tau\Phi_2^T & \tau\Phi_3^T \\ * & -2T & 0 \\ * & * & \Phi_4 \end{bmatrix} < 0 \tag{11}$$

where

$$\Phi_1 = \begin{bmatrix} \Phi_{11} & \Phi_{12} & \Phi_{13} & \Phi_{14} & \Phi_{15} & \Phi_{16} & \Phi_{17} \\ * & \Phi_{22} & \Phi_{23} & -\zeta_2^T & \Phi_{25} & 0 & D^T P_2 \\ * & * & \Phi_{33} & -\zeta_3^T & 0 & 0 & 0 \\ * & * & * & \Phi_{44} & -\zeta_5 & -\zeta_6 & -\zeta_7 \\ * & * & * & * & -2\Lambda & -G_1 B_2 & 0 \\ * & * & * & * & * & -\gamma^2 I & B_2^T P_2 \\ * & * & * & * & * & * & \Phi_{77} \end{bmatrix}$$

$$\Phi_2 = [\zeta_1 \quad \zeta_2 \quad \zeta_3 \quad \zeta_4 \quad \zeta_5 \quad \zeta_6 \quad \zeta_7]$$

$$\Phi_3 = [-P_1 A - G_2 C \quad -G_2 D \quad 0 \quad 0 \quad P_1 \quad P_1 B_1 - G_2 B_2 \quad -G_3]$$

$$\Phi_4 = -2P_1 + S + \frac{T}{2}$$

$$\Phi_{11} = -A^T P_1 - P_1 A - C^T G_2^T - G_2 C + Q_1 + Q_2$$
$$\quad + \tau^2 R + H_1^T H_1 - S + \tau \zeta_1 + \tau \zeta_1^T$$

$$\Phi_{12} = -G_2 D - V^T + S + \tau \zeta_2 + H_1^T H_2$$

$$\Phi_{13} = V^T + \tau \zeta_3$$

$$\Phi_{14} = \tau \zeta_4 - \zeta_1^T$$

$$\Phi_{15} = P_1 + \tau \zeta_5 - C^T G_1^T$$

$$\Phi_{16} = P_1 B_1 - G_2 B_2 + \tau \zeta_6$$

$$\Phi_{17} = C^T P_2 - G_3 + \tau \zeta_7$$

$$\Phi_{22} = -(1 - \mu) Q_1 + H_2^T H_2 - 2S + V + V^T$$

$$\Phi_{23} = S - V^T$$

$$\Phi_{25} = W^T L \Lambda - D^T G_1$$

$$\Phi_{33} = -Q_2 - S$$

$$\Phi_{44} = -R - \zeta_4 - \zeta_4^T$$

$$\Phi_{77} = -P_2.$$

Furthermore, the gain matrices $K_1$, $K_2$ and $K_3$ can be designed as

$$K_1 = (\Lambda L)^{-1} G_1, \quad K_2 = P_1^{-1} G_2, \quad K_3 = P_1^{-1} G_3.$$

*Proof.* It is shown that (7) holds for all nonzero $\omega(t)$ when the condition (10) and (11) are satisfied under zero-initial conditions. We design a Lyapunov functional candidate as

$$V(t) = \sum_{i=1}^{4} V_i(t) \tag{12}$$

and

$$V_1(t) = e^T(t) P_1 e(t) + e^{-t} \xi^T(t) P_2 \xi(t)$$

$$V_2(t) = \int_{t-\tau(t)}^{t} e^T(s) Q_1 e(s) ds + \int_{t-\tau}^{t} e^T(s) Q_2 e(s) ds$$

$$V_3(t) = \tau \int_{-\tau}^{0} \int_{t+\theta}^{t} e^T(s) R e(s) ds d\theta + \tau \int_{-\tau}^{0} \int_{t+\theta}^{t} \dot{e}^T(s) S \dot{e}(s) ds d\theta$$

$$V_4(t) = \int_{-\tau}^{0} \int_{t+\theta}^{t} (s - t - \theta) \dot{e}^T(s) T \dot{e}(s) ds d\theta.$$

Now, we take the derivative of $V(t)$ along the trajectories of (5) yield

$$\dot{V}_1(t) \leq e^T(t)(-A^T P_1 - P_1 A - C^T K_2^T P_1 - P_1 K_2 C)e(t) - 2e^T(t)P_1 K_2 De(t - \tau(t))$$
$$+ 2e^T(t)P_1 g(t) + 2e^T(t)(P_1 B_1 - P_1 K_2 B_2)\omega(t) - 2e^T(t)P_1 K_3 \bar{\xi}(t)$$
$$- \bar{\xi}^T(t)P_2 \bar{\xi}(t) + 2e^T(t)C^T P_2 \bar{\xi}(t) + 2e^T(t - \tau(t))D^T P_2 \bar{\xi}(t)$$
$$+ 2\omega^T(t)B_2^T P_2 \bar{\xi}(t) \tag{13}$$

$$\dot{V}_2(t) \leq e^T(t)(Q_1 + Q_2)e(t) - e^T(t - \tau)Q_2 e(t - \tau)$$
$$- (1 - \mu)e^T(t - \tau(t))Q_1 e(t - \tau(t)) \tag{14}$$

$$\dot{V}_3(t) = \tau^2 e^T(t)Re(t) + \tau^2 \dot{e}^T(t)S\dot{e}(t) - \tau \int_{t-\tau}^t e^T(s)Re(s)ds$$
$$- \tau \int_{t-\tau}^t \dot{e}^T(s)S\dot{e}(s)ds \tag{15}$$

$$\dot{V}_4(t) = \frac{\tau^2}{2}\dot{e}^T T\dot{e}(t) - \int_{-\tau}^0 \int_{t+\theta}^t \dot{e}^T(s)T\dot{e}(s)dsd\theta. \tag{16}$$

It is known from (13) that $V(t)|_{t=0} = 0$ and for $t > 0$ $V(t) \geq 0$ under the zero-initial conditions. Now, we define

$$J = \int_0^\infty [\tilde{z}^T(t)\tilde{z}(t) - \gamma^2 \omega^T(t)\omega(t)]dt. \tag{17}$$

Then, we have the following inequality

$$J \leq \int_0^\infty [\tilde{z}^T(t)\tilde{z}(t) - \gamma^2 \omega^T(t)\omega(t)]dt + V(t)|_{t\to\infty} - V(t)|_{t=0}$$
$$= \int_0^\infty [\tilde{z}^T(t)\tilde{z}(t) - \gamma^2 \omega^T(t)\omega(t) + \dot{V}(t)]dt. \tag{18}$$

Let $\rho(t) = [e^T(t), e^T(t - \tau(t)), e^T(t - \tau), \int_{t-\tau}^t e^T(s)ds, g^T(t), \omega(t), \bar{\xi}(t)]^T$. Then taking the derivative of $V(t)$ along the trajectories of (5) and employing Lemmas 1 and 2 yield

$$\tilde{z}^T(t)\tilde{z}(t) - \gamma^2 \omega^T(t)\omega(t) + \dot{V}(t)$$
$$\leq \rho^T(t)[\bar{\Phi}_1 + \frac{\tau^2}{2}\Phi_2^T T^{-1}\Phi_2 + \bar{\Phi}_3^T(\tau^2 S + \frac{\tau^2}{2}T)\bar{\Phi}_3]\rho(t) \tag{19}$$

where $\bar{\Phi}_1, \bar{\Phi}_3$ are obtained from $\Phi_1, \Phi_3$ by replacing $G_1$ with $\Lambda L K_1$, $G_2$ with $P_1 K_2$, $G_3$ with $P_1 K_3$. Noting $-P(S + T/2)^{-1}P \leq -2P + S + T/2$, by (11), the following matrix inequality holds

$$\begin{bmatrix} \Phi_1 & \tau\Phi_2^T & \tau\Phi_3^T \\ * & -2T & 0 \\ * & * & -P(S + \frac{T}{2})^{-1}P \end{bmatrix} < 0. \tag{20}$$

Then, pre- and postmultiplying (20) by $\text{diag}\{I, I, (S + \frac{T}{2})P^{-1}\}$ and its transpose, and noting $K_1 = (\Lambda L)^{-1}G_1, K_2 = P_1^{-1}G_2$, and $K_3 = P_1^{-1}G_3$, the following matrix inequality holds

$$
\begin{bmatrix}
\bar{\Phi}_1 & \tau\Phi_2^T & \tau\bar{\Phi}_3(S+\frac{T}{2}) \\
* & -2T & 0 \\
* & * & -S-\frac{T}{2}
\end{bmatrix} < 0 \tag{21}
$$

By Schur complement, the matrix inequality (21) is equivalent to

$$
\bar{\Phi}_1 + \frac{\tau^2}{2}\Phi_2^T T^{-1}\Phi_2 + \bar{\Phi}_3^T(\tau^2 S + \frac{\tau^2}{2}T)\bar{\Phi}_3 < 0 \tag{22}
$$

Then, for any $\rho(t) \neq 0$

$$
\rho^T(t)[\bar{\Phi}_1 + \frac{\tau^2}{2}\Phi_2^T T^{-1}\Phi_2 + \bar{\Phi}_3^T(\tau^2 S + \frac{\tau^2}{2}T)\bar{\Phi}_3]\rho(t) < 0. \tag{23}
$$

Noting (17), (18), (19), and (23), one has

$$
J \leq \int_0^\infty [\tilde{z}^T(t)\tilde{z}(t) - \gamma^2\omega^T(t)\omega(t) + \dot{V}(t)]dt < 0. \tag{24}
$$

Therefore, $\|\tilde{z}(t)\|_2 < \gamma\|\omega(t)\|_2$. According to Lyapunov stability theory, the error system (5) is globally asymptotically stable. This completes the proof.

**Remark 2.** It can be seen from our result that the $H_\infty$ performance state estimation of delayed static neural networks is guaranteed. There are three gain matrices to adjust the parameters of (4). It is obvious that much better $H_\infty$ performance can be achieved by our method.

**Remark 3.** In this paper, our new state estimator has more accurate effect than [13] and [14] because of our adding integral term $K_3\bar{\xi}(t)$. It can be shown from the following simulation. The significance of adding $e^{-t}$ to the integral term of our new estimator (4) is that when we deal with $e^{-t}\xi^T(t)P_2\xi(t)$ of Lyapunov functional, the derivative of it is $-e^{-t}\xi^T(t)P_2\xi(t) + e^{-t}\xi^T(t)P_2\dot{\xi}(t)$, which is no greater than $-e^{-t}\xi^T(t)P_2\xi(t)e^{-t} + e^{-t}\xi^T(t)P_2\dot{\xi}(t)$. Thus, (11) is solvable.

## 4   Numerical Example

In this section, we will take an example from [13] and [14] to show the effectiveness of the presented method. Consider the system (1) with the following parameters:

$$
A = diag\{1.06, 1.42, 0.88\}, \quad H_2 = 0,
$$

$$
H_1 = \begin{bmatrix} 1 & 0 & 0.5 \\ 1 & 0 & 1 \\ 0 & -1 & 1 \end{bmatrix}, W = \begin{bmatrix} -0.32 & 0.85 & -1.36 \\ 1.1 & 0.41 & -0.5 \\ 0.42 & 0.82 & -0.95 \end{bmatrix},
$$

$$
C = \begin{bmatrix} 1 & 0.5 & 0 \\ 0 & -0.5 & 0.6 \end{bmatrix}, D = \begin{bmatrix} 0 & 1 & 0.2 \\ 0 & 0 & 0.5 \end{bmatrix},
$$

$$
B_1 = [0.2 \quad 0.2 \quad 0.2]^T, B_2 = [0.4 \quad -0.3]^T.
$$

**Table 1.** Comparison of the $H_\infty$ performance index $\gamma_{min}$ for different $L$.

| Methods | I | 1.1I | 1.2I | 1.3I | 1.4I | 1.5I |
|---|---|---|---|---|---|---|
| Theorem 1 in [13] | 0.3868 | 0.5446 | 0.8220 | 1.5311 | 6.4813 | - |
| Theorem 1 in [14] | 0.3822 | 0.4923 | 0.6536 | 0.9805 | 2.2387 | - |
| Theorem 1 | 0.3529 | 0.4270 | 0.4986 | 0.6014 | 0.7588 | 1.0305 |

It is supposed that $L = \lambda I(\lambda > 0)$ and $\tau = 0.8$, $\mu = 0.6$. Thus, the $H_\infty$ performance index $\gamma_{min}$ can be obtained by our method for different values of $\lambda$. The results are listed in Table 1. It is obvious that the $\gamma_{min}$ is increasing when $\lambda$ is from 1 to 1.5. Due to our adding integral term in the new estimator, it is clear from the Table 1 that the $H_\infty$ indices are much better than the ones in [13] and [14]. However, the $H_\infty$ performance index can be calculated until $L = 1.7I$, when $L = 1.5I$, the three gain matrices are calculated as

$$K_1 = \begin{bmatrix} -0.0523 & -0.4464 \\ 0.2217 & -0.0813 \\ 0.2020 & -0.7930 \end{bmatrix}, K_2 = \begin{bmatrix} 10.8202 & 14.5750 \\ 0.1259 & -9.4775 \\ 16.0261 & 35.9349 \end{bmatrix}, \bar{K}_3 = \begin{bmatrix} 0.0759 & 0.0845 \\ 0.2522 & 0.2762 \\ 0.1563 & 0.1721 \end{bmatrix},$$

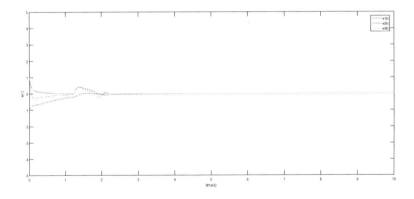

**Fig. 1.** Error system: $e_1(t), e_2(t), e_3(t)$.

where $K_3 = 10^{-5} \bar{K}_3$. It is seen from Table 1 that the $H_\infty$ performance index $\gamma_{min} = 1.0305$. Then, for our simulation, we choose the time-varying delay as $\tau(t) = 0.4 + 1.6cos(5t/6)$. Figure 1 is the response of error system (5). It is clear that Theorem 1 can guarantee the $H_\infty$ performance state estimation of time-varying delayed static neural networks.

## 5 Conclusion

In this paper, a new state estimator is constructed to deal with the problem of $H_\infty$ performance state estimation of static neural networks with time-varying delay. By designing a appropriate Lyapunov functional, a delay criterion is derived. As a result, the error system is globally asymptotically stable. In addition, the gain matrices and $H_\infty$ performance index are obtained. It is obviously that much better performance has been gained by our method. An example is given to show the effectiveness of our method.

**Acknowledgments.** This work was supported in part by the National Natural Science Foundation of China (Grant Nos. 61473070, 61433004, 61627809), and in part by SAPI Fundamental Research Funds (Grant No. 2018ZCX22).

## References

1. Liu, H., Wang, Z., Shen, B., Liu, X.: Event-triggered $H_\infty$ state estimation for delayed stochastic memristive neural networks with missing measurements: the discrete time case. IEEE Trans. Neural Netw. Learn. Syst. **29**, 3726–3737 (2018)
2. Duan, Q., Su, H., Wu, Z.G.: $H_\infty$ state estimation of static neural networks with time-varying delay. Neurocomputing **97**, 16–21 (2012)
3. Zhang, X.-M., Han, Q.-L.: Global asymptotic stability for a class of generalized neural networks with interval time-varying delay. IEEE Trans. Neural Netw. **22**, 1180–1192 (2011)
4. Qiao, H., Peng, J., Xu, Z.-B., Zhang, B.: A reference model approach to stability analysis of neural networks. IEEE Trans. Syst. Man Cybern. B Cybern. **33**, 925–936 (2003)
5. Huang, H., Feng, G., Cao, J.: Guaranteed performance state estimation of static neural networks with time-varying delay. Neurocomputing **74**, 606–616 (2011)
6. Zhang, H., Wang, Z., Liu, D.: A comprehensive review of stability analysis of continuous-time recurrent neural networks. IEEE Trans. Neural Netw. Learn. Syst. **25**, 1229–1262 (2014)
7. Zhang, X.-M., Han, Q.-L.: Global asymptotic stability analysis for delayed neural networks using a matrix-based quadratic convex approach. Neural Netw. **54**, 57–69 (2014)
8. Zhang, C.K., He, Y., Jiang, L., Wu, M.: Stability analysis for delayed neural networks considering both conservativeness and complexity. IEEE Trans. Neural Netw. Learn. Syst. **27**, 1486–1501 (2016)
9. Wang, Z., Ho, D.W.C., Liu, X.: State estimation for delayed neural networks. IEEE Trans. Neural Netw. **16**, 279–284 (2005)
10. Yao, Y.X., Radun, A.V.: Proportional integral observer design for linear systems with time delay. IET Control Theory Appl. **1**, 887–892 (2007)
11. Liu, B., Ma, X., Jia, X.-C.: Further results on $H_\infty$ state estimation of static neural networks with time-varying delay. Neurocomputing **285**, 133–140 (2018)
12. Zhang, X.-M., Han, Q.-L., Wang, Z.-D., Zhang, B.-L.: Neural state estimation for neural networks with two additive time-varying delay components. IEEE Trans. Cybern. **47**, 3184–3194 (2017)

13. Huang, H., Huang, T., Chen, X.: Guaranteed $H_\infty$ performance state estimation of delayed static neural networks. IEEE Trans. Circ. Syst. II Exp. Briefs. **60**, 371–375 (2013)
14. Huang, H., Huang, T., Chen, X.: Further result on guranteed $H_\infty$ performance state estimation of delayed static neural networks. IEEE Trans. Neural Netw. Learn. Syst. **26**, 1335–1341 (2015)
15. Zhang, X.M., Han, Q.L.: State estimation for static neural networks with time-varying delays based on an improved reciprocally convex inequality. IEEE Trans. Neural Netw. Learn. Syst. **29**, 1376–1381 (2018)
16. He, Y., Wang, Q.-G., Wu, M., Lin, C.: Delay-dependent state estimation for delayed neural networks. IEEE Trans. Neural Netw. **17**, 1077–1081 (2006)
17. Park, P.G., Ko, J.W., Jeong, C.: Reciprocally convex approach to stability of systems with time-varying delays. Automatica **47**, 235–238 (2011)
18. Zemouche, A., Boutayeb, M.: Comments on "a note on observers for discrete-time Lipschitz nonlinear systems". IEEE Trans. Circ. Syst. II Exp. Briefs. **60**, 56–60 (2013)

# Uncertainty Estimation via Stochastic Batch Normalization

Andrei Atanov[1], Arsenii Ashukha[2], Dmitry Molchanov[1,2],
Kirill Neklyudov[1,2(✉)], and Dmitry Vetrov[1,2]

[1] National Research University Higher School of Economics,
Samsung-HSE Laboratory, Moscow, Russia
andrewatanov@yandex.ru, dmolch111@gmail.com, k.necludov@gmail.com
[2] Samsung AI Center in Moscow, Moscow, Russia
ars.ashuha@gmail.com, vetrovd@yandex.ru

**Abstract.** In this work, we investigate Batch Normalization technique and propose its probabilistic interpretation. We propose a probabilistic model and show that Batch Normalization maximizes the lower bound of its marginal log-likelihood. Then, according to the new probabilistic model, we design an algorithm which acts consistently during train and test. However, inference becomes computationally inefficient. To reduce memory and computational cost, we propose Stochastic Batch Normalization – an efficient approximation of proper inference procedure. This method provides us with a scalable uncertainty estimation technique. We demonstrate the performance of Stochastic Batch Normalization on popular architectures (including deep convolutional architectures: VGG-like and ResNets) for MNIST and CIFAR-10 datasets.

**Keywords:** Uncertainty estimation · Deep Learning ·
Batch Normalization

## 1 Introduction

Deep Neural Networks have demonstrated state-of-the-art performance on many problems and are successfully integrated in real-life scenarios: semantic segmentation, object detection and scene recognition, to name but a few. Usually the quality of a model is measured in terms of accuracy, however, accurate uncertainty estimation is also crucial for real-life decision-making applications, such as self-driving systems and medical diagnostic. Despite high accuracy rate, DNNs are prone to overconfidence even on out-of-domain data.

The Bayesian framework lends itself well to uncertainty estimation [8], but exact Bayesian inference is intractable for large models such as DNNs. To address this issue, a number of approximation inference techniques have been proposed recently [3,11]. It has been shown that Dropout, a well-known regularization technique [10], can be treated as a special case of stochastic variational inference [5,9]. Also [1] showed that stochasticity induced by Dropout can provide well-calibrated uncertainty estimation for DNNs. Multiplicative Normalizing Flows

© Springer Nature Switzerland AG 2019
H. Lu et al. (Eds.): ISNN 2019, LNCS 11554, pp. 261–269, 2019.
https://doi.org/10.1007/978-3-030-22796-8_28

[7] is another approximation technique that produces great uncertainty estimation. However, such complex method is hard to scale to very deep convolutional architectures. Moreover, recently proposed Residual Network [2] with more than a hundred layers does not have any noise inducing layers such as Dropout. This type of layer leads to a significant accuracy degradation [2]. This problem can be addressed by non-Bayesian Deep Ensembles method [6], which provides competitive uncertainty estimation, but it requires to store several separate models and perform forward passes through all of them to make prediction.

Batch Normalization [4] is an essential part of very deep convolutional architectures. In our work, we treat Batch Normalization as a stochastic layer and propose a way to ensemble batch-normalized networks. The straightforward technique, however, ends up with high memory and computational cost. We, therefore, propose Stochastic Batch Normalization (SBN)—an efficient and scalable approximation technique. We show the performance of our method on out-of-domain uncertainty estimation problem for deep convolutional architectures including VGG-like, ResNet and LeNet-5 on MNIST and CIFAR10 datasets. We also demonstrate that SBN successfully extends Dropout and Deep Ensembles methods.

## 2   Method

We consider a supervised learning problem, with a dataset $D = \{(x_i, y_i)\}_{i=1}^{N}$. The goal is to train the parameters $\theta$ of the predictive likelihood $p_\theta(y \mid x)$, modelled by a neural network. To solve this problem stochastic optimization methods with a mini-batch gradient estimator usually are used.

**Batch Normalization.** Batch Normalization attempts to preserve activations of all layers with zero mean and unit variance. In order to do that it uses the mean $\mu(\mathcal{B})$ and variance $\sigma^2(\mathcal{B})$ over the mini-batch $\mathcal{B}$ during training and accumulated statistics on the inference phase:

$$\mathrm{BN}_{\gamma,\beta}^{\mathrm{train}}(x_i) = \frac{x_i - \mu(\mathcal{B})}{\sqrt{\sigma^2(\mathcal{B}) + \epsilon}} \cdot \gamma + \beta \qquad \mathrm{BN}_{\gamma,\beta}^{\mathrm{test}}(x_i) = \frac{x_i - \hat{\mu}}{\sqrt{\hat{\sigma}^2 + \epsilon}} \cdot \gamma + \beta \quad (1)$$

where $\gamma, \beta$ are the trainable Batch Normalization parameters (scale and shift) and $\epsilon$ is a small constant, needed for numerical stability. Note that during training mean and variance are computed over a randomly picked batch ($\mu(\mathcal{B})$, $\sigma(\mathcal{B})$), while during testing the exponentially smoothed statistics ($\hat{\mu}$, $\hat{\sigma}^2$) are used. We further address this inconsistency by proposed probabilistic model.

**Batch Normalization: Probabilistic View.** Note from (1) that forward pass through the batch-normalized network depends not only on $x_i$ but on the entire batch $\mathcal{B}$ as well. This dependency can be reinterpreted in terms of mini-batch statistics $\mu(\mathcal{B}), \sigma(\mathcal{B})$:

$$p_\theta(y_i \mid x_i, \mathcal{B}_{\setminus i}) = p_\theta(y_i | x_i, \mu(\mathcal{B}), \sigma(\mathcal{B})), \qquad (2)$$

where $\mathcal{B}_{\backslash i}$ is a batch without $x_i$. Due to the stochastic choice of mini-batches during training, for a fixed $x_i$ $\mathcal{B}_{\backslash i}$ is a random variable, so mini-batch statistics can be treated as a random variables. The conditional distribution $p_\theta(\mu, \sigma \,|\, x_i, \mathcal{B}_{\backslash i})$ is the product of two Dirac delta functions, centered at $\mu(\mathcal{B})$ and $\sigma(\mathcal{B})$, since statistics are deterministic functions of the mini-batch, and the distribution of mean and variance given $x_i$ is an expectation over mini-batch distribution. During inference we average the distribution $p_\theta(y|x, \mu, \sigma^2)$ over the normalization statistics:

$$p_\theta(\mu, \sigma | x_i) = \mathbb{E}_{\mathcal{B}_{\backslash i}} \, \delta_{\mu(\mathcal{B})}(\mu) \delta_{\sigma(\mathcal{B})}(\sigma) \qquad p_\theta(y|x) = \mathbb{E}_{p_\theta(\mu, \sigma|x)} p(y|x, \mu, \sigma) \quad (3)$$

**Connection to Batch Normalization.** In Sect. 3 we show that during training Batch Normalization (1) performs the unbiased one-sample MC estimation of a gradient of a lower bound to the marginal likelihood (3). Thus, such probabilistic model corresponds to Batch Normalization during training. However, on test phase Batch Normalization uses exponentially smoothed statistics $\mathbb{E}\mu \approx \hat{\mu}, \mathbb{E}\sigma \approx \hat{\sigma}$, which can be seen as a biased approximation of (3):

$$\mathbb{E}_{p_\theta(\mu, \sigma|x_i)} p(y_i|x_i, \mu, \sigma) \approx p_\theta(y|x, \mathbb{E}\mu, \mathbb{E}\sigma)$$

Straightforward MC averaging can be used for better unbiased estimation of (3), however, it is inefficient in practice. Indeed, to draw one sample from the distribution over statistics (3) we need to pass an entire mini-batch through the network. So, to make MC averaging for single test object, we need to perform several forward passes with different mini-batches sampled from the training data. To address this drawback we propose Stochastic Batch Normalization.

**Stochastic Batch Normalization.** To address memory and computational cost of straightforward MC estimation, we propose to approximate the distribution of Batch Normalization statistics $p_\theta(\mu, \sigma \,|\, x_i)$ with a fully-factorized parametric approximation $p_\theta(\mu, \sigma \,|\, x_i) \approx r(\mu)r(\sigma)$. We parameterize $r(\mu)$ and $r(\sigma)$ in the following way:

$$r(\mu) = \mathcal{N}(\mu | \mathrm{m}_\mu, \mathrm{s}_\mu^2) \qquad r(\sigma) = \mathrm{Log}\mathcal{N}(\sigma | \mathrm{m}_\sigma, \mathrm{s}_\sigma^2) \qquad (4)$$

Such approximation works well in practice. In Sect. 4 we show that it accurately fits the real marginals. Since approximation no longer depends on the training data, samples for each layer can be computed without passing the entire batch through the network and it is possible to make prediction in an efficient way.

To adjust parameters $\{m_\mu, s_\mu, m_\sigma, s_\sigma\}$ we minimize the KL-divergence between distribution induced by Batch Normalization (3) and our approximation $r(\mu)r(\sigma)$ for each object:

$$D_{\mathrm{KL}} \left( 1/N \sum_{i=1}^{N} p_\theta(\mu, \sigma \,|\, x_i) \,\|\, r(\mu)r(\sigma) \right) \longrightarrow \min_{m_\mu, s_\mu, m_\sigma, s_\sigma}$$

Since $r$ belongs to the exponential family, this minimization problem is equal to moment matching and does not require gradients computation. In our implementation we simply use exponential smoothing to approximate the sufficient statistics of mean and variance distributions. It can be done for any pre-trained batch-normalized network.

## 3   Lower Bound on Marginal Log-Likelihood

In Sect. 2 we propose the probabilistic view on Batch Normalization which models marginal likelihood $p_\theta(y|x)$. In this section we show that conventional Batch Normalization actually optimizes a lower bound on marginal log-likelihood in such probabilistic model. So the goal is to train the model parameters $\theta$ given training dataset $D = \{(x_i, y_i)\}_{i=1}^N$. Using Maximum Likelihood approach we need to maximize the following objective $\mathcal{L}(\theta)$:

$$\mathcal{L}(\theta) = \sum_{i=1}^N \log p_\theta(y_i|x_i) = \sum_{i=1}^N \log \mathbb{E}_{\mu,\sigma \sim p_\theta(\mu,\sigma|x_i)} \, p_\theta(y_i|x_i, \mu, \sigma) \tag{5}$$

However, the term $\log \mathbb{E}_{\mu,\sigma} \, p_\theta(y_i|x_i, \mu, \sigma)$ is intractable due to the expectation over statistics. We, therefore, construct a lower bound of $\mathcal{L}(\theta)$ using the Jensen-Shannon inequality:

$$\mathcal{L}_{\mathrm{BN}}(\theta) = \sum_{i=1}^N \mathbb{E}_{\mu,\sigma} \log p_\theta(y_i|x_i, \mu, \sigma) \le \sum_{i=1}^N \log \mathbb{E}_{\mu,\sigma} \, p_\theta(y_i|x_i, \mu, \sigma) = \mathcal{L}(\theta) \tag{6}$$

To use gradient-based optimization methods we need to compute gradient of $\mathcal{L}_{\mathrm{BN}}(\theta)$ w.r.t. parameters $\theta$. Unfortunately, distribution over $\mu$, $\sigma$ depends on $\theta$ and, therefore, we cannot propagate gradient through the expectation. However, we can use the definition of $p_\theta(\mu, \sigma|x_i)$ from Eq. (3) and reparametrize expectation in terms of mini-batch distribution:

$$
\begin{aligned}
\mathbb{E}_{\mu,\sigma} \log p_\theta(y_i|x_i, \mu, \sigma) &= \int p_\theta(\mu, \sigma|x_i) \log p_\theta(y_i|x_i, \mu, \sigma) d\mu d\sigma \\
&= \int \left( \int \delta_{\mu(\mathcal{B})}(\mu)\delta_{\sigma(\mathcal{B})}(\sigma) p(\mathcal{B}_{\setminus i}) d\mathcal{B}_{\setminus i} \right) \log p_\theta(y_i|x_i, \mu, \sigma) d\mu d\sigma \\
&= \int \left( \int \delta_{\mu(\mathcal{B})}(\mu)\delta_{\sigma(\mathcal{B})}(\sigma) \log p_\theta(y_i|x_i, \mu, \sigma) d\mu d\sigma \right) p(\mathcal{B}_{\setminus i}) d\mathcal{B}_{\setminus i} \\
&= \int \log p_\theta(y_i|x_i, \mu(\mathcal{B}), \sigma(\mathcal{B})) p(\mathcal{B}_{\setminus i}) d\mathcal{B}_{\setminus i} \\
&= \mathbb{E}_{\mathcal{B}_{\setminus i}} \log p_\theta(y_i|x_i, \mu(\mathcal{B}), \sigma(\mathcal{B}))
\end{aligned}
$$

Since distribution over mini-batches does not depend on $\theta$, we now can propagate the gradient through the expectation and use MC approximation for an unbiased estimation. During training Batch Normalization draws mini-batch $\mathcal{B}$ of size $M$ and approximate the full gradient $\nabla \mathcal{L}_{\mathrm{BN}}(\theta)$ in the following way:

$$\nabla \hat{\mathcal{L}}_{\mathrm{BN}}(\theta) = \frac{N}{M} \sum_{i=1}^M \nabla \log p_\theta(y_i|x_i, \mu(\mathcal{B}))$$

Note that Batch Normalization uses the same mini-batch $\mathcal{B}$ to calculate statistics as for gradient estimation. Taking an expectation over mini-batch $\mathcal{B}$, we can actually see that such procedure performs an unbiased estimation of $\nabla\mathcal{L}(\theta)$:

$$
\begin{aligned}
\mathbb{E}_{\mathcal{B}}\nabla\hat{\mathcal{L}}_{\mathrm{BN}}(\theta) &= \frac{N}{M}\sum_{i=1}^{M}\nabla\mathbb{E}_{\mathcal{B}}\log p_{\theta}(y_i|x_i,\mu(\mathcal{B})) \\
&= N\cdot\nabla\mathbb{E}_{\mathcal{B}}\log p_{\theta}(y_i|x_i,\mu(\mathcal{B}),\sigma(\mathcal{B})) \\
&= N\cdot\nabla\mathbb{E}_{x_i}\mathbb{E}_{\mathcal{B}\setminus i}\log p_{\theta}(y_i|x_i,\mu(\mathcal{B}),\sigma(\mathcal{B})) \\
&= \nabla\sum_{i=1}^{N}\mathbb{E}_{\mathcal{B}\setminus i}\log p_{\theta}(y_i|x_i,\mu(\mathcal{B}),\sigma(\mathcal{B})) \\
&= \nabla\mathcal{L}_{\mathrm{BN}}(\theta)
\end{aligned}
$$

So Batch Normalization produces an unbiased gradient estimation of $\nabla\mathcal{L}(\theta)$ during training and can be seen as an approximation for inference in proposed probabilistic model.

## 4  Statistics Distribution Approximation

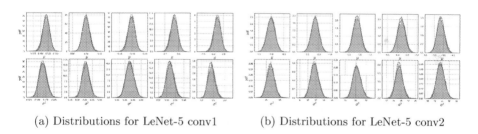

(a) Distributions for LeNet-5 conv1          (b) Distributions for LeNet-5 conv2

**Fig. 1.** The empirical marginal distribution over statistics (blue) for convolutional LeNet-5 layers and proposed approximation (green). Top row for mean distribution and bottom for variance. (Color figure online)

For computational and memory efficiency we propose the following approximation for the real distribution over the batch statistics, induced by Batch Normalization:

$$
r(\mu) = \mathcal{N}(\mu|\mathrm{m}_\mu,\mathrm{s}_\mu^2) \qquad\qquad r(\sigma) = \mathrm{Log}\mathcal{N}(\sigma|\mathrm{m}_\sigma,\mathrm{s}_\sigma^2) \tag{7}
$$

According to our observation, the real distributions are unimodal Fig. 1. Also the Central Limit Theorem implies that the means converge in distributions to Gaussians, therefore we model this distribution using a fully-factorized Gaussian. While the common choice for the variance is Gamma distribution, we choose the

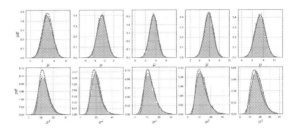

**Fig. 2.** The empirical marginal distribution over statistics (blue) for fully-connected LeNet-5 layer and the proposed approximation (green). Top row corresponds to the means, and the bottom row corresponds to the variances. (Color figure online)

log-normal distribution, as it allows for a more tractable moment-matching. Also as we show in Figs. 1 and 2, the log-normal distribution fits the data well.

To verify the right choice of parametric family we estimate an empirical marginal distributions over $\mu$ and $\sigma^2$ for LeNet-5 architecture on MNIST dataset. To sample statistics from the real distribution we pass different mini-batches from training data through the network. We use Kernel Density Estimation to plot the empirical distribution. The results for convolutional and fully-connected layers of LeNet-5 can be seen in Figs. 1 and 2. It can be seen that the approximation (7) fits the real marginal distributions over $\mu$, $\sigma$ accurately.

## 5  Experiments

**Table 1.** Test errors (%) and NLL scores for known classes. MNIST for LeNet-5 and CIFAR5 for VGG-11 and ResNet-18. SBN column correspond to methods with all Batch Normalization layers replaced by ours SBN.

| Network | Method | Error% | | NLL | |
|---|---|---|---|---|---|
| | | No SBN | SBN | No SBN | SBN |
| LeNet-5 MNIST | SBN | — | $0.53 \pm 0.05$ | — | $0.025 \pm 0.003$ |
| | Deep ensembles | $0.43 \pm 0.00$ | $0.43 \pm 0.00$ | $0.015 \pm 0.001$ | $0.014 \pm 0.001$ |
| | Dropout | $0.51 \pm 0.00$ | $0.49 \pm 0.00$ | $0.016 \pm 0.000$ | $0.015 \pm 0.000$ |
| VGG-11 CIFAR5 | SBN | — | $5.76 \pm 0.00$ | — | $0.302 \pm 0.002$ |
| | Deep ensembles | $\mathbf{5.18 \pm 0.00}$ | $5.23 \pm 0.00$ | $0.177 \pm 0.004$ | $\mathbf{0.154 \pm 0.002}$ |
| | Dropout | $\mathbf{5.32 \pm 0.00}$ | $5.38 \pm 0.00$ | $0.155 \pm 0.001$ | $\mathbf{0.149 \pm 0.001}$ |
| ResNet-18 CIFAR5 | SBN | — | $4.35 \pm 0.17$ | — | $0.255 \pm 0.018$ |
| | Deep ensembles | $\mathbf{3.37 \pm 0.00}$ | $3.34 \pm 0.00$ | $0.138 \pm 0.005$ | $\mathbf{0.110 \pm 0.004}$ |

We evaluate uncertainties on MNIST and CIFAR10 datasets using convolutional architectures. In order to apply Stochastic Batch Normalization to existing architectures we only need to update parameters of our approximation $r(\mu), r(\sigma)$ (4), which does not affect the training process at all. We show that SBN improves both Dropout and Deep Ensembles techniques in terms of out-of-domain uncertainty and test Negative Log-Likelihood (NLL), and maintains the same level of accuracy.

**Experimental Setup.** We compare our method with Dropout and Deep Ensembles. Since [2] showed that ResNet does not perform well with any Dropout layer and suffers from instability, we did not include this method into consideration for ResNet architecture. For Deep Ensembles we trained 6 models for all architectures and did not use adversarial training (as suggested by [6]) since this technique results in lower accuracy.

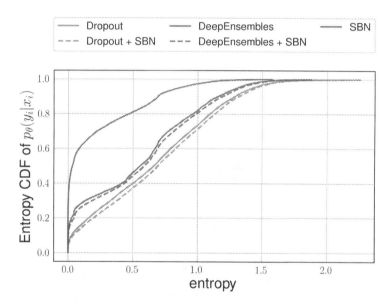

**Fig. 3.** Results for LeNet-5 on notMNIST. Empirical CDF of entropy for out-of-domain data. SBN corresponds to model with all Batch Normalization layers replaced by Stochastic Batch Normalization. The more to the right and the lower, the better.

**Uncertainty Estimation on notMNIST.** For this experiment we trained LeNet-5 model on MNIST and evaluated the entropy of the predictive distribution on notMNIST, which is out-of-domain data for MNIST, and plot the empirical CDF on Fig. 3. We also report the test set accuracy and NLL scores, the results can be seen at Table 1.

**Uncertainty Estimation on CIFAR10.** To show that our method scales to deep convolutional architectures well, we perform experiments on VGG-like and ResNet architectures. We split CIFAR10 dataset into two datasets (CIFAR5), and plot the empirical CDF in Fig. 4. We trained networks on randomly chosen 5 classes and evaluated predictive uncertainty on the remaining.

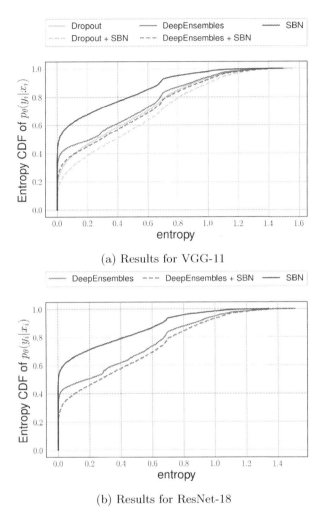

(a) Results for VGG-11

(b) Results for ResNet-18

**Fig. 4.** Empirical CDF of entropy for out-of-domain data. (a) VGG-11 and (b) ResNet-18 on five classes of CIFAR10, hidden during training. SBN corresponds to model with all Batch Normalization layers replaced by Stochastic Batch Normalization. The more to the right and the lower, the better.

We observed that Stochastic Batch Normalization improves both Dropout and Deep Ensembles in terms of out-of-domain uncertainties and NLL score on test data (from the same domain) at the same level of accuracy. However, SBN itself ends up with the more overconfident predictive distribution in comparison to baselines Dropout and Deep Ensembles.

# 6   Conclusion

In this paper, we propose a probabilistic interpretation of Batch Normalization technique. We study a probabilistic point of view and design a new algorithm that behaves consistently during training and test stages. We compare the performance of the proposed algorithm with concurrent techniques on image classification and uncertainty estimation tasks.

**Acknowledgments.** This research is in part based on the work supported by Samsung Research, Samsung Electronics.

# References

1. Gal, Y., Ghahramani, Z.: Dropout as a Bayesian approximation: representing model uncertainty in deep learning. arXiv:1506.02142 (2015)
2. He, K., Zhang, X., Ren, S., Sun, J.: Deep residual learning for image recognition. CoRR abs/1512.03385 (2015)
3. Hoffman, M.D., Blei, D.M., Wang, C., Paisley, J.: Stochastic variational inference. J. Mach. Learn. Res. **14**, 1303–1347 (2013)
4. Ioffe, S., Szegedy, C.: Batch normalization: accelerating deep network training by reducing internal covariate shift. CoRR abs/1502.03167 (2015)
5. Kingma, D.P., Salimans, T., Welling, M.: Variational dropout and the local reparameterization trick. In: Cortes, C., Lawrence, N.D., Lee, D.D., Sugiyama, M., Garnett, R. (eds.) Advances in Neural Information Processing Systems 28, pp. 2575–2583. Curran Associates, Inc. (2015)
6. Lakshminarayanan, B., Pritzel, A., Blundell, C.: Simple and scalable predictive uncertainty estimation using deep ensembles. In: Guyon, I., et al. (eds.) Advances in Neural Information Processing Systems 30, pp. 6405–6416. Curran Associates, Inc. (2017)
7. Louizos, C., Welling, M.: Multiplicative normalizing flows for variational Bayesian neural networks. In: Proceedings of the 34th International Conference on Machine Learning, ICML 2017, Sydney, NSW, Australia, 6–11 August 2017, pp. 2218–2227 (2017)
8. MacKay, D.J.C.: A practical Bayesian framework for backpropagation networks. Neural Comput. **4**(3), 448–472 (1992). https://doi.org/10.1162/neco.1992.4.3.448
9. Molchanov, D., Ashukha, A., Vetrov, D.: Variational dropout sparsifies deep neural networks. arXiv preprint arXiv:1701.05369 (2017)
10. Srivastava, N., Hinton, G., Krizhevsky, A., Sutskever, I., Salakhutdinov, R.: Dropout: a simple way to prevent neural networks from overfitting. J. Mach. Learn. Res. **15**(1), 1929–1958 (2014)
11. Welling, M., Teh, Y.W.: Bayesian learning via stochastic gradient Langevin dynamics. In: Getoor, L., Scheffer, T. (eds.) ICML, pp. 681–688. Omnipress (2011)

# A Hybrid Neurodynamic Algorithm to Multi-objective Operation Management in Microgrid

Chunliang Gou[1,2], Xing He[1,2(⊠)], and Junjian Huang[1,2]

[1] Chongqing Key Laboratory of Nonlinear Circuits and Intelligent Information Processing, College of Electronic and Information Engineering, Southwest University, Chongqing 400715, China
524634865@qq.com, hexingdoc@swu.edu.cn, hmomu@sina.com
[2] Key Laboratory of Machine Perception and Children's Intelligence Development, Chongqing University of Education, Chongqing 400067, China

**Abstract.** In this paper, we consider a microgrid framework consisting of four power generation units, such as gas turbine, fuel cell, diesel generator and photovoltaic power generation. We focus on the minimum power generation cost under the lowest environmental pollution, combining with particle swarm optimization (PSO) and projection neural network. In this framework, we consider the two objectives simultaneously, both economic cost and pollution emission. The projection neural network is used to find the local optimal value, and then the PSO algorithm is used to update the weight to increase the solution diversify and seek global optimization. The convergence and stability of the projection neural network algorithm are reflected in the simulation.

**Keywords:** Multi-objective optimization ·
Particle swarm optimization · Microgrid · Projection neural network

## 1 Introduction

A microgrid, which refers to a small power distribution system consisting of distributed power supply (micro-turbines, PV), energy storage device(fuel cells), load and so on [1]. At present, the scheduling strategy of the microgrid is divided into fixed strategy and optimization strategy. Since dynamic optimization considers the coordination between multi-period equipment operations, the optimization effect is better for microgrids that usually contain time-coupled characteristic components such as energy storage and generators [2]. How to calculate the lowest economic cost under multiple power generation units and consider the minimum environmental pollution emissions is the primary problem of microgrid economic benefits. In the multi-objective microgrid, to achieve maximum

This work is supported by the Natural Science Foundation Project of Chongqing CSTC (Grant no. cstc2018jcyjAX0583, cstc2018jcyjAX0810).

H. Lu et al. (Eds.): ISNN 2019, LNCS 11554, pp. 270–277, 2019.
https://doi.org/10.1007/978-3-030-22796-8_29

economic benefits of microgrid is to find the optimal solution under multiple objectives [3].

Multi-objective optimization, is an algorithm that optimizing two or more objective with a number of constraints, and aims to find the best pareto point in multi-objective problems, which has been widely used in various fields, such as civil, commercial, and military [4]. There are many methods for solving multi-objective optimization problems, such as weight sum method, NSGA-II, MOEA/D and indicator-based selection. In this paper, we combine the weighted Chebyshev method with the microgrid model to calculate some practical optimization problems. Based on the above proposed, the method is used to sacalarize multi-objective to single objective, and the projection neural network is used to find individual best solutions for PSO. Meanwhile, PSO is employed to reinitialization and optimize weight vectors [5].

The remainder of this paper is organized as follows. In Sect. 2, the modeling and problem formulation of microgrid is given. In Sect. 3, the method about goal conversion and the algorithm based on projection neural network and PSO is proposed. Section 4 presents the simulation example to verify the proposed hybrid system. The conclusions is given in Sect. 5.

## 2    Problem Formulation and Model Description

### 2.1    Electricity Cost Function

**Fuel Generator Function.** Fuel power generation is the most basic and traditional part of electrical energy. The whole fuel generator cost in terms of microgrid model is described as follows:

$$F_{MT}\left(P_{MT}\right) = Q \cdot P_{MT} \cdot \Delta t / \left(\eta_{MT} \cdot C\right) \tag{1}$$

where $Q$ represents the natural gas prices, $\Delta t$ indicates generator running time, $C$ is natural gas calorific value, $P_{MT}$ and $\eta_{MT}$ is the fuel generator power and power generation efficiency, respectively.

**Fuel Cell Function.** Fuel cells are also an important part of the microgrid. As an energy storage unit, it also plays an important role. The real fuel cell cost is described as follows:

$$F_{FC}\left(P_{FC}\right) = \rho \cdot \frac{1}{\varsigma} \cdot \frac{P_{FC}}{\eta_{FC}} \tag{2}$$

where $\rho$ represents actual price parameter and $\varsigma$ denotes gas calorific value. $P_{FC}$ and $\eta_{FC}$ is the fuel cell power and power generation efficiency, respectively.

**Diesel Generators Function.** Diesel generator as a supplemental unit of smart grid, the cost function is described as follows:

$$F_{DE}\left(P_{DE}\right) = \nu \cdot P_{DE}^2 + \tau \cdot P_{DE} + \iota \tag{3}$$

where $\nu$, $\tau$ and $\iota$ are the generator specifications, $P_{DE}$ is the diesel generators power and assume that the efficiency of the diesel generator is 80%.

**Photovoltaic (PV) Function.** Photovoltaic power generation as the cleanest source of energy has been increasingly used in people's daily lives. The microgrid power output of PV is described as follows:

$$F_{PV}(P_{PV}) = \varrho \cdot P_{PV} \cdot \Delta t - k_{PV} \cdot P_{PV} \cdot \Delta t \tag{4}$$

where $k_{PV}$ is the operation and maintenance costs of PV, $P_{PV}$ is the PV power and $\varrho$ is the photovoltaic unit parameter. In summary, the problem above the total cost function can be expressed as follows:

$$Min \quad F_c = F_{MT} + F_{FC} + F_{DE} + F_{PV} \tag{5}$$

## 2.2 Emission Function

Industrial production is always accompanied by this pollutant discharge, and the common industrial pollutants are Sox, NOx, CO2, etc. Their emissions have a mathematical relationship with power generation. The pollutant emissions of the entire microgrid can be expressed by the following function:

$$Min \ E\left(P_{out,q}\right)$$
$$= \sum_{q=1}^{N} \left(\alpha_q + \beta_q \cdot P_{out,q} + \gamma_q \cdot P_{out,q}^2 + \delta_q \cdot \exp\left(d_q \cdot p_q\right)\right) \tag{6}$$

where $P_{out,q}$ are the power of fuel generator, fuel cell, diesel generators, PV units ($q = 1, 2, 3, 4$), respectively. $\alpha_q$, $\beta_q$ and $\gamma_q$ are the emission parameters of $q^{th}$ generator unit and $d_q$ and $p_q$ are the exponential parameters of $q^{th}$ generator unit.

## 2.3 Constraints

$$\sum_{q=1}^{4} \left(P_{out,q}\right) = P_L \tag{7}$$

$$P_{out,q,\min} \leqslant P_{out,q} \leqslant P_{out,q,\max} \tag{8}$$

where $P_{out,q}$ represents $q^{th}$ power generation unit, $P$ denotes the total load, and $P_{out,q,\min}$, $P_{out,q,\max}$ means the maximum and the minimum output power, respectively.

## 2.4 Multi-objective Problem Model

In this section, we focus two important goals of power generation cost and pollutant emission in the microgrid, simultaneously. Then, the multi-objective problem is described as follows:

$$Min \ [F_c, E] \tag{9}$$

subject to the constraints (7)–(8).

## 3    Algorithm Analysis

In this section, we consider using the weighted Chebyshev method [6] to transform the multi-objective problem into two single objective problem. Projective neural network is adopted to solve the local best point of the single objective. Figure 1 shows the iterative process of the algorithm. If the obtained solution satisfies the stopping criterion, then the algorithm will converge to the optimal output solutions.

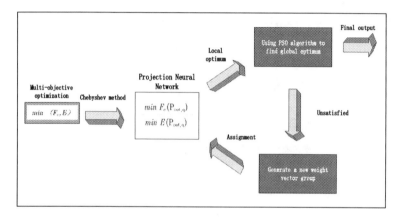

**Fig. 1.** Overview of the proposed hybrid projection neural network algorithm for seeking pareto optimal solution.

### 3.1    Problem Transformation of Multi-objective

This Chebyshev method considers an group ideal parameter $r^* = (r_1^*, r_2^*)^T$, $r_1^* < \min(F_c(P_{out,q}) | P_{out,q} \in \Omega)$, $r_2^* < \min(E(P_{out,q}) | P_{out,q} \in \Omega)$ and where $\Omega = \{P_{out,q} | l_i \leqslant P_{out,q} \leqslant u_i, i = 1,...,n\}$, $l_i$ and $u_i$ are maximum and minimum point respectively, $j = 1, 2$. The transformation equation is described as follows:

$$Min \ f(x|\lambda_i)$$
$$= Min \ \max_{1 \leqslant j \leqslant 2} \left\{ \lambda_1^i |F_c(P_{out,q}) - r_1^*|, \lambda_2^i |E(P_{out,q}) - r_2^*| \right\} \tag{10}$$

where $i = 1, 2, ..., M$, $\lambda_j^i > 0$ and $\sum_{j=1}^m \lambda_j^i = 1$, $j$ denotes the $j^{th}$ objective function.

### 3.2    Projection Neural Network

In general, for a single-objective problem, it will contain a single target and an inequality constraint [7]. Combined with the Chebyshev method we discussed in the previous section, the problem can be transformed into two separate goals: electricity cost and emission. In this way, projection neural networks are used to solve these subproblems [8]. The electricity cost derivation equation can be expressed as follows:

$$\frac{dP_{out,q}}{dt} = -P_{out,q} + H\left(P_{out,q} - \nabla F_c\left(P_{out,q}\right) + \lambda\right)$$

$$\frac{d\lambda}{dt} = \sum_{q=1}^{N}\left(P_{out,q}\right) - P_L \tag{11}$$

and the emission derivation equation is described as follows:

$$\frac{dP_{out,q}}{dt} = -P_{out,q} + H\left(P_{out,q} - \nabla E\left(P_{out,q}\right) + \lambda\right)$$

$$\frac{d\lambda}{dt} = \sum_{q=1}^{N}\left(P_{out,q}\right) - P_L \tag{12}$$

where $P_{out,q}$ represents $q^{th}$ power generation unit; $H$ indicates projection operator; $\nabla F_c$, $\nabla E$ are the gradients of $F_c$ and $E$, respectively.

$$H\left(x\right) = \begin{cases} u, x > u \\ x, l \leqslant x \leqslant u \\ l, x < l \end{cases} \tag{13}$$

### 3.3   Particle Swarm Optimization

In this section, PSO is used to accomplish following goals. The first one is to obtain global optimal solution, and the secondary goal is update initial value and weight vector. Let the position and velocity of $i^{th}$ particle denote as $P_{out,q}^i = \left(P_{out,1}^i, P_{out,2}^i, ..., P_{out,n}^i\right)^T$ and $v^i = \left(v_1^i, v_2^i, ..., v_n^i\right)^T$, respectively. Basis on iteration of the particle change, its velocity and position are redefined according to

$$\begin{cases} v^q \leftarrow \omega v^q + c_1 r_1\left(\tilde{P}_{out,q}^i - P_{out,q}^i\right) + c_2 r_2\left(\hat{P}_{out,q} - P_{out,q}^i\right) \\ P_{out,q}^i \leftarrow P_{out,q}^i + v^i \end{cases} \tag{14}$$

where $\omega$ denotes inertia weight, its value range is 0.8 to 0.2; $c_1$ and $c_2$ are the individual optimal weight value and global optimal weight value; the parameter value of $r_1$ and $r_2$ are generally taken as 0.1 to 0.8; $\tilde{P}_{out,q}^i$ and $\hat{P}_{out,q}$ are the Individual optimal point and global optimal point of the $i^{th}$ particle. $c_1$ and $c_2$ are generally set to 1.0 to 2.0 or controlled adaptively.

### 3.4   Hybrid Neurodynamic Optimization Algorithm

According to the weighted Chebyshev scalarization mentioned before, the multi-objective based on microgrid is transformed into multiple single-objective optimization problems to solve. However, for $i^{th}$ subproblem, the projection neural network algorithm [9] is considered. The $i^{th}$ projection neural network with weighted Chebyshev is described as follows:

$$\frac{dP^i_{out,q}}{dt} = -P^i_{out,q} + H\left(P^i_{out,q} - \nabla F\left(z^i\right) + \lambda^i\right)$$

$$\frac{d\lambda^i}{dt} = \sum_{q=1}^{N}\left(z^i\right) - P_L \tag{15}$$

where $P^i_{out,q}$ is the current value of the $i^{th}$ projective neural network; $F = \{F_c, E\}$; $\lambda^i$ is hidden state vector; $z^i = \left(P^i_{out,q}, \delta\right)$; $\delta = F_c\left(z^i\right)$. When the projection neural network converges to stopping condition: $||\hat{P}_{out}\left(t\right) - \hat{P}_{out}\left(t-1\right)|| \leqslant \varepsilon_1$, the procedures end. The value of $\varepsilon_1$ is a positive number close to zero.

### 3.5   Weight Update Method

In the scalarization approach, the distribution of the multi-objective solution largely depends on the selection of the weight vector. In this section, PSO is adopted to update the weight value. The updated method of PSO $\lambda^j$ is

$$\begin{cases} \varphi^j \leftarrow \omega\varphi^j + c_1 r_1 \left(\tilde{\lambda}^j - \lambda^j\right) + c_2 r_2 \left(\hat{\lambda}^l - \lambda^j\right) \\ \lambda^j \leftarrow \lambda^j + \varphi^j \end{cases} \tag{16}$$

where $\varphi^j$ denotes velocity information; $\tilde{\lambda}^j$ represents individual optimal value; $\hat{\lambda}^l$ is the global optimal value, and before an iteration update is completed(16), we select the nearest weight vector $\left\{\hat{\lambda}^1, \hat{\lambda}^2, ..., \hat{\lambda}^M\right\}$ as the assignment of index $l$, $\lambda^j \in R^m$ and $j = 1,...,M \cdot p$, $p$ represents the number of groups of $\tilde{\lambda}$, and $M$ represents the number of $\tilde{\lambda}$ in each group. The weight vector converges: $||\lambda\left(k\right) - \lambda\left(k-1\right)|| \leqslant \varepsilon_3$, $\varepsilon_3$ is a positive number close to zero. Algorithm starts from execution to meet conditional termination (i.e., $S\left(A\right) \leqslant \varepsilon_2$, where $A$ is the solution set obtained by the above method, and $\varepsilon_2$ is a positive number close to zero) [10]. When the value in the solution set A satisfies the well-distributed, the algorithm terminates and the result is the final solutions of the algorithm.

## 4   Simulation Result

By simulating the above problem, the results are presented to illustrate the feasibility of the hybrid projection neural network algorithm. Two goals of electricity cost and environmental cost are considered in simulation to find the optimal value and verify the stability of the algorithm. Figure 2 shows bi-objective pareto solutions via projection neural network algorithm. We consider the actual situation two cost function that involve power balance constraint and generation capacity limit. If we take more group weight, dots in the graph will be more densely distributed, we can obtain more Pareto-optimal solutions.

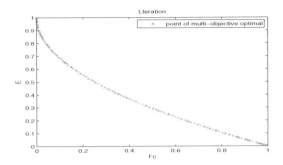

**Fig. 2.** Areto-optimal solutions of bi-objective.

In order to verify the convergence and stability of the hybrid neurodynamic algorithm, we recorded the iteration process of each variable in the objective function in Fig. 3. The four lines represent the power of four generators. The curve tends to stabilize after several iterations. We obtained the desired result by simulation experiment.

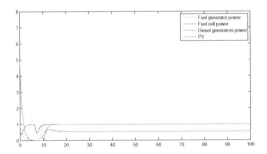

**Fig. 3.** Variable convergence chart.

## 5    Conclusions

In this paper, a hybrid projection neural network algorithm is used to solve microgrid multi-objective optimization. Two important issues of power generation cost and pollution emission in the microgrid are considered. Simulation result shows the hybrid neruodynamic algorithm has high stability and convergence speed, and the diversity of solutions is guaranteed based on PSO algorithm. Future works may focus on extending to more objectives optimization in microgrid, reducing the complexity and improving stability and accuracy of the algorithm.

# References

1. Dimeas, A., Hatziargyriou, N.: A multi-agent system for microgrids. In: Vouros, G.A., Panayiotopoulos, T. (eds.) SETN 2004. LNCS (LNAI), vol. 3025, pp. 447–455. Springer, Heidelberg (2004). https://doi.org/10.1007/978-3-540-24674-9_47
2. Nagata, T., Sasaki, H.: A multi-agent approach to power system restoration. IEEE Trans. Power Syst. **3**, 1551–1556 (2002)
3. Tapia, M.G.C., Coello, C.A.C.: Applications of multi-objective evolutionary algorithms in economics and finance: a survey. In: IEEE Congress on Evolutionary Computation (CEC), pp. 532–539, September 2007
4. Marler, R.T., Arora, J.S.: Survey of multi-objective optimization methods for engineering. Struct. Multidiscip. Optim. **26**(6), 369–395 (2004)
5. Agrawal, S., Panigrahi, B.K., Tiwari, M.K.: Multiobjective particle swarm algorithm with fuzzy clustering for electrical power dispatch. IEEE Trans. Evol. Comput. **12**(5), 529–541 (2008)
6. Leung, M.F., Wang, J.: A collaborative neurodynamic approach to multiobjective optimization. IEEE Trans. Neural Netw. Learn. Syst. (2018, in press). https://doi.org/10.1109/TNNLS.2806481
7. Xia, Y., Leung, H., Wang, J.: A projection neural network and its application to constrained optimization problems. IEEE Trans. Circ. Syst. **49**(4), 447–458 (2002)
8. Liu, Q., Wang, J.: A projection neural network for constrained quadratic minimax optimization. IEEE Trans. Neural Netw. Learn. Syst. **26**(11), 2891–2900 (2015)
9. He, X., Huang, T., Li, C., Che, H., Dong, Z.: A recurrent neural network for optimal real-time price in smart grid. Neurocomputing **149**, 608–612 (2015)
10. Van Veldhuizen, D.A., Lamont, G.B.: On measuring multiobjective evolutionary algorithm performance. IEEE Congr. Evol. Comput. **1**, 204–211 (2000)

# Simulation of a Chaos-Like Irregular Neural Firing Pattern Based on Improved Deterministic Chay Model

Zhongting Jiang[1], Dong Wang[1,2(✉)], Jin Sun[1], Hengyue Shi[1],
Huijie Shang[1], and Yuehui Chen[1,2]

[1] School of Information Science and Engineering,
University of Jinan, Jinan 250022, China
ise_wangd@ujn.edu.cn
[2] Key Laboratory of Medicinal Plant and Animal Resources of Qinghai-Tibet
Plateau in Qinghai Province, Qinghai Normal University, Xining, China

**Abstract.** In this paper, the deterministic Chay model was improved considering the generation mechanism of an action potential, with special relevance to the opening of potassium channel after depolarization. Then a chaos-like irregular non-periodic neural firing pattern, which was lying between period n and period (n + 1) bursting in a period-adding bifurcation and composed of alternating period n and period (n + 1) bursts, was also simulated by this improved Chay model. The nonlinear time series analysis results suggest this pattern display both deterministic and stochastic dynamic characteristics, as same as those results in the previous studies. This pattern was always simulated by stochastic neuron models and considered to be coherence resonance near the bifurcation points induced by the inner noise. However, there was no noise in this improved deterministic Chay model. This present paper attempted to discuss and preliminarily explain the generation mechanism of this firing pattern from the standpoint of the unification of certainty and randomness.

**Keywords:** Neural discharge activity · Deterministic Chay model ·
Action potential · Chaos-like · Neural firing pattern

## 1 Introduction

The biological nervous system shows strong non-linearity from a nerve unit to a neural network such as the brain. As the main content of neurodynamics, using the mathematical model to simulate the real neural discharge activity, and using the nonlinear theory to analyze the dynamic features under simulation and experimental data, can deepen our understanding to some phenomena of the nervous system [1–4].

As the basic unit of nervous system, related research on a neuron is the foundation of complex biomechanical of neural activities at different disciplines and levels [5]. The information encoding mode of neuron is to change the time and rate of action potential (AP) [6]. The transmembrane ionic current, which is caused by the opening and closing of ion channels, is the material basis to the generation and change of the AP, as shown in Fig. 1 [7]. Then, the correct description, reasonable assumption and simplicity for

© Springer Nature Switzerland AG 2019
H. Lu et al. (Eds.): ISNN 2019, LNCS 11554, pp. 278–287, 2019.
https://doi.org/10.1007/978-3-030-22796-8_30

the ion channel behavior at the AP level have been very important for the construction and improvement of neuron models [6].

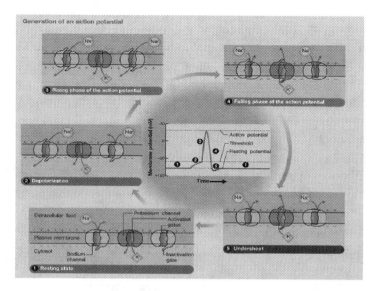

**Fig. 1.** The transmembrane ionic current in different phases of AP [7]

With the in-depth research, the viewpoint that the neural information is coded by not only frequency but also rhythm is gradually and widely accepted [8, 9]. By analysis on interspike intervals (ISIs) only for one neuron, abundant complex discharge rhythms and transitions patterns display nonlinear phenomena, such as periodic, chaotic firing patterns, period adding bifurcation, and so on [10–12]. The underlying unification between confirmation and randomicity from these rhythms is presented, especially in multimodal neural firing patterns, like chaotic discharge, integer multiple firing pattern and so forth. Chaos has been attracting a great deal of attention and identified as the typical deterministic firing pattern with some stochastic characteristics by a series of mature dynamical analysis methods developed in the past few decades [10]. Others like on-off firing and integer multiple firing were once considered as the deterministic chaos, which seemed to be better explained via the introduction of stochastic resonance (SR) mechanisms [13, 14]. However, the newly observed chaos-like irregular neural firing patterns in both model simulation and biological experiment, such like the patterns lying between periodic patterns in period adding bifurcation, still face how to be effectively identified [15, 16].

This present paper dealt with the chaos-like irregular patterns simulated by the improved Chay model, which was added a peak constraint term without noise according to the reasonable assumption for the AP. Then a preliminary explanation was tried to give to the analysis results. By the work done in this paper, we hope to enrich the theoretical connotation for the research of nonlinear dynamics and neuroscience, and provide some practical methods.

## 2 Theoretical Chay Model

Chay model [17] is a typical realistic model which describes the firing behavior of neurons based on the dynamic behavior of ion channels. It can simulate abundant discharging behavior observed in the real experiment, not only rest and discharge, but also complex rhythms, such as periodic rhythms and chaotic rhythm, as well as doubling period of rhythm and period-adding bifurcation (PAB). It can also well simulate and explain transitions between different complex rhythms [11, 16].

Deterministic Chay model (DCM) is formulated as follows:

$$\frac{dV}{dt} = g_I m_\infty^3 h_\infty (V_I - V) + g_{K,v} n^4 (V_K - V) + g_{K,C} \frac{C}{1+C} (V_K - V) + g_L (V_L - V) \quad (1)$$

$$\frac{dn}{dt} = \frac{n_\infty - n}{\tau_n} \quad (2)$$

$$\frac{dC}{dt} = \rho \left( m_\infty^3 h_\infty (V_C - V) - K_C C \right) \quad (3)$$

$V$, $n$ and $C$ represent cell membrane potential, $K^+$ channel activation probability, intracellular $Ca^{2+}$ concentration, respectively. $\tau_n$ is relaxation time. $V_C$ is the equilibrium potential of $Ca^{2+}$ channel. In many previous researches, a noise term was always joined in the first formula to make up the stochastic Chay model (SCM), which was used to study the influence of noise in neural activities.

Here, we made a reasonable hypothesis considering the mechanism of AP generation. When the voltage of AP reaches the spike peak, that is the joint point between phase 2 and 3 in Fig. 1, the $K^+$ channel will open instantaneously and completely accompanied by the end of depolarization and the beginning of repolarization. Consistent with the physiological process, a coefficient $w_K$ was joined in the Eq. (2), as Eq. (4).

$$\frac{dn}{dt} = w_K \frac{n_\infty - n}{\tau_n} \quad (4)$$

When the opening probability of $K^+$ channel opens instantaneously and completely at the spike peak (the junction of phase 3 and phase 4 in Fig. 1), the value of the second formula of the Chay model should be 1. At this time, the value of $w_K$ should be $\tau_n/(n_\infty - n)$. This constraint does not work except at the peak, where $w_K = 1$. Thus, Eq. (4) and Eqs. (1), (3) formed the improved Chay model based on peak constraints. It should be noted that this improved model was an deterministic neuron model without any noise, here named improved DCM in this paper. The parameter settings in this paper can be referred to in [15].

## 3  Time Series Analysis Methods

As mentioned above, the dynamical features are supposed to be implied in neural discharge rhythms. Thus, the ISI time series transformed from the spike trains are usually used for further analysis. The time series analysis methods in this paper included nonlinear analysis methods based on phase space reconstruction theory, such as first return map (FRM), nonlinear prediction (NPE), approximate entropy (ApEn). Others as complexity, surrogate data (SD) and autocorrelation coefficient (ACC) analysis were also covered. All the methods were as same as those described in [16].

## 4  Numerical Simulation Results and Analysis

A series of results suggested that the improved DCM could well numerically simulate abundant neural firing patterns and bifurcations as same as those in previous studies. All these similar activities will be particularly introduced in another article.

Here, much more attention will be paid to a kind of chaos-like irregular firing pattern lying at the bifurcation points in the PAB. This cannot be described in the DCM according to references results [16], which was also confirmed in this paper. In the DCM, when $\lambda n = 240$ with $Vc$ decreased from 486 mV to 162 mV, the rhythm is PAB scenarios illustrated in Fig. 2(a) from period 1 to period 5 bursting, and there is no other rhythm near the bifurcation points.

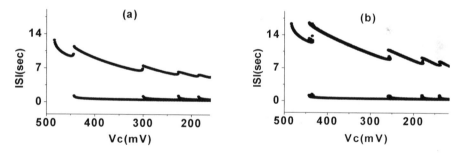

**Fig. 2.** (a) PAB without any other non-periodic patterns in the DCM ($\lambda n = 240$). (b) PAB with irregular firing patterns among neighbouring burstings in the improved DCM ($\lambda n = 240$)

However, the results were quite different in the improved DCM. The irregular firing between period $n$ and period $(n + 1)$ bursting ($n = 1, 2, 3, 4$) in PAB was numerically simulated, which illustrated in Fig. 2(b). This pattern could be considered as the transitions between period $n$ and period $(n + 1)$ bursts ($n = 1, 2, 3, 4$), which illustrated in Fig. 3. It was intuitively different from chaos because of no other bursts. Obvious single burst was indicated by oblique arrows.

From the generated location and the composition of spike trains, this irregular non-periodic pattern imitated by the improved DCM is similar to the stochastic multimodal firing imitated by the SCM reported before. However, the analysis of ISI shows that it has definite properties like chaos.

The firing mode among period 1 and period 2 bursting when $\lambda n = 240$, $Vc = 438$ mV in the improved DCM was taken for example. It has 3 types of ISIs, the smallest of which is the main ISI shared by period 1 and 2 bursting, illustrated in Fig. 4 (a) and (b). The FRM of ISI has a deterministic structure, which is different from the SD of the ISI, which illustrated in Fig. 4(c) and (d), respectively. The ACC of the ISI with one step lag was $-0.8370$, then vibrated and decreased slowly from 1 while the lag was within 12, while only vibrating around between $-0.1$ and $0.1$ by with the lag extended, which illustrated in Fig. 4(e). The comparison of NPE between ISI and its SD suggested it can be predicted in a short step while not in a long step, which illustrated in Fig. 4(f), indicating the chaos-like features for the pattern. ApEn and complexity was 0.315213 and 0.325043 respectively, which shows that the model has lower complexity and higher orderliness.

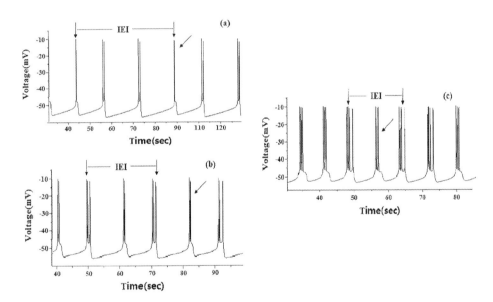

**Fig. 3.** Spike trains of irregular non-periodic bursting among period n and period (n + 1) in the improved DCM ($\lambda$n = 240), the interval among two vertical arrows indicates IEI. (a) n = 1, Vc = 438 mV. (b) n = 2, Vc = 259 mV. (c) n = 2, Vc = 182 mV

The above results suggested that the non-periodic had similar dynamical features, at least about the chaos-like features, to the stochastic multimodal firing simulated through the SCM and observed in pacemaker experiment. To further investigate the randomness of the patterns, a detailed analysis on "event" dynamics is carried out according to the methods described before [15]. Since there were only two alternating bursts in this pattern, each burst could be defined as an event, and the time interval among two consecutive events could be defined inter-event intervals (IEI), which

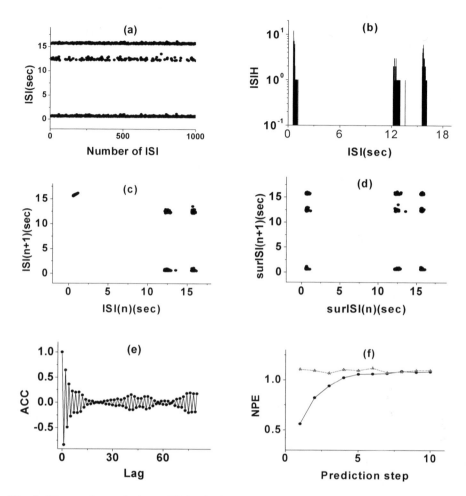

**Fig. 4.** Time series analysis on ISI in the improved DCM (a) ISI series. (b) ISI histogram. (c) FRM. (d) FRM of SD of ISI. (e) ACC. (f) NPE (circle—raw data, triangle—SD)

illustrated in Fig. 3. In the above example, period 2 burst is considered the event. Thus, the analysis on period 2 IEI was shown in Fig. 5.

The analysis results of IEI exhibited typical features of the integer multiple patterns with ApEn 0.688489 and complexity 0.858858 respectively. That was, the chaos-like irregular firing patterns simulated through the improved DCM indeed had characteristics of randomness, which was reflected by the stochastic transition among period 1 and 2 bursts. Analysis on other bursting between period $n$ and period $(n + 1)$ bursting ($n = 2, 3, 4$) got similar results.

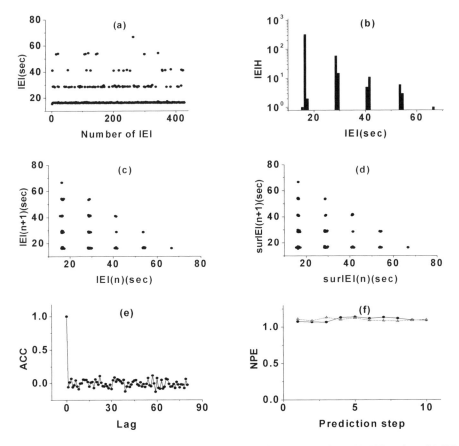

**Fig. 5.** Time series analysis on the IEI series in the improved DCM. (a) IEI series. (b) IEI histogram. (c) FRM. (d) FRM of SD of IEI. (e) ACC. (f) NPE

## 5   Discussion and Conclusion

In this paper, the improved DCM was established considering the $K^+$ channel opening probability during the generation of AP, and had a powerful simulation capability compared to the original model. The improvement with the coefficient $w_K$ adding was carried on mainly for two key reasons as follow.

Firstly, as known, original DCM is a continuous dynamical system as similar to HH model. This is the one reason that the two models could simulate closely to the experiments [18]. But the introduction of $w_K$ turn the improved DCM into discrete with significant sensitive to the $K^+$ channel opening situation. Meanwhile, another factor about the sensory systems of higher animals is that it is very important to define the sensitive intensity of self-perceived information. However, there is no individual coding element that can encode the whole range of sensitive organisms [19]. Maybe, it is obviously useful to use discrete variable in a smaller range to construct an invariant representation in a larger intensity range, such as the intervention of $w_K$.

Secondly, despite it is important for each neuron to encode information through changing the time and rate of its AP, mechanisms that how to control the time and rate of AP are not completely understood. Meanwhile, spiking of individual neuron, especially for the cortical neurons, is given little information. One possible solution for neurons adjust the rate of information transmission through varying membrane potential, so as to adapt to the continuous discharge activities [6, 7]. The improvement to the original Chay model in this paper was also a rewarding attempt in this respect.

The outstanding and interesting work in this paper was that a chaos-like irregular non-periodic firing was simulated through the improved DCM, which was only able to be simulated through the SCM and identified as the stochastic multimodal neural firing before. This pattern was explained using the SR mechanism caused by the noise in the PAB. However, noise plays a role in the corresponding interval. The size of noise and the interval of control parameters need to be controlled artificially. The operation of parameter adjustment is time-consuming and labor-consuming, and it is easy to omit intervals. The improved DCM can be seen that when the noise is not artificially added, it can be simulated. This part has been added to the discussion for explanation. Thus, generated by a deterministic model but with the same dynamic and bifurcation as that in a stochastic model, how should this pattern be understand? We might try to get it from the standpoint of inner dynamics in the nonlinear system.

As same as other biological phenomena, neural discharge activities is caused by the influence of internal dynamics and random forces of organisms, which are random volatility caused by external inputs or other effects [20]. Chaos is the typical phenomenon with deterministic and stochastic features simultaneously, which was observed and validated in both real neural experiment and neuron model. In fact, with typical fractal structure, a neuron itself is a nonlinear system reflecting the unification between confirmation and randomicity, also order and disorder, the unification of certainty and randomness was also reflected in both the firing patterns simulated through the improved DCM and the stochastic multimodal patterns studied before.

Indeed, an important feature of real neuron system is to extract meaningful internal and external signals under various qualifications, in order to successfully catch prey, avoid predators and so on. After a long period of evolution, this capability of picking up related information systematically under exceeding variable environment are called robust, even in the face of fragment information or signal degradation [21]. A typical example is integer multiple firing pattern encodes different signal information in different animals. So, to the chaos-like irregular non-periodic firing pattern here, it may be a special pattern for organism to perceive, encode and process a large class of special information as same as the stochastic multimodal neural firing pattern. This form of information processing is irrelevant to whether noise affect the system, at least partly. Certainly, the hypothesis mentioned above still needs further experimental, stimulation and analytical verification.

**Acknowledgments.** This research was supported by the Shandong Provincial Natural Science Foundation, China (No. ZR2018LF005), the National Key Research and Development Program of China (No. 2016YFC0106000), the Natural Science Foundation of China (Grant No. 61302128, 61573166, 61572230), and the Youth Science and Technology Star Program of Jinan City (201406003).

# References

1. Khurana, V., Kumar, P., Saini, R., Roy, P.P.: EEG based word familiarity using features and frequency bands combination. Cogn. Syst. Res. **49**, 33–48 (2018)
2. Huaguang, G., Zhiguo, Z., Bing, J., Shenggen, C.: Dynamics of on-off neural firing patterns and stochastic effects near a sub-critical Hopf bifurcation. PLoS ONE **10**(4), e0121028 (2015)
3. Li, C.H., Yang, S.Y.: Eventual dissipativeness and synchronization of nonlinearly coupled dynamical network of Hindmarsh-Rose neurons. Appl. Math. Model. **39**(21), 6631–6644 (2015)
4. Shi, R., Hu, G., Wang, S.: Reconstructing nonlinear networks subject to fast-varying noises by using linearization with expanded variables. Commun. Nonlinear Sci. Numer. Simul. **72**, 407–416 (2019)
5. Zhao, Z., Gu, H.: Identifying time delay-induced multiple synchronous behaviours in inhibitory coupled bursting neurons with nonlinear dynamics of single neuron. Procedia IUTAM **22**, 160–167 (2017)
6. Azarfar, A., Calcini, N., Huang, C., Zeldenrust, F., Celikel, T.: Neural coding: a single neuron's perspective. Neurosci. Biobehav. Rev. **94**, 238–247 (2018)
7. Fletcher, A.: Action potential: generation and propagation. Anaesth. Intensive Care Med. **17**(4), 204–208 (2016)
8. Ren, W., Hu, S.J., Zhang, B.J., Wang, F.Z., Gong, Y.F., Xu, J.: Period-adding bifurcation with chaos in the interspike intervals generated by an experimental neural pacemaker. Int. J. Bifurcat. Chaos **7**(08), 1867–1872 (1997)
9. Schoch, A., Pahle, J.: Requirements for band-pass activation of $Ca^{2+}$-sensitive proteins such as NFAT. Biophys. Chem. **245**, 41–52 (2019)
10. Huang, S., Zhang, J., Wang, M., Hu, C.: Firing patterns transition and desynchronization induced by time delay in neural networks. Physica A **499**, 88–97 (2018)
11. Jia, B., Gu, H., Xue, L.: A basic bifurcation structure from bursting to spiking of injured nerve fibers in a two-dimensional parameter space. Cogn. Neurodyn. **11**(2), 1–12 (2017)
12. Bao, B.C., Wu, P.Y., Bao, H., Xu, Q., Chen, M.: Numerical and experimental confirmations of quasi-periodic behavior and chaotic bursting in third-order autonomous memristive oscillator. Chaos. Soliton. Fract. **106**, 161–170 (2018)
13. Shang, H., Xu, R., Wang, D., Zhou, J., Han, S.: A stochastic neural firing generated at a Hopf bifurcation and its biological relevance. In: Liu, D., Xie, S., Li, Y., Zhao, D., El-Alfy, E.S. (eds.) ICONIP 2017. LNCS, vol. 10637, pp. 553–562. Springer, Cham (2017). https://doi.org/10.1007/978-3-319-70093-9_58
14. Shang, H., et al.: Dynamical analysis of a stochastic neuron spiking activity in the biological experiment and its simulation by $I_{Na,P} + I_K$ model. In: Huang, T., Lv, J., Sun, C., Tuzikov, Alexander V. (eds.) ISNN 2018. LNCS, vol. 10878, pp. 850–859. Springer, Cham (2018). https://doi.org/10.1007/978-3-319-92537-0_96
15. Shang, H., Xu, R., Wang, D.: Dynamic analysis and simulation for two different chaos-like stochastic neural firing patterns observed in real biological system. In: Huang, D.-S., Bevilacqua, V., Premaratne, P., Gupta, P. (eds.) ICIC 2017. LNCS, vol. 10361, pp. 749–757. Springer, Cham (2017). https://doi.org/10.1007/978-3-319-63309-1_66
16. Shang, H., Jiang, Z., Xu, R., Wang, D., Wu, P., Chen, Y.: The dynamic mechanism of a novel stochastic neural firing pattern observed in a real biological system. Cogn. Syst. Res. **53**, 123–136 (2019)
17. Chay, T.R.: Chaos in a three-variable model of an excitable cell. Physica D **16**(2), 233–242 (1985)

18. Drukarch, B., et al.: Thinking about the nerve impulse: a critical analysis of the electricity-centered conception of nerve excitability. Prog. Neurobiol. **169**, 172–185 (2018)
19. Sun, W., Marongelli, E.N., Watkins, P.V., Barbour, D.L.: Decoding sound level in the marmoset primary auditory cortex. J. Neurophysiol. **118**(4), 2024–2033 (2017)
20. James, A., Karl, J., Michael, B.: Clinical applications of stochastic dynamic models of the brain, part I: a primer. Neurosci. Neuroimaging **2**(3), 216–224 (2017)
21. Bakay, W.M.H., Anderson, L.A., Garcia-Lazaro, J.A., McAlpine, D., Schaette, R.: Hidden hearing loss selectively impairs neural adaptation to loud sound environments. Nat. Commun. **9**(1), 4298 (2018)

# A New Complex Hyper-chaotic System and Chaotic Synchronization of Error Feedback with Disturbance

Weidong Guan[1,2,3], Dengwei Yan[1,2,3], Lidan Wang[1,2,3(✉)],
and Shukai Duan[1,2,3]

[1] Southwest University, Chongqing 400715, China
ldwang@swu.edu.cn
[2] Chongqing Key Laboratory of Brain Inspired Computing and Intelligent
Control, Chongqing 400715, China
[3] National and Local Joint Engineering Laboratory of Intelligent Transmission
and Control Technology, Chongqing 400715, China

**Abstract.** In this paper, a new complex hyper-chaotic system is proposed. Through the separation of real and imaginary parts, the basic dynamics such as symmetry, dissipation, equilibrium stability, Lyapunov exponent spectrum and power spectrum are studied. Then, according to the Lyapunov stability theory, using the error feedback synchronization method, we design a complex feedback controller to realize the chaotic synchronization of the proposed chaotic system with both parameters and external disturbances. Theoretical analysis shows that the controller can make the synchronization error gradually towards zero point. In addition, the numerical simulation of the complex chaotic synchronization system is carried out. The simulation results further verify the effectiveness of the proposed method.

**Keywords:** Complex hyper-chaotic system · Parameter perturbation ·
External disturbance · Feedback synchronization

## 1 Introduction

In the past few decades, chaos control and synchronization have flourished. In 1990, scientists at the University of Maryland in the United States, Ott, Grebogi and Yorke, first proposed a method for implementing chaotic control using two-dimensional discrete mapping [1]. In a broad sense, chaotic synchronization belongs to the category of chaos control, and chaotic synchronization is a specific chaos control. In 1990, two scientists from the US Naval Laboratory, Pecora and Carrol, proposed a chaotic self-synchronization method [2]. They first used the drive-response method to synchronize the two chaotic systems. Since then, chaotic synchronization has caused a wide range of interests and an in-depth study of this [3].

Complex chaotic systems have more complex dynamic behaviors than real chaotic systems, such as complex Chen systems [4], complex Lu systems [5], complex Lorenz systems [6], and other complex systems [7]. With the development of complex systems, the research on synchronization of complex systems has gradually been widely

© Springer Nature Switzerland AG 2019
H. Lu et al. (Eds.): ISNN 2019, LNCS 11554, pp. 288–296, 2019.
https://doi.org/10.1007/978-3-030-22796-8_31

carried out. In 2004, Mahmoud and Bountis studied the dynamic behavior of a complex nonlinear oscillator for the first time [8]. Since then, the results of synchronization of complex systems have been constantly emerging. In 2008, Zhu and Zhang designed controllers based on passive control principle to achieve complete synchronization of complex chaotic systems [9]; In 2009, Mahmoud and others proposed complete synchronization of chaotic complex Chen system and chaotic complex Lü system [10]. In 2013, Zhang and Zhao proposed the modified function projective synchronization of different complex chaotic systems [11], and realized the modified function projective synchronization of two different complex chaotic systems. In 2014, Skardal et al. studied the optimal synchronization of complex networks [12]. It is noteworthy that the influence of disturbance on the system is not taken into account in the research work of these literatures. In fact, complex systems do not exist in isolation, especially external disturbances of complex systems. Therefore, the study of synchronization of complex systems with disturbances has theoretical and practical significance.

In Sect. 2, a new complex chaotic system is proposed. Separate real part and imaginary part, and two real systems are derived. A hyper-chaotic attractor is obtained by adjusting the parameters. Subsequently, Sect. 3 analyses its dynamic characteristics in detail. In Sect. 4, based on Lyapunov stability theory and error feedback synchronization method, a complex feedback controller is designed to realize chaotic synchronization when the parameters and external of the complex hyper-chaotic system are disturbed. Section 5 is a conclusion.

# 2   A New Complex Hyper-chaotic System

## 2.1   A Subsection Sample

In this paper, a new complex hyper-chaotic system is proposed. The dynamic equation is as follows.

$$
\begin{aligned}
\dot{x} &= ax - x + yz; \\
\dot{y} &= -xz + yz - bx; \\
\dot{z} &= c - \frac{1}{2}(\bar{x}y + x\bar{y});
\end{aligned}
\tag{1}
$$

Where, $x = x_1 + ix_2, y = x_3 + ix_4$ are complex variables, the '$-$' represents complex conjugate variable, $z = x_5, i = \sqrt{-1}$. In view of the fact that complex variables have real and imaginary parts, we separated the system with complex variables into two real systems, imaginary part system and real part system, combined into a new real system. The combined real system has the same chaotic characteristics as the complex system. The mathematical form of the complex hyper-chaotic system is obtained as follows,

$$\begin{aligned}
\dot{x}_1 &= ax_1 - x_1 + x_3x_5; \\
\dot{x}_2 &= ax_2 - x_2 + x_4x_5; \\
\dot{x}_3 &= -x_5x_1 + x_3x_5 - bx_1; \\
\dot{x}_4 &= -x_5x_2 + x_4x_5 - bx_2; \\
\dot{x}_5 &= c - x_1x_3 - x_2x_4;
\end{aligned} \tag{2}$$

When the system parameters $a = 0.5$, $b = 0.8$, $c = 20$, the *Lyapunov* exponents of the system are $L_1 = 0.325$, $L_2 = 0$, $L_3 = 0.007$, $L_4 = -0.068$, $L_5 = -0.402$ respectively. There exist two positive *Lyapunov* exponents, and the sum of *Lyapunov* exponents is less than zero, which indicates that the system is in hyper-chaotic state. The Lyapunov exponent spectrum of the system is shown in Fig. 1. It can be seen that the five Lyapunov exponents of the system tend to a fixed constant over time. On this basis, the dimension of Lyapunov exponent of the new system can be produced.

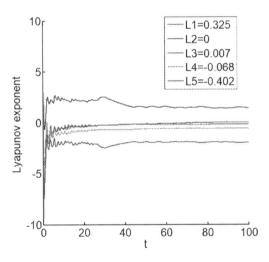

**Fig. 1.** Lyapunov exponents of the system (2).

$$D_L = k + \frac{S_k}{|L_k|} = 3 + \frac{L_1 + L_2 + L_3}{|L_4 + L_5|} = \frac{0.325 + 0 + 0.007}{|-0.068 - 0.402|} = 3.706 \tag{3}$$

Observing the result of formula (3), $D_L$ is exactly not an integer, which further determined that the complex system with hyper-chaotic characteristics. The hyperchaotic attractor of the system is vividly shown in Fig. 2.

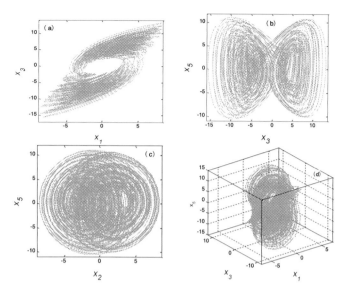

**Fig. 2.** The hyper-chaotic attractors of the system (2): (a) $x_1 - x_3$, (b) $x_3 - x_5$, (c) $x_2 - x_5$, (d) $x_1 - x_3 - x_5$.

## 3  Dynamic Analysis

### 3.1  Initial Sensitivity and Power Spectrum Characteristics

Set the initial values of $x_1$, $x_2$, $x_3$, $x_4$ and $x_5$ to be $-0.1$, change $x_3$ only with the difference of 0.00001, and get the sequence diagram of variables as shown in Fig. 3.

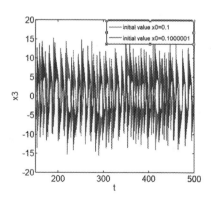

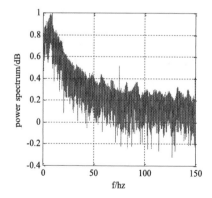

**Fig. 3.** Sensitivity of variable $x_3$ to initial value.

**Fig. 4.** Sensitivity of variable $x_3$ to initial value.

It can be seen that the initial value of the system changes very little, the time-domain waveforms are completely different, and there is no overlapping trend. On the contrary, the difference is disorderly, which indicates that the system is very sensitive to the initial value. In addition, the power spectrum of the system (2) is continuous, as shown in Fig. 4. There are no obvious peaks in the drawing, and the spectrum of the sequence is very wide, which also shows that the system (2) satisfies hyper-chaotic characteristics.

## 3.2  Symmetry and Dissipation

Replaced $(x_1, x_2, x_3, x_4, x_5)$ with $(-x_1, -x_2, -x_3, -x_4, x_5)$, the system has not changed, so it is axisymmetric about $x_5$. Replaced $(x_1, x_2, x_3, x_4, x_5)$ with $(-x_1, -x_2, x_3, x_4, -x_5)$, the system has not changed, so it is axisymmetric about $x_3, x_4$.

$$\nabla V = \frac{\partial \dot{x}_1}{\partial x_1} + \frac{\partial \dot{x}_2}{\partial x_2} + \frac{\partial \dot{x}_3}{\partial x_3} + \frac{\partial \dot{x}_4}{\partial x_4} + \frac{\partial \dot{x}_5}{\partial x_5} = 2a - 2 < 0 \qquad (4)$$

From Eq. (4), it is clear that the system is a dissipative system. It converges exponentially to a set of zero measures, $\frac{dV}{dt} = e^{-2(1-a)}$, that is $V(t) = V_0 e^{-2(1-a)t}$, where $V_0$ is the initial value of the system volume. This is a necessary feature for the existence of chaotic or hyper-chaotic attractors.

## 3.3  Equilibrium Point and Stability Analysis

Let the right side of the nonlinear Eq. (2) be equal to zero, Two equilibrium points of the system can be calculated by solving the problem, which are $S_0 = [-0.3, 0.167 + 5.785i, 0.2, 0.1 - 3.471i, 0]$, $S_1 = [-0.3, 0.167 - 5.785i, 0.2, 0.1 + 3.471i, 0]$. At the equilibrium point $S_0 = [-0.3, 0.167 + 5.785i, 0.2, 0.1 - 3.471i, 0]$, the Jacobian matrix of the system can be obtained by linearizing the system.

$$J = \begin{bmatrix} a-1 & 0 & 0 & 0 & x_3 \\ 0 & a-1 & 0 & x_5 & x_4 \\ -x_5-b & 0 & x_5 & 0 & -x_1+x_3 \\ 0 & -x_5-b & 0 & x_5 & -x_2+x_4 \\ -x(3) & -x(4) & -x(1) & -x(2) & 0 \end{bmatrix} \qquad (5)$$

Through $|\lambda I - J| = 0$, the eigenvalues appears $\lambda_1 = 0.084 + 6.450i$, $\lambda_2 = -0.324 - 6.457i$, $\lambda_3 = -0.5$, $\lambda_4 = -0.01$, $\lambda_5 = -0.260 + 0.007i$. According to Routh-Hurwitz condition, it can be concluded that the equilibrium point $S_0$ is an unstable saddle point. Similarly, the corresponding eigenvalues of $S_1$ equilibrium point can be obtained. $\lambda_1 \lambda_2 \lambda_3 \lambda_4 \lambda_5 = 0.084 - 6.450i, -0.324 + 6.457i, -0.5, -0.01, -0.260 -0.007i$.

Obviously, at least one real part of the eigenvalue corresponding to each equilibrium point is positive and at least one real part is negative, so all the equilibrium points of the system are unstable saddle focus.

## 4  Feedback Synchronization of Complex Hyper-chaotic Systems with Perturbations

### 4.1  Theoretical Derivation

The above-proposed system with perturbation is used as the driving system.

$$\begin{aligned}
\dot{x}_1 &= (a + \xi_a)x_1 - x_1 + x_2 x_3 + d_1; \\
\dot{x}_2 &= -x_1 x_3 + x_2 x_3 - (b + \xi_b)x_1 + d_2; \\
\dot{x}_3 &= c + \xi_c - \tfrac{1}{2}(\bar{x}_1 x_2 + x_1 \bar{x}_2) + d_3;
\end{aligned} \tag{6}$$

The response system is as follows,

$$\begin{aligned}
\dot{y}_1 &= (a + \xi'_a)y_1 - y_1 + y_2 y_3 + d'_1 - k_1(y_1 - x_1); \\
\dot{y}_2 &= -y_1 y_3 + y_2 y_3 - (b + \xi'_b)y_1 + d'_2 - k_2(y_2 - x_2); \\
\dot{y}_3 &= c + \xi'_c - \tfrac{1}{2}(\bar{y}_1 y_2 + y_1 \bar{y}_2) + d'_3 - k_3(y_3 - x_3);
\end{aligned} \tag{7}$$

Among them, $\xi_a \xi_b \xi_c \xi'_a \xi'_b \xi'_c$ are parameter perturbations, $d_1 d_2 d_3 d'_1 d'_2 d'_3$ are external perturbations and $k_1 k_2 k_3$ are feedback intensity coefficients. $x_1 = u_{11} + iu_{21}$, $x_2 = u_{31} + iu_{41}$, $x_3 = u_{51}$, $y_1 = u_{12} + iu_{22}$, $y_2 = u_{32} + iu_{42}$, $y_3 = u_{52}$. The synchronization error between drive system (6) and response system (7) is $e_1 = y_1 - x_1$, $e_2 = y_2 - x_2$, $e_3 = y_3 - x_3$. Error dynamics equation is $\dot{e}_1 = \dot{y}_1 - \dot{x}_1$, $\dot{e}_2 = \dot{y}_2 - \dot{x}_2$, $\dot{e}_3 = \dot{y}_3 - \dot{x}_3$ and $e_1 = e_{u1} + ie_{u2}, e_2 = e_{u3} + ie_{u4}, e_3 = e_{u5}$.

By introducing the variables of Formula (6) and (7) into the error dynamics equation, the following forms of it can be described:

$$\begin{aligned}
\dot{e}_1 = \dot{e}_{u1} + i\dot{e}_{u2} &= [(a - 1)e_{u1} + \xi'_a u_{12} - \xi_a u_{11} + u_{52}e_{u1} + u_{31}e_{u5} - k_1 e_{u1} + d'_1 \\
&\quad - d_1 - k_1 e_{u1}] + i[(a - 1)e_{u2} + \xi'_a u_{22} - \xi_a u_{21} + u_{52}e_{u2} + u_{41}e_{u5} - k_1 e_{u2}] \\
\dot{e}_2 = \dot{e}_{u3} + i\dot{e}_{u4} &= [-u_{52}e_{u1} - e_{u5}u_{11} + u_{52}e_{u3} + e_{u5}u_{31} - be_{u1} - \xi'_b u_{12} + \xi_b u_{11} + d'_2 - d_2 \\
&\quad - k_2 e_{u3}] + i[-be_{u2} - u_{52}e_{u2} + \xi_b u_{21} - e_{u5}u_{21} + u_{52}e_{u4} - k_2 e_{u4} + e_{u5}u_{41} - \xi'_b u_{22}] \\
\dot{e}_3 = \dot{e}_{u5} &= -(u_{32}e_{u1} + u_{42}e_{u2} + u_{11}e_{u3} + u_{21}e_{u4}) + \xi'_c - \xi_c + d'_3 - d_3 - k_3 e_{u5}
\end{aligned} \tag{8}$$

According to Formula (8), it can be obtained.

$$\begin{aligned}
\dot{e}_{u1}e_{u1} &\le (a - 1)e_{u1}^2 + |u_{52}|e_{u1}^2 - k_1 e_{u1}^2 + \tfrac{|u_{31}|}{2}(e_{u5}^2 + e_{u1}^2) + l_1 e_{u1}^2 \\
\dot{e}_{u2}e_{u2} &\le (a - 1)e_{u2}^2 + |u_{52}|e_{u2}^2 - k_1 e_{u2}^2 + \tfrac{|u_{41}|}{2}(e_{u5}^2 + e_{u2}^2) + l_2 e_{u2}^2 \\
\dot{e}_{u3}e_{u3} &\le \tfrac{|u_{52}| + |b|}{2}(e_{u1}^2 + e_{u3}^2) + \tfrac{|u_{11}|}{2}(e_{u3}^2 + e_{u5}^2) \\
&\quad + e_{u3}^2(u_{52} - k_2) + \tfrac{|u_{31}|}{2}(e_{u3}^2 + e_{u5}^2) + l_3 e_{u3}^2 \\
\dot{e}_{u4}e_{u4} &\le \tfrac{|u_{52}| + |b|}{2}(e_{u2}^2 + e_{u4}^2) + \tfrac{|u_{21}|}{2}(e_{u4}^2 + e_{u5}^2) \\
&\quad + e_{u4}^2(u_{52} - k_2) + \tfrac{|u_{41}|}{2}(e_{u4}^2 + e_{u5}^2) + l_4 e_{u4}^2 \\
\dot{e}_{u5}e_{u5} &\le \tfrac{|u_{32}|}{2}(e_{u1}^2 + e_{u5}^2) + \tfrac{|u_{42}|}{2}(e_{u2}^2 + e_{u5}^2) + \\
&\quad \tfrac{|u_{11}|}{2}(e_{u3}^2 + e_{u5}^2) + \tfrac{|u_{21}|}{2}(e_{u4}^2 + e_{u5}^2) + l_5 e_{u5}^2 - k_3 e_{u5}^2
\end{aligned} \tag{9}$$

Where

$$l_1 = \frac{1}{\gamma}\left(|u_{12}||\xi'_a| + |u_{11}||\xi_a| + |d'_1| + |d_1|\right)$$

$$l_2 = \frac{1}{\gamma}\left(|u_{22}||\xi'_a| + |u_{21}||\xi_a|\right)$$

$$l_3 = \frac{1}{\gamma}\left(|u_{12}||\xi'_b| + |u_{11}||\xi_b| + |d'_2| + |d_2|\right)$$

$$l_4 = \frac{1}{\gamma}\left(|u_{22}||\xi'_b| + |u_{21}||\xi_b|\right)$$

$$l_5 = \frac{1}{\gamma}\left(|\xi_c| + |\xi'_c| + |d'_3| + |d_3|\right)$$

Construction of Lyapunov function

$$V(t) = \frac{1}{2}\left(e_{u1}^2 + e_{u2}^2 + e_{u3}^2 + e_{u4}^2 + e_{u5}^2\right) \tag{10}$$

Bring Formula (9) into Formula (10), through inequality transformation, get:

$$
\begin{aligned}
\dot{V}(t) \leq & \left[(a-1) + |u_{52}| - k_1 + \frac{|u_{31}|}{2} + \frac{|u_{52}|}{2} + \frac{|b|}{2} + \frac{|u_{32}|}{2} + l_1\right]e_{u1}^2 \\
& + \left[(a-1) + |u_{52}| - k_1 + \frac{|u_{41}|}{2} + \frac{|u_{52}|}{2} + \frac{|b|}{2} + \frac{|u_{42}|}{2} + l_2\right]e_{u2}^2 \\
& + \left[\frac{|u_{52}| + |b|}{2} + \frac{|u_{11}|}{2} + |u_{52}| - k_2 + \frac{|u_{31}|}{2} + \frac{|u_{11}|}{2} + l_3\right]e_{u3}^2 \\
& + \left[\frac{|u_{52}| + |b|}{2} + \frac{|u_{21}|}{2} + |u_{52}| - k_2 + \frac{|u_{41}|}{2} + \frac{|u_{21}|}{2} + l_4\right]e_{u4}^2 \\
& + \left[\frac{|u_{32}|}{2} + \frac{|u_{42}|}{2} + |u_{11}| + |u_{21}| + |u_{31}| + |u_{41}| + l_5 - k_3\right]e_{u5}^2
\end{aligned}
\tag{11}
$$

$$
k_1 > \max\left\{
\begin{aligned}
& (a-1) + |u_{52}| + \frac{|u_{31}|}{2} + \frac{|u_{52}|}{2} + \frac{|b|}{2} + \frac{|u_{32}|}{2} + l_1; \\
& (a-1) + |u_{52}| + \frac{|u_{41}|}{2} + \frac{|u_{52}|}{2} + \frac{|b|}{2} + \frac{|u_{42}|}{2} + l_2
\end{aligned}
\right\}
$$

$$
k_2 > \max\left\{
\begin{aligned}
& \frac{|u_{52}| + |b|}{2} + \frac{|u_{11}|}{2} + |u_{52}| + \frac{|u_{31}|}{2} + \frac{|u_{11}|}{2} + l_3; \\
& \frac{|u_{52}| + |b|}{2} + \frac{|u_{21}|}{2} + |u_{52}| + \frac{|u_{41}|}{2} + \frac{|u_{21}|}{2} + l_4
\end{aligned}
\right\}
\tag{12}
$$

$$k_3 > \frac{|u_{32}|}{2} + \frac{|u_{42}|}{2} + |u_{11}| + |u_{21}| + |u_{31}| + |u_{41}| + l_5$$

According to formula (11), if the above formula (12) are satisfied, $\dot{V}(t) < 0$, which indicates error dynamics system (8) is asymptotically stable. Let all perturbation values be less than 0.5, $\gamma = 0.1$, and bring in formulas (12). Observe the maximum absolute value in the range of each variable, we can get, $k_1 > 237.5; k_2 > 233; k_3 > 170$.

### 4.2 Simulation Experiment

Here we choose $\xi_a = \xi_b = \xi_c = \xi'_a = \xi'_b = \xi'_c = \sin(t)$ and $d_1 d_2 d_3 d'_1 d'_2 d'_3$ for 0.1. In the simulation experiment, Range-Kutta method is chosen to decouple the drive system (6) and the response system (7), and the time step is chosen as $\tau = 0.001$ (s). Arbitrary selection of a set of initial conditions to verify the correctness of theoretical derivation. The selection of variable initial values for $(1, 6, 5, 2, 3)(6, 1, 3, 5, 0)$ to get the error initial value for $(e_{u1}, e_{u2}, e_{u3}, e_{u4}, e_{u5}) = (5, -5, -2, 3, -3)$. Then, Fig. 5 shows the synchronization error curve when the driving system (6) and the response system (7) select the initial values mentioned. It can be seen from the graph that when the time tends to infinity, the synchronization error gradually approaches zero and stabilizes at zero.

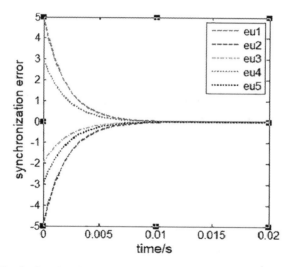

**Fig. 5.** Synchronization error curve of the set of initial values.

## 5   Conclusion

The state variables of complex chaos, in the complex domain, are more complex than real chaotic systems. Complex variables not only increase the content of transmitted information but also improve the security. At present, complex chaos has been widely used in many disciplines such as communication, finance, biology and so on. In this paper, a new complex chaotic system is proposed, and its abundant dynamic characteristics are verified by a comprehensive analysis method. Based on Lyapunov stability theory and error feedback synchronization method, a complex feedback controller is designed to realize chaotic synchronization when the parameters of complex chaos and external disturbances are disturbed. The theory proves that the controller can make the synchronization error of complex system gradually tend to zero and stabilize to zero. The numerical simulation results show that the method is simple, easy to operate and has good control effect. The speed of convergence is fast and has wide application value in practical engineering.

# References

1. Ott, E., Grebogi, C., Yorke, J.A.: Controlling chaos. Phys. Rev. Lett. **64**(11), 1196–1199 (1990)
2. Pecora, L.M., Caeeoll, T.L.: Synchronization in chaotic systems. Phys. Rev. Lett. **64**(8), 8210–8224 (1990)
3. Yang, X.S.: On the existence of generalized synchronizer in unidirectionally coupled systems. Appl. Math. Comput. **122**(1), 71–79 (2001)
4. Elabbasv, E.M., Agiza, H.N.: Synchronization of modified Chen system. Int. J. Bifurcat. Chaos **14**(11), 3969–3979 (2004)
5. Mahmound, G.M., Bountis, T.: Active control and global synchronization of the complex Chen system and Lu system. Int. J. Bifurcat. Chaos **17**(12), 4295–4308 (2007)
6. Mahmound, G.M.: Modified projective lag synchronization of two nonidentical hyperchaotic complex nonlinear systems. Int. J. Bifurcat. Chaos **21**(8), 2369–2379 (2011)
7. Gamal, M.M., Mansour, E.A., Nabil, S.: On autonomous and nonautonomous modified hyperchaotic complex Lu systems. Int. J. Bifurcat. Chaos **21**(7), 1913–1936 (2011)
8. Mahmoud, G.M., Bountis, T.: The dynamics of systems of complex nonlinear oscillators: a review. Int. J. Bifurcat. Chaos **14**(11), 3821–3846 (2004)
9. Zhu, H.L., Zhang, X.B.: Dynamical analysis of a new complex chaotic system and its synchronization. J. Dyn. Control **6**(4), 307–311 (2008)
10. Mahmoud, G.M., Bountis, T., AbdEl-Latif, G.M., Mahmoud, E.E.: Chaos synchronization of two different chaotic complex Chen and Lii systems. Nonlinear Dyn. **55**(1–2), 43–53 (2009)
11. Zhang, X.B., Zhao, H.G.: Modified function projective synchronization of different chaotic systems. J. Chongqing Normal Univ. **30**(2), 65–68 (2013)
12. Skardal, P.S., Taylor, D., Sun, J.: Optimal synchronization of complex networks. Phys. Rev. Lett. **113**(14), 144101 (2014)

# Model Optimization, Bayesian Learning, and Clustering

# A New Adaptive Hybrid Algorithm for Large-Scale Global Optimization

Ninglei Fan[1], Yuping Wang[1(✉)], Junhua Liu[1], and Yiu-ming Cheung[2]

[1] School of Computer Science and Technology, Xidian University,
Xi'an 710071, China
ningleifan@qq.com, ywang@xidian.edu.cn, msliujunhua@163.com
[2] Department of Computer Science, Hong Kong Baptist University,
Kowloon, Hong Hong, China
ymc@comp.hkbu.edu.hk

**Abstract.** Large-scale global optimization (LSGO) problems are one of most difficult optimization problems and many works have been done for this kind of problems. However, the existing algorithms are usually not efficient enough for difficult LSGO problems. In this paper, we propose a new adaptive hybrid algorithm (NAHA) for LSGO problems, which integrates the global search, local search and grouping search and greatly improves the search efficiency. At the same time, we design an automatic resource allocation strategy which can allocate resources to different optimization strategies automatically and adaptively according to their performance and different stages. Furthermore, we propose a self-learning parameter adjustment scheme for the parameters in local search and grouping search, which can automatically adjust parameters. Finally, the experiments are conducted on CEC 2013 LSGO competition benchmark test suite and the proposed algorithm is compared with several state-of-the-art algorithms. The experimental results indicate that the proposed algorithm is pretty effective and competitive.

**Keywords:** Large scale global optimization ·
Parameter automatical adjustment · Global search · Local search ·
Grouping search · Resource allocation · Self-learning

## 1 Introduction

Many real world problems in the areas of biology, physics, military and engineering applications can be modeled as large scale global optimization problems, which are quite difficult. There are two main reasons for this difficulty. One is that the search space is huge, and the other is that there are a lot of local optimal solutions. For LSGO problems, one of the most effective algorithms is the cooperative co-evolution algorithms, which decompose the high-dimensional problem into several low dimensional problems, and then uses evolutionary algorithm to optimize these problems. In general, cooperative co-evolution algorithms are

© Springer Nature Switzerland AG 2019
H. Lu et al. (Eds.): ISNN 2019, LNCS 11554, pp. 299–308, 2019.
https://doi.org/10.1007/978-3-030-22796-8_32

quite effective for separable and partially separable functions. However, for fully nonseparable functions, the effectiveness and efficiency will decrease greatly. To enhance the effectiveness and efficiency of this kind of algorithms, we design effective local search, global search, grouping search strategies and cooperative mechanism, and combine these strategies and mechanism to propose a new adaptive hybrid algorithm. We also design an effective scheme to automatically allocate the resources (the proper number of function evaluations) to local search strategy, global search strategy, and grouping search strategy based on not only the characteristics of different functions, but also the performance of different strategies in different stages. This scheme ensures that resources are highly utilized during the search process. Besides, the parameters in the local search and the grouping search are also dynamically adjusted. In this way, the proposed algorithm has strong learning ability and can adapt to different scenarios.

The remainder of this paper is organized as follows. Section 2 briefly summarizes the related works on large-scale global optimization problems. In the Sect. 3, we describe in details of the proposed algorithm. Section 4 gives the experimental results of the proposed algorithm on CEC 2013 competition benchmarks and compares the proposed algorithm with several state-of-the-art algorithms. At last, we give the conclusion in Sect. 5.

## 2   Related Work

The framework of cooperative co-evolution algorithm was first proposed by Potter and De Jong in [1]. Its basic idea is to decompose an $n$-dimensional problem into $n$ one-dimensional problems. Then these one-dimensional problems are solved cooperatively by an evolutionary algorithm. However, the variables in a problem are often interactive (or correlated). The work [2] shows that the correlation of variables can greatly affect the performance of such algorithms. So it is very important to take the correlation of variables into account. Some researchers have conducted the research on this direction and proposed some grouping methods which classify the variables into several groups. In each group, the variables are interacted and the variables in two different groups are independent. By using group method, a cooperative evolutionary algorithm is designed by Yang, Tang, and Yao in [3]. And it uses fixed group size and adaptive weights in the evolution process. In addition, Van den Bergh and Engelbrecht also combine random grouping with particle swarm optimization [4]. This simple grouping method effectively improves the performance of the above algorithm for high-dimensional problems. Later, Yang, Tang, and Yao make some improvements on their previous work and treat the size of the group as a parameter to adaptively determine it according to the historical performance in [5]. This practice is more realistic and more reasonable, and the flexibility and performance of the algorithm has been improved. However, literature [6] points out that the more variables there are, the more difficult the random grouping is to group related variables together. To overcome the disadvantages of random grouping, Omidvar, Li, Mei, and Yao propose a cooperative co-evolution with differential grouping [7]. This differential grouping method greatly improves the accuracy of random grouping. But it still does not recognize the correlation between some variables. To overcome the shortcoming, a global differential grouping method is proposed in [8]. In addition, Omidvar, Yang, Mei, Li, and Yao introduce a new faster and more accurate differential grouping in [9]. In recent years, Sun, Kirley and Halgamuge introduce a recursive differential decomposition method in [10]. These methods have effectively improved the performance of differential grouping. Furthermore, a formula-based variable grouping method is proposed in [11] by Wang, Liu, Wei, Zong and Li. Unlike differential grouping which is for black box problems, it is for white box problems and uses expressions of functions to automatically identify correlations between variables. These works have greatly enhance cooperative co-evolution algorithms.

Another important technique is hybrid algorithms which combine global search and local search efficiently. Antonio LaTorre MUELAS S and PENA J-M propose a MOS-based hybrid algorithm in [12]. This algorithm combines three individual algorithms whose function evaluations are assigned to each individual algorithm dynamically in the process of optimization. In [13], Wang and Li design a two-stage based ensemble optimization algorithm. In the first stage, the search tries to enter the potential area as quickly as possible. And in the second stage, the search tries to focus on a limited area to find as good solution as possible. Recently, a two phase hybrid algorithm with a new decomposition method [14] is proposed by Liu, Wang, Liu and Li. It uses different strategies to optimize fully separable functions, partially separable functions, and completely nonseparable functions in two phases. The experimental results have indicated the performance has been further improved compared to the original two-stage algorithm. Moreover, an algorithm based on local search chains is proposed in [15] by Molina, Lozano, and Herrera. The algorithm uses global search and local search in turn, and provides corresponding parameters for the local search algorithm through local search chains. What is more, Brest, Zamuda, Fister, et al. introduce a self-adaptive differential evolution algorithm in [16] for large-scale global optimization problems. This algorithm utilizes four differential evolution strategies combined with population size selection mechanisms. And Molina and Herrera propose an iterative hybridization of differential evolution with local search in [17]. These attempts have made a great contribution to the LSGO problem solving. However, the existing algorithms still face many challenges for the large-scale global optimization problems, and more research works are necessary.

## 3  Proposed Algorithm

In this section, we shall propose a new adaptive hybrid algorithm (briefly NAHA) which combines global search, local search and grouping search properly so that the exploration and the exploitation can be well balanced. Most importantly, we design a mechanism to adjust these three types of searches adaptively during the search process. Also, we integrate three individual algorithms automatically to be suitable to different situations.

## 3.1  NAHA

---

**Algorithm 1.** New adaptive hybrid algorithm(NAHA)

---

1: First, we use recursive differential grouping to group variables.
2: Randomly generate a population (population size is *NP*).
3: Initialize parameter:*gen, itsr, iter, neva, keva, geva, peva*.
4: **while** *termination condition is not satisfied* **do**
   5:Use the *DE* to optimize the population for one cycle, then*gen* = *gen*+1.
   6:**if** *mod(gen, itsr)=0* **then**
      7:Randomly select an individual *vec* from the population, and apply local search algorithm (Solis Wets algorithm [18]) to *vec* with *neva* function evaluations to get *vec\**. Finally, substitute *vec* with *vec\**;
   **end**
   8:**if** *mod(gen, iter)=iter/3* **then**
      9:Apply overall local search to *bestvec* with *keva* function evaluations to get *bestvec\** and replace *bestvec* with *bestvec\**;
   **end**
   10:**if** *mod(gen, iter)=2iter/3* **then**
      11:Select some potential dimensions *Dim*, and use grouping search to *Dim* of *bestvec* with *geva* function evaluations to get *bestvec\**;
   **end**
   12:**if** *mod(gen, iter)=0* **then**
      13:Use one-dimensional search (MTS-LS1 [19]) to *bestvec* with *peva* function evaluations to get *bestvec\** and substitute *bestvec* with *bestvec\**;
      14:According to the performance of the above steps in the previous round, Reassign *itsr, iter, neva, keva, geva* and *peva*;
      15:*gen*=0;
   **end**
**end**

---

Because large-scale global optimization (LSGO) problems are quite complex and their characteristics vary greatly from one problem to another problem, a single search strategy will have significant limitations to different problems. Without knowing the characteristics of the problem considered, it is hard to know which kind of search strategies is effective in advance. It is better for an algorithm to automatically explore the potential area and to adaptively adjust search strategy. Based on this consideration, a new adaptive hybrid algorithm (NAHA) is designed. In this algorithm, on the one hand, the global search and the local search cooperate with each other. On the other hand, overall search, grouping search and one-dimensional search also work together. The framework of the algorithm is given in Algorithm 1.

It is worth noting that steps 12–14 are very critical and have a great impact on the performance of the algorithm. When the algorithm make the function value decrease faster, we do not change the parameters *iter, itsr, neva*. Otherwise, we will increase *iter, itsr, neva* accordingly to push the search to enter a more promising area. In addition, the values of *keva, geva* and *peva* are taken in the following way. Assume that the total number of function evaluations for the three strategies is fixed and denoted as *totalnum*. Here, we assign the number of function evaluations to each strategy in the next round search based on the contribution of each strategy. Suppose that before the $i$-th strategy search, the best individual's fitness is $exefit_i$, after the search, it is $houfit_i$. Suppose that the number of function evaluations by the $i$-th strategy in the previous round is $exeva_i$. The number of function evaluations is arranged for the $i$-th strategy in the next generation as $[r_i * totalnum]$, where $[r_i * totalnum]$ is the integer part of $r_i * totalnum$ and $r_i$ is computed by

$$con_i = \frac{exefit_i - houfit_i}{exefit_i} \quad (i = 1, 2, 3) \tag{1}$$

$$r_i = \sigma + \frac{con_i/exeva_i + \eta}{\sum_{i=1}^{s}(con_i/exeva_i) + s\eta}(1 - s\sigma) \quad (i = 1, 2, 3) \tag{2}$$

where $\sigma$ is a threshold and $\sigma = 0.1$ in the experiments. The chosen method on three search strategies in the algorithm is in step 9, step 11 and step 13. In order to avoid the case where the denominator is zero, we also add a small threshold $\eta$ to the numerator and denominator. In our experiment $\eta$ is assigned a value of 1e−20. The number of function evaluations for strategies 1, 2, and 3, denoted by *keva*, *geva* and *peva*, respectively, are equal to *totalnum* multiplied by corresponding $r_i$ ($i = 1, 2, 3$). Moreover, in the local search algorithm, it is generally required to pass the search range $sr$. In order to solve this problem, we design three ways to set the search range. These three ways consider many possibilities and can be suitable to different situations. The first one defines the search range of the current individual $x$ by $[x - std(x), x + std(x)]$, where $std(x)$ is the standard deviation of $x$ in the current population. The second one defines the search range of the current individual $x$ by $[x - abd(x), x + abd(x)]$, where $abd(x)$ is the absolute difference between the updated individual of $x$ and $x$ in the two adjacent generations. The third one is to generate a random vector by using the standard Gaussian distribution with mean zero and variance one first, and then compute the absolute value of each component of the vector and use the absolute values as the components of a new vector $abG(x)$. The search range is defined by $[x - abG(x), x + abG(x)]$. Since we do not know which way is appropriate in advance, So we want to design a simple scheme to realize this. For the $i$-th way, record $gum_i$ by setting $gum_i = startval$ initially for ($i = 1, 2, 3$), where *startval* is a parameter. In our experiments, *startval* is set to 3. When the $con_i$ is below our preset threshold, $gum_i$ will be decremented by 1, that is, $gum_i = gum_i - 1$. On the contrary, When $con_i$ is higher than our preset threshold, $gum_i$ will be increased by 1, that is, $gum_i = gum_i + 1$. In next round, we choose one way among three ways according to the probability $pro_i$ ($i = 1, 2, 3$). That is, the probability that the $i$-th way is selected is $pro_i$.

$$pro_i = \frac{max(gum_i, 1)}{\sum_{i=1}^{3} max(gum_i, 1)} \quad (i = 1, 2, 3) \tag{3}$$

In this way, the more appropriate way to determine the search range will have the greater probability to be selected in subsequent iterations. Note that from formula (3), using $max(gum_i,1)$ instead of $gum_i$ is mainly to avoid completely losing the chance of being selected of the $i$-th way to determine the search range.

Finally, Let us explain step 11 in Algorithm 1 in detail. In step 11, we randomly select one of the groups based on the result in the first step. If the problem is completely non-separable or fully separable, we will use random grouping to pick a set of variables. Then we use algorithm *SaNSDE* [20] to optimize this set of variables. Other variables are fixed to the same values of the corresponding variables of the current best individual *bestvec*. Through this we hope the search can reach a promising area. Then we apply local search (Solis Wets algorithm) on the solution obtained by *SaNSDE* to optimize this sub-problem. In this round of optimization, we will record the variables of the group and the corresponding $con_i$. In the next round of optimization, we will first compare $con_i$ in the previous round with the threshold we set. When $con_i$ is less than the threshold we set, then another sub-problem (another group of variables) is optimized, otherwise, this set of variables is continued to be optimized.

## 4    Experimental Results and Analysis

In order to identify the efficiency of our algorithm, we conduct the experiments on the CEC 2013 benchmark problems and compare the proposed algorithm NAHA with three state-of-the-art algorithms MOS, CMAESCC-RDG and MA-SW-Chains. CEC 2013 benchmark suite is designed specifically for large-scale global optimization problems. It includes the fully separable functions, partially separable functions, and completely nonseparable functions. Many functions have a lot of local optimal solutions. Furthermore, it takes into account many situations which might happen in real world problems. This benchmark suite is very difficult and challenging enough to test the performance of algorithms. For a detailed description of this benchmark suite, please refer to [21].

In the experiments, we conducted the proposed algorithm on each test problem in 25 independent runs, and recorded the mean and median values in 25 runs. In addition, We also recorded the standard deviation of these test values in 25 runs. Finally, all these results are shown in Table 1.

In order to illustrate the statistical difference between our algorithm and each of compared algorithms, we did a hypothesis testing. Our null hypothesis is that there is no difference between them. The results of the hypothesis test are given in Table 2, where A represents acceptance of the null hypothesis and R indicates rejection of the null hypothesis.

Moreover, we use the scoring criteria in [22] to make an overall evaluation for these four algorithms. Let $SR$ denote the sum of the ranks defined as follows.

$$SR = \sum_{i=1}^{N} rank_i \tag{4}$$

where $N$ represents the total number of functions, and $rank_i$ means the rank of the performance of the algorithm on the $i$-th function. The final score and rank are shown in Table 3.

**Table 1.** The result of the NAHA, MOS, CMAESCC-RDG and MA-SW-Chains algorithms applied to the CEC 2013 benchmark suite. The best performing results are marked in bold.

| Function | Stats | NAHA | MOS | CMAESCC-RDG | MA-SW-Chians |
|---|---|---|---|---|---|
| f1 | Median | **0.00e+00** | **0.00e+00** | 2.84e+05 | 7.12e−13 |
|  | Mean | **0.00e+00** | **0.00e+00** | 2.89e+05 | 1.34e−12 |
|  | Std | **0.00e+00** | **0.00e+00** | 3.27e+04 | 2.45e−12 |
| f2 | Median | **4.01e+00** | 8.36e+02 | 4.66e+03 | 1.24e+03 |
|  | Mean | **3.66e+00** | 8.32e+02 | 4.68e+03 | 1.25e+03 |
|  | Std | **1.53e+00** | 4.48e+01 | 1.77e+02 | 1.05e+02 |
| f3 | Median | 2.00e+01 | 9.10e−13 | 2.03e+01 | **6.83e−13** |
|  | Mean | 2.00e+01 | 9.17e−13 | 2.03e+01 | **6.85e−13** |
|  | Std | 9.16e−05 | 5.12e−14 | 4.96e−02 | **2.12e−13** |
| f4 | Median | 6.33e+07 | 1.56e+08 | **5.83e+06** | 2.75e+09 |
|  | Mean | 5.53e+07 | 1.74e+08 | **5.90e+06** | 3.81e+09 |
|  | Std | 3.09e+07 | 7.87e+07 | **6.56e+05** | 2.73e+09 |
| f5 | Median | 4.84e+06 | 6.79e+06 | **2.19e+06** | **2.03e+06** |
|  | Mean | 4.58e+06 | 6.94e+06 | **2.20e+06** | **2.25e+06** |
|  | Std | 5.03e+05 | 8.85e+05 | **3.76e+05** | **1.30e+06** |
| f6 | Median | 1.03e+06 | 1.39e+05 | 9.95e+05 | **6.33e+02** |
|  | Mean | 1.03e+06 | 1.48e+05 | 9.95e+05 | **1.86e+04** |
|  | Std | 6.31e+03 | 6.43e+04 | 2.88e+01 | **2.54e+04** |
| f7 | Median | 5.78e−03 | 1.62e+04 | **2.94e−20** | 4.03e+06 |
|  | Mean | 1.17e−02 | 1.62e+04 | **8.12e−17** | 3.85e+06 |
|  | Std | 1.40e−02 | 9.10e+03 | **2.17e−16** | 6.34e+05 |
| f8 | Median | 5.81e+10 | 8.08e+12 | **8.71e+06** | 4.60e+13 |
|  | Mean | 4.84e+10 | 8.00e+12 | **9.74e+06** | 4.62e+13 |
|  | Std | 3.19e+10 | 3.07e+12 | **5.83e+06** | 9.02e+12 |
| f9 | Median | 2.85e+08 | 3.87e+08 | 1.57e+08 | **1.42e+08** |
|  | Mean | 2.90e+08 | 3.83e+08 | 1.65e+08 | **1.44e+08** |
|  | Std | 3.11e+07 | 6.29e+07 | 4.16e+07 | **1.55e+07** |
| f10 | Median | 9.13e+07 | 1.18e+06 | 9.04e+07 | **3.34e+02** |
|  | Mean | 9.12e+07 | 9.02e+05 | 9.12e+07 | **3.72e+04** |
|  | Std | 7.03e+05 | 5.07e+05 | 1.53e+06 | **6.25e+04** |
| f11 | Median | **3.76e+06** | 4.48e+07 | 1.64e+07 | 2.10e+08 |
|  | Mean | **3.23e+06** | 5.22e+07 | 1.62e+07 | 2.10e+08 |
|  | Std | **1.70e+06** | 2.05e+07 | 6.11e+05 | 2.35e+07 |
| f12 | Median | 1.13e+03 | **2.46e+02** | 1.01e+03 | 1.25e+03 |
|  | Mean | 1.21e+03 | **2.47e+02** | 9.81e+02 | 1.24e+03 |
|  | Std | 3.55e+02 | **2.54e+02** | 7.30e+01 | 8.33e+01 |
| f13 | Median | **2.18e+06** | 3.30e+06 | **2.49e+06** | 1.91e+07 |
|  | Mean | **2.21e+06** | 3.40e+06 | **2.47e+06** | 3.58e+07 |
|  | Std | **6.21e+05** | 1.06e+06 | **3.83e+05** | 4.30e+07 |
| f14 | Median | **8.66e+06** | 2.42e+07 | 2.74e+07 | 1.43e+08 |
|  | Mean | **8.85e+06** | 2.56e+07 | 2.76e+07 | 1.45e+08 |
|  | Std | **2.45e+06** | 7.94e+06 | 1.49e+06 | 1.60e+07 |
| f15 | Median | 4.44e+06 | 2.38e+06 | **2.18e+06** | 5.80e+06 |
|  | Mean | 4.64e+06 | 2.35e+06 | **2.19e+06** | 5.98e+06 |
|  | Std | 1.23e+06 | 1.94e+05 | **2.28e+05** | 1.42e+06 |

Now let's compare the overall performance of four compared algorithms. From Table 3 we can see that the total score of our NAHA algorithm on all test

**Table 2.** Results obtained by hypothesis testing

| Function | MOS | CMAESCC-RDG | MA-SW-Chains |
|---|---|---|---|
| f1 | 1.00e+00, A | 1.66e−24, R | 1.15e−02, R |
| f2 | 3.80e−32, R | 7.27e−36, R | 1.49e−27, R |
| f3 | 7.1e−130, R | 1.27e−20, R | 7.1e−130, R |
| f4 | 2.94e−07, R | 3.21e−08, R | 4.12e−07, R |
| f5 | 2.54e−11, R | 6.07e−16, R | 1.44e−08, R |
| f6 | 5.30e−29, R | 9.63e−20, R | 7.96e−40, R |
| f7 | 4.53e−09, R | 3.35e−04, R | 1.16e−20, R |
| f8 | 2.54e−12, R | 8.00e−08, R | 6.28e−19, R |
| f9 | 7.44e−07, R | 1.18e−11, R | 5.85e−17, R |
| f10 | 3.69e−50, R | 1.00e+00, A | 2.12e−52, R |
| f11 | 1.47e−11, R | 2.26e−22, R | 1.96e−24, R |
| f12 | 6.99e−11, R | 4.20e−02, A | 6.84e−02, A |
| f13 | 6.18e−05, R | 8.74e−02, A | 6.68e−04, R |
| f14 | 4.21e−10, R | 2.04e−21, R | 5.37e−24, R |
| f15 | 2.46e−09, R | 7.39e−10, R | 1.60e−02, R |

**Table 3.** Total score and ranking

| Algorithm | SR | Place |
|---|---|---|
| NAHA | 34 | 1 |
| MOS | 38 | 3 |
| CMAESCC-RDG | 34.5 | 2 |
| MA-SW-Chains | 43.5 | 4 |

algorithms is best and the total rank of NAHA is the first. This indicates that the proposed NAHA performs best among four compared algorithms. CMAESCC-RDG gets the second best total score and the total rank is second. This means that CMAESCC-RDG performs worse than NAHA, but performs better than MOS and MA-SW-Chains. Also, MOS gets the third best score and rank, Thus, MOS performs worse than NAHA and CMAESCC-RDG, but performs better than MA-SW-Chains. The overall performance of MA-SW-Chains is the worst. It can be seen from the $t$-test results in Table 2 that for most test functions the performance difference between NAHA and any other compared algorithm is obvious. This conclusion and the results in Table 3 mean that the overall performance of NAHA is the best and really better than that of any compared algorithm.

In the experiment we found that NAHA can make the functions decrease very fast in the early stage, but quite slowly in the later stage. This indicates the ability of jumping out the local optima of NAHA is not enough in the late stage. In addition, CMAESCC-RDG and NAHA have difficulty to find the optimal solution on Ackley functions f3, f6 and f10. This indicates CMAESCC-RDG and NAHA are ineffective to Ackley functions. Thus, it is necessary to further improve NAHA by avoiding it to trap into local optima.

## 5   Conclusion

The proposed algorithm NAHA effectively combines the local search, grouping search and global search together and fully make use of the advantages of the three strategies. It also can automatically adjust parameters during the optimization process. This greatly expands its application fields and improves search efficiency. Experimental results have shown that NAHA performs well on fully separable, partially separable and completely nonseparable functions. However, the proposed algorithm still has a lot of rooms for improvement. In the future, we will improve resource allocation scheme, enhance the grouping search, local search and global search and avoid trapping into local optima.

**Acknowledgments.** This work is supported by National Natural Science Foundation of China under the Project 61872281.

## References

1. Potter, M.A., De Jong, K.A.: A cooperative coevolutionary approach to function optimization. In: Davidor, Y., Schwefel, H.-P., Männer, R. (eds.) PPSN 1994. LNCS, vol. 866, pp. 249–257. Springer, Heidelberg (1994). https://doi.org/10.1007/3-540-58484-6_269
2. Salomon, R.: Re-evaluating genetic algorithm performance under coordinate rotation of benchmark functions. A survey of some theoretical and practical aspects of genetic algorithms. Bio Syst. **39**(3), 263–278 (1996)
3. Yang, Z., Tang, K., Yao, X.: Large scale evolutionary optimization using cooperative coevolution. Inf. Sci. **178**(15), 2985–2999 (2008)
4. Van den Bergh, F., Engelbrecht, A.P.: A cooperative approach to particle swarm optimization. IEEE Trans. Evol. Comput. **8**(3), 225–239 (2004)
5. Yang, Z., Tang, K., Yao, X.: Multilevel cooperative coevolution for large scale optimization. In: 2008 IEEE Congress on Evolutionary Computation, Hong Kong, pp. 1663–1670. IEEE Press (2008)
6. Omidvar, M.N., Li, X., Yang, Z., Yao, X.: Cooperative co-evolution for large scale optimization through more frequent random grouping. In: 2010 IEEE Congress on Evolutionary Computation, Barcelona, pp. 1754–1761. IEEE Press (2010)
7. Omidvar, M.N., Li, X., Mei, Y., Yao, X.: Cooperative co-evolution with differential grouping for large scale optimization. IEEE Trans. Evol. Comput. **18**(3), 378–393 (2014)
8. Mei, Y., Omidvar, M.N., Li, X., Yao, X.: A competitive divide-and-conquer algorithm for unconstrained large-scale black-box optimization. ACM Trans. Math. Softw. **42**(2), 1–24 (2016)

9. Omidvar, M.N., Yang, M., Mei, Y., Li, X., Yao, X.: DG2: a faster and more accurate differential grouping for large-scale black-box optimization. IEEE Trans. Evol. Comput. **21**(6), 929–942 (2017)

10. Sun, Y., Kirley, M., Halgamuge, S.K.: A recursive decomposition method for large scale continuous optimization. IEEE Trans. Evol. Comput. **22**(5), 647–661 (2018)

11. Wang, Y., Liu, H., Wei, F., Zong, T., Li, X.: Cooperative co-evolution with formula-based variable grouping for large-scale global optimization. Evol. Comput. **26**(4), 569–596 (2017)

12. Latorre, A., Muelas, S., Pena, J.M.: Large scale global optimization: experimental results with MOS-based hybrid algorithms. In: 2013 IEEE Congress on Evolutionary Computation, Cancun, pp. 2742–2749. IEEE Press (2013)

13. Wang, Y., Li, B.: Two-stage based ensemble optimization for large-scale global optimization. In: 2010 IEEE Congress on Evolutionary Computation, Barcelona, pp. 4488–4495. IEEE Press (2010)

14. Liu, H., Wang, Y., Liu, L.: A two phase hybrid algorithm with a new decomposition method for large scale optimization. Integr. Comput.-Aided Eng. **25**(4), 349–367 (2018)

15. Molina, D., Lozano, M., Herrera, F.: MA-SW-Chains: memetic algorithm based on local search chains for large scale continuous global optimization. In: 2010 IEEE Congress on Evolutionary Computation, Barcelona, pp. 1–8. IEEE Press (2010)

16. Brest, J., Boskovic, B., Zamuda, A., Fister, I.: Self-adaptive differential evolution algorithm with a small and varying population size. In: 2012 IEEE Congress on Evolutionary Computation, Brisbane, pp. 2827–2834. IEEE Press (2012)

17. Molina, D., Herrera, F.: Iterative hybridization of DE with local search for the CEC'2015 special session on large scale global optimization. In: 2015 IEEE Congress on Evolutionary Computation, Sendai, pp. 1974–1978. IEEE Press (2015)

18. Solis, F.J., Wets, R.J.B.: Minimization by random search techniques. Math. Oper. Res. **6**(1), 19–30 (1981)

19. Tseng, L.Y., Chen, C.C.C.: Multiple trajectory search for large scale global optimization. In: 2008 IEEE Congress on Evolutionary Computation, Hong Kong, pp. 3052–3059. IEEE Press (2008)

20. Yang, Z., Tang, K., Yao, X.: Self-adaptive differential evolution with neighborhood search. In: 2010 IEEE Congress on Evolutionary Computation, Barcelona, pp. 1110–1116. IEEE Press (2010)

21. Li, X., Tang, K., Omidvar, M.N.: Benchmark functions for the CEC 2013 special session and competition on large-scale global optimization. Gene **7**(33), 8–30 (2013)

22. Maucec, M.S., Brest, J.: A review of the recent use of differential evolution for large-scale global optimization: an analysis of selected algorithms on the CEC 2013 LSGO benchmark suite. Swarm Evol. Comput. 1–18 (2018)

# An Effective Hybrid Approach for Optimising the Learning Process of Multi-layer Neural Networks

Seyed Jalaleddin Mousavirad[1], Azam Asilian Bidgoli[1],
Hossein Ebrahimpour-Komleh[1], Gerald Schaefer[2(⊠)], and Iakov Korovin[3]

[1] Department of Computer and Electrical Engineering,
University of Kashan, Kashan, Iran
[2] Department of Computer Science, Loughborough University, Loughborough, UK
gerald.schaefer@ieee.org
[3] Southern Federal University, Taganrog, Russia

**Abstract.** Finding the optimal connection weights in a neural network is one of the most challenging tasks in machine learning and pattern recognition. The main disadvantage of conventionally used algorithms such as back-propagation is that they show a tendency of getting trapped in local rather than global optima. To address this, population-based metaheuristic algorithms can be employed. In this paper, we propose a novel approach to optimise the weights of a neural network. Our method integrates an imperialist competitive algorithm, a powerful metaheuristic algorithm, with chaos theory and back-propagation for neural network learning. Experimental results on the three-bit parity problem and several function approximation tasks confirm that our proposed algorithm significantly outperforms several state-of-the-art methods for neural network weight optimisation.

**Keywords:** Machine learning · Neural networks ·
Weight optimisation · Imperialist competitive algorithm ·
Chaos theory · Back-propagation

## 1 Introduction

Artificial neural networks (ANNs) are mathematical models that have been extensively used for applications such as speech processing [5], pattern recognition [12], and image processing [6]. Useful characteristics of ANNs include adaptability, learning capability, and generalisation ability [2]. One of the most widely used ANN architectures is the multi-layer perceptron (MLP), a feed-forward neural network that consists of simple neurons called perceptrons.

Learning in ANNs means finding connection weights between nodes so as to minimise the prediction error. Generally, learning algorithms can be divided into two groups: gradient-based algorithms and stochastic search algorithms.

© Springer Nature Switzerland AG 2019
H. Lu et al. (Eds.): ISNN 2019, LNCS 11554, pp. 309–317, 2019.
https://doi.org/10.1007/978-3-030-22796-8_33

Gradient-based methods such as back-propagation (BP) are traditional algorithms for neural network training but suffer from drawbacks such as high dependency on initial values and local optima entrapment [3]. To overcome these limitations, stochastic search algorithms such as metaheuristic algorithms can be employed. In recent years, these algorithms have been extensively used for neural network learning and have been shown to outperform gradient-based approaches [3,7].

The imperialist competitive algorithm (ICA) [4] is a metaheuristic algorithm based on socio-political evolution that has advantages over traditional optimisation algorithms in not needing to calculate the gradient and reducing the probability of getting stuck in local minima. It has been shown to yield competitive performance for various applications [9]. The main ICA strategies supporting this efficacy are imperialist competition and assimilation.

When developing metaheuristic algorithms, two contradictory criteria need to be balanced: exploration, which refers to identifying promising regions in search space, and exploitation, which is the ability to search around a solution with the aim of improving it.

Chaos is related to the study of chaotic dynamical systems. One of the main characteristic here is the sensitivity to initial conditions. To improve both exploration and exploitation [13], random numbers in a chaotic-based metaheuristic algorithms can be created using chaotic sequences instead of random variables.

In this paper, we propose an enhanced ICA incorporating chaos theory and BP for enhanced neural network learning. The proposed method integrates the global search strategy of ICA with the local search ability of the BP algorithm and uses chaotic maps to maintain the diversity of the population. Experimental results on the three-bit parity problem and several function approximation tasks confirm that our proposed algorithm significantly outperforms state-of-the-art methods for neural network weight optimisation.

The remainder of the paper is organised as follows: Sect. 2 briefly explains the ICA, while Sect. 3 introduces our proposed algorithm. Experimental results are given in Sect. 4, and Sect. 5 concludes the paper.

## 2     Imperialist Competitive Algorithm

The imperialist competitive algorithm (ICA) is a population-based metaheuristic algorithm that has been shown to yield good performance to solve various global optimisation problems [4]. Inspired by colonial competition among countries, each candidate solution in ICA is a country. Some of the superior countries are imperialist, while the remaining countries are their colonies. Colonies are distributed among imperialists according to their power.

After the formation of the initial empires, colonies move towards their empire by

$$Position_{i+1} = Position_i + \gamma \times \delta \oplus d, \qquad (1)$$

where $Position_i$ is the colony's position in the $i$-th iteration, $\gamma$ is the assimilation coefficient, $\delta$ is a random vector normally distributed in $[0;1]$, and $d$ is the distance between colony and its imperialist.

Revolution is a sudden change in the position of colonies in search space and is employed to improve the exploration capability of ICA.

The total power of an empire depends on both the power of the imperialist country and of its colonies, i.e.

$$TC_n = Cost(imperialist_n) + \xi mean\{Cost(Colonies\ of\ empires)\}, \quad (2)$$

where $TC_n$ is the total cost of the $n$-th empire and $\xi$ is a positive number.

In imperialist competition, all empires try to take possession of colonies of other empires to increase their own power while decreasing the power of others. This is modelled by selecting some of the weakest colonies of the weakest empires and having the remaining empires compete for them. Eventually, all empires except the most powerful one will annihilate, and all colonies will converge to the same position.

# 3    Proposed Algorithm

In this paper, we propose a novel ICA-based algorithm for neural network learning. In particular, we improve the ICA algorithm for neural network learning using chaos theory and back-propagation. In the following, we explain our algorithm in detail.

## 3.1    Representation and Objective Function

In our proposed algorithm, each country consists of three parts: the connection weights between input layer and hidden layer $w_{i,j}$, the connection weights between hidden layers and output layer $\varphi_{i,j}$, and the bias weights $\beta_{i,j}$. The resulting representation of a country is illustrated in Fig. 1.

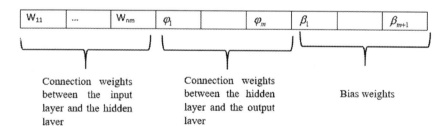

**Fig. 1.** Representation of a country. $n$ is the number of inputs and $m$ the number of neurons in the hidden layer.

As objective function we employ, as is commonly done, the mean squared error (MSE) defined as

$$MSE = \frac{1}{P} \sum_{i=1}^{P} (d_i - o_i)^2, \quad (3)$$

where $d_i = (d_{i,1}, d_{i,2}, ..., d_{i,k})$ is the desired output vector and $o_i = (o_{i,1}, o_{i,2}, ..., o_{i,k})$ is the actual output. The aim of the learning process is to find optimal values of $x$ (the connection weights and biases) so as to minimise the difference between $d$ and $o$.

## 3.2   Chaos-Enhanced ICA

An important requirement for a metaheuristic algorithm is the generation of random numbers with a sensible uniformity. Chaos is a pseudo-random process in non-linear dynamics that can be used as a source of randomness [1]. Using randomness characteristics of chaos, a candidate solution moves in a chaotic way enabling it to escape from a local optima. In our approach, we use a sinusoidal chaotic map to generate random numbers. It generates chaotic sequences in the range $[0; 1]$ by

$$x_{n+1} = ax^2{}_n sin(\pi.x_n), \tag{4}$$

which simplifies to

$$x_{n+1} = sin(\pi.x_n) \tag{5}$$

when $a = 2.3$ and $x_0 = 0.7$.

Chaotic sequences are applied in two ways: generating the initial empires and during the assimilation process. To this end, initial countries are generated by iterating the sinusoidal chaos map, while during assimilation $\delta$ is substituted, leading to

$$Position_{i+1} = Position_i + \gamma \times cm \oplus d, \tag{6}$$

where $cm$ is a vector generated from the sinusoidal chaotic map.

## 3.3   Back-Propagation

Gradient descent back-propagation (GDBP) is a classic algorithm to train an MLP by modifying the weights in the direction corresponding to the negative gradient of an error function. In our algorithm, we adopt GDBP in each iteration as a local search operator to improve the best found solutions.

GDBP proceeds in three stages:

1. Forward propagation: input values pass from hidden layers to generate the outputs;
2. Backward propagation: the error in the layers of the MLP is back-propagated;
3. Weight update: finally, weights are updated in a direction that corresponds to the negative gradient of the error function.

The process ends when a stopping condition has been satisfied.

## 4    Experimental Results

We have evaluated the performance of our proposed algorithm on two types of problems: the three-bit parity problem and three function approximation problems. Lower and upper bounds of all weights are $-1$ and $+1$, the maximal iteration number was set to 500, the numbers of countries to 40, the number of imperialists to 5, and revolution rate to 0.1.

To put our obtained results into context, we compare our algorithm with the basic ICA algorithm and with GDBP, and with several state-of-the-art optimisation algorithms [10,11], namely genetic algorithm (GA), particle swarm optimisation (PSO), ant colony optimisation (ACO), evolutionary strategy (ES), population-based incremental learning (PBIL), biogeography-based optimisation (BBO), and grey wolf optimiser (GWO). For all algorithms, we report statistical results in terms of average MSE and its standard deviation on the test dataset.

### 4.1    Three-Bit Parity Problem

An important way to evaluate neural network training algorithms is the parity problem [8]. A well-known non-linear benchmark, it is a mapping problem defined on distinct binary vectors with a result of 1 if the number of 1s in the vector is odd, and 0 otherwise. As we trained MLPs with structure 3-7-1 to solve a 3-bit parity problem, 36 variables must be optimised.

The results for all algorithms are given in Table 1. From there, we can see that our proposed algorithm clearly achieves the lowest MSE confirming it has the best ability to avoid local minima. It also gives a very low standard deviation which shows that the method is robust.

**Table 1.** Results on three-bit parity problem

| Algorithm | Avg. MSE | Std. dev. MSE |
|---|---|---|
| ICA | 1.68E−02 | 5.42E−04 |
| GDBP | 9.87E−04 | 3.63E−03 |
| GA | 1.81E−04 | 4.13E−04 |
| PSO | 8.41E−02 | 3.59E−02 |
| ACO | 1.80E−01 | 2.53E−02 |
| ES | 1.19E−01 | 1.16E−02 |
| PBIL | 3.02E−02 | 3.97E−02 |
| BBO | 3.65E−07 | **0** |
| GWO | 9.41E−03 | 2.95E−02 |
| Proposed algorithm | **3.14E−08** | 5.83E−08 |

## 4.2  Function Approximations

We used three popular function approximation datasets, given in Table 2, to evaluate our proposed algorithm. We trained MLPs of structure 1-15-1 to approximate these functions.

**Table 2.** Function approximation datasets.

| Function | Training samples | Test samples |
| --- | --- | --- |
| Sigmoid | 61: $x \in [-3{:}0.1{:}3]$ | 121: $x \in [-3{:}0.05{:}3]$ |
| Cosine | 31: $x \in [1.25{:}0.05{:}2.75]$ | 38: $x \in [1.25{:}0.04{:}2.75]$ |
| Sine | 105: $x \in [0{:}0.03{:}\pi]$ | 153: $x \in [0.1{:}0.2{:}\pi]$ |

**Sigmoid Function.** The sigmoid function, $y = 1/1(1 + e^{-x})$, is the simplest function applied in the experiments. Figure 2 shows that our proposed algorithm is able to very accurately estimate the real curve. This is further confirmed in Table 3 which shows results for all tested algorithms. From there, it is evident that our algorithm yields superior results compared to all other approaches.

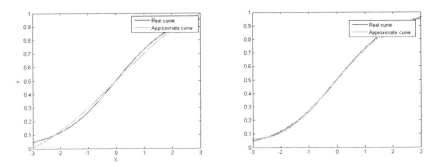

**Fig. 2.** Approximated curves for *sigmoid* function. Left: ICA, right: proposed algorithm.

**Cosine Function.** This dataset, based on $y = cos(x\pi/2)$, is more challenging than the *sigmoid* function. Figure 3 shows the approximated curves for ICA and the proposed algorithm and demonstrates the efficacy of our approach. Table 4 compares our method with other algorithms and confirms that it outperforms all other approaches.

**Table 3.** Results on *sigmoid* function.

| Algorithm | Avg. MSE | Std. dev. MSE |
|---|---|---|
| ICA | 5.53E−04 | 1.55E−04 |
| GDBP | 2.75E−01 | 2.32E−03 |
| GA | 1.09E−03 | 9.16E−04 |
| PSO | 2.30E−02 | 9.43E−03 |
| ACO | 2.35E−02 | 1.00E−02 |
| ES | 7.56E−02 | 1.64E−02 |
| PBIL | 4.05E−03 | **2.74E−17** |
| BBO | 1.33E−05 | 3.57E−04 |
| GWO | 2.03E−04 | 2.26E−04 |
| Proposed algorithm | **1.60E−11** | 2.00E−11 |

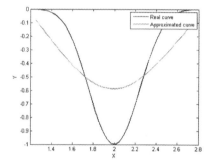 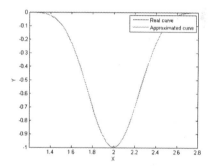

**Fig. 3.** Approximated curves for *cosine* function. Left: ICA, right: proposed algorithm.

**Table 4.** Results on *cosine* function.

| Algorithm | Avg. MSE | Std. dev. MSE |
|---|---|---|
| ICA | 4.23E−02 | 4.83E−03 |
| GDBP | 1.75E−01 | 6.56E−02 |
| GA | 1.09E−02 | 6.32E−03 |
| PSO | 5.90E−02 | 2.10E−02 |
| ACO | 5.09E−02 | 1.08E−02 |
| ES | 8.66E−02 | 2.22E−02 |
| PBIL | 9.43E−02 | 1.85E−02 |
| BBO | 1.37E−02 | **1.83E−18** |
| GWO | 3.11E−03 | 2.16E−03 |
| Proposed algorithm | **1.02E−09** | 3.26E−10 |

**Sine Function.** This dataset, based on $y = sin(2x)$, is the most difficult one. Nevertheless, as can be see from Fig. 4, once again our approach yields a very close function approximation, while Table 5 again impressively shows the superiority of our algorithm compared to all other methods.

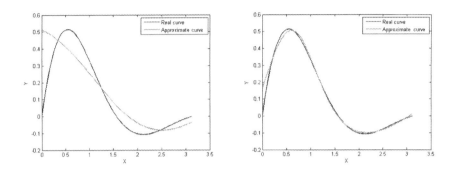

**Fig. 4.** Approximated curves for *sine* function. Left: ICA, right: proposed algorithm.

**Table 5.** Results on *sine* function.

| Algorithm | Avg. MSE | Std. dev. MSE |
|---|---|---|
| ICA | 4.72E−01 | 1.42E−03 |
| GDBP | 1.2 | 1.11E−01 |
| GA | 4.21E−01 | 6.12E−02 |
| PSO | 5.27E−01 | 7.29E−02 |
| ACO | 5.30E−01 | 5.33E−02 |
| ES | 7.07E−01 | 7.74E−02 |
| PBIL | 4.83E−01 | 7.94E−03 |
| BBO | 1.03E−01 | 0 |
| GWO | 2.62E−01 | 1.15E−01 |
| Proposed algorithm | **1.06E−08** | 3.48E−09 |

# 5    Conclusions

Effective learning is crucial for successful application of neural networks. Since conventional algorithms based on gradient descent methods suffer from a tendency to get trapped in local optima, metaheuristic algorithms can be used to tackle this problem. The imperialist competitive algorithm (ICA) is a population-based metaheuristic algorithm inspired by imperialist competition. In this paper, we have proposed an improved ICA for neural network training where a sinusoidal chaotic map is used to maintain the diversity of the population, while back-propagation is employed to enhance exploitation of the algorithm. The proposed algorithm has been evaluated on the three-bit parity and several function approximation problems and has been demonstrated to provide excellent results, significantly outperforming existing state-of-the-art metaheuristics. In future, we plan to extend our approach to other uses of neural networks such as classification tasks.

**Acknowledgements.** The paper is published due to the financial support of the Ministry of Science and Higher Education of Russia, contract 14.575.21.0152, signed 26/09/2017, unique identifier RFMEFI57517X0152.

# References

1. Alatas, B.: Chaotic bee colony algorithms for global numerical optimization. Expert Syst. Appl. **37**(8), 5682–5687 (2010)
2. Alba, E., Chicano, J.F.: Training neural networks with GA hybrid algorithms. In: Deb, K. (ed.) GECCO 2004. LNCS, vol. 3102, pp. 852–863. Springer, Heidelberg (2004). https://doi.org/10.1007/978-3-540-24854-5_87
3. Amirsadri, S., Mousavirad, S.J., Ebrahimpour-Komleh, H.: A levy flight-based grey wolf optimizer combined with back-propagation algorithm for neural network training. Neural Comput. Appl. **30**, 1–14 (2017)
4. Atashpaz-Gargari, E., Lucas, C.: Imperialist competitive algorithm: an algorithm for optimization inspired by imperialistic competition. In: IEEE Congress on Evolutionary Computation, pp. 4661–4667 (2007)
5. Chan, W., Jaitly, N., Le, Q., Vinyals, O.: Listen, attend and spell: a neural network for large vocabulary conversational speech recognition. In: IEEE International Conference on Acoustics, Speech and Signal Processing, pp. 4960–4964 (2016)
6. Egmont-Petersen, M., de Ridder, D., Handels, H.: Image processing with neural networks - a review. Pattern Recogn. **35**(10), 2279–2301 (2002)
7. Heidari, A.A., Faris, H., Aljarah, I., Mirjalili, S.: An efficient hybrid multilayer perceptron neural network with grasshopper optimization. Soft Comput. 1–18 (2018)
8. Liu, D., Hohil, M.E., Smith, S.H.: N-bit parity neural networks: new solutions based on linear programming. Neurocomputing **48**(1–4), 477–488 (2002)
9. Mahmoudi, M.T., Taghiyareh, F., Forouzideh, N., Lucas, C.: Evolving artificial neural network structure using grammar encoding and colonial competitive algorithm. Neural Comput. Appl. **22**(1), 1–16 (2013)
10. Mirjalili, S.: How effective is the grey wolf optimizer in training multi-layer perceptrons. Appl. Intell. **43**(1), 150–161 (2015)
11. Mirjalili, S., Mirjalili, S.M., Lewis, A.: Let a biogeography-based optimizer train your multi-layer perceptron. Inf. Sci. **269**, 188–209 (2014)
12. Mousavirad, S.J., Rezaee, K., Nasri, K.: A new method for identification of Iranian rice kernel varieties using optimal morphological features and an ensemble classifier by image processing. Majlesi J. Multimedia Process. **1**, 1–8 (2012)
13. Saremi, S., Mirjalili, S., Lewis, A.: Biogeography-based optimisation with chaos. Neural Comput. Appl. **25**(5), 1077–1097 (2014)

# A Collaborative Neurodynamic Optimization Approach to Bicriteria Portfolio Selection

Man-Fai Leung[1,3] and Jun Wang[1,2,3(✉)]

[1] Department of Computer Science, City University of Hong Kong,
Kowloon, Hong Kong
manfleung7-c@my.cityu.edu.hk, jwang.cs@cityu.edu.hk
[2] School of Data Science, City University of Hong Kong, Kowloon, Hong Kong
[3] Shenzhen Research Institute, City University of Hong Kong, Shenzhen, China

**Abstract.** In this paper, a collaborative neurodynamic optimization approach is applied for bicriteria portfolio selection in the Markowitz mean-variance framework. The bicriteria portfolio selection problem consists of two objectives (risk and return) which are scalarized using a weighted Chebyshev function. Multiple neurodynamic optimization models are used to generate a set of Pareto-optimal solutions. Particle swarm optimization is used to diversify the Pareto-optimal solutions by optimizing the weights of the scalarized objective functions. Experimental results show the superiority of the applied approach.

**Keywords:** Portfolio selection · Multiobjective optimization ·
Neurodynamic optimization approach

## 1 Introduction

Since the pioneering work by Markowitz [1], portfolio selection has received great attention from academic and economic points of view. After that, a substantial amount of research works were investigated (e.g., [2–4]). The Markowitz mean-variance (M-V) model is formulated as a bicriteria optimization problem which considers maximizing the expected return while minimizing the risk of a portfolio. A set of Pareto-optimal (PO) solutions is obtained by optimizing the problem. The set of solutions is called an Pareto frontier.

Multiobjective optimization involves optimizing problems with at least two objective functions simultaneously, under a set of constraints. A set of non-dominated solutions is aimed to be generated. A non-dominated solution has

This work was supported in part by the Research Grants Council of the Hong Kong Special Administrative Region of China, under Grants 11208517 and 11202318, in part by the National Natural Science Foundation of China under Grant 61673330 and in part by International Partnership Program of Chinese Academy of Sciences under Grant GJHZ1849.

© Springer Nature Switzerland AG 2019
H. Lu et al. (Eds.): ISNN 2019, LNCS 11554, pp. 318–327, 2019.
https://doi.org/10.1007/978-3-030-22796-8_34

the property that none of its objective values can be improved further without worsening its other objective values. The non-dominated solution is regarded as a PO solution if no other solution dominates it within the feasible region. One main advantage of considering portfolio selection as a bicriteria optimization problem is portfolio managers or investors can choose one optimal asset alloca- tion strategy based on different risk aversion [5]. For example, [6] formulates the portfolio selection as a multiobjective linear programming model.

Neurodynamic optimization is a brain-like approach based on neural net- works by Hopfield and Tank [7]. Various neural network models are proposed for solving various optimization problems with different properties (e.g., [8–29]). Recently, neurodynamic approaches are proposed to solve multiobjective opti- mization problems [30,31].

Based on the above discussions, a collaborative neurodynamic optimization approach is applied for portfolio optimization. A set of scarlarized optimization problems is generated using weighted Chebyshev technique. Multiple recurrent neural networks are used to search for PO solutions. Besides, particle swarm optimization is used to diversify the solutions by optimizing the weights of the scalarized objective functions. Experimental results show the effectiveness of the applied algorithm.

## 2 Preliminaries

### 2.1 Bicriteria Portfolio Optimization

A typical bicriteria portfolio selection problem is formulated as:

$$\min \begin{cases} f_1(x) = \sum_{i=1}^{n} \sum_{j=1}^{n} \sigma_{ij} x_i x_j \\ f_2(x) = - \sum_{i=1}^{n} x_i r_i \end{cases} \text{s.t.} \begin{cases} \sum_{i=1}^{n} x_i = 1 \\ 0 \leq x_i \leq 1 \end{cases} \tag{1}$$

where $r = (r_1, r_2, \dots, r_n)^T$ and $x = (x_1, x_2, \dots, x_n)^T$ are the expected returns and allocations of a portfolio respectively; $n$ is the number of invested assets, $\sigma_{ij}$ is the covariance matrix between assets $i$ and $j$, $\sum_{i=1}^{n} x_i = 1$ is the budget con- straint and $x_i \geq 0$ implies no short sell is allowed. The term $\sum_{i=1}^{n} \sum_{j=1}^{n} \sigma_{ij} x_i x_j$ measures the portfolio risk (variance) and it should be minimized; the term $\sum_{i=1}^{n} x_i r_i$ models the expected return to be maximized.

Suppose that there exists two solutions $x'$ and $x''$, $x'$ dominates $x''$ if and only if $f_k(x') \leq f_k(x'')$, $k = 1, 2$. A solution is called Pareto-optimal (denoted by $x^*$) if no solution dominates it. The PO solutions form a Pareto frontier.

Weighted Chebyshev scalarization [32] is popular for multiobjective optimiza- tion. It formulates as a set of scalarized optimization problems (subproblems) with various weights. A set of PO solutions is obtained by optimizing the sub- problems. Let $f^* = (f_1^*, f_2^*)^T$, where $f_k^* = \min\{f_k(x) | \sum_{i=1}^{n} x_i = 1, 0 \leq x_i \leq 1\}$, $k = 1, 2$. The weighted Chebyshev function for a bicriteria optimization problem is defined as:

$$\min_{x} \max \{\lambda(f_1(x) - f_1^*), (1 - \lambda)(f_2(x) - f_2^*)\}. \tag{2}$$

where $0 \leq \lambda \leq 1$.

In view that (2) is non-smooth, it is reformulated as a constrained optimization problem: let $\zeta = \max\{\lambda(f_1(x) - f_1^*), (1 - \lambda)(f_2(x) - f_2^*)\}$, the optimization problem (1) is reformulated as:

$$\begin{aligned}
\min \quad & \zeta \\
\text{s.t.} \quad & f_1(x) - f_1^* - \zeta \le 0, \ f_2(x) - f_2^* - \zeta \le 0, \\
& \sum_{i=1}^{n} x_i = 1 \text{ and } 0 \le x_i \le 1.
\end{aligned} \tag{3}$$

### 2.2   Particle Swarm Optimization

Particle swarm optimization (PSO) [33] is a metaheuristic method for solving optimization problems. Members in PSO search for the global optimum of an optimization problem. Each of the members moves according to its past experience and a global best leader. The leader is regarded as the final output when the termination condition is reached. The velocity $\lambda$ and position $v$ of each member is updated as:

$$\begin{cases}
v \leftarrow \psi v + c_1 r_1(\tilde{\lambda} - \lambda) + c_2 r_2(\hat{\lambda} - \lambda) \\
\lambda \leftarrow \lambda + v
\end{cases} \tag{4}$$

where $\psi$ is an inertia weight; $r_1$ and $r_2$ are random variables within the range of 0.0 and 1.0; $c_1$ and $c_2$ are learning factors; $\tilde{\lambda}$ is the past best of each member; $\hat{\lambda}$ is the global best leader.

### 2.3   Performance Measures

Spacing is a popular indicator to measure the distribution of the solutions in multiobjective optimization [34]:

$$S(\mathcal{A}) = \frac{1}{\bar{d}} \sqrt{\frac{1}{|\mathcal{A}|} \sum_{i=1}^{|\mathcal{A}|} (d_i - \bar{d})^2} \tag{5}$$

where $\bar{d}$ is the mean value over $d_i$; $\mathcal{A}$ denotes a set of PO solutions, $|\mathcal{A}|$ is the cardinality of $\mathcal{A}$; $d_i$ is the Euclidean distance between it and its nearest member in $\mathcal{A}$. A smaller value of the indicator indicates the more equally spaced of the solutions (i.e., the better).

Hypervolume is another popular indicator which measures the precision and distribution of the solutions [35]. For the bicriteria case, let $f' = (f_1', f_2')^T$ be a reference vector dominated by a set of PO solutions. The HV value of ($\mathcal{A}$) is the non-overlapping region of all the hypercubes formed by $f'$ and member $a$ in $\mathcal{A}$:

$$HV(\mathcal{A}, f') = L\left( \bigcup_{a \in \mathcal{A}} [f_1(a), f_1'] \times [f_2(a), f_2'] \right) \tag{6}$$

where $L$ is the Lebesgue measure. Larger value of the indicator means the better set of the solutions.

## 3    Collaborative Neurodynamics

The main goal of solving a bicriteria portfolio optimization problem is to generate a set of well-distributed PO solutions. To ensure optimality of the solutions, Neurodynamic approach is used to solve the subproblems scalarized by the weighted Chebyshev approach. For each of the subproblems, a projection neural network is used to search for the PO solution and the solution is stored in an external archive. As a result, the solutions of the subproblems constitute a Pareto front. The projection neural network [36] for optimization problem (3) is described as:

$$\begin{cases} \epsilon\dfrac{dy}{dt} = -y + g[y - e_{n+1} - \nabla h(y)z - \nabla p(y)] \\ \epsilon\dfrac{dz}{dt} = -z + [z + h(y)]^+ \\ \epsilon\dfrac{du}{dt} = -p(y) \end{cases} \tag{7}$$

where $\epsilon$ is a positive time constant, $y = (x^T, \zeta)^T \in \mathbb{R}^{n+1}$ is the state vector, $z \in \mathbb{R}^2$ and $u \in \mathbb{R}$ are hidden state vectors, $e_{n+1} = (0, 0, \ldots, 0, 1)^T \in \mathbb{R}^{n+1}$, $h(x, \zeta) = (f_1(x) - f_1^* - \zeta, f_2(x) - f_2^* - \zeta)^T$, $p(x) = \sum_{i=1}^n x_i - 1$, $g(\cdot)$ is a piecewise-linear activation function defined as follows:

$$g(y) = \begin{cases} 1, & y > 1 \\ y, & 0 \le y \le 1 \\ 0, & y < 0 \end{cases}$$

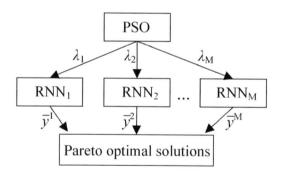

**Fig. 1.** Diagram of the collaborative neurodynamic approach.

The distribution of the solutions generated by the projection neural networks is associated with the values of the weights. To generate a set of well-distributed solutions, PSO weight optimization is used [37] by optimizing the metric stated in (6). Note that the global best leader $\hat{\lambda}$ is selected randomly to enhance turbulence. The weight optimization terminates until the S metric of the external archive is smaller than a threshold (i.e., $S(\mathcal{A}) \le \varepsilon$, where $\varepsilon$ is a sufficiently small positive number). Figure 1 and Procedure 1 show the diagram and pseudo codes of the collaborative neurodynamic approach (CNA) to bicriteria portfolio optimization.

---

**Procedure 1**

---

1: **Initialization:**
    a) Set $\varepsilon > 0$;
    b) Set $\mathcal{A} = \emptyset$;
    c) Initialize $y^m$ randomly in $[0,1]^{n+1}$, $m = 1, 2, \ldots, M$;
    d) Generate $\lambda_m$ for all $m$ uniformly;
2: **for** $m = 1, 2, \ldots, M$ **do**
3:     Compute steady states $\bar{y}^m$ using (7);
4: **end for**
5: Update external archive $\mathcal{A}$;
6: **if** $S(\mathcal{A}) \leq \varepsilon$ **then**
7:     exit;
8: **else**
9:     **for** $m = 1, 2, \ldots, M$ **do**
10:         Determine global best weight vector $\hat{\lambda}$;
11:         Determine personal best weight vectors $\tilde{\lambda}_m$;
12:         Update weight $\lambda_m$ using (4);
13:     **end for**
14:     **go to** 2;
15: **end if**

---

## 4    Experimental Results

Four datasets are constructed based on four major stock markets around the world: DAX, FTSE, HSI and S&P. The datasets are derived from weekly adjusted

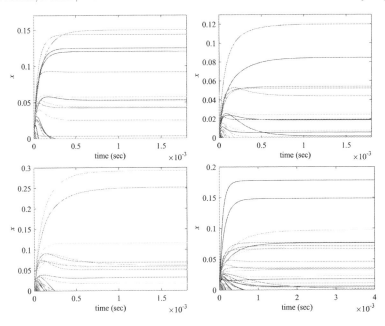

**Fig. 2.** Transient state x in neurodynamic model (7) based on DAX (top-left subplot), FTSE (top-right subplot), HSI (bottom-left subplot) and S&P (bottom-right subplot) datasets.

closing prices of the stocks from January 2000 to January 2018. Suspended and newly enlisted stocks within the period are excluded [38]. The number of stocks are 48, 58, 90 and 384 respectively. Figure 2 shows the transient behaviors of the state vectors corresponding to the portfolio of each dataset with maximum portfolio return.

Two state-of-the-art multiobjective evolutionary algorithms are used to compare the performance with CNA. They are NSGAII [39] and MOEA/D [40]. The population size of each approach is set to 20. Figure 3 shows the Pareto frontiers generated by the compared approaches, where the horizontal axis is variance and the vertical axis is portfolio return. It can be seen that the generated solutions by CNA are more evenly distributed.

**Table 1.** Performance comparison of the compared approaches in terms of mean HV

|      | NSGAII | MOEA/D | CNA |
|------|--------|--------|-----|
| DAX  | 9.9978e−1 (2.96e−5) | 9.9929e−1 (1.16e−4) | **9.9981e−1 (3.60e−5)** |
| FTSE | 9.9838e−1 (1.48e−4) | 9.9786e−1 (1.54e−4) | **9.9845e−1 (8.59e−5)** |
| HSI  | 9.9782e−1 (1.95e−4) | 9.9723e−1 (2.30e−4) | **9.9846e−1 (1.07e−4)** |
| S&P  | 9.9846e−1 (1.26e−4) | 9.9799e−1 (1.29e−4) | **9.9864e−1 (1.64e−4)** |

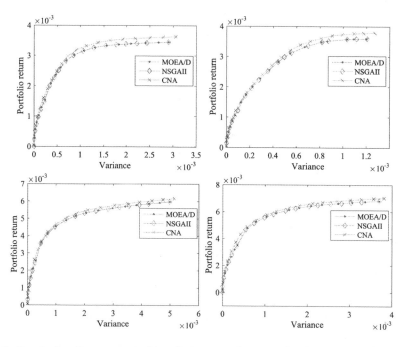

**Fig. 3.** Pareto frontiers generated by the compared approaches based on DAX (top-left subplot), FTSE (top-right subplot), HSI (bottom-left subplot) and S&P (bottom-right subplot) datasets.

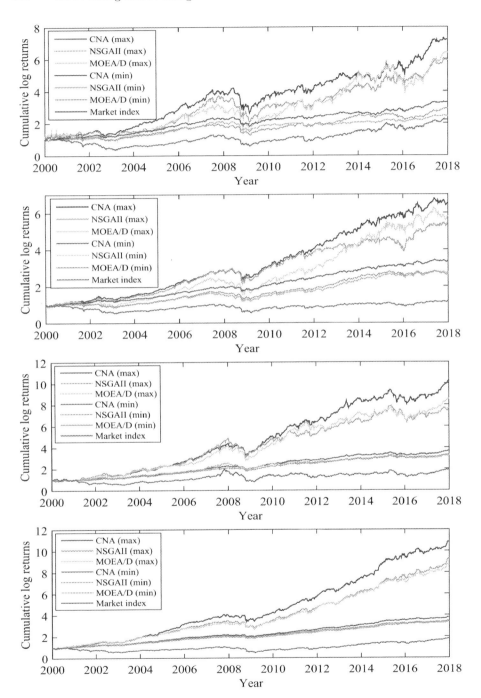

**Fig. 4.** Cumulative log returns of different portfolios based on DAX (the first subplot), FTSE (the second subplot), HSI (the third subplot) and S&P (the last subplot) datasets.

Table 1 tabulates the mean HV values of the solutions obtained by the compared approaches with 50 independent runs, and the standard deviations are in round brackets. The best mean HV value is in bold. It can be seen that CNA outperforms NSGAII and MOEA/D for all the test instances in terms of HV. Figure 4 shows the cumulative log returns of portfolios based on the four datasets from year 2000 to 2018. CNA (max) and CNA (min) denote respectively the optimal portfolios by the collaborative neurodynamic approach with the maximum and minimum annualized returns, NSGAII (max) and NSGAII (min) denote respectively the optimal portfolios by NSGAII with the maximum and minimum annualized returns, and MOEA/D (max) and MOEA/D (min) denote respectively the optimal portfolios by MOEA/D with the maximum and minimum annualized returns. It can be seen that CNA (max) outperforms others in all the cases.

## 5   Concluding Remarks

A collaborative neurodynamic optimization approach is applied for bicriteria portfolio optimization in this paper. By using multiple projection neural networks with different weights, PO solutions are generated and also optimized by using PSO. Experimental results show that the approach outperforms the compared approaches. Further investigations focus on neurodynamic approaches to portfolio selection with various problem formulations and the use of alternative risk measures.

## References

1. Markowitz, H.: Portfolio selection. J. Financ. **7**(1), 77–91 (1952)
2. Steinbach, M.C.: Markowitz revisited: mean-variance models in financial portfolio analysis. SIAM Rev. **43**(1), 31–85 (2001)
3. Cui, X., Zhu, S., Li, D., Sun, J.: Mean-variance portfolio optimization with parameter sensitivity control. Optim. Methods Softw. **31**(4), 755–774 (2016)
4. Cui, X., Li, D., Li, X.: Mean-variance policy for discrete-time cone-constrained markets: time consistency in efficiency and the minimum-variance signed supermartingale measure. Math. Financ. **27**(2), 471–504 (2017)
5. Kolm, P.N., Tütüncü, R., Fabozzi, F.J.: 60 years of portfolio optimization: practical challenges and current trends. Eur. J. Oper. Res. **234**(2), 356–371 (2014)
6. Ogryczak, W.: Multiple criteria linear programming model for portfolio selection. Ann. Oper. Res. **97**(1–4), 143–162 (2000)
7. Hopfield, J.J., Tank, D.W.: Computing with neural circuits: a model. Science **233**(4764), 625–633 (1986)
8. Liu, S., Wang, J.: A simplified dual neural network for quadratic programming with its KWTA application. IEEE Trans. Neural Netw. **17**(6), 1500–1510 (2006)
9. Hu, X., Wang, J.: Design of general projection neural networks for solving monotone linear variational inequalities and linear and quadratic optimization problems. IEEE Trans. Syst. Man Cybern. B **37**(5), 1414–1421 (2007)

10. Hu, X., Wang, J.: An improved dual neural network for solving a class of quadratic programming problems and its k-winners-take-all application. IEEE Trans. Neural Netw. **19**(12), 2022 (2008)
11. Liu, Q., Wang, J.: A one-layer recurrent neural network with a discontinuous activation function for linear programming. Neural Comput. **20**(5), 1366–1383 (2008)
12. Liu, Q., Wang, J.: A one-layer recurrent neural network with a discontinuous hard-limiting activation function for quadratic programming. IEEE Trans. Neural Netw. **19**(4), 558–570 (2008)
13. Xia, Y., Feng, G., Wang, J.: A novel recurrent neural network for solving nonlinear optimization problems with inequality constraints. IEEE Trans. Neural Netw. **19**(8), 1340–1353 (2008)
14. Guo, Z., Liu, Q., Wang, J.: A one-layer recurrent neural network for pseudoconvex optimization subject to linear equality constraints. IEEE Trans. Neural Netw. **22**(12), 1892–1900 (2011)
15. Liu, Q., Wang, J.: A one-layer recurrent neural network for constrained nonsmooth optimization. IEEE Trans. Syst. Man Cybern. B **41**(5), 1323–1333 (2011)
16. Liu, Q., Wang, J.: Finite-time convergent recurrent neural network with a hard-limiting activation function for constrained optimization with piecewise-linear objective functions. IEEE Trans. Neural Netw. **22**(4), 601–613 (2011)
17. Liu, Q., Guo, Z., Wang, J.: A one-layer recurrent neural network for constrained pseudoconvex optimization and its application for dynamic portfolio optimization. Neural Netw. **26**, 99–109 (2012)
18. Hosseini, A., Wang, J., Hosseini, S.M.: A recurrent neural network for solving a class of generalized convex optimization problems. Neural Netw. **44**, 78–86 (2013)
19. Liu, Q., Wang, J.: A one-layer projection neural network for nonsmooth optimization subject to linear equalities and bound constraints. IEEE Trans. Neural Netw. Learn. Syst. **24**(5), 812–824 (2013)
20. Li, G., Yan, Z., Wang, J.: A one-layer recurrent neural network for constrained nonsmooth invex optimization. Neural Netw. **50**, 79–89 (2014)
21. Liu, Q., Huang, T., Wang, J.: One-layer continuous-and discrete-time projection neural networks for solving variational inequalities and related optimization problems. IEEE Trans. Neural Netw. Learn. Syst. **25**(7), 1308–1318 (2014)
22. Li, G., Yan, Z., Wang, J.: A one-layer recurrent neural network for constrained nonconvex optimization. Neural Netw. **61**, 10–21 (2015)
23. Liu, Q., Wang, J.: A projection neural network for constrained quadratic minimax optimization. IEEE Trans. Neural Netw. Learn. Syst. **26**(11), 2891–2900 (2015)
24. Xia, Y., Wang, J.: A bi-projection neural network for solving constrained quadratic optimization problems. IEEE Trans. Neural Netw. Learn. Syst. **27**(2), 214–224 (2016)
25. Le, X., Wang, J.: A two-time-scale neurodynamic approach to constrained minimax optimization. IEEE Trans. Neural Netw. Learn. Syst. **28**(3), 620–629 (2017)
26. Qin, S., Le, X., Wang, J.: A neurodynamic optimization approach to bilevel quadratic programming. IEEE Trans. Neural Netw. Learn. Syst. **28**(11), 2580–2591 (2017)
27. Liu, Q., Yang, S., Wang, J.: A collective neurodynamic approach to distributed constrained optimization. IEEE Trans. Neural Netw. Learn. Syst. **28**(8), 1747–1758 (2017)
28. Yan, Z., Fan, J., Wang, J.: A collective neurodynamic approach to constrained global optimization. IEEE Trans. Neural Netw. Learn. Syst. **28**(5), 1206–1215 (2017)

29. Che, H., Wang, J.: A two-timescale duplex neurodynamic approach to biconvex optimization. IEEE Trans. Neural Netw. Learn. Syst. (2018, in press). https://doi.org/10.1109/TNNLS.2018.2884788
30. Yang, S., Liu, Q., Wang, J.: A collaborative neurodynamic approach to multiple-objective distributed optimization. IEEE Trans. Neural Netw. Learn. Syst. **29**(4), 981–992 (2018)
31. Leung, M.F., Wang, J.: A collaborative neurodynamic approach to multiobjective optimization. IEEE Trans. Neural Netw. Learn. Syst. **29**(11), 5738–5748 (2018)
32. Miettinen, K.: Nonlinear Multiobjective Optimization. Kluwer, Boston (1999)
33. Shi, Y., Eberhart, R.C.: Empirical study of particle swarm optimization. In: Congress on Evolutionary Computation, pp. 1945–1950 (1999)
34. Van Veldhuizen, D.A., Lamont, G.B.: On measuring multiobjective evolutionary algorithm performance. In: Congress on Evolutionary Computation, pp. 204–211 (2000)
35. Bader, J., Zitzler, E.: HypE: an algorithm for fast hypervolume-based many-objective optimization. Evol. Comput. **19**(1), 45–76 (2011)
36. Xia, Y.: An extended projection neural network for constrained optimization. Neural Comput. **16**(4), 863–883 (2004)
37. Leung, M.F., Wang, J.: A neurodynamic approach to multiobjective linear programming. In: 15th International Symposium on Neural Networks, pp. 11–18 (2018)
38. Chang, T.J., Meade, N., Beasley, J.E., Sharaiha, Y.M.: Heuristics for cardinality constrained portfolio optimisation. Comput. Oper. Res. **27**(13), 1271–1302 (2000)
39. Deb, K., Pratap, A., Agarwal, S., Meyarivan, T.A.M.T.: A fast and elitist multi-objective genetic algorithm: NSGA-II. IEEE Trans. Evol. Comput. **6**(2), 182–197 (2002)
40. Zhang, Q., Li, H., Maringer, D., Tsang, E.: MOEA/D with NBI-style Tchebycheff approach for portfolio management. In: Congress on Evolutionary Computation, pp. 1–8 (2010)

# A Learning-Based Approach for Perceptual Models of Preference

Junhui Mei[1], Xinyi Le[1(⊠)], Xiaoting Zhang[2], and Charlie C. L. Wang[3]

[1] Shanghai Jiao Tong University, Shanghai 200240, China
junhui_mei@126.com, lexinyi@sjtu.edu.cn
[2] Boston University, Boston, MA, USA
xiaoting@bu.edu
[3] The Chinese University of Hong Kong, Shatin, Hong Kong
cwang@mae.cuhk.edu.hk

**Abstract.** This paper introduces a novel data-driven approach based on subjective constraints and feature learning for training perceptual models of preference. Fuzzy evaluation is applied to describe the subjective opinions from a large set of data collected from user study. Combined with the objective attributes of the training models and the subjective preferences, an optimization method is developed successfully for training and learning perceptual models. Two applications are given in details for the selection of "best" viewpoint of 3D objects and the optimized direction of 3D printing, which verify the effectiveness of our approach. This work also demonstrate a good human-computer interaction practice that draws supporting knowledge from both the machine side and the human side.

**Keywords:** Perceptual model · Feature learning ·
Viewpoint selection · 3D printing direction

## 1 Introduction

Recently, it is really prevailing of learning algorithms on the study of classification and identification of patterns, signals, and other objective information [17,24]. However, the preference, opinions and views from the human side are also essential in decision making and reasoning [1]. The questions of what are good views, which one looks better, why it is more popular have been addressed in wide areas, such as art, media, literature appreciation and architecture design.

A highly reliable and effective performance evaluation rule is essential in handling cases like subjectivity and imprecise information [4,9]. Fuzzy set theory that was first introduced by Zadeh [30] is a suitable tool to evaluate the preference from the human side. The theory of fuzzy set has been applied in different

The work described in the paper was jointly sponsored by Natural Science Foundation of Shanghai (18ZR1420100) and National Natural Science Foundation of China (61703274). This work was partially supported by NSFC 61628211.

© Springer Nature Switzerland AG 2019
H. Lu et al. (Eds.): ISNN 2019, LNCS 11554, pp. 328–339, 2019.
https://doi.org/10.1007/978-3-030-22796-8_35

evaluation systems for many daily applications [3], such as pattern classification [11], shipping performance evaluation [5], feature extraction [25] and personnel selection [8], etc. As classical logic only permits conclusions which are either true or false, fuzzy degree between black and white was proposed to describe the expression of partial truth, tall, small etc. However, as the presence of imprecision, vagueness and subjectivity in fuzzy set theory, it is always necessary to further train and fine-tune a fuzzy model to improve its veracity.

Past decades have witnessed the development of numerous learning methods, such as support vector machines [7], echo state networks [12,32], convolutional deep neural networks [15], and deep Boltzmann machines [19]. These techniques mainly depend on a huge number of figures to learn the features and have shown the effectiveness in classification tasks. With respect to those preference related problems, it is always difficult to collect a large set of data for training. In addition, because of the bias and imprecision of personal preference, generalization ability is of great importance for such kind of problem. Extreme learning machine (ELM) proposed by [10] has shown very good generalization ability and robustness performance. Recently, as feature selection has drawn increasingly attentions [28], some extreme learning machine approaches with auto-encoders for multi-layer perception have been developed [22]. The architecture composes of self-taught feature extraction and supervised feature classification, which are bridged by random hidden weights. Multi-layer ELM is able to achieve more compact and meaningful feature representations compared to shallow ELM. Nowadays, ELM has been applied in a wider range of areas, such as vigilance estimation [21] and time sequence classification [16]. Especially, some ELM approaches have been successfully employed in the graphics area for classification and optimization [27,29,31]. An ELM classifier was trained in [29] for 3D shape segmentation. Unlike most works focusing on learning features, a shallow ELM for optimization has been applied for finding the "best" 3D printing direction [31].

In this paper, we proposed a fuzzy ELM approach for perceptual models of preference. The proposed approach is able to solve the evaluation problem by combing both the subjective knowledge and the objective evaluation. Fuzzy set theory is applied for dealing with subjective preference information, and multi-layer ELM is used to extract good features without supervised labels. This approach has been tested in practical applications to substantiate its effectiveness. Two applications of perception models in computer graphics area are introduced in this paper. Zhang et al. [31] optimized the printing direction based on four metrics. However, the trained model may be less effective when it is applied to 3D models with different size and style. Similarly, Secord et al. [20] proposed a data-driven approach for viewpoint preference. However, their model exists multicollinearity in different attributes, which will lead to instability caused by coupled parameters. To overcome the shortcomings of the existing methods, we propose a novel fuzzy extreme learning approach to improve the results of these two applications. Our method combines the advantages of fuzzy theory, feature extraction and ELM. Furthermore, the method can be easily modified to optimize its performance in different applications. Experiment results show that the perceptual models learned by our method of fuzzy ELM are effective and easy-to-implement.

The rest of this paper is organized as follows. Section 2 introduces the preliminary work, including the fundamental concepts and the theories of ELM and fuzzy model. Section 3 describes the proposed framework for preference learning problems. Section 4 optimizes the performance of 3D printed models through the orientation selection. Section 5 presents the optimization results of perceptual models of viewpoint preference of 3D models. Simulations and experiments have been conducted to verify the effectiveness of our method. Finally, our paper ends with the conclusion in Sect. 6.

## 2    Preliminaries

### 2.1    Extreme Learning Machine

Unlike the traditional function approximation theories which require to adjust input weights and the hidden layer biases, the input weights and hidden layer biases of *Extreme Learning Machine* (ELM) can be randomly assigned if only the activation function is infinitely differentiable. A basic ELM neural network is composed of one input layer, one hidden layer and one output layer. The input nodes depend on the input data. In the hidden layer, all the nodes are randomly generated and independent of training data. In the output layer, all the weights $\beta_i$ are problem-based and could be adjusted to solve problems such as feature learning, clustering, regression and classification.

Consider there are $N$ input-output samples $(x_i, t_i) \in \Re^n \times \Re^m$ for training, where $x_i$ is an input vector with dimension $n$ and $t_i$ is a target vector with dimension $m$. The output of an single layer feedforward neural network with $L$ hidden nodes can be represented by

$$f_L(x) = \sum_{i=1}^{L} \beta_i h_i(a_i, b_i, x) = h(x)\beta, \tag{1}$$

where $a_i$ and $b_i$ are learning parameters of hidden nodes and $\beta_i$ is the weight connecting the $i$th hidden node to the output node. $h_i(a_i, b_i, x)$ is the output of the $i$th hidden node with respect to the input $x$. $\beta = [\beta_1, ..., \beta_L]^T$ is the vector of the output weights between the hidden layer of $L$ nodes and the output node, and $h(x) = [h_1(x), ..., h_L(x)]$ is the output (row) vector of the hidden layer with respect to the input $x$. $h_i(a_i, b_i, x)$ could be composed of additive hidden nodes with different activation functions, such as sigmoid functions, hardlimit functions, gaussian functions, wavelet, hyperbolic functions, etc.

For a multiclass classifier or a regression problem, we assume there are $m$ output nodes. If the original class label is $p$, the expected output vector of the $m$ output nodes is $t_i = [0, ..., 0, 1^p, 0, ..., 0]^T$. In this case, only the $p$th element of

$t_i = [t_{i,1}, ..., t_{i,m}]$ is one, while the rest elements are set to zero. The optimization problem for ELM with multi-output nodes can be formulated as

$$\min \frac{1}{2}\|\beta\|^2 + \frac{1}{2}\lambda \sum_{i=1}^{N} \|\xi_i\|^2$$
$$\text{s.t. } h(x_i)\beta = t_i^T - \xi_i^T, \ i = 1, ..., N \tag{2}$$

where $\xi_i = [\xi_{i,1}, ..., \xi_{i,m}]^T$ is the training error vector of the $m$ output nodes with respect to the training sample $x_i$.

After the hidden nodes' parameters are chosen randomly, ELM neural network can be considered as a linear system and the output weights can be optimized according to the optimization procedure (2). The universal approximation capability has been analyzed in [10] that ELM neural network with randomly generated additive and a wide range of activation functions can universally approximate any continuous target functions in any compact subset of the Euclidean space. There are two phases in the training process of ELM: (1) feature mapping and (2) output weights solving. As solving the output weight is based on the optimization problem, we can design different feature mapping approaches to improve the ELM algorithm.

## 2.2 Fuzzy Model for Pairwise Comparison

For some preference and decision-making problems, it is very difficult to collect the opinions such as "which one is the best?" or "What is the optimal solution?". However, it is possible to collect the information for pairwise comparison, such as "which one is better, A or B?". People are more likely to answer these *2-Alternative Forced Choice* (2AFC) questions. In this paper, pairwise comparison information is assumed to be collected randomly, consistently and with a roughly uniform distribution.

According to the study of pairwise comparisons in [18], the value $s_{ij}$ is assigned of the comparison pair of $i$ and $j$, which represents a relative preference of $i$ over $j$. If the element $i$ is preferred to $j$ then $s_{ij} > 1$. Let us consider a prioritisation problem with $N$ unknown priorities, then the reciprocal property $s_{ji} = 1/s_{ij}$ for $i, j = 1, 2, ..., N$ always holds. A positive reciprocal matrix of pairwise comparisons $S = \{s_{ij}\} \in \Re^{N \times N}$ is constructed through $N(N-1)/2$ judgements. Then a priority vector $v = (v_1, v_2, ..., v_N)^T$ may be derived from the matrix. Moreover, we can make them satisfy the partition-of-unity as

$$v_1 + v_2 + \cdots + v_N = 1, \ v_i \geq 0, i = 1, 2, ..., N \tag{3}$$

When all elements $s_{ij}$ have perfect values, then

$$s_{ij} = v_i/v_j, \ s_{ij} = s_{ik}s_{kj}, \ i, j, k = 1, 2, .., N \tag{4}$$

However, the evaluations, $\{s_{ij}\}$, are usually not perfect – i.e., they only approximately estimate the exact ratios $v_i/v_j$. In addition, the comparisons are not complete for all the judgements when $N$ is very large.

A fuzzy approach to priorities derivation is then derived for dealing the imperfect pairwise comparisons based on inexact and incomplete judgments. In fuzzy set theory, the membership function of a fuzzy set (with a range covering the interval $(0, 1)$) represents the degree of truth as an extension of valuation. The values between 0 and 1 characterize fuzzy members, which belong to the fuzzy set only partially [13]. Fuzzy membership function thus becomes a suitable approach to evaluate perceptual models of preference. A normal fuzzy set $\tilde{s}$ is a triangular fuzzy number, defined by three real numbers $l \leq m \leq u$, and has a linear piecewise continuous membership function $\mu_{\tilde{s}}(\cdot)$ with the following characteristics [14]:

- $\mu_{\tilde{s}}(\cdot)$ is a continuous mapping from $\Re$ to the closed interval $[0, 1]$
- For all $x \in [-\infty, l]$ and $x \in [u, \infty]$, $\mu_{\tilde{s}}(x) = 0$.
- $\mu_{\tilde{s}}(x)$ is strictly linearly increasing on $[l, m]$ and strictly linearly decreasing on $[m, u]$.
- For $x = m$, $\mu_{\tilde{s}}(x) = 1$.

Let the pairwise comparison judgements $\{s_{ij}\}$ be represented by fuzzy numbers $s_{ij} = (l_{ij}, m_{ij}, u_{ij})$, and consider a set of $m$ ($m < N(N-1)/2$) incomplete pairwise comparisons. If the judgements are inconsistent, there is no priority vector that satisfies all interval judgements simultaneously. However, it is reasonable to try and find a vector that satisfies all judgements "as well as possible". This implies that a solution vector has to satisfy all interval judgements approximately in order to become "good enough".

## 3    Model Description

The proposed fuzzy extreme learning machine model combines convolutional ELM and fuzzy pairwise learning as optimization constraints, as shown in Fig. 1. The training procedure consists of two phases: feature mapping and optimization.

### 3.1    ELM Feature Mapping

Convolution is the process of multiplying each element of the image with its local neighbors, weighted by the kernel. Different kinds of kernels can cause a wide range of effects, such as blurring, sharpening, embossing, edge detection, and more. Convolution have been widely used to find the features, especially on images. Inspired by these prior works, the procedure of ELM feature mapping can be described as follows:

- First, choose kernels randomly to obtain convolutional feature maps with input data and kernels.
- Second, pooling operation is performed to maintain rotation invariance and minimize data size.
- Then, generating random and sparse weights from convoluted and pooled data.
- Finally, feature map have been prepared for the exaction and optimization in the next phase.

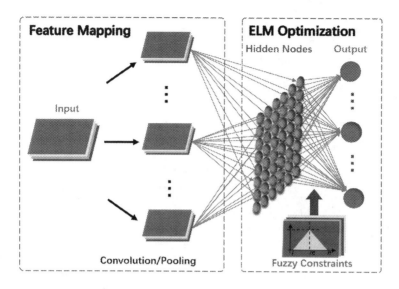

**Fig. 1.** Block diagram of fuzzy extreme learning machine

## 3.2 ELM Optimization

Based on the preliminaries about the extreme learning machine and fuzzy constraints, the optimization procedure of fuzzy extreme learning machine will be formulated in this subsection. Assume $v$ is the feature mapping of input data. Denote $H$ as the feature mapping operator from input data to hidden layer output $H : \{x_i\} \mapsto \{z_i\}$.

$$z_i = H(x_i), i = 1, 2, 3, ..., N. \tag{5}$$

According to (2), the optimization problem for fuzzy ELM can be formulated as follows:

$$\min \frac{1}{2}\|\beta\|^2 + \frac{1}{2}\lambda\|H(x)\beta - v\|^2 + \frac{1}{2}\rho\|\mu\|^2$$
$$\text{s.t.} Rv - d\mu \leq 0, 0 \leq \mu \leq 1,$$
$$\sum_{i=1}^{N} v_i = 1, v_i > 0, i = 1, 2, ..., N. \tag{6}$$

where $d = [d_1, d_2, ..., d_m]^T$ is a tolerance parameter vector, $\lambda$ and $\rho$ are multipliers satisfying $\lambda > 0$ and $\rho > 0$. As the optimization problem (6) is a convex quadratic problem with linear equality and inequality constraints, common toolbox is applicable to work it out in a very short time [2].

To conclude, a fuzzy extreme learning machine procedure for preference optimization can be concluded as follows:

- Given a training set $\{x_i\}$, and $m$ pairwise comparisons $\{s_{ij}\}$ within the training set $\{x_i\}$, $i, j = 1, 2, ..., N$.

- Define triangular membership functions $(l_{ij}, m_{ij}, u_{ij})$ for all the pairwise comparisons.
- Denote hidden node number as $L$ and randomly generate kernels for convolution.
- Using convolution and pooling for feature mapping and consider the feature mapping relationship as $H : \{x_i\} \mapsto \{z_i\}$.
- Add fuzzy constraints when minimization the mismatching and solve the convex optimization problem (6)
- Calculate the output weight $\beta$.
- Find the ranking priority result as $\{v_i\}$.
- Obtain the best choice of training set satisfying $v_{i^*} = \max(v_i)$.

## 4    Perceptual Model for 3D Printing Orientations

Additive manufacturing methods often require robust branching support structures to prevent material collapse at overhangs, resulting in unsightly surface artifacts after the supports have been removed. Figure 2 shows the 3D printed model containing artifacts from different support structures when printing a 3D model along different directions. Improper support will damage the small features of the model and have influence on visual artifacts. This section improves the perceptual model for determining 3D printing orientations already discussed in [31]. Four metrics including contact area, visual saliency, viewpoint preference, and smoothness entropy were considered. Despite the effectiveness of the proposed method in [31], the algorithm still had some drawbacks. First, these metrics have different kinds of dimensions. But the evaluation function is not dimension invariant (i.e., $F(d) \neq F(10d)$), the model should be normalized before training. Second, metrics in [31] cannot be proved to be ergodic. There possibly exist some other metrics undetected. Thirdly, graphics computation is really time consuming. For some complex models, it may cause several hours for calculating the metrics. In order to overcome these shortages, we propose a fuzzy ELM approach in this section to improve the algorithm.

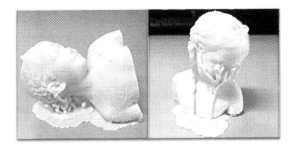

**Fig. 2.** 3D printed model from different support structure placement.

## 4.1   Collecting Human Preferences

The service of Amazon Mechanical Turk (AMT) is applied to collect human preferences. Pairwise comparisons are conducted through investigation on Amazon
Mechanical Turk to select between two printing directions. Several models are
employed to generate tasks as follows. We uniformly sample the Gauss sphere
to obtain 1448 possible printing directions. For each model, 500 pairs of printing
directions are randomly picked. Each pair was shown to 8 people. Among them,
16 random pairs of each model occur more than once for testing the reliability.
For each picked pair $s_{ij}$, let's assume that the direction $i$ was picked $p_i$ times,
and $j$ was chosen $p_j$ times for all comparisons between directions $i$ and $j$ for all
$i, j = 1, 2, 3..., N$. For some comparisons, if $p_i = p_j$ means the preference on $i$
and $j$ is very contradictive, the uncertainty of $s_{ij}$ is comparatively large. However, when $p_i = 0$ the direction $i$ is preferred definitely; vice versa for $p_j = 0$.
Moreover, as $p_i$ and $p_j$ increase, the vagueness of preference will decrease. Based
on the above facts, the pairwise comparison judgements $\{s_{ij}\}$ can be represented
by fuzzy numbers $s_{ij} = (l_{ij}, m_{ij}, u_{ij})$. Then, we can evaluate human preferences
and apply these information to train the perceptual model.

## 4.2   Input Representation

Given a set of 3D shapes, in OBJ (or .OBJ) format or STL (STereoLithog-
raphy) format. Both formats are widely used for rapid prototyping, 3D printing and computer-aided manufacturing [6]. STL files describe only the surface
geometry of a three-dimensional object without any representation of color, texture or other common CAD model attributes. OBJ (or .OBJ) is also a universally accepted geometry definition file format representing 3D geometry alone—
namely, the position of each vertex, and the faces that make each polygon defined
as a list of vertices.

In order to find the "best" 3D printing results of models, some information
should be added to the 3D model itself. Distribution of contact area ($f_1$) should
be first considered. For different printing orientations, the surface area of regions
connecting to supporting structures is determined by considering overhangs as
well as potential connections at the base of supports. According to the graphics
study in [23], surface regions need support only if the angle between its tangent
plane and the printing direction is larger than the critical angle $\alpha$. Faces with
angle less than $\alpha$ are self-supported. In this section, set $\alpha = 25°$ according to
[26]. A distribution of contact area $f_1$ then can be computed and visualized as
one of the important features.

In addition, subjective preference is also very important. For example, we
prefer there are no supports on eyes and faces. For some man-made models,
we prefer there's no defect in working plane. An interactive tool is developed
to incorporate the subjective preference into a model. Denote distribution of
subjective important feature as $f_2$. Select one of important directions $d_{a1}$, a
Harmonic field $H_1(d)$ is computed as

$$\nabla^2 H(d) = 0$$
$$\text{s.t. } H(d_{a1}) = 1, H(-d_{a1}) = 0, \tag{7}$$

Similarly, select one least important direction (e.g., base) as $d_{a2}$, a Harmonic field $H_2(d)$ can be similarly calculated. The distribution of subjective importance feature can be emerged as

$$f_1 = \min(H_1, H_2, ...) \tag{8}$$

Moreover, the geometry feature, such as coordinates and relationship between vertexes, also should be considered. Denote geometry feature as $f_3$. As a result, the input data for perceptual 3D printing model, which is a combination of contacted area distribution feature($f_1$), subjective importance feature($f_2$), and structure geometry feature ($f_3$).

## 5    Perceptual Models of Viewpoint Preference

In this section, we will demonstrate the better performance on three perceptual models for 3D printing orientations by placing fewer support structures at visually important regions. Our results are also compared with Autodesk MeshMixer obtained by considering only a single factor.

Using the proposed approach, Fig. 3 show the "optimal", "default", and "meshmixer" printing directions results for "Kitten". Figures 4 and 5 show the "optimal" and "default" printing directions results for "Bunny" and "Armadillo", respectively, which further substantiate the effectiveness of proposed method. We can see the additive support is limited and no support is added on the regions that are important for viewpoint preference. Therefore, the negative influence caused by supporting structures can be reduced as much as possible.

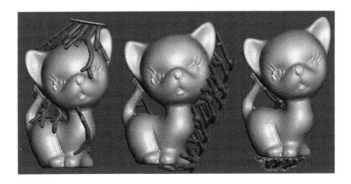

**Fig. 3.** "Optimal", "default" and "Meshmixer" printing directions for "Kitten".

**Fig. 4.** "Optimal"and "default" printing directions for "Bunny".

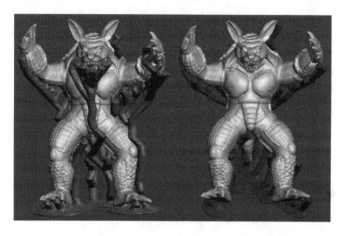

**Fig. 5.** "Optimal"and "default" printing directions for "Armadillo".

## 6   Conclusion

This paper introduces a fuzzy extreme learning machine approach for training perceptual models of preference. A novel fuzzy extreme learning machine
approach was presented for obtaining preference priority from a set of inconsistency pairwise comparisons. Our approach combines the advantage of feature
mapping, ELM optimization, and fuzzy membership formulation. Two applications are given in details for the selection of "best" viewpoint of 3D objects and
the optimization of 3D printing direction. Compare to the previous results, our
approach has better robustness and generalization capabilities as the input comparisons are inaccurate and inconsistency. Moreover, our model can be improved
if more models has been trained. Not only in graphics area, our approach can
be applied in a wide range of area such as parametric design for clothing patterns, mechanical structure optimization, etc. Our approach demonstrates a good

human-computer interaction practice. Future work will focus on improving our algorithm for better performance and extend our approach for more applications.

# References

1. Acampora, G., Cadenas, J.M., Loia, V., Ballester, E.M.: A multi-agent memetic system for human-based knowledge selection. IEEE Trans. Syst. Man Cybern. Part A: Syst. Hum. **41**(5), 946–960 (2011)
2. Boyd, S., Vandenberghe, L.: Convex Optimization. Cambridge University Press, Cambridge (2004)
3. Chakraborty, S., Konar, A., Jain, L.C.: An efficient algorithm to computing max;min inverse fuzzy relation for abductive reasoning. IEEE Trans. Syst. Man Cybern. Part A: Syst. Hum. **40**(1), 158–169 (2010)
4. Chen, S.M., Tsai, B.H.: Autocratic decision making using group recommendations based on intervals of linguistic terms and likelihood-based comparison relations. IEEE Trans. Syst. Man Cybern. Syst. **45**(2), 250–259 (2015)
5. Chou, T.Y., Liang, G.S.: Application of a fuzzy multi-criteria decision-making model for shipping company performance evaluation. Marit. Policy Manag. **28**(4), 375–392 (2001)
6. Chua, C.K., Leong, K.F.: Rapid Prototyping: Principles and Applications, vol. 1. World Scientific, Singapore (2003)
7. Cortes, C., Vapnik, V.: Support-vector networks. Mach. Learn. **20**(3), 273–297 (1995)
8. Güngör, Z., Serhadlıoğlu, G., Kesen, S.E.: A fuzzy AHP approach to personnel selection problem. Appl. Soft Comput. **9**(2), 641–646 (2009)
9. Gören, S., Baccouche, A., Pierreval, H.: A framework to incorporate decision-maker preferences into simulation optimization to support collaborative design. IEEE Trans. Syst. Man Cybern. Syst. **47**(2), 229–237 (2016)
10. Huang, G.B., Zhu, Q.Y., Siew, C.K.: Extreme learning machine: theory and applications. Neurocomputing **70**(1), 489–501 (2006)
11. Ishibuchi, H., Nakashima, T., Murata, T.: Performance evaluation of fuzzy classifier systems for multidimensional pattern classification problems. IEEE Trans. Syst. Man Cybern. Part B Cybern. **29**(5), 601–618 (1999)
12. Jaeger, H.: Echo state network. Scholarpedia **2**(9), 2330 (2007)
13. Klir, G., Yuan, B.: Fuzzy sets and fuzzy logic, theory and applications (2008)
14. Klir, G.J., Yuan, B.: Fuzzy Sets, Fuzzy Logic, and Fuzzy Systems. World Scientific, Singapore (1996)
15. Krizhevsky, A., Sutskever, I., Hinton, G.E.: Imagenet classification with deep convolutional neural networks. In: Advances in Neural Information Processing Systems, pp. 1097–1105 (2012)
16. Liu, H., Yu, L., Wang, W., Sun, F.: Extreme learning machine for time sequence classification. Neurocomputing **174**, 322–330 (2016)
17. Michalski, R.S., Carbonell, J.G., Mitchell, T.M.: Machine Learning: An Artificial Intelligence Approach. Springer, Heidelberg (2013)
18. Mikhailov, L.: Deriving priorities from fuzzy pairwise comparison judgements. Fuzzy Sets Syst. **134**(3), 365–385 (2003)
19. Salakhutdinov, R., Hinton, G.E.: Deep Boltzmann machines. In: AISTATS, vol. 1, p. 3 (2009)

20. Secord, A., Lu, J., Finkelstein, A., Singh, M., Nealen, A.: Perceptual models of viewpoint preference. ACM Trans. Graph. (TOG) **30**(5), 109 (2011)
21. Shi, L.C., Lu, B.L.: EEG-based vigilance estimation using extreme learning machines. Neurocomputing **102**, 135–143 (2013)
22. Tang, J., Deng, C., Huang, G.B.: Extreme learning machine for multilayer perceptron. IEEE Trans. Neural Netw. Learn. Syst. **27**(4), 809–821 (2016)
23. Vanek, J., Galicia, J.A.G., Benes, B.: Clever support: efficient support structure generation for digital fabrication. Comput. Graph. Forum **33**(5), 117–125 (2014)
24. Vapnik, V.: The Nature of Statistical Learning Theory. Springer, Heidelberg (2013)
25. Wang, C.C., Chang, T.K., Yuen, M.M.: From laser-scanned data to feature human model: a system based on fuzzy logic concept. Comput.-Aided Des. **35**(3), 241–253 (2003)
26. Wang, C.C., Leung, Y.S., Chen, Y.: Solid modeling of polyhedral objects by layered depth-normal images on the GPU. Comput.-Aided Des. **42**(6), 535–544 (2010)
27. Wang, Y., Xie, Z., Xu, K., Dou, Y., Lei, Y.: An efficient and effective convolutional auto-encoder extreme learning machine network for 3D feature learning. Neurocomputing **174**, 988–998 (2016)
28. Wu, Q., Wang, Z., Deng, F., Chi, Z., Feng, D.D.: Realistic human action recognition with multimodal feature selection and fusion. IEEE Trans. Syst. Man Cybern. Syst. **43**(4), 875–885 (2013)
29. Xie, Z., Xu, K., Liu, L., Xiong, Y.: 3D shape segmentation and labeling via extreme learning machine. In: Computer Graphics Forum, vol. 33, pp. 85–95. Wiley (2014)
30. Zadeh, L.A.: Fuzzy sets. Inf. Control **8**(3), 338–353 (1965)
31. Zhang, X., Le, X., Panotopoulou, A., Whiting, E., Wang, C.C.: Perceptual models of preference in 3D printing direction. ACM Trans. Graph. (TOG) **34**(6), 215 (2015)
32. Zhang, X., Le, X., Wu, Z., Whiting, E., Wang, C.C.: Data-driven bending elasticity design by shell thickness. In: Computer Graphics Forum, vol. 35, pp. 157–166 (2016)

# Resistant Neural Network Learning via Resistant Empirical Risk Minimization

Zaur M. Shibzukhov[1,2(✉)]

[1] Mathematics and Informatics Institute of Moscow Pedagogical State University,
Krasnoprudnaya 14, Moscow, Russia
[2] Institute of Applied Mathematics and Automation KBSC RAS,
Shortanova 89a, Nalchik, Russia
szport@gmail.com

**Abstract.** The article proposes an extended version of the principle of minimizing the empirical risk for training neural networks that is stable with respect to a large number of outliers in the training data. It is based on the use of -averaging and -averaging functions instead of arithmetic mean for estimating empirical risk. An iteratively re-weighted scheme is proposed for minimizing differentiable resistant estimates of mean loss functions. This schema allows to use weighted version of traditional back-propagation algorithms for neural networks learning in presence of large number of outliers.

**Keywords:** Neural networks · Robust estimation ·
Resistant averaging function

## 1  Introduction

This article is devoted to the problem of approximation of functions of several variables using multilayer neural networks (NN), which is resistant to a large number of outliers in the training data. NNs are known as universal approximators [1], which can approximate any continuous functions of many variables, which is also confirmed when solving applied problems. The data in applied problems, as a rule, are distorted or erroneous. The first type of data distortion is noise. It is effectively eliminated with the help of modern methods of approximation of continuous functions (including NN), based on minimization of the sum of squares of errors between observed and predicted values. The second type of distortion is emissions—significant deviations associated with both large errors in the data and the fact that the data may reflect a mixture of different processes. Usually such distortions can cover from 1% to 10% of the data, but can also cover up to 40–50%. In order to properly restore the functional dependencies of the data, it is advisable to identify and remove the outliers from the data that are used for the approximation. However, this approach is not

This work have been supported by the RFBR grant N18-01-00050.

H. Lu et al. (Eds.): ISNN 2019, LNCS 11554, pp. 340–350, 2019.
https://doi.org/10.1007/978-3-030-22796-8_36

always possible. In such cases, it is difficult to do without a prior attempt to bring the desired functional relationship closer, which would ensure the detection of outliers in accordance with the distribution of errors or losses or other criteria. Approximation methods based on minimizing arithmetic mean or sum of squares of errors are not resistant to outliers. Outliers can cause such a significant shift in the desired parameters that their detection by the distribution of errors becomes impossible. This problem is well studied in the case of linear regression, when a relatively small amount of emissions can significantly distort the coefficients [2–4].

To overcome this problem, a number of linear regression recovery methods have been developed that are resistant to a relatively large number of errors. Among them, the most popular are LMedS (*Least Median of Squares*) and LTS (*Least Trimmed Squares*) [5]. They provide resistant (stable) results in the presence of up to 50% of outliers. Such results are virtually impossible with M-estimator methods [6]. This is possible because the median and the truncated average can be considered as reliable estimates for the mean. Nonlinear regression has also been studied in the presence of outliers based on the M-regression method [7]. NN training based on the M-regression approach has also been studied in [9–11]. NN training based on LTS is considered in [12,13]. LMedS based NN training is considered in [14]. An interesting PCLTS algorithm is proposed in [15].

LMedS and LMedA methods are individual cases for solving a regression problem via minimizing the functional

$$Q(\mathbf{w}) = \mathrm{med}\{\varrho(r_1(\mathbf{w})), \ldots, \varrho(r_1(\mathbf{w}))\},$$

where $\varrho(r)$ is the nonnegative quasi-convex function, $r_k(\mathbf{w}) = f(\mathbf{x}_k, \mathbf{w}) - y_k$, $\{\mathbf{x}_1, \ldots, \mathbf{x}_N\} \subset \mathbb{R}^n$ and $\{y_1, \ldots, y_N\} \subset \mathbb{R}$ are given input vectors and expected output, $f(\mathbf{x}, \mathbf{w})$ is the transformation function of the parametrized model for the restored dependence. LTS and LTA estimators are individual cases for solving a regression problem based on minimizing the functional

$$Q(\mathbf{w}) = \frac{1}{N - p} \sum_{k=1}^{N-p} \varrho(r_{(k)}(\mathbf{w})),$$

where $r_{(1)}, \ldots, r_{(N)}$ is the sequence $r_1, \ldots, r_N$ in ascending order, $p > N/2$. For LMedS and LTS estimators $\varrho(r) = r^2$, and for LMedA and LTA $\varrho(r) = |r|$.

Robust non-linear regression recovery with a large number of outliers in the data (up to 40–50%) is an important task in cases where the detection of outliers is very difficult. However, if you remove the outliers in the source data, and the remaining amount of data is enough to restore the dependency, then resistant regression methods can be effective. They will allow us, at least, to find such an approximation of the desired dependence, so that by the distribution of errors, it will be possible to detect outliers, and at best, to restore the desired relationship, despite outliers.

This paper proposes generalized approach that covers LMedS and LMedA, LTS and LTA estimations. It is based on minimizing robust and differentiable

M-averaged and WM-averages. Thus, we can employ gradient minimization and algorithms similar to IRLS (*Iteratively Re-weighted Least Squares*) [8] to search for optimal values for dependence parameters in linear regression model. The main idea is to employ robust, but smooth estimating functions instead of robust and non-smooth versions of the latter. Minimizing robust mean or sum estimates is more preferable, since, it is the instability of mean (as a rule, the arithmetic average) or the sum estimation methods in relation with outliers is the main reason why the required parameters shift in the problem of minimizing the approximate functional dependencies employing empirical risk minimization principle.

## 2    Classical Empirical Risk Minimization Principle and Robust Learning

Regression and classification problems are often formulated as empirical risk minimization problem:

$$\mathbf{w}^* = \arg \min_{\mathbf{w}} Q(\mathbf{w})$$

where

$$Q(\mathbf{w}) = \frac{1}{N} \sum_{k=1}^{N} \varrho(r_k(\mathbf{w})). \tag{1}$$

The problem of outliers arises when the empirical distribution of losses $z_1 = \varrho(r_1(\mathbf{w}))$, ..., $z_N = \varrho(r_N(\mathbf{w}))$ contains outliers. They cause distortion of the values of the desired parameters. First robust approaches to suppress outliers are based on a choice of the function $\varrho(r)$, which grows slower than $r^2$, i.e. $|\varrho'(r)| \leqslant 1$. However, this approach does not always able to suppress the influence of outliers. In order to explain let's consider the sum:

$$S\{r_1, \ldots, r_N\} = \varrho(r_1) + \cdots + \varrho(r_N).$$

The following equality

$$|S\{r_1, \ldots, r_N + \Delta\} - S\{r_1, \ldots, r_N\}| = \varrho'(\tilde{r})\Delta,$$

where $\tilde{r} \in [r_N, r_N + \Delta]$, can explain the reason: when $\Delta$ is large or the number of outliers is large the distortion of $S$ become large too. So minimization of the distorted function $Q(\mathbf{w})$ usually led to distortion of the desired parameters $\mathbf{w}^*$.

The alternative is based on using functions of summing or averaging, which are resistant to outliers. Well-known examples are median and quantile. But they are not continuously differentiable. Therefore, gradient based approaches to training are not possible. However differentiable averaging functions, which are resistant to outliers can be defined and constructed.

## 3 Differentiable M-averages

Thus, for estimating the average we propose to use the M-averages, which are defined as follows:

$$\mathsf{M}_\rho\{z_1,\ldots,z_N\} = \arg\min_u \sum_{k=1}^N \rho(z_k - u),$$

where $\rho(r)$ is the strictly convex and twice continuously differentiable function. Well known averages are examples of M-averages:

- arithmetical mean: $\rho(r) = r^2$;
- median: $\rho(r) = |r|$;
- $\alpha$-quantile: $\rho(r) = \begin{cases} \alpha r, & \text{if } r \geqslant 0 \\ (\alpha - 1)r, & \text{if } r < 0, \end{cases}$ where $\alpha \in (0,1)$.

In the cases when $\rho'$ and $\rho''$ are exist the $\mathsf{M}_\rho$ is differentiable:

$$\frac{\partial \mathsf{M}_\rho}{\partial z_k} = \frac{\rho''(z_k - \bar{z}_\rho)}{\rho''(z_1 - \bar{z}_\rho) + \cdots + \rho''(z_N - \bar{z}_\rho)},$$

where $\bar{z}_\rho = \mathsf{M}_\rho\{z_1,\ldots,z_N\}$. Wherein $\frac{\partial \mathsf{M}_\rho}{\partial z_k} \geqslant 0$ and $\sum_{k=1}^N \frac{\partial \mathsf{M}_\rho}{\partial z_k} = 1$.

## 4 Learning via Minimizing M-averages of Losses

So now we able to formulate the problem of searching of the optimal parameter values $\mathbf{w}^*$ as a problem of minimization of the following function

$$Q(\mathbf{w}) = \mathsf{M}_\rho\{\ell_1(\mathbf{w}),\ldots,\ell_N(\mathbf{w})\},$$

where $\ell_k(\mathbf{w}) = \varrho(r_k(\mathbf{w}))$. It can be solved numerically using gradient descent method and it's variants. Gradient of $Q(\mathbf{w})$ has the following form:

$$\nabla Q(\mathbf{w}) = \sum_{k=1}^N v_k(\mathbf{w})\nabla \ell_k(\mathbf{w})$$

where

$$v_k(\mathbf{w}) = \frac{\partial \mathsf{M}_\rho\{\ell_1(\mathbf{w}),\ldots,\ell_N(\mathbf{w})\}}{\partial z_k}.$$

In order to minimize the overhead of calculation of averaging value an iteratively re-weighted schema is applied. The averaging value is calculated iteratively:

$$\mathbf{w}_{t+1} \leftarrow \arg\min_{\mathbf{w}} \sum_{k=1}^N v_k \ell_k(\mathbf{w}),$$

where with $z_k = \ell_k(\mathbf{w})$

$$v_k = \frac{\rho''(z_k - \bar{z}_t)}{\rho''(z_1 - \bar{z}_t) + \cdots + \rho''(z_N - \bar{z}_t)}.$$

So, to minimize the functional, we can apply the IR-ERM learning algorithm (*Iteratively Re-weighted Empirical Risk Minimization*) [16] which in pseudocode language can be expressed as:

> **procedure** IR-ERM($\mathbf{w}_0$)
> $\quad t \leftarrow 0$
> $\quad$**repeat**
> $\qquad z_1 = \ell_1(\mathbf{w}_t), \ldots, z_N = \ell_N(\mathbf{w}_t)$
> $\qquad \bar{z}_t = \mathsf{M}_\rho\{z_1, \ldots, z_N\}$
> $\qquad$**for** $k = 1, \ldots, N$ **do**
> $$v_k = \frac{\rho''(z_k - \bar{z}_t)}{\rho''(z_1 - \bar{z}_t) + \cdots + \rho''(z_N - \bar{z}_t)}$$
> $\qquad$**end**
> $\qquad \mathbf{w}_{t+1} \leftarrow \arg\min_{\mathbf{w}} \sum_{k=1}^{N} v_k \ell_k(\mathbf{w})$
> $\qquad t \leftarrow t + 1$
> $\quad$**until** $\{\bar{z}_t\}$ and $\{\mathbf{w}_t\}$ are stabilized.
> **end**

## 5  Resistant Differentiable M-averages

We give examples of differentiable averages, to be used as an adequate replacement for the median:

1. $\rho_{\text{sqrt},\varepsilon}(r) = \sqrt{\varepsilon^2 + r^2} - \varepsilon;$
2. $\rho_{\ln,\varepsilon}(r) = |r| - \varepsilon \ln(\varepsilon + |r|) - \varepsilon \ln \varepsilon;$
3. $\rho_{\arctan,\varepsilon}(r) = r \arctan \frac{r}{\varepsilon} - \frac{1}{2}\varepsilon \ln[1 + \left(\frac{r}{\varepsilon}\right)^2].$

In order to define differentiable replacements for quantiles let's define

$$\rho_\alpha(r) = \begin{cases} \alpha \rho(r), & \text{if } r \geqslant 0 \\ (1-\alpha)\rho(r), & \text{if } r < 0, \end{cases}$$

where $\rho(r)$ is a function for the definition of the replacement for the median (for example, $\rho_{\text{sqrt},\varepsilon}$, $\rho_{\ln,\varepsilon}$, $\rho_{\arctan,\varepsilon}$).

The key to understanding the empirical mean resistance to outliers can be the following inequality:

$$|\mathsf{M}_\rho\{r_1, \ldots, r_N + \Delta\} - \mathsf{M}_\rho\{r_1, \ldots, r_N\}| = \varrho''(\tilde{r} - u_{\tilde{r}})\Delta,$$

where $\rho(r)$ is the convex function, $\rho''(r)$ is the continuous function, $\Delta > 0$ is the distortion value, $\tilde{r} \in [r_N, r_N + \Delta]$, $u_{\tilde{r}} = \mathsf{M}_\rho\{r_1, \ldots, r_N\}$.

We give the following estimates for $\rho''(r)\Delta$:

(1) $\rho''_{\text{sqrt},\varepsilon}(r) = \frac{\varepsilon^2}{(\varepsilon^2+r^2)^{3/2}}$, $\rho''_{\text{sqrt},\varepsilon}(r)\Delta < \frac{\varepsilon^2\Delta}{r^3}$;

(2) $\rho''_{\ln,\varepsilon}(r) = \frac{\varepsilon}{(\varepsilon+|r|)^2}$, $\rho''_{\ln,\varepsilon}(r)\Delta < \frac{\varepsilon^2\Delta}{r^2}$;

(3) $\rho''_{\arctan,\varepsilon}(r) = \frac{\varepsilon}{\varepsilon^2+r^2}$, $\rho''_{\arctan,\varepsilon}(r)\Delta < \frac{\varepsilon\Delta}{r^2}$.

It is easy to see that for the sufficiently small $\varepsilon$ the M-averages with $\rho_{\text{sqrt},\varepsilon}$, $\rho_{\ln,\varepsilon}$, $\rho_{\arctan,\varepsilon}$ and can be resistant to relatively high number of outliers:

$$|M_\rho\{r_1,\ldots,r_N+\Delta\} - M_\rho\{r_1,\ldots,r_N\}| < \varepsilon.$$

## 6   Learning via WM-averages Minimizing

Along with the LTS and LTA, we can also employ LWS (*Least Winsorized Squares*) and LWA (*Least Winsorized Absolutes*) estimations, which are based on *Winsorized Mean* (WM) method:

$$\text{WM}_\rho\{z_1,\ldots,z_N\} = \frac{1}{N}\sum_{k=1}^{N}\min\{z_k,\bar{z}_\rho\}. \tag{2}$$

To generalize this method of robust estimation of the mean, we introduce a differentiable function $\sigma(z,u)$, which satisfies the following conditions:

(1) $\lim_{z\to+\infty}\sigma(z,u) = u$;

(2) $\lim_{z\to-\infty}\sigma(z,u)/z = 1$.

So

$$\text{WM}_\rho\{z_1,\ldots,z_N\} = \frac{1}{N}\sum_{k=1}^{N}\sigma(z_k,\bar{z}_\rho).$$

Partial derivatives of $\text{WM}_\rho$ can be written as follows:

$$\frac{\partial\text{WM}_\rho}{\partial z_k} = \frac{1}{N}\sigma'_z(z_k,\bar{z}_\rho) + \frac{1}{N}\frac{\partial\text{M}_\rho}{\partial z_k}\sum_{l=1}^{N}\sigma'_u(z_l,\bar{z}_\rho).$$

In the case of (2)

$$\frac{\partial\text{WM}_\rho}{\partial z_k} = \begin{cases} \frac{1}{N} + \frac{m}{N}\frac{\partial\text{M}_\rho}{\partial z_k}, & \text{if } z_k \leqslant \bar{z}_\rho \\ \frac{m}{N}\frac{\partial\text{M}_\rho}{\partial z_k}, & \text{if } z_k > \bar{z}_\rho, \end{cases}$$

where $m = |\{z_k : z_k > \bar{z}_\rho\}|$.

For example, there is also smooth variant of $\min\{z,u\}$:

$$\sigma(z,u) = \frac{1}{2}(z + u - \rho(z-u)), \tag{3}$$

where $\rho$ is a function, which is used for construction M-average as replacement of median (for example, $\rho_{\text{sqrt},\varepsilon}$, $\rho_{\ln,\varepsilon}$, $\rho_{\arctan,\varepsilon}$). In this case

$$\frac{\partial\text{WM}_\rho}{\partial z_k} = \frac{1}{2N}(1 - \rho'(z_k-\bar{z}_\rho)) + \frac{1}{2}\frac{\partial\text{M}_\rho}{\partial z_k}.$$

It can be easily shown that $\sum\limits_{k=1}^{N} \frac{\partial \mathsf{WM}_\rho}{\partial z_k} = 1$. So that $\mathsf{WM}_\rho$-averaging has the same property as the M-averaging.

We can further generalize this approach by the way of using an arbitrary M-average instead of the arithmetic mean. Let's $\mathsf{M}_\varphi$ is an M-averaging function for replacement of the arithmetic mean. Then we can define the following averaging function:

$$\mathsf{WM}_{\varphi\rho}\{z_1, \ldots, z_N\} = \mathsf{M}_\varphi\{\sigma(z_1, \bar{z}_\rho), \ldots, \sigma(z_N, \bar{z}_\rho)\}.$$

Partial derivatives of $\mathsf{WM}_{\varphi\rho}$ can be calculated according to following formula:

$$\frac{\partial \mathsf{WM}_\rho}{\partial z_k} = \frac{\partial \mathsf{M}_\varphi}{\partial v_k}\sigma_z'(z_k, \bar{z}_\rho) + \frac{\partial \mathsf{M}_\rho}{\partial z_k}\sum_{l=1}^{N}\frac{\partial \mathsf{M}_\varphi}{\partial v_l}\sigma_u'(z_l, \bar{z}_\rho).$$

The search for the optimal values of the parameters $\mathbf{w}^*$ that minimize the function

$$Q(\mathbf{w}) = \mathsf{M}_\rho\{\ell_1(\mathbf{w}), \ldots, \ell_N(\mathbf{w})\}$$

can be preformed using the IR-WERM learning algorithm (*Iteratively Reweighted Winsorized Empirical Risk Minimization*):

> **procedure** IR-ERM($\mathbf{w}_0$)
> $\quad t \leftarrow 0$
> $\quad$**repeat**
> $\qquad z_1 = \ell_1(\mathbf{w}_t), \ldots, z_N = \ell_N(\mathbf{w}_t)$
> $\qquad \bar{z}_t = \mathsf{M}_\rho\{z_1, \ldots, z_N\}$
> $\qquad$**for** $k = 1, \ldots, N$ **do**
> $$v_k = \frac{\partial \mathsf{M}_\varphi}{\partial v_k}\sigma_z'(z_k, \bar{z}_\rho) + \frac{\partial \mathsf{M}_\rho}{\partial z_k}\sum_{l=1}^{N}\frac{\partial \mathsf{M}_\varphi}{\partial v_l}\sigma_u'(z_l, \bar{z}_\rho).$$
> $\qquad$**end**
> $\qquad \mathbf{w}_{t+1} \leftarrow \arg\min\limits_{\mathbf{w}} \sum\limits_{k=1}^{N} v_k \ell_k(\mathbf{w})$
> $\qquad t \leftarrow t+1$
> $\quad$**until** $\{\bar{z}_t\}$ and $\{\mathbf{w}_t\}$ are stabilized.
> **end**

## 7    Results and Discussion

Here we present some examples of using the IR-WERM algorithm for teaching NN with one, two, and three hidden layers, using two sets of data for training that contain outliers (25% and 45%, respectively). We considered NN, which are represent functions $F : \mathbb{R}^2 \to \mathbb{R}$. All NN have two inputs $x_1, x_2$. Each layer of the NN contains sigmoidal neurons with the following transformation model: $v = \arctan(w_0 + w_1 u_1 + \cdots + w_n u_n)$, where $u_1, \ldots, u_n$ are the inputs of the layer, $v$ is the output, $\arctan(s)$ is the activation function. On the top of last layer there is a linear function $y = w_0 + w_1 v_1 + \cdots + w_m v_m$. It calculates the output value of the NN.

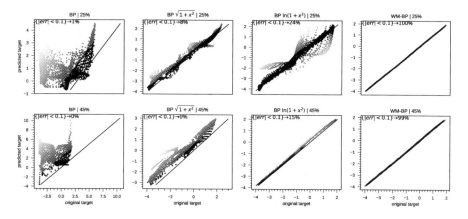

**Fig. 1.** Y-Y plots for NN with one hidden layer (7 neurons), 25% and 45% of outliers in the training data.

In order to generate dataset for learning we randomly generate weights of the original NN and calculates it's values of the grid $X^2 \subset \mathbb{R}^2$, where $X = \{-2 + 0.08t : t = 0, \ldots, 50\}$ (totally 2500 data points). Then a noise with the amplitude 0.1 is added. Two datasets was then constructed by adding the outliers: 25% and 45% of data points of the original dataset are chosen, correspondently, and for each chosen data point $(x_1, x_2)$ the value $y$ is increased by $10|y|$.

These examples are used to demonstrate the ability to reduce the absolute value of a large number of errors and the resistance to large number of the outliers. First, the NN learns to reduce the sum of squared errors. Next, NN is trained by minimizing (1) with $\varrho(r) = \sqrt{1 + r^2}$ using BP. Third, NN learns to reduce the sum of squared errors. Next, NN is trained by minimizing (1) with $\varrho(r) = \ln(1 + r^2)$ using BP. And finally, NN trained using IR-WERM with $\rho_{\alpha,\varepsilon}$, where $\rho_\varepsilon(r) = \sqrt{\varepsilon^2 + r^2} - \varepsilon$, $\alpha = 0.9$, $\varepsilon = 0.001$. Each figure (Y-Y plot) shows the distribution of target value pairs: the predicted value (along the Y axis) and the known value (along the X axis). It demonstrates the ability of the proposed approach and the IR-WERM learning algorithm to reduce the absolute errors on training data and the resistance to large portion of the outliers in the training data.

**One Hidden Layer.** This NN has one hidden layer, which is contains 7 neurons. Y-Y plots are presented on the Fig. 1. BP algorithm for minimization of (1) with $\varrho(r) = r^2$, $\varrho(r) = \sqrt{1 + r^2}$ and $\varrho(r) = \ln(1 + r^2)$ show poor results. Only IR-WERM algorithm for minimization (2) demonstrates resistance to large number of outliers.

**Two Hidden Layers.** This NN has two hidden layers: first layer contains 7 neurons, second layer contains 3 neurons. Y-Y plots are presented on the Fig. 2. BP algorithm for minimization of (1) with $\varrho(r) = r^2$, $\varrho(r) = \sqrt{1 + r^2}$ shows poor results. For $\varrho(r) = \ln(1 + r^2)$ it shows good result for 25%, not good result for

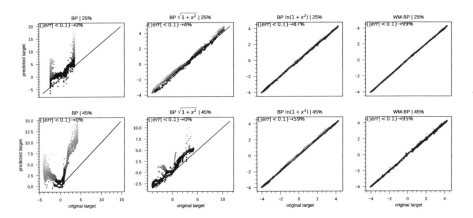

**Fig. 2.** Y-Y plots for NN with two hidden layers (7, 3 neurons), 25% and 45% of outliers in the training data.

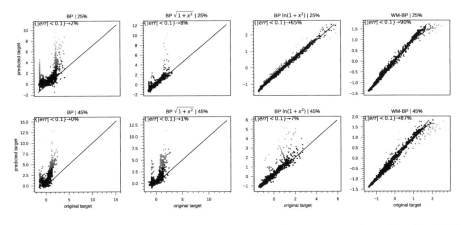

**Fig. 3.** Y-Y plots for NN with three hidden layers (7, 7, 3 neurons), 25% and 45% of outliers in the training data.

45%. Only IR-WERM algorithm for minimization (2) demonstrates resistance to large number of outliers.

**Three Hidden Layers.** This NN has three hidden layers: first layer contains 7 neurons, second layer contains 7 neurons, third layer contains 3 neurons. Y-Y plots are presented on the Fig. 3. BP algorithm for minimization of (1) with $\varrho(r) = r^2$, $\varrho(r) = \sqrt{1 + r^2}$ show poor results. For $\varrho(r) = \ln(1 + r^2)$ it shows good result for 25% and poor result for 45%. IR-WERM algorithm for minimization (2) demonstrates good resistance property to large number of outliers.

Algorithm IR-WERM was used from the project mlgrad[1]. The examples considered above are represented in the Jupyter[2] notebooks (see /examples folder in the project mlgrad).

# 8    Conclusion

The paper considers a new approach to improving the reliability of learning neural networks for training data that contain a significant amount of outliers. It's based on minimization of differentiable analogues of median, quantiles and winsorized means of loss functions. The above techniques are more desirable in cases where gradient-based minimization algorithms are preferable. For example, these methods allow the use of a series of weighted variants of back-propagation algorithms for training NN, resistant to a large number of outliers. In particular, a kind of iteratively re-weighted least squares (IRLS) procedures can be used. In these procedures, a weighted version of the back-propagation algorithm is used for each step. The examples presented above have clearly shown the resistance property to a large number of outliers.

# References

1. Hornik, K., Stinchcombe, M., White, H.: Multilayer feedforward networks are universal approximators. Neural Netw. **2**, 359–366 (1989)
2. Maronna, R., Martin, R., Yohai, V.: Robust Statistics: Theory and Methods. Wiley, New York (2006)
3. Rousseeuw, P.J., Leroy, A.M.: Robust Regression and Outlier Detection. Wiley, New York (2003)
4. Rousseeuw, P.J.: Least median of squares regression. J. Am. Stat. Assoc. **79**, 871–880 (1984)
5. Huber, P.J.: Robust Statistics. Wiley, New York (1981)
6. Stromberg, A.J., Ruppert, D.: Breakdown in nonlinear regression. J. Am. Stat. Assoc. **87**, 991–997 (1992)
7. Holland, P.W., Welsch, R.E.: Robust regression using iteratively reweighted least-squares. Commun. Stat. Theory Methods **6**, 813–827 (1977)
8. Chen, D.S., Jain, R.C.: A robust back-propagation learning algorithm for function approximation. IEEE Trans. Neural Netw. **5**, 467–479 (1994)
9. Liano, K.: Robust error measure for supervised neural network learning with outliers. IEEE Trans. Neural Netw. **7**, 246–250 (1996)
10. El-Melegy, M.T., Essai, M.H., Ali, A.A.: Robust training of artificial feedforward neural networks. In: Hassanien, A.E., Abraham, A., Vasilakos, A.V., Pedrycz, W. (eds.) Foundations of Computational, Intelligence Volume 1. Studies in Computational Intelligence, vol. 201, pp. 217–242. Springer, Heidelberg (2009). https://doi.org/10.1007/978-3-642-01082-8_9

---

[1] https://bitbucket.org/intellimath/mlgrad.
[2] https://jupyter.org.

11. Rusiecki, A.: Robust LTS backpropagation learning algorithm. In: Sandoval, F., Prieto, A., Cabestany, J., Graña, M. (eds.) IWANN 2007. LNCS, vol. 4507, pp. 102–109. Springer, Heidelberg (2007). https://doi.org/10.1007/978-3-540-73007-1_13

12. Jeng, J.-T., Chuang, C.-T., Chuang, C.-C.: Least trimmed squares based CPBUM neural networks. In: International Conference on System Science and Engineering, pp. 187–192 (2011)

13. Rusiecki, A.: Robust learning algorithm based on iterative least median of squares. Neural Process. Lett. **36**, 145–160 (2012)

14. Lina, Y.-L., Hsiehb, J.-G., Jenga, J.-H., Cheng, W.-C.: On least trimmed squares neural networks. Neurocomputing **61**, 107–112 (2015)

15. Beliakov, G., Kelarev, A., Yearwood, J.: Robust artificial neural networks and outlier detection (2012). https://arxiv.org/abs/1110.0169

16. Shibzukhov, Z.M.: On the principle of empirical risk minimization based on averaging aggregation functions. Dokl. Math. **96**(2), 494–497 (2017)

# Robotic Path Planning Based on Improved Ant Colony Algorithm

Tingting Liu[1], Chuyi Song[1], and Jingqing Jiang[1,2(✉)]

[1] College of Mathematics, Inner Mongolia University for Nationalities,
Tongliao 028000, China
317647944@qq.com, songchuyi@sina.com,
jiangjingqing@aliyun.com
[2] College of Computer Science and Technology, Inner Mongolia University
for Nationalities, Tongliao 028000, China

**Abstract.** Ant colony algorithm is an intelligent bionic optimization algorithm. Its self-organization and intelligence provide guiding for studying the global path planning problem. Based on this, an improved ant colony algorithm is proposed to solve the problem of robotic path planning and improved the convergence speed. The environment model is established by grid method and the traditional ant colony algorithm is improved. The heuristic factor and pheromone updating strategy of the algorithm are improved to enhance the precision of the algorithm and the ability of later convergence. Simulation experiments show that the improved algorithm has a faster convergence speed to achieve the optimal path compared with other algorithms. It shows that the improved algorithm is effective and reliable.

**Keywords:** Ant colony algorithm · Path planning · Pheromone · Mobile robot

## 1 Introduction

Path planning means that in an unknown or known environment, mobile robots need to find a path that safely bypasses obstacles from the starting point to the target point and does not collide. At present, mobile robot path planning is a research hotpot in the field of robots. The traditional mobile robot path planning methods include Dijkstra [1], Artificial potential field method [2], A* algorithm [3] etc. There are some intelligent optimization algorithms such as genetic algorithms, particle swarm optimization algorithms, and ant colony algorithms. Dijkstra is a classical shortest path search algorithm with the advantage of being able to get the shortest path. The disadvantage is that the computational complexity is higher and the efficiency is lower. The artificial potential field method has the characteristic of high efficiency, but it is easy to generate oscillations in front of obstacles and fall into local optima. So it is difficult for mobile robots to reach the target point. A* is a heuristic algorithm. Because the heuristic function is difficult to determine, it is easy to fall into the dead cycle and result in unsatisfactory planning path.

Ant colony algorithm is a bionic algorithm designed to simulate the behavior of the ant searching for food at the shortest path. Ant colony algorithm has certain advantages

© Springer Nature Switzerland AG 2019
H. Lu et al. (Eds.): ISNN 2019, LNCS 11554, pp. 351–358, 2019.
https://doi.org/10.1007/978-3-030-22796-8_37

in solving discrete problems and sensitivity to path. The traditional ant colony algorithm is mainly to solve the travelling salesman problem. In recent years, researchers have also achieved good results in the path planning using ant colony algorithms. The paper [2] introduces the barrier exclusion weight and the new heuristic factor into the path selection probability to improve the quality of global search ability. The improved A * algorithm based on two-dimensional grid map is proposed in paper [3], which introduces direction vector and parallel search, so that robot path search has both directionality and parallelism. Paper [4] removes some "useless" obstacles in the viewable method, and selects the direction of the path planning through the principle of the shortest line between two points. The aim is reducing the complexity of the path planning and the operation time, and increasing the scope of application of the visual graph method.

Because the traditional ant colony algorithm has certain defects, this paper proposes an improved ant colony algorithm to plan the mobile robot path, which can get the global optimal path under different complexity circumstance. The efficiency of the path search is improved, and the convergence speed of the algorithm is accelerated. Finally, the feasibility of the algorithm is verified by the simulation results.

## 2   Problem Description and Modeling

Video graphic method [4], topological method [5], Free Space Law [6], grid method [8] are methods for environmental modeling. The grid method is simple in structure and easy to practice. Therefore, this paper will use the grid method to construct the environment space. According to the knowing robot environment information, the robot in reality is converted into plane understandable data by extracting and analyzing. It effectively reduces the trouble in the path search process. The conditions for this study are:

(1)   All locations of the obstacle are known and the height of the obstacle is ignored.
(2)   The starting and ending positions of the robot are known.

The grid model is from left to right, from top to bottom, and the black part represents the barrier grid. When the obstacle is smaller than a grid, it is treated according to a grid that occupies a grid. The white part is a grid that does not have obstacles. The side length of a grid is the length of the robot's movement step, which is a unit length. The grids are numbered from the left lower corner. The corresponding relationship between the serial number and the coordinates are as follows:

$$x_i = mod\left(\frac{N_i}{N}\right) - 0.5 \tag{1}$$

$$y_i = N - ceil\left(\frac{N_i}{N}\right) + 0.5 \tag{2}$$

Where mod means modular, ceil means ceiling, $N_i$ represents the number of the grid, and $N$ represents the number of grids in each column.

## 3  Ant Colony Algorithm

The ant colony algorithm is a typical path planning algorithm. Each ant leaves pheromones in the process of finding food, and can perceive pheromones within a certain range to guide its own movement. Therefore, the ant colony will exhibit a positive feedback phenomenon. The more ants pass through this path, the more pheromones there will be on this path. Then, the path is selected according to the amount of pheromones. The food will be found.

In the initial state, when the ant begins to forage, the amount of pheromone on each path is equal. From the grid located in the node i, the ant selects the next node j based on the node's pheromone concentration and heuristic information. The transfer probability formula is shown in Eq. (3):

$$P_{ij}^k(t) = \begin{cases} \dfrac{\tau_{ij}^{\alpha}(t)\eta_{ij}^{\beta}(t)}{\sum_{s\in Q}\tau_{ij}^{\alpha}(t)\eta_{ij}^{\beta}(t)} & j \in Q \\ 0 & other \end{cases} \tag{3}$$

Where $\eta_{ij}(t)$ represents the heuristic information at the moment $t$ on path $ij$. $\eta_{ij}(t) = \frac{1}{d_{ij}}$. $\alpha$ represents the pheromone enhancement coefficient, and the larger the value, the deeper influence on the initial random pheromone. $\beta$ is the expected heuristic factor, which reflects the importance of the heuristic information in the ant's path selection process. $P_{ij}^k$ is the probability that ant k is transferred from node i to node j at moment t. $\tau_{ij}^{\alpha}(t)$ is the pheromone concentration at moment t on the path from node i to node j.

The pheromone update principle on each path is updated by local pheromones (4) from node i to j.

$$\tau_{ij}(t) = (1 - \lambda)\tau_{ij}(t) + \lambda\tau_0, \lambda \in (0,1) \tag{4}$$

Where $\lambda$ is the rate at which the local pheromone is evaporated, and $\tau_0$ is a small positive real number.

When all ants reach the target grid, an iteration search completes. The global pheromone is updated according to the formula (5).

$$\tau_{ij}(t+1) = (1 - \rho)\tau_{ij}(t) + \rho\Delta\tau_{ij}(t) \tag{5}$$

Where $\rho$ is the rate at which the global pheromone evaporated. $\Delta\tau_{ij}(t)$ is the change of global pheromone concentration, as shown in formula (6).

$$\Delta\tau_{ij}(t) = \begin{cases} \frac{Q}{L^k}, & (ij \in)L^k \\ 0, & other \end{cases} \tag{6}$$

Where $Q$ is constant and greater than 0, $L^k$ is the shortest path established by ant K.

## 4  Improved Ant Colony Algorithms

### 4.1  Adjust the Heuristic Function

In the traditional ant colony algorithm, the heuristic function is the reciprocal of the Euclidean distance between two nodes. Due to the requirements of search time, this paper uses improved strategies to heuristic functions according to literature [8]:

$$\eta_{ij} = (\frac{1}{d_{ij} + d_{jG}})^2 \tag{7}$$

Using formula 7 as an heuristic function increases the effectiveness of the algorithm and reduces the convergence time. At the same time, it ensures that the search will toward to the target point at a faster speed.

### 4.2  Strategy for Updating Pheromones

In the traditional ant colony algorithm, only the pheromone on the better path is updated, which cause the pheromone adjustment to be delayed. So the ant cannot immediately find the optimal path and may lead to the wrong path from the beginning. Therefore, in iteration, find the local best solution $L^{best}$ and the local worst solution $L^{worst}$. It increases the amount of pheromone releasing on the best path, and reduces the amount of pheromone release on the worst path. The solution will not fall into local optimization for convergence faster. The pheromone will be updated according to the following formula [9].

$$\tau_{ij}^k(t + \Delta t) = (1 - \rho) \cdot \tau_{ij}(t) + \Delta\tau_{ij}(t) + \frac{Q}{L^{best}} - \frac{Q}{L^{worst}} \tag{8}$$

The $L^{best}$ represents the optimal path length in this iteration, and $L^{worst}$ represents the worst path length in this iteration.

### 4.3  Range of Pheromones

In the pre-search period, if the pheromone on a certain path grows too fast, all ants may search only one path from beginning to end, which is not the optimal path. To avoid this, set a range for pheromones $[\tau_{min}, \tau_{max}]$. The value of the initial pheromones of the traditional ant colony algorithm is 0, and the search path is limited. After given a range, the pheromone diversity on different paths can be reduced [10].

### 4.4  Algorithm Flow

The heuristic function (7) and pheromone updating strategy (8) are introduced to the ant colony algorithm simultaneously. The proposed algorithm is applied to the path planning. The algorithm steps are as follows:

Step 1: Initializes the parameters, generates obstacles in a known environment, sets the starting point and the target point, and calculates the distance between the two nodes.

Step 2: Puts M ants at the starting point, establishes a TABU table, and then looks for the next feasible node according to (3).

Step 3: Determines whether the ant has reached its destination. If there is a dead-lock, terminate the search and set the current path length to infinity until all ants reach the end, save the path and path length that the ant has passed.

Step 4: Updates the pheromones for the first time according to the update rules (7) and (8).

Step 5: Judges whether the maximum number of iterations is reached. If the maximum number of iterations is reached, the current shortest path is output and the process ends. Otherwise, go to step 2.

## 5   Simulation Results and Analysis

In this paper, Matlab 2016a is used to simulate the feasibility and validity of this algorithm. Through the same parameters and environment, the traditional and improved ant colony algorithms are compared. Experimental parameters are as follows: $m = 50$, $\alpha = 1$, $\beta = 7$, $\rho = 0.3$, $Q = 12$.

Figure 1 show the optimal path of traditional ant colony algorithm on $20 \times 20$ grids. Figure 2(a) and (b) show the convergence curve of traditional and improved ant colony algorithm on $20 \times 20$ grids respectively. The optimal path can be obtained by traditional ant colony algorithm is 29.21. It can be obtained at 16 iterations. Since the target node is introduced into the heuristic function in this paper, the directionality of the search is increased. Table 1 compares the optimal path length and the number of iterations on $20 \times 20$ grids for three algorithms. From the Fig. 2(b), it can be seen that the improved algorithm can converge faster than the traditional ant colony algorithm. The search efficiency of the algorithm is improved effectively.

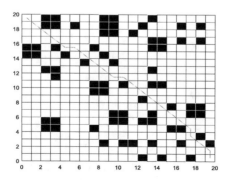

**Fig. 1.** Optimal path of traditional ant colony algorithm on $20 \times 20$ grids

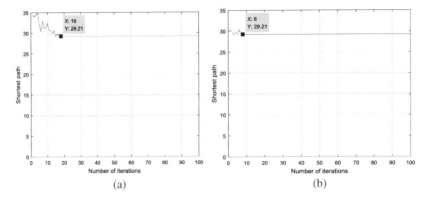

Fig. 2. Convergence curve of traditional (a) and improved (b) ant colony algorithm on $20 \times 20$ grids

Table 1. Comparasion of optimal path length and number of iterations on $20 \times 20$ grids

| Algorithm | Optimal path length | Number of iterations |
|---|---|---|
| Ant colony algorithm | 29.21 | 18 |
| Improved ant colony algorithm | 29.21 | 8 |
| Literature [10] | 29.21 | 14 |

The experiment conducts on a $30 \times 30$ grid environment. Figure 3 is the optimal path of the improved ant colony algorithm. Figure 4(a) and (b) are the convergence curve of the traditional and improved ant colony algorithm respectively. Table 2 compares the optimal path length and the number of iterations on $30 \times 30$ grids for three algorithms. By comparing with the traditional ant colony algorithm, the same path is obtained. Under the same circumstance, the improved algorithm achieves the optimal path 43.36 only need 11 iterations.

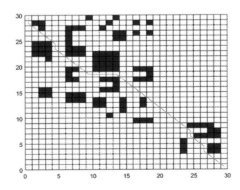

Fig. 3. Optimal path of improved algorithm on $30 \times 30$ grids

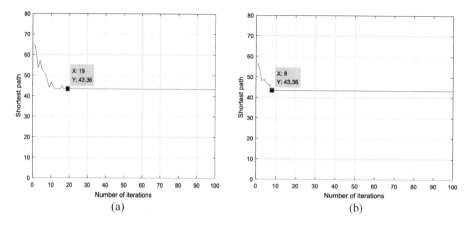

**Fig. 4.** Convergence curve of traditional (a) and improved (b) ant colony algorithm on $30 \times 30$ grids

**Table 2.** Comparasion of optimal path length and number of iterations on $30 \times 30$ grids

| Algorithm | Optimal path length | Number of iterations |
|---|---|---|
| Ant colony algorithm | 43.36 | 19 |
| Improved ant colony algorithm | 43.36 | 8 |
| Literature [10] | 43.36 | 16 |

Therefore, from the above simulation results, it can be seen that the improved algorithm in this paper reaches the shortest path length as the traditional ant colony algorithm. However the convergence speed is higher than the traditional algorithm. Theses indicate that the proposed algorithm has some certain advantages.

## 6 Conclusions

In this paper, we improve the traditional ant colony algorithm in the path planning problem of mobile robot. The distance between nodes, the adjustment of the heuristic function, and the updating mode of pheromones are improved simultaneously. The experiment is carried out under the same environment. Through experiments, the algorithm proposed in this paper can find the optimal path through less iteration, and the search efficiency is improved. The rational path can be obtained by experiments on different scales. The feasibility and validity of the proposed algorithm are demonstrated.

**Acknowledgement.** This work was supported by The National Natural Science Foundation of China (Project No. 61662057, 61672301) and Higher Educational Scientific Research Projects of Inner Mongolia Autonomous Region (Project No. NJZC17198).

# References

1. Wang, X., Liu, Y., Li, G.: Path planning for mobile robots based on improved Dijkstra algorithm. J. Tianjin Univ. Urban Constr. **24**, 378–381 (2018)
2. Chen, J., Dong, J., Zhu, X.: Improving the path planning of mobile robots by artificial potential field method. Command Control Simul. 1–6 (2019)
3. Chen, H., Li, Y., Luo, J.: Research on path planning of mobile robots based on improved A* algorithm optimization. Autom. Instrum. **12**, 1–4 (2018)
4. Shao, W., Luo, Z.: The application of improved visual map method in path planning. J. Nanyang Norm. Univ. **17**, 38–42 (2018)
5. Zhuang, H., Du, S., Wu, T.: Robot path planning and related algorithms research. Sci. Technol. Bull. **3**, 210–215 (2004)
6. Zou, J., Xu, X.: Unmanned vehicle path planning based on improved mixed genetic algorithm. J. Jiamusi Univ. (Nat. Sci. Ed.) **36**, 573–577 (2018)
7. Zhang, Y., Zhang, Z.: Robotic path planning based on improved multi-step ant colony algorithm. Comput. Eng. Des. **39**, 3829–3834+3866 (2018)
8. Zhan, W., Qu, J., Lu, X., Hou, L.: Mobile robot global path planning based on improved ant colony algorithm. Mod. Electron. Technol. **41**, 170–173 (2018)
9. He, Y., Zhang, Z., Han, M., Chen, Q., Huang, X.: Research on ant colony algorithm based on omnidirectional mobile robot path planning. J. Test Technol. **32**, 374–380 (2018)
10. Zhou, J., Zheng, X.: Path planning method based on improved ant colony algorithm. J. Hubei Univ. Technol. **33**, 49–52+101 (2018)
11. Zhu, H., Sun, Z., Woody: Three-dimensional path planning for mobile robots based on improved ant colony algorithm. J. Huazhong Norm. Univ. (Nat. Sci. Ed.) **50**, 812–817 (2016)
12. Wang, X., Yang, L., Zhang, Y., Meng, S.: Robot path planning based on improving potential field ant colony algorithm. Control Decis. Mak. **33**, 75–81 (2018)

# An Early Warning Method for Basic Commodities Price Based on Artificial Neural Networks

Jesús Silva[1(✉)], Noel Varela[2], Hugo Martínez Caraballo[3],
Jesús García Guiliany[3], Luis Cabas Vásquez[4], Jorge Navarro Beltrán[4],
and Nadia León Castro[4]

[1] Universidad Peruana de Ciencias Aplicadas, Lima, Peru
`jesussilvaUPC@gmail.com`
[2] Universidad de la Costa, St. 58 #66, Barranquilla, Atlántico, Colombia
`nvarela2@cuc.edu.co`
[3] Universidad Simón Bolívar, Barranquilla, Colombia
`{hugo.martinez,jesus.garcia}@unisimonbolivar.edu.co`
[4] Corporación Universitaria Latinoamericana, Barranquilla, Colombia
`{lcabas,cinpro}@ul.edu.co`, `jorgeelbacan05@gmail.com`

**Abstract.** The prices of products belonging to the basic family basket are an important component in the income of producers and consumer spending; its excessive variations constitute a source of uncertainty and risk that affects producers, since it prevents the realization of long-term investment plans, and can refuse lenders to grant them credit. His study to identify these variations, as well as to detect their sources, is then of great importance. The analysis of the variations of the prices of the basic products over time, include seasonal patterns, annual fluctuations, trends, cycles and volatility. Because of the advance in technology, applications have been developed based on Artificial Neural Networks (ANN) which have helped the development of massive sales forecast on consumer products, improving the accuracy of traditional forecasting systems. This research uses the RNA to develop an early warning system for facing the increase in basic agricultural products, considering seasonal factors.

**Keywords:** Support vector machines · Cyclic variation · Predictive model · Multilayer perceptron · Multiple Input Multiple Output · Forecast

## 1 Introduction

In the markets of agricultural products, the quantities offered and demanded in each period (month or week) are disparate, which causes variations in prices. During the harvest periods, a large quantity of the product is offered in the markets, greater than the amount that is usually demanded for consumption. In these cases, the prices determined by the market, through supply and demand, are relatively low, lower than the average price of the year. Conversely, during periods when there is no harvest and

© Springer Nature Switzerland AG 2019
H. Lu et al. (Eds.): ISNN 2019, LNCS 11554, pp. 359–369, 2019.
https://doi.org/10.1007/978-3-030-22796-8_38

therefore supply is low, less than the quantity normally demanded, market prices are high, reflecting the relative scarcity of the product [1–3].

To measure the price fluctuations in the different months of the year, it is very useful to build seasonality indexes [4]. If you take the average annual price of a product, it is obvious that the prices of some months would be higher than this average, while prices in other months will be lower. Percentage, the price average in relation to itself will be equal to the unit; prices above the average, in relation to this average, will result in coefficients greater than 1; and prices below the annual average will result in coefficients less than 1 [5].

On the other hand in terms of predictions, there are probabilistic, deterministic, and hybrids, such as [4, 5]: Simple Moving Average, Weighted Moving Average, Exponential Smoothing, Regression Analysis, Box-Jenkins method (ARIMA), trend projections, etc., which have been used to generate forecasts, providing certain advantages and disadvantages compared to the others. However, these models are still unable to offer good results in an environment of high uncertainty and constant changes. To this end, new paradigms based on numeric modeling of nonlinear systems are necessary, such as the Artificial Neural Networks (ANN), and the Support Vector Regression (SVR) [7].

The present study proposes to Multi-Layer Perceptron with Multiple-Input Multiple-Output (MLP and MIMO) as a model for the prediction of prices of the products that belong to the family basket from Colombian State as the warning level criterion. Due to the nature of the products, the seasonal factor is integrated.

## 2   Theoretical Review

### 2.1   Artificial Neural Networks

Artificial Neural Networks (ANNs) can learn from data and can be used to construct reasonable input-output mapping, with no prior assumptions are made on the statistical model of the input data [6]. ANNs have nonlinear modeling capability with a data-driven approach so that the model is adaptively formed based on the features presented from the data [7].

An introduction to ANNs model specifications and implementation and their approximation properties has been provided from an econometric perspective [8]. Numerous studies have shown that ANNs can solve a variety of challenging computational problems, such as pattern classification, clustering or categorization, function approximation, prediction or forecasting, optimization (travelling salesman problem), retrieval by content, and control [9].

Some studies of ANN application related to financial early warning models have been conducted by [10] as well as [11] who used ANN as a classifier with a categorical output. Other authors used ANNs as financial forecasting models with continuous value. Some of them are [12] as well as [13], who implemented ANNs with a single-step prediction output. A previous study on ANNs forecasting model was also proposed by [14] for a multi-step prediction with a direct strategy, so the number of models is equal to the number of the prediction horizon. In the context of basic commodities price, the need for prediction is not limited to one-step forward but could be extended to include multi-step ahead predictions. Three strategies to tackle the multi-step forecasting problem can be considered, namely recursive, direct, and multiple output strategies [15]. The Multiple Input Multiple Output (MIMO) techniques train a single prediction model f that produces vector outputs of future prediction values [16].

The present study proposes to Multi-Layer Perceptron with Multiple Input and Multiple Output (MLP-MIMO) as to agricultural products price prediction model coupled with the coefficient of variation from the Colombian state price reference to the criteria of warning level.

## 2.2 Garson's Algorithm to Determine the Level of Importance

Garson's algorithm was developed to determine the degree or level of importance of an entry indicator in an ANN. In many cases related to the measurement of the variables, the weights in the hidden layer and their interactions in the output network are considered. A measure proposed by Garson [17] consists of dividing the weights of the hidden layer into components associated with each input node and then assigning each of them a percentage of the total weights.

Several studies show the effectiveness of the Garson algorithm to evaluate the importance of an entry in the RNA [18–20]. The certainty of the algorithm of Garson was experimentally determined, concluding that the measure is applied successfully under a wide variety of conditions. As a result of this analysis, the Garson's algorithm, on a scale from 0 to 1, determines a unique value for each explanatory variable that describes the relationship with the response variable in the model.

# 3    Materials and Methods

## 3.1    Data

In this study, the data were obtained from the National Administrative Department of Statistics of Colombia (DANE - National Administrative Department of Statistics), which provided a sales database of 1054 distributors of products of the basic basket from the main regions of Colombia in the time period from 2016 to 2018 [21]. The macroeconomic variables considered in this study range from food inflation, GDP, employment rate, minimum wage to commercial balance and capital flow of the nation [22]. Internal factors such as demand, and substitute and complementary products were also analyzed. Seasonal factors were incorporated to adjust the predictions obtained [23].

## 3.2    Methods

The early warning model consists of three main components, namely preprocessing, predictive model, and post-processing, as depicted in Fig. 1 [24].

Preprocessing: Before all raw data about commodity prices are presented to the predictive model, the preprocessing operations are applied on the data. The price surveys were conducted by local government at working days, so the commodity price

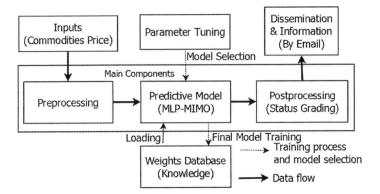

**Fig. 1.** Model of the early warning system [24].

data represent a daily basis with missing values in weekends and holidays. The data are therefore added to weekly data with the mean function to reduce the volume of the data for computational efficiency [24].

Predictive models: The predictive models are built from the trained final MLP-MIMO models obtained from the parameter tuning. Every single commodity in every city has its own model parameter structure. The weights after training are stored in a weight database so the model can be reloaded at any time [7].

Post-processing: The output of the predictive model is a normalized price prediction for eight weeks ahead. The post-processing is responsible for denormalizing the predicted price and determining early warning status based on the maximum predicted price according to Section 'Price Spike and Early Warning Status Leveling'. An alert will be sent to the stakeholders when the price is above a given threshold (on status 'watch' or 'monitoring') by an email service [10].

Finally, the Garson's algorithm for determining the level of importance was developed to determine the degree or level of importance of an input indicator in ANN. In many cases related to the measurement of the variables, the weights in the hidden layer and their interactions in the output network are considered. A measure proposed by [25] consists of dividing the weights of the hidden layer into components linked to each input node and then assigning each of them a percentage of the total weights.

### 3.3   Seasonality and Unitary Seasonal Roots

When working with time series, one of the most important questions that the researcher should ask about is: what is the data generating process (DGP, in English, Data Generating Process) from which the studied sample comes? The conventional approach is to try to detect the different components of the DGP. Typically, 4 components are considered: the trend (stochastic or not), the cyclical part, the purely random component and the seasonal component. Precisely, the seasonal component can be of a different nature: deterministic or stochastic. The most common ways of modeling seasonality involve: using dummy variables, seasonal autoregressive models (ARMA)

(SARMA) or seasonal integration, and later modeling with a SARMA or ARMA model [10, 13, 20].

If the DGP implies that the stationary behavior of the series is purely deterministic, then it can be expressed as follows (Ec. 1) [26]:

$$y_r = \mu + \sum_{i=1}^{11} \alpha_i D_{i,t} + e_t \tag{1}$$

Where $e_t$ is a Gaussian white noise error term, and $D_{i,t}$ is a dummy variable that takes the value of 1 if the observation corresponds to month i, and 0 otherwise.

Now, if the stationary behavior is stochastic, it is possible that it is stationary or not. In other words, the behavior can be such that in the event of disturbances in the series, the system tends to return to its seasonal but non-deterministic behavior (stationary stochastic seasonality) or that such disturbances, on the contrary, imply a permanent change in seasonal behavior (non-stationary stochastic seasonality) [10].

The case of stationary seasonal behavior can be represented with SARMA (p, q) × (P, Q) s models that take the following structure (Ec. 2) [27]:

$$\Phi(L^S)\phi(L)y_t = \Theta(L^S)\theta(L)e_t \tag{2}$$

where L represents the lag operator and $\Phi(\cdot)$, $\varphi(\cdot)$, $\Theta(\cdot)$ and $\theta(\cdot)$ represent polynomials in the lag operator.

## 4   Results and Discussion

### 4.1   Product Selection

According to the United Nations Organization for Agriculture and Food (FAO) the basic products are divided into [2]:

- Food and non-alcoholic beverages
- Alcohol and tobacco
- Restaurants and hotels
- Dress and shoes
- Rental housing
- Housing services
- Furniture, home equipment
- Health
- Transport
- Communications
- Recreation and culture
- Personal care
- Educational services
- Financial services
- Others

Taking into account these categories, it is easy to identify each month how much the value of products and services increases and if inflation remains stable. For the purposes of the following investigation, the group of foods belonging to the perishable category will be assumed [10].

Perishable foods are those that are likely to spoil, break down or become unsafe to consume. They should be stored refrigerated at 40 °F (4.4 °C) or less to remain safe or prolong the time they will remain healthy because refrigeration slows bacterial growth. There are two completely different families of bacteria that can be found in food: "pathogenic bacteria", the class that causes food poisoning disease, and "spoilage bacteria", the class of bacteria that causes food spoilage and develops odors, unpleasant flavors and textures. Examples of foods that should be kept refrigerated for safety include meats, poultry, fish, dairy products, soft cheeses, cheesecake, most cakes, all cooked leftovers and any foods purchased refrigerated or labeled "keep refrigerated" ("keep refrigerated"). Very few fresh fruits and vegetables will remain safe at room temperature for a long time, so most should be stored in the refrigerator to prevent spoilage or mold growth. Some condiments that are safe at room temperature (such as ketchup, mustard, and soy sauce) can be kept chilled to preserve texture or flavor, but it is not necessary [7, 9, 20].

To select the products on which the forecasts were made, the f1-score criterion is used [24]. The ordering by this factor considered both the quantity and the value of sales for the selection of the most important products. Table 1 presents the values of the f1-score factor for each selected product.

**Table 1.** Prioritization and selection of products

| Code | Product | Quantity | F1-SCORE |
|------|---------|----------|----------|
| P1 | Dairy products | 815 | 145879.957 |
| P2 | Vegetables | 1025 | 68264.5132 |
| P3 | Fruits | 1254 | 67848.756 |
| P4 | Condiments | 626 | 81443.7233 |
| P5 | Red meats | 458 | 58743.3935 |
| P6 | White meats | 1478 | 58489.3872 |
| P7 | Fish | 325 | 32722.393 |

## 4.2    Model of the Early Warning System

According to [16], the increase in the price is considered normal when it is below a certain threshold. The threshold is derived from the government reference price established by the Ministry of Commerce and the variation coefficient noted (CVtarget). There are four degrees of warning status: normal, advisory, monitoring, and warning, whose criteria are presented in Table 2.

**Table 2.** Levels of warning status and their criteria.

| Level | Status | Interval price increase related to the reference price |
|-------|--------|--------------------------------------------------------|
| I | Normal | ≤ 1.85CVtarget |
| II | Advisory | (1.85CVtarget, 2CVtarget] |
| III | Monitoring | (2CVtarget, 3.14CVtarget] |
| IV | Warning | >3.14CVtarget |

Table 3 shows the descriptive statistics for this market in each of the months. It can be deduced from this table that the distributions of the prices in all the months present an asymmetry towards the right, besides that the distribution in all the months is leptocurtic. Moreover, the descriptive statistics show an apparently different behavior in both the average and volatility in each of the months.

**Table 3.** Descriptive statistics for the price of the products under study

| Month | Min | Max | Average | Variance | Standard error | Coefficient asymmetry | Kurtosis |
|-------|-----|-----|---------|----------|----------------|-----------------------|----------|
| January | 111,35 | 443,78 | 210,2227 | 4983,6393 | 70,5949 | 1,6387 | 6,6491 |
| February | 112,6 | 577,95 | 225,2414 | 9455,707 | 97,2405 | 2,2091 | 8,9857 |
| March | 95,45 | 451,52 | 206,8382 | 5273,9794 | 72,6222 | 1,6627 | 6,9786 |
| April | 108,01 | 538,88 | 224,37 | 8820,5716 | 93,9179 | 1,7576 | 6,8356 |
| May | 93,34 | 444,99 | 203,8559 | 5609,5177 | 74,8967 | 1,4112 | 6,0556 |
| June | 112,78 | 474,36 | 230,9486 | 7859,2319 | 88,6523 | 1,0148 | 3,8402 |
| July | 111,48 | 434,4 | 209,1532 | 5387,7657 | 73,4014 | 1,2029 | 5,0726 |
| August | 105,58 | 532,06 | 230,8282 | 9129,2901 | 95,5473 | 1,4682 | 5,6772 |
| September | 113,92 | 356,33 | 202,5277 | 3189,4362 | 56,4751 | 0,6726 | 3,6953 |
| October | 105,07 | 567,6 | 230,7041 | 9843,1243 | 99,2125 | 1,8015 | 7,1485 |
| November | 114,3 | 489,76 | 213,6318 | 5959,7064 | 77,1991 | 2,094 | 8,6065 |
| December | 111,26 | 621,72 | 222,0477 | 11843,8742 | 108,8296 | 2,3321 | 9,3665 |

Given the above, it is necessary to determine the characteristics of the DGP of each of the series. For this purpose, the existence or not of seasonal unitary roots in the series must be identified. For this, as mentioned above, the HEGY seasonal unit root test of [12] for the monthly series.

Following [21], the need to consider dummy variables that capture the non-stochastic seasonality in the series is initially taken into account. Therefore, the test must be performed together with these dummy variables. The results of the test for the series of prices of the sample under study are presented in Table 4.

The seasonal dichotomous variables are not significant, so they are excluded to perform the unit root test.

To determine if the non-stochastic seasonality detected contributes to forecast the price of the sample, 3 ARIMA models will be used to generate price forecasts and compare them with the real values. In particular, after integrating the series to find

**Table 4.** Test of Hylleberg, Engle, Granger and Yoo for the future of the sample

| Null hypothesis | Test statistic |
|---|---|
| $\pi 1 = 0$ (non-seasonal unit root) | $-0,866$ |
| $\pi 2 = 0$ (bi-monthly root) | $-1,194$ |
| $\pi 3 = \pi 4 = 0$ (unit root for periods of 4 months) | 692,315 *** |
| $\pi 5 = \pi 6 = 0$ (unit root quarterly) | 421,117 *** |
| $\pi 7 = \pi 8 = 0$ (semi-annual unit root) | 608,359 *** |
| $\pi 9 = \pi 10 = 0$ (unit root at the frequency $5\pi/6$) | 145,757 *** |
| $\pi 11 = \pi 12 = 0$ (annual unit root) | 406,645 *** |
| $\pi 2 = \pi 3 = \cdots = 0$ (all the unit roots are present seasonal) | 454,508 *** |
| $\pi 1 = \pi 3 = \cdots = 0$ (all unit roots are present, seasonal and non-seasonal) | 419,209 *** |

Note: rejects Ho with a level of significance of: 10% (*), 5% (**), 1% (***).

stationary processes and perform a control for possible problems of heteroskedasticity and autocorrelation of the series, we proceed to model these filtered series (stationary series) in 2 different ways: with an ARMA model and a model SARMA. On the other hand, a SARIMA model is estimated for the unfiltered series; that is, without taking into account the non-stationary seasonality [15]. The best models for each case are reported in Table 5.

**Table 5.** Estimated models

| Filtered series | Best model SARIMA | Best ARIMA model |
|---|---|---|
| Crude sugar | SARIMA (9,0,2) (0,0,1) with mean 0 | ARIMA (10,0,2) with mean 0 |
| Refined sugar | SARIMA (2,0,2) (3,0,0) with mean 0 | ARIMA (10,0,2) with mean 0 |
| Unfiltered series | Best model SARIMA | |
| Crude sugar | SARIMA (4,1,2) (2,0,1) with mean 0 | |
| Refined sugar | SARIMA (1,1,1) (2,0,0) with mean 0 | |

Finally, Table 6 shows the monthly seasonality indices of real prices. The period of increase in prices began between the months of April-May-June and ended in November. For the males of 1 and 11/4 of the year, prices reached increases of up to 5% and 3.5% in the month of August and decreased to $-6.7$ and $-4.6\%$ points in February. For males aged 11/2 and 13/4, the high price season began in April and May with price increases of up to 3% and 2% in the month of June, respectively. In the low price season real prices fell to $-4\%$ in the month of February.

The reference price of each product and the threshold for each warning status used in this study are presented in Table 7, according to [16] together with DANE [21].

According to the Garzon's coefficient, the most relevant internal variables are the price and year, while in the macroeconomic policies are the foreign investment and the range of corruption, due to that both destabilize the price of the dollar.

**Table 6.** Seasonality of real monthly prices

| Months | P1 | P2 | P3 | P4 | P5 | P6 | P7 |
|---|---|---|---|---|---|---|---|
| January | 0,957 | 0,977 | 0,982 | 0,995 | 0,852 | 0,745 | 0,794 |
| February | 0,933 | 0,954 | 0,964 | 0,961 | 0,8584 | 0,847 | 0,876 |
| March | 0,963 | 0,966 | 0,967 | 0,971 | 0,741 | 0,725 | 0,723 |
| April | 0,986 | 0,986 | 0,995 | 1,002 | 0,8154 | 0,832 | 0,812 |
| May | 0,999 | 0,999 | 1,009 | 1,017 | 0,9584 | 0,921 | 0,941 |
| June | 1,041 | 1,035 | 1,030 | 1,019 | 1,1124 | 1,190 | 1,190 |
| July | 1,027 | 1,026 | 1,020 | 1,010 | 1,312 | 1,3412 | 1,306 |
| August | 1,050 | 1,035 | 1,020 | 1,010 | 1,214 | 1,257 | 1,268 |
| September | 1,034 | 1,023 | 1,019 | 1,013 | 1,154 | 1,178 | 1,137 |
| October | 1,022 | 1,014 | 1,013 | 1,008 | 1,175 | 1,199 | 1,188 |
| November | 1,004 | 1,010 | 1,002 | 1,004 | 1,185 | 1,124 | 1,163 |
| December | 0,987 | 0,989 | 0,981 | 0,991 | 0,851 | 0,812 | 0,840 |

**Table 7.** Reference price, interval price increase, and the levels of early warning status.

| Commodity | Interval percentage of price increase relative to reference price | | | |
|---|---|---|---|---|
| | Normal | Advisory | Watch | Warning |
| P1 | ≤5% | (5%, 10%] | (10%, 15%] | >15% |
| P2 | ≤25% | (25%, 50%] | (50%, 75%] | >75% |
| P3 | ≤10% | (10%, 20%] | (20%, 30%] | >30% |
| P4 | ≤5% | (5%, 10%] | (5%, 10%] | >15% |
| P5 | ≤5% | (5%, 10%] | (10%, 15%] | >15% |
| P6 | ≤25% | (25%, 50%] | (50%, 75%] | >75% |
| P7 | ≤25% | (25%, 50%] | (50%, 75%] | >75% |

## 5  Conclusions

Using monthly data corresponding to the sale of perishable products of the family basket during the 2016–2018 period, it is found that there is no deterministic seasonal pattern in the series. However, this study finds the existence of seasonal unitary roots. In other words, a "summer" can turn into a "winter" due to unforeseen shocks. This result is used to generate forecasts outside the sample for the 12 months of each year. Said forecasts with models with filtered series that take into account the non-stationary stochastic seasonality behave better in terms of measures such as the Mean Absolute Error, the Mean Absolute Percentage Error and the Root Mean Square Error. That is, the finding of the existence of non-stationary stochastic seasonality allows us to improve the performance of forecast models.

Thus, the results imply that although there is seasonality, this is not deterministic. In this order of ideas, the proposed model presents an improvement over others available in the literature that do not take into account the "stochastic" seasonality due to the presence of seasonal roots.

## References

1. Fonseca, Z., et al.: Encuesta Nacional de la Situación Nutricional en Colombia 2010. Da Vinci, Bogotá (2011)
2. Instituto Colombiano de Bienestar Familiar (ICBF): Ministerio de Salud y Protección Social, Instituto Nacional de Salud (INS), Departamento Administrativo para la Prosperidad Social, Universidad Nacional de Colombia. The National Survey of the Nutritional Situation of Colombia (ENSIN) (2015)
3. Food and Agriculture Organization of the United Nations (FAO): Pan American Health Organization (PAHO), World Food Programme (WFP), United nations International Children's Emergency Fund (UNICEF). Panorama of Food and Nutritional Security in Latin America and the Caribbean, Inequality and Food Systems, Santiago (2018)
4. Frank, R.J., Davey, N., Hunt, S.P.: Time series prediction and neural networks. J. Intell. Rob. Syst. **31**(3), 91–103 (2001)
5. Haykin, S.: Neural Networks and Learning Machines. Prentice Hall International, Upper Saddle River (2009)
6. Jain, A.K., Mao, J., Mohiuddin, K.M.: Artificial neural networks: a tutorial. IEEE Comput. **29**(3), 1–32 (1996)
7. Kulkarni, S., Haidar, I.: Forecasting model for crude oil price using artificial neural networks and commodity future prices. Int. J. Comput. Sci. Inf. Secur. **2**(1), 81–89 (2008)
8. McNelis, P.D.: Neural networks in finance: gaining predictive edge in the market, vol. 59, no. 1, pp. 1–22. Elsevier Academic Press, Massachusetts (2005)
9. Mombeini, H., Yazdani-Chamzini, A.: Modelling gold price via artificial neural network. J. Econ. Bus. Manag. **3**(7), 699–703 (2015)
10. Sevim, C., Oztekin, A., Bali, O., Gumus, S., Guresen, E.: Developing an early warning system to predict currency crises. Eur. J. Oper. Res. **237**(1), 1095–1104 (2014)
11. Zhang, G.P.: Time series forecasting using a hybrid ARIMA and neural network model. Neurocomputing **50**(1), 159–175 (2003)
12. Horton, N.J., Kleinman, K.: Using R For Data Management, Statistical Analysis, and Graphics. CRC Press, Clermont (2010)
13. Chang, P.C., Wang, Y.W.: Fuzzy Delphi and backpropagation model for sales forecasting in PCB industry. Expert Syst. Appl. **30**(4), 715–726 (2006)
14. Lander, J.P.: R for Everyone: Advanced Analytics and Graphics. Addison-Wesley Professional, Boston (2014)
15. Chopra, S., Meindl, P.: Supply Chain Management: Strategy, Planning and Operation. Prentice Hall, Upper Saddle River (2001)
16. Izquierdo, N.V., Lezama, O.B.P., Dorta, R.G., Viloria, A., Deras, I., Hernández-Fernández, L.: Fuzzy logic applied to the performance evaluation. Honduran coffee sector case. In: Tan, Y., Shi, Y., Tang, Q. (eds.) ICSI 2018. LNCS, vol. 10942, pp. 164–173. Springer, Cham (2018). https://doi.org/10.1007/978-3-319-93818-9_16
17. Babu, C.N., Reddy, B.E.: A moving-average filter based hybrid ARIMA–ANN model for forecasting time series data. Appl. Soft Comput. **23**(1), 27–38 (2014)
18. Cai, Q., Zhang, D., Wu, B., Leung, S.C.: A novel stock forecasting model based on fuzzy time series and genetic algorithm. Procedia Comput. Sci **18**(1), 1155–1162 (2013)

19. Egrioglu, E., Aladag, C.H., Yolcu, U.: Fuzzy time series forecasting with a novel hybrid approach combining fuzzy c-means and neural networks. Expert Syst. Appl. **40**(1), 854–857 (2013)
20. Kourentzes, N., Barrow, D.K., Crone, S.F.: Neural network ensemble operators for time series forecasting. Expert Syst. Appl. **41**(1), 4235–4244 (2014)
21. Departamento Administrativo Nacional de Estadística-DANE: Manual Técnico del Censo General. DANE, Bogotá (2018)
22. Fajardo-Toro, C.H., Mula, J., Poler, R.: Adaptive and hybrid forecasting models—a review. In: Ortiz, Á., Andrés Romano, C., Poler, R., García-Sabater, J.-P. (eds.) Engineering Digital Transformation. LNMIE, pp. 315–322. Springer, Cham (2019). https://doi.org/10.1007/978-3-319-96005-0_38
23. Deliana, Y., Rum, I.A.: Understanding consumer loyalty using neural network. Pol. J. Manag. Stud. **16**(2), 51–61 (2017)
24. Chang, O., Constante, P., Gordon, A., Singana, M.: A novel deep neural network that uses space-time features for tracking and recognizing a moving object. J. Artif. Intell. Soft Comput. Res. **7**(2), 125–136 (2017)
25. Scherer, M.: Waste flows management by their prediction in a production company. J. Appl. Math. Comput. Mech. **16**(2), 135–144 (2017)
26. Sekmen, F., Kurkcu, M.: An early warning system for Turkey: the forecasting of economic crisis by using the artificial neural networks. Asian Econ. Financ. Rev. **4**(1), 529–543 (2014)
27. Ke, Y., Hagiwara, M.: An English neural network that learns texts, finds hidden knowledge, and answers questions. J. Artif. Intell. Soft Comput. Res. **7**(4), 229–242 (2017)

# Overcoming Catastrophic Interference with Bayesian Learning and Stochastic Langevin Dynamics

Mikhail Leontev[1,2], Alexander Mikheev[3], Kirill Sviatov[3], and Sergey Sukhov[1(✉)] (iD)

[1] Kotel'nikov Institute of Radio Engineering and Electronics of Russian Academy of Sciences (Ulyanovsk branch), Ulyanovsk, Russia
ulstaer@gmail.com, s.sukhov@gmail.com
[2] S.P. Kapitsa Technological Research Institute, Ulyanovsk State University, Ulyanovsk, Russia
[3] Ulyanovsk State Technical University, Ulyanovsk, Russia
a.miheev@simcase.ru, k.svyatov@gmail.com

**Abstract.** Neural networks encounter serious catastrophic forgetting when information is learned sequentially. Although simply replaying all previous data alleviates the problem, it may require large memory to store all previous training examples. Even with enough memory, joint training can be infeasible if access to past data is limited. We developed generative methods for preventing catastrophic forgetting that do not require the presence of previously used data. Developed methods are based on activation maximization of output neurons and on sampling of posterior probability of data distribution. The methods can work for regular feedforward networks. The proof of concept experiments were performed on publicly available datasets.

**Keywords:** Catastrophic interference · Feedforward neural network · Generative replay

## 1 Introduction

When an artificial neural network (ANN) is trained on one set of data and then later is introduced to another set of examples, catastrophic forgetting (or catastrophic interference) may occur [1]. Catastrophic interference (CI) can be very harmful if one wants to build truly intelligent systems, as with CI ANNs can solve only isolated problems [2]. With respect to this, intensive research is devoted to the development of methods for preventing CI. In spite of some progress, the problem of catastrophic forgetting is still a big challenge for the Artificial Intelligence community [3].

One of the simplest ways to overcome CI in ANNs is to interleave new training patterns with rehearsal of previously learned ones [4, 5]. However, storing all

The work was supported by Russian Foundation for Basic Research and the government of Ulyanovsk region (Grant No. 18-47-732006).

previously used data may require huge amount of memory. Also, in some cases of continuous learning, the previously used data may be no longer available. This may require developing methods that do not rely on the presence of formerly used training patterns.

Catastrophic interference can be prevented if one defines a different sub-network for each task to be learned. One way to do this is by differently regularizing the network's parameters during training on each new task. In particular, it was suggested that regularization methods such as dropout and L2 regularization help to reduce the interference of new learning [5]. More recent examples of this approach, Elastic Weight Consolidation [6] and Synaptic Intelligence [7], estimate the importance of ANN's weights for the previously learned tasks and slow down the changes of important weights during learning. The shortcoming of regularization approaches is that they usually require larger network than normally needed. Also, the information about the importance of the weights (such as Fisher information matrix [6]) should be stored separately and may occupy a significant amount of memory. Calculation of Fisher information matrix is required also in another recent approach for overcoming CI by matching the moment of the posterior distribution of the neural network trained on consecutive tasks [8].

One another set of methods is based on generation of missing training data. The newly generated examples can be jointly used with training samples of new tasks. With recent development of generative ANNs, these methods become more and more popular. The following section describes the generative methods in more details.

## 2   Related Work

One of the methods to prevent CI is to create artificial data and to use them for subsequent training. In early implementation of this strategy, pseudo-data were constructed from a noise signal in a method called *pseudo-rehearsal* [9–11]. A pseudo-sample in this method is constructed by generating random input vector and passing it forward through the network in a standard way; the corresponding output vector becomes the associated target output [9]. The technique of pseudo-rehearsal was used successfully on simple ANNs [10, 11]. However, pseudo-rehearsal did not demonstrate scalability to deeper networks due to the difficulty of generating meaningful pseudo-inputs without supervision.

Recently, novel approaches based on generative models were suggested to prevent catastrophic interference. The generative models differ from pseudo-rehearsal techniques in that the virtual inputs are generated from learned past input distribution. Mocanu et al. [12] employed generative capabilities of Restricted Boltzmann Machines to implement online learning of ANN without forgetting. In deep generative replay framework suggested by Shin et al. [13], the authors train a generative adversarial network (GAN) to mimic past data. In yet another approach, the generative model was integrated into the main model by equipping it with generative feedback connections using the concept of Variational Auto-Encoder (VAE) [14]. The virtual samples generated by VAE or GAN were shown to help in preventing catastrophic interference on a number of tasks.

On the contrary to the above generative approaches, in this paper, we suggest several methods that do not require additional generative sub-network and can be used for regular feedforward ANNs. One method uses maximization of output activity to produce artificial samples. Another method uses Markov chain Monte Carlo (MCMC) approach to sample the probability distribution of input data. The generation of new samples is performed through a process well-known in physics – Brownian motion. The details of these methods are discussed in the following sections.

# 3  Method

## 3.1  Activation Maximization

One of the disadvantages of using random noise for pseudo-samples is that some output neurons are activated with very low probability. For a target activation of a desired output neuron, one can use activation maximization method [15–17]. This method gained popularity both in machine learning and neuroscience as a mean to study internal representations of neural networks. The corresponding patterns can be found by backpropagation starting from arbitrary random initial pattern. The algorithm for generating such 'creations' was first proposed by Lewis [18]. Recently, the original idea of Ref. [18] acquired larger popularity with respect to 'dreams synthesis' for deep convolutional networks. For example, Erhan et al. [16] visualized deep models by finding an input image which maximizes the neuron activity of interest by carrying out an optimization using gradient ascent in the image space. In present work, the procedure of dreams synthesis has different purpose. Namely, 'dreams' are used as virtual dataset that contains the information about previous tasks.

## 3.2  Reconstruction of Probability Distribution of Input Data

Previous method finds input patterns corresponding to the maximum activity of the specific output neuron. However, for proper generalization in ANNs, one needs not only the samples corresponding to the maximum activity, but also all other input patterns that can be referred to the same class. In other words, one needs to reconstruct a posteriori probability function $p(x|y = t, \theta)$, where $x$ corresponds to the input feature, $y$ is the output vector, $\theta$ are the parameters (weights) of the neural network, and $t$ is the target vector usually represented in one-hot encoding. In general, the calculation of this prability is an infeasible task. The common current approach is to sample $p(x|y, \theta)$ using Bayes' rule and MCMC techniques. According to Bayes' rule,

$$p(x|y, \theta) \propto p(y|x, \theta)p(x),$$

where $p(y|x, \theta)$ is the probability of a given input pattern $x$ to belong to a certain class, $p(x)$ is the *a priori* probability of input data that can be chosen to have diagonal normal distribution. Probabilities $p(y|x, \theta)$ are obtained at the output of ANNs using softmax activation function.

One of the possible methods for sampling $p(y|x, \theta)$ came from physics and uses stochastic Langevin dynamics that describes Brownian motion in input data space [19].

Probability $p(y|x, \theta)$ can be matched with "potential energy" $U = -\ln p(y|x, \theta)$, which coincides with common cross-entropy loss function. Brownian motion in potential energy landscapes is being intensively studied both theoretically and experimentally in different physical models [20]. Here we employ the random walk of Brownian motion to generate artificial input data. It is well-known that following the Langevin dynamics, the trajectory of the parameters converges to the probability distribution $p(y|x, \theta)$, which solves the problem in question.

For a Brownian motion in overdamped regime in an external field, the random walk is determined by the following Langevin equation [19]:

$$\dot{x} = -\nabla U(x) + \nabla \ln p(x) + \xi, \tag{1}$$

where $\xi \propto N(0, 2\mathbf{I}\epsilon)$ is the Gaussian noise corresponding to the time step (learning rate) $\epsilon$, $\mathbf{I}$ is the identity matrix. Numerical solution of Eq. (1) can be performed with predictor-corrector algorithm [21]:

$$\bar{x} = x_n + \epsilon F(x_n) + \xi, \tag{2}$$

$$x_{n+1} = x_n + \frac{\epsilon}{2}[F(x_n) + F(\bar{x})] + \xi, \tag{3}$$

where $F = -\nabla U(x) + \nabla \ln p(x)$ includes the gradient of the loss function and L2 regularization term and serves as a "force" acting on a Brownian particle, $x_n$, $x_{n+1}$ are two consecutive generated samples.

One shortcoming of the method based on Brownian motion is that it may produce highly correlated consecutive samples $x_n$, $x_{n+1}$ (slow mixing). Also, it may take indefinitely long to sample multimodal distributions as the sampler may get stuck in a local minimum for a long time. Also, strictly speaking, one needs to collect not all the samples, but only part of them that result in correct target class. In other words, we need not the whole $p(y|x, \theta)$, but only the part of it satisfying the condition $\arg\max y = \arg\max t$. For certain shapes of multidimensional surface $U(x)$, the majority of generated samples may produce very low activity on a target output neuron. All these shortcomings should be addressed in generative algorithm.

### 3.3 Restricted Brownian Motion

To overcome the problems of Brownian sampler described in previous section, ideally, one needs another sampling strategy with fine sampling occurring only near the local minima of the loss function with very coarse sampling in other areas. Such explorative behavior is known in nature and is called Levy flight [22]. However, the development of corresponding sampling methods requires further research. Instead, to improve the classification result in current paper, we use an approach that combines the activation maximization and Langevin dynamics. The proposed algorithm consists of the following steps:

1. Activation maximization corresponding to a particular target output $t$ is performed towards local minimum of the loss function starting from random input pattern.

2. Brownian motion in the local minimum is performed with only those random steps allowed that maintain the condition $\arg\max y = \arg\max t$.
3. The steps 1 and 2 are repeated several times starting from different random input samples for every target vector $t$.

The limitation of the proposed method is that this restricted Brownian motion cannot sample the whole space $U(x)$. Instead, the algorithm should be restarted several times in a hope that all essential minima of the loss function will be found. The number of restarts becomes the new hyperparameter that should be fine-tuned in experiments.

## 4    Experiment

Three-layer fully connected feedforward network was sequentially trained on three sets (A, B, C) of permuted MNIST data [23] of hand-written digits. Permuted MNIST datasets became a standard for testing of CI prevention algorithms [5, 6]. The input features were normalized to be within the interval [0, 1]. The MNIST dataset was divided into 50000 training samples and 10000 validation samples. The neural network had the following parameters: 784 neurons in input layer corresponding to the number of pixels in MNIST images, 400 neurons in each of two hidden layers, and 10 neurons in output layer corresponding to the number of classes (digits). ReLU activation function was used for hidden layers. The output layer had softmax activation function. Softmax activation transformed the output of the last hidden layer into probabilities $p(y|x, \theta)$. The cross-entropy was used as a loss function. The training was performed with stochastic gradient descend using learning rate 0.001 and batches of size 32. L2 regularization with coefficient $10^{-4}$ was used during training. The code was implemented in Keras with TensorFlow backend. The network was first trained for 20 epochs on dataset A, then for 20 epochs on dataset B and, finally, another 20 epochs on dataset C.

In joint training, new training data (B and C) were added while keeping previous training data (A). With joint training, the network constantly improved its classification accuracy on test data (Fig. 1a). When network is trained on new data with old training data thrown away, the catastrophic interference occurs even with L2 regularization (Fig. 1a). These two cases correspond to two extremes (upper and lower bounds) of possible learning outcomes. All the methods aimed at prevention of catastrophic forgetting should have their outcomes in between those extremes.

One of the simplest methods to prevent catastrophic interference is pseudo-rehearsal. This method uses noise samples together with samples from new dataset for the subsequent training. In experiment, 50000 random samples were used. The result of using pseudo-rehearsal is depicted in Fig. 1b. As one can see, pseudo-rehearsal partially helps in overcoming catastrophic interference. However, at the end of training process, the test accuracy for dataset A still drops below 90% (see Table 1). The reason of this modest performance is unpredictability of random samples, which do not activate all the output neurons. For example, no samples with $\arg\max y = 1$ were generated in our experiments.

In the next experiment we generated artificial samples (dreams) with activation maximization method. To generate the dreams, the original network was supplemented

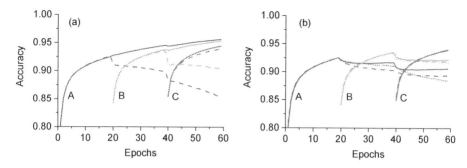

**Fig. 1.** Average test accuracy of the network when learning with different approaches. (a) The upper and lower bounds are determined by joint training (solid line) and by training without retaining previous datasets (dashed line). (b) The results of training with different methods for overcoming catastrophic interference: pseudo-rehearsal (dotted line), sampling with activation maximization (dashed line), a posteriori function sampling over restricted regions (solid line)

**Table 1.** Average test accuracy of classification on permuted MNIST datasets after 60 epochs of training. Each experiment was performed 10 times. Reported is the mean ($\pm$standard deviation).

| Methods | Dataset A | Dataset B |
|---|---|---|
| Catastrophic forgetting – *lower bound* | 0.855 ($\pm$0.026) | 0.905 ($\pm$0.015) |
| Pseudo-rehearsal | 0.885 ($\pm$0.010) | 0.917 ($\pm$0.006) |
| Activation maximization | 0.894 ($\pm$0.008) | 0.919 ($\pm$0.004) |
| Restricted Brownian motion | 0.906 ($\pm$0.006) | 0.921 ($\pm$0.006) |
| Joint training – *upper bound* | 0.956 ($\pm$0.001) | 0.953 ($\pm$0.002) |

by additional layer with 784 neurons and sigmoid activation function (Fig. 2). This layer had an input from a single node and output of the layer was fed into the original network. The constant value of 1 was supplied as an input of this modified network (Fig. 2). The modification of ANN has a dual purpose: (1) the sigmoid function effectively restricts all generated values $\{x_i\}$ to the range (0, 1) required by original normalization; (2) additional layer converts the problem of input samples generation into a standard problem of Bayesian learning of weights $W$ [19] (Fig. 2).

All the weights of the original network were frozen. The weights $W$ of modified ANN were initialized randomly from a normal distribution with zero mean and variance 2. The modified network (Fig. 2) was trained with stochastic gradient descent using 0.01 learning rate and $L2 = 0.01$ regularization coefficient. The learning stopped when the change of the loss function during one training epoch was less than 0.0003. The artificial pattern (dream) was reconstructed from $W$ by applying sigmoid function $\sigma$: $x_i = \sigma(W)$. The procedure was repeated several times starting from random initialization of $W$ for every class (every target vector $t$). The examples of generated dreams are shown in Fig. 3. The generated patterns do not show the resemblance to the images of digits, but nevertheless induce high activity of a corresponding output neuron.

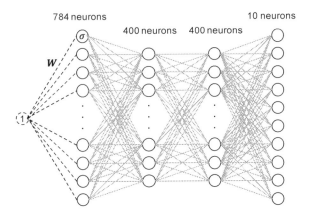

**Fig. 2.** Modification of neural network for samples generation with Bayesian learning. The initial networks with weights represented as solid lines is supplemented with additional weights *W* represented with dashed lines.

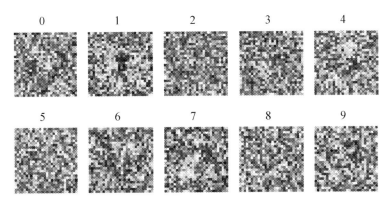

**Fig. 3.** Artificial input samples (dreams) corresponding to original MNIST dataset are obtained by activation maximization approach.

Totally, 2000 dreams were generated per class. The experiments have shown that further increasing the number of dreams does not improve the classification accuracy. The evolution of classification accuracy with this method is shown in Fig. 1b. One can see that activation maximization method does not perform significantly better than pseudo-rehearsal method.

Finally, we tested the generative method based on restricted Brownian motion. The same architecture of the network as in previous experiment (Fig. 2) was used for samples generation. Following the procedure indicated in Sect. 3.3, first, local minimum was found following activation maximization method. Next, artificial samples in the vicinity of this minimum were collected following Brownian random walk: after one step of gradient descend, noise $\xi$ was added to the weights *W* according to Eqs. (2)

and (3). Learning rate $\epsilon$ used in Langevin dynamics (2) and (3) was set to be $\epsilon = 0.1$, L2 regularization coefficient was chosen to be 0.01. L2 regularization in this case accounts for the prior distribution of learned weights $W$. To decrease the correlations between collected samples, only every 10-th sample was saved; 50 restarts from different random locations were performed for each target class and 100 samples were collected in the vicinity of every local minima. One can see from Fig. 1b and Table 1 that current method much better helps in preventing the catastrophic interference on dataset A. One reason that method is still not ideal is that the sampled loss function could have large number of local minima that may not correspond to the original data. One can observe this situation in Fig. 3 where generated samples do not resemble original MNIST images. Searching for proper local minima may be one the future methods for improvement of the current method.

## 5   Conclusion

In this paper we propose and test generative methods aimed at overcoming catastrophic interference during consecutive training. These generative methods do not require separate storage of previous training data, do not need separate generative parts inside ANNs, and can be used for regular feedforward networks. Proof of concept experiments were performed with fully connected ANNs on dataset of hand-written digits MNIST. The generalization of this method for deep convolutional networks is also straightforward. The method based on generation of artificial training samples with activation maximization did not demonstrate the performance significantly better that the method based on random samples. The MCMC sampling of posterior probability using restricted Brownian motion and predictor-corrector algorithm demonstrated superior performance as compared to pseudo-rehearsal and activation maximization methods.

## References

1. McCloskey, M., Cohen, N.-J.: Catastrophic interference in connectionist networks: the sequential learning problem. In: Bower, G.H. (ed.) Psychology of Learning and Motivation, vol. 24, pp. 109–165. Academic Press, San Diego (1989)
2. Liu, B.: Lifelong machine learning: a paradigm for continuous learning. Front. Comput. Sci. **11**(3), 359–361 (2017)
3. Silver, D.-L., Yang, Q., Li, L.: Lifelong machine learning systems: beyond learning algorithms. In: AAAI Spring Symposium Lifelong Machine Learning, p. 5. AAAI, Stanford (2013)
4. Choy, M.-C., Srinivasan, D., Cheu, R.-L.: Neural networks for continuous online learning and control. IEEE Trans. Neural Netw. **17**(6), 1511–1531 (2006)
5. Goodfellow, I.-J., Mirza, M., Xiao, D., Courville, A., Bengio, Y.: An empirical investigation of catastrophic forgetting in gradient-based neural networks. arXiv preprint arXiv:1312.6211 (2013)
6. Kirkpatrick, J., et al.: Overcoming catastrophic forgetting in neural networks. Proc. Natl. Acad. Sci. U.S.A. **114**(13), 3521–3526 (2017)

7. Zenke, F., Poole, B., Ganguli S.: Continual learning through synaptic intelligence. In: Proceedings of the 34th International Conference on Machine Learning, vol. 70, pp. 3987–3995. PMLR, Sydney (2017)

8. Lee, S.-W., Kim, J.-H., Jun, J., Ha, J.-W., Zhang, B.-T.: Overcoming catastrophic forgetting by incremental moment matching. In: Advances in Neural Information Processing Systems, pp. 4652–4662. Curran Associates Inc., Long Beach (2017)

9. Robins, A.: Catastrophic forgetting, rehearsal and pseudorehearsal. Connect. Sci. 7(2), 123–146 (1995)

10. French, R.-M.: Pseudo-recurrent connectionist networks: an approach to the 'sensitivity-stability' dilemma. Connect. Sci. 9(4), 353–380 (1997)

11. Ans, B., Rousset, S.: Avoiding catastrophic forgetting by coupling two reverberating neural networks. Comptes Rendus de l'Académie des Sciences-Series III-Sciences de la Vie 320 (12), 989–997 (1997)

12. Mocanu, D.-C., Vega, M.-T., Eaton, E., Stone, P., Liotta, A.: Online contrastive divergence with generative replay: Experience replay without storing data. arXiv preprint arXiv:1610.05555 (2016)

13. Shin, H., Lee, J.-K., Kim, J., Kim, J.: Continual learning with deep generative replay. In: Advances in Neural Information Processing Systems, pp. 2990–2999. Curran Associates Inc., Long Beach (2017)

14. van de Ven, G.-M., Tolias, A.-S.: Generative replay with feedback connections as a general strategy for continual learning. arXiv preprint arXiv:1809.10635 (2018)

15. Simonyan, K., Vedaldi, A., Zisserman, A.: Deep inside convolutional networks: Visualising image classification models and saliency maps. arXiv preprint arXiv:1312.6034 (2013)

16. Erhan, D., Bengio, Y., Courville, A., Vincent, P.: Visualizing higher-layer features of a deep network. Univ. Mont. 1341(3), 1 (2009)

17. Yosinski, J., Clune, J., Fuchs, T., Lipson, H.: Understanding neural networks through deep visualization. arXiv preprint arXiv:1506.06579 (2015)

18. Lewis, J.: Creation by refinement: a creativity paradigm for gradient descent learning networks. In: IEEE 1988 International Conference on Neural Networks, vol. 2, pp. 229–233. IEEE, San Diego (1988)

19. Welling, M., Teh, Y.-W.: Bayesian learning via stochastic gradient Langevin dynamics. In: Proceedings of the 28th International Conference on Machine Learning (ICML 2011), pp. 681–688. Omnipress, Bellevue (2011)

20. Douglass, K.-M., Sukhov, S., Dogariu, A.: Superdiffusion in optically controlled active media. Nat. Photon. 6(12), 834–837 (2012)

21. Romero, A.-H., Sancho, J.-M.: Brownian motion in short range random potentials. Phys. Rev. E 58(3), 2833 (1998)

22. Kamaruzaman, A.-F., Zain, A.-M., Yusuf, S.-M., Udin, A.: Levy flight algorithm for optimization problems-a literature review. Appl. Mech. Mater. 421, 496–501 (2013)

23. Lecun, Y., Bottou, L., Bengio, Y., Haffner, P.: Gradient-based learning applied to document recognition. Proc. IEEE 86(11), 2278–2324 (1998)

# An Improved Selection Operator
# for Multi-objective Optimization

Hong Zhao[1], Zhi-Hui Zhan[1(✉)], Wei-Neng Chen[1], Xiao-Nan Luo[2],
Tian-Long Gu[2], Ren-Chu Guan[3], Lan Huang[3], and Jun Zhang[1,4]

[1] Guangdong Provincial Key Lab of Computational Intelligence and Cyberspace
Information, School of Computer Science and Engineering,
South China University of Technology, Guangzhou 510006, China
zhanapollo@163.com
[2] School of Computer Science and Engineering,
Guilin University of Electronic Technology, Guilin 541004, China
[3] College of Computer Science and Technology, Jilin University,
Changchun 130012, China
[4] Department of Computer Science, City University of Hong Kong,
Kowloon, Hong Kong

**Abstract.** Non-dominated sorting genetic algorithm II (NSGA-II) obtains a
great success for solving multi-objective optimization problems (MOPs). It uses
a tournament selection operator (TSO) to select the suitable individuals for the
next generation. However, TSO selects individuals based on the non-dominated
rank and the crowding distance of each individual, which exhausts a lot of
computational burden. In order to relieve the heavy computational burden, this
paper proposes an improved selection operator (ISO) that is based on two
selection schemes, i.e., a rank-based selection (S-Rank) and a random-based
selection (S-Rand). S-Rank is a scheme that selects individuals based on its non-
dominated ranks, in which if the individuals have the different non-dominated
ranks, the individuals with lower (better) ranks will be selected for the next
generation. On the contrary, if the individuals have the same rank, we first select
an objective randomly from all objectives, and then select the individual with the
better fitness on this objective to enter the next generation. This is the S-Rand
scheme that can increase the diversity of individuals (solutions) due to the
random selection of objective. The proposed ISO only calculates the crowding
distance of the last (selected) rank individual, and avoids the calculation of the
crowding distance of all individuals. The performance of ISO is tested on two
different benchmark sets: the ZDT test set and the UF test set. Experimental
results show that ISO effectively reduces the computational burden and enhance
the selection diversity by the aid of S-Rank and S-Rand.

**Keywords:** Multi-objective optimization problems (MOPs) ·
Tournament selection operator · Crowded comparison method (CCM) ·
No-dominated sorting genetic algorithm II (NSGA-II)

© Springer Nature Switzerland AG 2019
H. Lu et al. (Eds.): ISNN 2019, LNCS 11554, pp. 379–388, 2019.
https://doi.org/10.1007/978-3-030-22796-8_40

# 1   Introduction

Many real-world problems have multiple objectives, but they often conflict with each other and need to be optimized simultaneously, which are called multi-objective optimization problems (MOPs) and have attracted a lot of attentions in the past several years [1–4]. There are three mainly methods to deal with MOPs. The first is to transform the MOPs into a single objective optimization problem by means of weighting technique [5]. The second is to use multiple populations to deal with multiple objectives (MPMO) [6, 7]. The third is to use Pareto dominance relation to obtain a group of solutions that are not dominated by each other [8]. The first method needs to set the weight vector of each objective in advance, which has a great impact on the final result. Moreover, with weight changes, the algorithm needs to be rerun. In practice, most of the weights are unpredictable. Therefore, the second and the third methods are generally adopted at present.

Non-dominated sorting genetic algorithm (NSGA-II) [8] is one of the most famous MOEA variant that uses Pareto dominance relation to solve MOPs. Since objectives in MOPs often conflict with each other, it is difficult to select better solutions to enter the next generation during evolution. NSGA-II uses a fast non-dominated sorting (FNDS) with crowded comparison method (CCM), and a tournament selection operator (TSO) to deal with this difficulty. After generating a new population $Q_t$, NSGA-II combines the original population $P_t$ and the new generated population $Q_t$ together, forming the merged population $R_t = P_t \cup Q_t$. Then, NSGA-II divides the combined individuals (solutions) into a set of non-dominated ranks according to FNDS. At last, NAGA-II conducts the population $P_{t+1}$ by adding the individuals from lower (better) rank to higher rank, until the population reaches the population size $NP$. Note that the FNDS is with the CCM, so that not only the rank information of each individual is obtained, but also its crowding distance. Therefore, if the number of the last rank that entering the population exceeds the population size, NSGA-II uses the crowding distance information to select the still needed individuals. After the population $P_{t+1}$ forms, NSGA-II adopts TSO on the population $P_{t+1}$ to select individuals for the next generation, forming the population $Q_{t+1}$. That is, for a pair of individuals selected from population $P_{t+1}$, the individual with lower (better) rank will be selected to enter the next generation. In contrast, if the individuals have the same rank, the individual with less crowed (i.e., larger crowding distance means better) is selected to enter the next generation. The TSO selects individuals according to the rank and the crowding distance of each individual to enter the next generation. After the TSO selecting individuals to form the population $Q_{t+1}$, the $Q_{t+1}$ is further undergone the crossover and mutation operations to form the final offspring population $Q_{t+1}$.

Therefore, the TSO in NSGA-II can be regarded as rank-based and crowded-based. However, the operations may take a large computational burden because FNDS not only calculates the rank information of all individuals in the population, but also their crowding distance information. In fact, it is not necessary to calculate the crowding distance of individuals at the ranks before $F_l$ (the last selected non-dominated rank) in NSGA-II, because all the individuals at these ranks will be added into the next generation. We can only calculate the crowding distance of individuals on the rank $F_l$ to select the $NP$ individuals for the next generation.

In order to reduce the computational burden, we propose an improved selection operator (ISO) for NSGA-II in this paper to reduce the computational burden, resulting in the NSGA-II-ISO algorithm. The proposed ISO utilizes a rank-based selection (S-Rank) and a randomness-based selection (S-Rand) to select the individuals for the next generation. In S-Rank, if the selected individuals have different ranks, the individuals have a lower (better) rank will be selected to enter the next generation. In S-Rand, if the selected individuals have same ranks, we can first randomly select an objective, and then the better fitness individual of this objective can be selected to enter the next generation.

The difference of NSGA-II and NSGA-II-ISO is shown in Fig. 1. In Fig. 1, $R_t$ is merged by the parent population $P_t$ and the offspring population $Q_t$ in the $t^{th}$ generation evaluation. In Fig. 1(a), $NP$ individuals are selected from the population $R_t$ to enter the next generation $P_{t+1}$ by FNDS and CCM. Then, the TSO is performed on the population $P_{t+1}$ to generate $Q_{t+1}$. In TSO, $i$ and $j$ are two individuals that are randomly selected from the population $P_{t+1}$, $i_{rank}$ and $j_{rank}$ are the non-dominated ranks of $i$ and $j$ that calculated by FNDS, $i_{distance}$ and $j_{distance}$ are the crowding distance of $i$ and $j$ that calculated by CCM. The process of TSO can be divided into the following two situations: (i) If $i_{rank}$ is less than $j_{rank}$, the $i^{th}$ individual is selected to enter $Q_{t+1}$; if $j_{rank}$ is less than $i_{rank}$, the $j^{th}$ individual is selected to enter $Q_{t+1}$. (ii) If $i_{rank}$ is equal to $j_{rank}$, the crowding distance of $i$ and $j$ ($i_{distance}$ and $j_{distance}$) should be compared. That is, if $i_{distance}$ is greater than $j_{distance}$, the $i^{th}$ individual is selected to enter $Q_{t+1}$; if $j_{distance}$ is greater than $i_{distance}$, the $j^{th}$ individual is selected to enter $Q_{t+1}$. In conclusion, the NSGA-II need to calculate the crowding distance of each individual in $P_{t+1}$. The main process of NSGA-II-ISO is shown in Fig. 1(b). Similar to NSGA-II, firstly, $NP$ individuals are selected from the population $R_t$ to enter the next generation $P_{t+1}$ by FNDS and CCM*. The difference between CCM* in NSGA-II-ISO and CCM in NSGA-II is that CCM* only needs to calculate the crowding distance of the individuals at the rank $F_l$ while CCM calculate the crowding distance of all the individuals at all ranks. Then, the TSO is replaced by the ISO in NSGA-II-ISO. The ISO also can be divided into two situations, the first situation is the same as the TSO (i); and the second situation is that when $i_{rank}$ is equal to $j_{rank}$, the fitness value of $i$ and $j$ on the $m^{th}$ objective are compared, where $m$ is an objective that randomly select from all the objectives. $f_i(m)$ and $f_j(m)$ are the fitness values of $i$ and $j$ on the $m^{th}$ objective. The individual with better (smaller) fitness value will be selected to enter $Q_{t+1}$. That is, the crowding distance of the individuals before the rank $F_l$ are not needed to calculated, and so the ISO reduce the computational burden of NSGA-II.

(a) NSGA-II            (b) NSGA-II-ISO

**Fig. 1.** The difference between NSGA-II and NSGA-II-ISO. (a) NSGA-II. (b) NSGA-II-ISO.

The following parts of this paper are organized as follows. Related works of MOPs are introduced in Sect. 2. Section 3 elaborates on NSGA-II-ISO. Experimental results are shown in Sect. 4. Finally, Sect. 5 introduced the conclusions and future works.

## 2  Related Works of MOPs

There are many researches on MOPs. According to their different features, they can be divided into the following groups. Some researchers focus on the aggregation approach to solve MOPs [5, 9]. Parsopoulos *et al.* [5] utilized a weighted aggregation technique to help the PSO searching the PFs. That is, MOPs are transformed into a single objective optimization problem by using weights of objectives. However, the weight of each objective is hard to determine, and they have an impact on the performance of MOPs. To solve this problem, Zhang *et al.* [9] proposed a new decomposition-based method to determine the weight of each objective, namely, the multi-objective evolutionary algorithm based on decomposition (MOEA/D). MOEA/D decomposes the MOPs into several sub-problems, and optimizes the sub-problems simultaneously, which has low computational complexity during evolution.

One of the most significant researches about MOPs is the MPMO proposed by Zhan *et al.* [6]. In recently years, the MPMO framework has been widely studied by researchers to extend to many-objective optimization [7], differential evolution [11], group search [12], invasive weed optimization [13], and multi-objective cloud system [14]. Using the concept of Pareto dominance to deal with MOPs is very popular. Deb *et al.* [8] proposed NSGA-II that selected the next generation according to the ranking of each individual with PF. Subsequently, a many-objective optimization method (NSGA-III) [10] based on NSGA-II is proposed by Deb and Jain to solve problems with box constraints. NSGA-II has been used to the real-world problems, such as single-objective transmission planning [15], optimal feature selection [16], and multiobjective beampattern optimization [17].

## 3  NSGA-II-ISO

### 3.1  ISO

According to NSGA-II, TSO has two aspects to improve: (i) It takes a lot of computational burden due to the calculation process of the crowding distance of each individual. (ii) It may select many repetitive individuals and so that the diversity of population (solutions) is reduced. Correspondingly, in the processing of forming population $P_{t+1}$ from the population $R_t$, TSO can be improved from two aspects. On the one hand, if the rank $F_l$ is the last non-dominated rank to be added into the population $P_{t+1}$, the crowding distance of the individuals at ranks before $F_l$ are not necessary to be calculated, since all of them can be added into $P_{t+1}$. It should be noticed that the crowding distance of individuals in $F_l$ should be calculated to make sure $NP$ individuals in $P_{t+1}$. Therefore, the CCM can be only performed on the individuals at rank $F_l$.

Based on the above considerations, the detail of ISO is shown in Algorithm 1. Firstly, $Q_{t+1}$ is initialized to empty (Step 1). Then, when $|Q_{t+1}|$ is smaller than $NP$ (Step 2), we randomly select two individuals $i$ and $j$ from $P_{t+1}$ (Step 3), and compare their ranks ($i_{rank}$ and $j_{rank}$) (Step 4). If $i_{rank}$ is not equal to $j_{rank}$, the lower (better) rank individual will be selected to add into $Q_{t+1}$ (Step 5). Otherwise, the individual with a better fitness value for a random objective $m$ ($m$ is selected from all the objectives) will be selected to add into $Q_{t+1}$ to ensure the diversity of selection (Step 7–Step 8).

---

**Algorithm 1 ISO**

---

**Begin**
1. $Q_{t+1}=\Phi$;
2. **While** $|Q_{t+1}|\leq NP$
3.     Randomly select two individuals $i$ and $j$ from $P_{t+1}$;
4.     **If** $i_{rank}\neq j_{rank}$
5.         Select the lower (better) rank individual to add into $Q_{t+1}$;
6.     **Else**
7.         Randomly select an objective $m$ from all the objectives;
8.         Select the individual that has the smaller (better) fitness value on objective $m$ to add into
9.     **End If**
10. **End While**
**End**

---

**Algorithm 2 NSGA-II-ISO**

---

**Begin**
1. Randomly generate a population $P_t$ with $NP$ individuals, and set $t=0$;
2. Perform the tournament selection, crossover, and mutation operators on $P_t$, and generate an offspring population $Q_t$;
3. **While** $t<t_{max}$  // $t_{max}$ is the maximum evolution generation
4.     Merge $P_t$ and $Q_t$ into $R_t$;
5.     Sort $R_t$ by the FNDS, and get the Pareto front set $F$;
6.     Set $P_{t+1}$ to an empty and the $F_l$ is the first rank, i.e., $l=1$;
7.     **While** $|P_{t+1}|+|F_l|\leq NP$
8.         $P_{t+1}=P_{t+1}\cup F_l$;
9.         $l=l+1$;
10.    **End While**
11.    Perform the CCM* on $F_l$ and add $|NP-|P_{t+1}||$ individuals to the population $P_{t+1}$;
12.    Perform **Algorithm 1** on the population $P_{t+1}$ to obtain the offspring population $Q_{t+1}$;
13.    Perform the crossover and mutation operators on $Q_{t+1}$;
14.    $t=t+1$;
15. **End While**
16. Report the final population;
**End**

## 3.2    Implementation of ISO Based NSGA-II

The pseudocode of NSGA-II-ISO is shown in Algorithm 2. In Algorithm 2, $t$ denotes the current evolutionary generation, $t_{max}$ denotes the maximum evolutionary genera-tions. $P_t$ is the parent population of the $t^{th}$ generation while $Q_t$ is the offspring popu-lation at the $t^{th}$ generation. $F$ stores all the different ranks formed by FNDS, with $F_i$ denoting the $i^{th}$ rank. In initialization process, a parent population $P_t$ with $NP$ indi-viduals is randomly generated, and $t$ is initially set to 0 (Step 1). The NSGA-II-ISO not stopped until the generation equals to $t_{max}$ (Step 3). The $R_t$ stores the merged popu-lation by combining $P_t$ and $Q_t$, whose size is $2NP$ (Step 4). Then, FNDS is performed on the merged population $R_t$ to obtain the non-dominated ranks of individuals in $R_t$, whose ranks are stored in the set $F$ (Step 5). Next, the individuals with lower (better) ranks will be firstly selected to enter the next generation, until there are $NP$ individuals in the population. Particularly, NSGA-II-ISO uses the CCM* to select the individuals at the last non-dominated rank to enter the next generation (Step 6–Step 11). Finally, the ISO, crossover, and mutation operators are performed on $P_{t+1}$ (Step 12–Step 13). The whole evolutionary process not stops until $t_{max}$ generations is satisfied.

# 4    Experiments and Results

## 4.1    Experiments Setup

The experiments are conducted on 12 functions to test the effectiveness of ISO. These functions consist of two groups. The first group is selected from the ZDT test sets [18], i.e., ZDT1-ZDT4, and ZDT6. The second group is the UF test sets selected from CEC'2009 [19], which includes two-objective unconstrained problems UF1-UF7.

To verify the advantages of ISO, we consider other two section operators in NSGA-II as comparisons: (i) The original selection operator in NSGA-II, i.e., the TSO. (ii) The IND operator, that is, when the selected individuals have the same rank, we randomly select an individual to enter the next generation, resulting in NSGA-II-IND. The relevant parameters are set as follows. $NP$ is set to be 100 for all the algorithms on all the test problems. The maximum number of function evolutions is 25000 (referring to $t_{max}$= 250). The crossover probability $p_c$ is set to be 0.9. The mutation probability $p_m$ is set to be $1/D$, where $D$ is the dimensions of the problem. The distribution indexes for crossover and mutation operators are $\eta_c$= 20 and $\eta_m$= 20, respectively. Each algorithm runs 30 times independently.

## 4.2    Results Comparisons

(1)    The comparison of IGD results
Table 1 shows the mean and standard variance of the inverted generation distance (IGD) obtained by the algorithms (NSGA-II, NSGA-II-ISO, and NSGA-II-IND) with 30 independent runs on ZDT test sets and UF test sets. The best results of each algorithm are signed with **boldface**. The results show that NSGA-II-ISO outperforms NSGA-II and NSGA-II-IND in dealing with ZDT and UF test sets. For ZDT1 and

ZDT2 whose objectives are all unimodal, NSGA-II-ISO still performs better than NSGA-II and NSGA-II-IND. It indicates that NSGA-II-ISO performs better on the MOPs with simple objectives. Meanwhile, NSGA-II-ISO also obtained the best results on ZDT3, ZDT4, and ZDT6.

**Table 1.** The mean and standard deviation value of IGD with 30 Times

| Functions | | NSGA-II-ISO | NSGA-II-IND | NSGA-II | Func | NSGA-II-ISO | NSGA-II-IND | NSGA-II |
|---|---|---|---|---|---|---|---|---|
| ZDT1 | Mean | **3.08e−03** | 9.98e−03 | 9.96e−03 | UF2 | **2.42e−02** | 3.53e−02 | 2.93e−02 |
|  | Std | **2.97e−06** | 3.08e−05 | 3.09e−05 |  | **3.49e−04** | 3.45e−04 | 2.74e−04 |
| ZDT2 | Mean | **3.04e−03** | 9.44e−03 | 9.56e−03 | UF3 | **3.76e−03** | 2.74e−02 | 1.71e−01 |
|  | Std | **2.79e−06** | 2.80e−05 | 2.83e−05 |  | **3.54e−04** | 2.22e−04 | 9.06e−03 |
| ZDT3 | Mean | **8.16e−01** | 2.73e+00 | 2.75e+00 | UF4 | **5.72e−02** | 9.54e−02 | 9.71e−02 |
|  | Std | **2.08e−01** | 2.31e+00 | 2.35e+00 |  | **2.34e−03** | 2.83e−03 | 2.94e−03 |
| ZDT4 | Mean | **7.12e−03** | 3.67e−02 | 2.66e−02 | UF5 | **2.99e+00** | 5.25e+00 | 9.77e+00 |
|  | Std | **1.79e−05** | 1.10e−03 | 2.11e−04 |  | **1.10e+01** | 8.12e+00 | 3.28e+01 |
| ZDT6 | Mean | **1.50e−02** | 5.63e−02 | 3.17e−02 | UF6 | **1.32e−02** | 2.10e−02 | 8.45e−02 |
|  | Std | **7.83e−05** | 1.02e−03 | 3.19e−04 |  | **1.12e−04** | 2.43e−04 | 1.90e−03 |
| UF1 | Mean | **1.05e−03** | 6.70e−03 | 1.22e−02 | UF7 | **5.75e−03** | 7.82e−03 | 1.18e−02 |
|  | Std | **1.76e−05** | 2.43e−05 | 3.61e−05 |  | **1.59e−05** | 1.97e−05 | 5.16e−05 |

The Pareto optimal fronts of UF test sets are complicated. Table 1 demonstrates that NSGA-II-ISO takes the first place in dealing with the UF test sets, especially on UF1, UF3, and UF7 test set. For UF2, UF4, UF5, and UF6, NSGA-II-ISO shows its competitive performance.

(2) The non-dominated solutions of the final evolution
To further investigate the effectiveness of NSGA-II-ISO, we investigate the non-dominated solutions of ZDT test sets and UF test sets in the last generation, respectively shown in Figs. 2 and 3. As shown in Fig. 2, although all the algorithms can obtain the non-dominated solutions, NSGA-II-ISO provides the best solutions that approximating the true PFs and has the better diversity to spread along the PF.

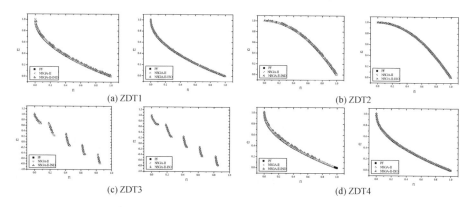

(a) ZDT1                           (b) ZDT2

(c) ZDT3                           (d) ZDT4

**Fig. 2.** Final non-dominated solutions of the ZDT problems in all the 30 runs. (a) ZDT1. (b) ZDT2. (c) ZDT3. (d) ZDT4.

Figure 3 shows the final non-dominated solutions obtained by different algorithms by averaging over 30 independent runs in dealing with UF test sets. From Fig. 3(a), (c), and (e), we find that the number of no-dominated solutions obtained by NSGA-II-ISO and the two comparison algorithms are few or far from the true PF values in optimizing UF1, UF3, and UF5. In this case, NSGA-II-ISO still provides better non-dominated solutions which are more approximate to the true PFs than other algorithms. For UF2, UF4, UF6, and UF7 test sets, both the NSGA-II-ISO and NSGA-II obtain promising solutions. In contrast, NSGA-II-IND obtains worse solutions than NSGA-II-ISO and NSGA-II.

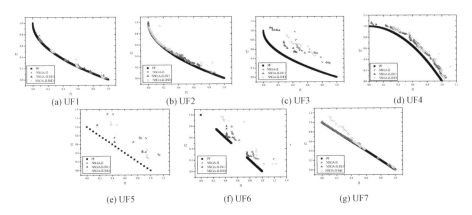

**Fig. 3.** Final non-dominated solutions of the UF problems in all the 30 runs. (a) UF1. (b) UF2. (c) UF3. (d) UF4. (e) UF5. (f) UF6. (g) UF7.

(3)  The average time of CPU used by each algorithm

To test the algorithm efficiency, we compare the average elapsed time of the algorithms running for 30 times, with the time unit represented by milliseconds (ms). The best results with minimum expenditure of time are signed in **boldface**. Results are shown in Table 2.

**Table 2.** The mean and time of CPU used by different algorithms runs 30 times

| Time (ms) | NSGA-II | NSGA-II-ISO | NSGA-II-IND | Time (ms) | NSGA-II | NSGA-II-ISO | NSGA-II-IND |
|---|---|---|---|---|---|---|---|
| ZDT1 | 2.77e+02 | 2.65e+02 | **2.55e+02** | UF1 | 2.39e+02 | **2.23e+02** | 2.28e+02 |
| ZDT2 | 2.68e+02 | 2.57e+02 | **2.47e+02** | UF2 | 2.79e+02 | 2.72e+02 | **2.66e+02** |
| ZDT3 | 2.66e+02 | **2.54e+02** | 2.60e+02 | UF3 | 2.77e+02 | 2.55e+02 | **2.51e+02** |
| ZDT4 | 1.56e+02 | **1.40e+02** | **1.40e+02** | UF4 | 2.78e+02 | **2.62e+02** | **2.62e+02** |
| ZDT6 | 2.79e+02 | 2.63e+02 | **2.36e+02** | UF5 | 2.47e+02 | **2.45e+02** | **2.45e+02** |
| UF1 | 2.39e+02 | 2.30e+02 | **2.25e+02** | UF6 | 2.50e+02 | 2.31e+02 | **2.26e+02** |
| UF2 | 2.79e+02 | 2.72e+02 | **2.66e+02** | UF7 | 2.48e+02 | 2.43e+02 | **2.38e+02** |

From the table, we find both NSGA-II-ISO and NSGA-II take more time than NSGA-II-IND. The elapsed time of NSGA-II-ISO is second to NSGA-II-IND. For some test problems, NSGA-II-ISO performs best in all the compared algorithms, such as ZDT3 and UF1. But for ZDT4, UF4 and UF5, the runtime of NSGA-II-ISO and NSGA-II-IND is the same. This verifies the effectiveness of ISO in reducing the computational burden than the original selection operator in NSGA-II.

## 5   Conclusion

An improved selection operation (ISO) is introduced in this paper to help NSGA-II efficiently solve MOPs. Based on the dominance relationship of the selected individuals, ISO combines S-Rank with S-Rand to reduce computational burdens and improve algorithm efficiency. Meanwhile, ISO can maintain the diversity of population by randomly selecting an objective to calculate the fitness value of individuals. We compare the performance of NSGA-II and NSGA-II-ISO on 12 test functions. Experimental results verify the effectiveness of ISO. In the future, we will consider a decision-making solution method to further improve the performance of NSGA-II-ISO in solving MOPs.

**Acknowledgments.** This work was partially supported by the Outstanding Youth Science Foundation with No. 61822602, the National Natural Science Foundations of China (NSFC) with No. 61772207 and 61873097, the Natural Science Foundations of Guangdong Province for Distinguished Young Scholars with No. 2014A030306038, the Project for Pearl River New Star in Science and Technology with No. 201506010047, the GDUPS (2016), the Science and Technology Planning Project of Guangdong Province, China, with No 2014B050504005, and the Fundamental Research Funds for the Central Universities.

## References

1. Chen, Z.G., et al.: Multiobjective cloud workflow scheduling: a multiple populations ant colony system approach. IEEE Trans. Cybern. (2018). https://doi.org/10.1109/tcyb.2018.2832640
2. Yu, X., et al.: Set-based discrete particle swarm optimization based on decomposition for permutation-based multiobjective combinatorial optimization problems. IEEE Trans. Crbern. **48**(7), 2139–2153 (2018)
3. Chen, N., et al.: An evolutionary algorithm with double-level archives multiobjective optimization. IEEE Trans. Cybern. **45**(9), 1851–1863 (2015)
4. Lin, Q.Z., et al.: A hybrid evolutionary immune algorithm for multiobjective optimization problems. IEEE Trans. Evol. Comput. **20**(5), 711–729 (2016)
5. Parsopoulos, K.E., Vrahatis, M.N.: Particles swarm optimization method in multiobjective problems. In: Proceedings ACM Symposium Applications Computation, pp. 603–607 (2002)
6. Zhan, Z.H., Li, J.J., Cao, J.N., Zhang, J., Chung, H.S., Shi, Y.H.: Multiple populations for multiple objectives: a coevolutionary technique for solving multiobjective optimization problems. IEEE Trans. Cybern. **43**(2), 445–463 (2013)
7. Liu, X.F., Zhan, Z.H., Gao, Y., Zhang, J., Kwong, S., Zhang, J.: Coevolutionary particle swarm optimization with bottleneck objective learning strategy for many-objective optimization. IEEE Trans. Evol. Comput. (2018). https://doi.org/10.1109/tevc.2018.2875430
8. Deb, K., Agrawal, S., Pratap, A., Meyarivan, T.: A fast and elitist multiobjective genetic algorithm: NSGA-II. IEEE Trans. Evol. Comput. **6**(2), 182–197 (2002)

9. Zhang, Q., Li, H.: MOEA/D: a multi-objective evolutionary algorithm based on decomposition. IEEE Trans. Evol. Comput. **11**(6), 712–731 (2007)
10. Deb, K., Jain, H.: An evolutionary many-objective optimization algorithm using reference-point-based nondominated sorting approach, Part I: solving problems with box constraints. IEEE Trans. Evol. Comput. **18**(4), 577–601 (2014)
11. Li, Y.Z., Wu, Q.H., Li, M.S., Zhan, J.P.: Mean-variance model for power system economic dispatch with wind power integrated. Energy **72**(1), 510–520 (2014)
12. Li, Y.Z., Li, M.S., Wen, B.J., Wu, Q.H.: Power system dispatch with wind power integrated using mean-variance model and group search optimizer. In: 2014 IEEE PES General Conference & Exposition, National Harbor, MD, pp. 1–5 (2014)
13. Ismail, H.A., Packianather, M.S., Grosvenor, R.I.: Multi-objective invasive weed optimization of the LQR controller. Int. J. Autom. Comput. **3**, 321–339 (2017)
14. Yao, G.S., Ding, Y.S., Jin, Y.C., Hao, K.G.: Endocrine-based coevolutionary multi-swarm for multi-objective workflow scheduling in a cloud system. Soft. Comput. **21**(15), 4309–4322 (2017)
15. Murugan, P., Kannan, S., Baskar, S.: Application of NSGA-II algorithm to single-objective transmission constrained generation expansion planning. IEEE Trans. Power Syst. **24**(4), 1790–1797 (2009)
16. Singh, U., Singh, S.N.: Optimal feature selection via NSGA-II for power quality disturbances classification. IEEE Trans. Ind. Inform. **14**(7), 2994–3002 (2018)
17. Jayaprakasam, S., Abdul Rahim, S.K., Leow, C.Y., Ting, T.O., Eteng, A.A.: Multiobjective beampattern optimization in collaborative beamforming via NSGA-II with selective distance. IEEE Trans. Antennas Propag. **65**(5), 2348–2357 (2017)
18. Zitzler, E., Deb, K., Thiele, L.: Comparison of multiobjective evolutionary algorithms: empirical results. Evol. Comput. **8**(2), 173–195 (2000)
19. Zhang, Q.F., Zhou, A., Zhao, S.Z., Suganthan, P.N., Liu, W.D., Tiwari, S.: Multiobjective optimization test instances for the CEC 2009 special session and competition. In: Proceedings of IEEE Congress Evolutionary Computation, pp. 1–30 (2009)

# Evolutionary Optimization of Liquid State Machines for Robust Learning

Yan Zhou[1], Yaochu Jin[1,2(✉)], and Jinliang Ding[1]

[1] State Key Laboratory of Synthetical Automation for Process Industry,
Northeastern University, Shenyang, China
[2] Department of Computer Science, University of Surrey, Guildford, UK
yaochu.jin@surrey.ac.uk

**Abstract.** Liquid State Machines (LSMs) are a computational model of spiking neural networks with recurrent connections in a reservoir. Although they are believed to be biologically more plausible, LSMs have not yet been as successful as other artificial neural networks in solving real world learning problems mainly due to their highly sensitive learning performance to different types of stimuli. To address this issue, a covariance matrix adaptation evolution strategy has been adopted in this paper to optimize the topology and parameters of the LSM, thereby sparing the arduous task of fine tuning the parameters of the LSM for different tasks. The performance of the evolved LSM is demonstrated on three complex real-world pattern classification problems including image recognition and spatio-temporal classification.

**Keywords:** Liquid State Machine · Evolution strategy · CMA-ES · Pattern recognition

## 1 Introduction

The human brain is a complex system which has the capabilities to learn multiple types of knowledge in dynamical scenarios. Although artificial neural networks and deep learning have recently made significant progresses in accomplishing many human-level cognitive tasks [8], their efficiency and flexibility remain weaker than the human brain. Most deep neural network models are based on the simplified and highly abstract neural model containing only matrix operations and non-linear activation functions, and the learning algorithm they use relies on error back-propagation through the feedforward structure. Therefore, there are still gaps need to be filled between the capability of artificial neural networks and human cognitive systems [10].

This work was supported by the National Natural Science Foundation of China under Grand 61525302, 61590922, the National Key Research and Development Program of China under Grant 2018YFB1701104, the Project of Ministry of Industry and Information Technology of China under Grand 20171122-6, and the Fundamental Research Funds for the Central Universities under Grand N160801001, N161608001.

© Springer Nature Switzerland AG 2019
H. Lu et al. (Eds.): ISNN 2019, LNCS 11554, pp. 389–398, 2019.
https://doi.org/10.1007/978-3-030-22796-8_41

As a biologically more plausible framework based on spiking neurons [3], Liquid State Machines (LSMs) were proposed to mimic actual neural activations in brain [9]. An LSM contains random and sparse recurrent connections in the reservoir to convert spatio-temporal properties into high-dimensional states and hence provides linear separability for classification, which could highly enhance the computational capabilities and flexibility [9]. LSMs are believed to be more easily trained than feedforward spiking neural networks since only the read-out weights need to be trained [9].

Although the topology as well as the weights in the reservoir of the original LSM is randomly generated and fixed, increasing efforts have been paid to optimize the topology and the weight of the reservoir using both unsupervised and supervised learning. Most unsupervised methods are inspired by the Hebbian learning rules [12], in particular spike time depends plasticity (STDP) rules [15]. Although the STDP has been widely used in training spiking neural network, such Hebbian learning rules are still not very efficient and their performance is sensitive to input encoding and feature extraction. Among various issues, it has been found that there is strong interference in learning the structures in the sensory input using plasticity rules [2], which means that the previously learned structure information will be overwritten when learning a new pattern. Meanwhile, synaptic plasticity can only adjust the strength of synapses while the structure is fixed in the initialization, which limited the learning capabilities of those learning rules.

In addition to supervised and unsupervised learning methods, evolutionary computation has also been adopted to improve the performance of spiking neural network (SNN). For example, a multi-objective evolutionary algorithm was used to optimize the complexity and parameters of a feedforward spiking neural network [5]. A genetic algorithm is employed to optimize a SNN [1] and has showed considerable performance enhancement in pattern learning accelerated using graphics processing units. The evolving spiking neural network (eSNN) is also one popular framework of neuroevolution in spiking neural network. The architecture of eSNN is to learn input pattern by creating output neurons, each of them being labelled with a certain class label [13]. However, the eSNN is only used for training the readout layer in eSNN, denoted reSNN [6]. A covariance matrix adaptation evolution strategy (CMA-ES) was used to optimize a gene regulatory network that regulates the parameters of a BCM plasticity rule, named GRN-BCM, to tune the structure of an SNN [11].

This paper proposes a framework for optimization of the topology and parameters of the LSM using the CMA-ES to generate an initial LSM before learning starts. Here, the structure of the LSM is described by a probability distribution, whose parameters are subject to optimization. The CMA-ES is adopted in this work as it has been shown to be efficient and effective for optimization of continuous, non-linear and non-convex problems [4]. We show that by avoiding fine tuning of the structure and the synaptic strength of the LSM, the proposed framework is able to achieve robust and high learning performance on complex real-world classification problems.

## 2    Liquid State Machines

### 2.1    Spike Train Encoding

Different from conventional neural networks, information processing between neurons is in the form of spike trains in SNNs, which consist of spikes with various intervals. Thus, depending on different scenarios, the data of sensory input needs to be transformed from continuous variables into discrete temporal spike times or spike rates. An appropriate encoding method is therefore significant for pattern recognition, which converts continuous signals into spike trains. Typical encoding schemes include rate coding, count coding, binary coding, timing coding, and rank order coding [3]. In this work, an efficient timing coding method, named square cosine encoding [17], and a differential encoding method for spatial-temporal data [2] are adopted.

The square cosine encoding utilizes several cosine encoding neurons to encode continuous input variables into spike times. Each real value will be converted into several values with limited range, and these values will determine when spikes appear in the spike train. The parameters of cosine are uniformly distributed so that the spike times generated by the encoding neurons are different.

For spatial-temporal data, the difference between two time-adjacent data is the most significant feature. Therefore, the encoding method in [2] is employed, where the data in each sequence will be processed according to Eqs. (1) and (2).

$$M = ||[\Delta(D_1, D_2), ..., \Delta(D_{N-1}, D_N)]||  \tag{1}$$

$$\Delta(D_{n-1}, D_n) = \begin{cases} 1 & if\, \Delta(D_{n-1}, D_n) \geq threshold \cdot max(M(\cdot)) \\ 0 & else \end{cases}  \tag{2}$$

where $M$ represents a sequence and $D_n$ represents an individual data in that sequence. If the difference is greater than a threshold, the encoding neuron will fire at that moment, otherwise, it will keep silent.

### 2.2    Network Topology

The topology of the LSM consists of three main components as shown in Fig. 1, which are the input layer with encoding neurons introduced above, a reservoir of spiking neurons and a readout layer that extracts liquid states from the reservoir. The encoding layer only contains excitatory neurons that convert the input data into the spike trains. The number of encoding neurons depends on the dimension of the input and the encoding strategy. The reservoir is a kind of recurrent neural network, where the synapses are randomly generated and the neurons are either excitatory or inhibitory. The reservoir contains $N_E = 400$ excitatory neurons and $N_I = 100$ inhibitory neurons, following the 4:1 ratio between excitatory and inhibitory neurons [9]. The readout layer also contains the same number of neurons as the reservoir, with each neuron receiving projections from only

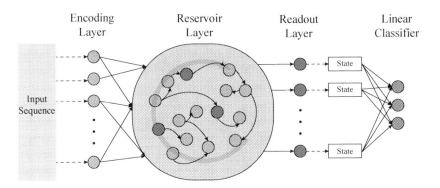

**Fig. 1.** The network topology of the LSM.

one specific neuron in the reservoir without a time delay. Therefore, the readout layer is composed of $N_{read} = 500$ readout neurons.

The synapses between the encoding layer and the reservoir are $S_{en-E}$ and $S_{en-I}$, representing the synapses of the encoding neurons to excitatory and inhibitory neurons, respectively. Both $S_{en-E}$ and $S_{en-I}$ only randomly choose 30% excitatory and inhibitory neurons in the reservoir as post-synaptic neurons, and the connection probability is $P_{en-E} = 0.01$ and $P_{en-I} = 0.01$, respectively. Note that the delay of $S_{en-E}$ and $S_{en-I}$ are both 0 ms.

There are four types of synapses in the reservoir according to pre- and post-synaptic neuron types, which are $S_{EE}$, $S_{EI}$, $S_{IE}$ and $S_{II}$. The synapses in the reservoir are recurrent and randomly generated in the initialisation and remain unchanged throughout the simulations. The connectivity of the reservoir adopted in this work is the model introduced in [9], which randomly allocates the neurons in a 'column' $(5 \times 5 \times 10)$. The connection probability of two neurons is dependent on the Euclidean distance of each two neurons, and the policy is described in the following:

$$P_{ij} = C \cdot e^{-(D(i,j)/\lambda)^2} \tag{3}$$

where $D(a, b)$ is the Euclidean distance between two neurons, $\lambda = 2$ is a important parameter that can control the density of the connection. The value of $C$ for each type of synapses are set as $C_{EE} = 0.3$, $C_{EI} = 0.2$, $C_{IE} = 0.4$, and $C_{II} = 0.1$.

The initial strength of each synapse will be set by picking values randomly from the gamma distribution after all synapses are generated. This work uses the original setting, the mean of the gamma distribution is set to $A_{EE} = 30$ mA, $A_{EI} = 60$ mA, $A_{IE} = -19$ mA and $A_{II} = -19$ mA. For the input synapses, parameter $A_{en-E} = 18$ mA for a projection onto an excitatory neuron and $A_{en-I} = 9$ mA for a projection onto an inhibitory neuron. The variance of the gamma distribution is set to the same to its mean. In this work, a scaling parameter $f = 1$ has been added to all synapses to generally adjust strength in the initialisation. Additionally, delays also exist when the spikes transmit through synapses in the reservoir, which is $t_{delay-E} = 0.8\,\text{ms}$ for $S_{EE}$ and $S_{EI}$, and $t_{delay-I} = 1.5\,\text{ms}$ for others. To simplify the readout process, no delay exists between the reservoir and the readout layer.

## 2.3 Neurons and Synapses

This work employs I&F neural model [3] in the reservoir. The dynamics of the post-synaptic neuron is described as in Eq. (4)

$$\tau_n \frac{du_i(t)}{dt} = u_{rest} - u_i + \sum_j I_{ij}(t) \cdot R \tag{4}$$

where $u_i$ is the trajectory of the membrane potential, $u_{rest}$ is resting potential, $\tau_n$ is membrane time constant, $R = 1\Omega$ is the impedance. The term $I_{ij}(t)$ describes the synaptic model and the synaptic dynamics is given in Eq. (5)

$$I_{ij}(t) = \sum_f w_{ij} \cdot e^{(-\frac{t}{\tau_s})} \cdot \delta_{ij}(t - t_j^{(f)}) + I_0 \tag{5}$$

where $\delta_{ij}(t - t_j^{(f)})$ is the Dirac function of post-synaptic neuron $i$ to pre-synaptic neuron $j$, $t_j^{(f)}$ is the firing time of the pre-synaptic neuron, $\tau_s$ is decay time constant and $I_0$ is the inherent injection to the pre-synaptic neuron.

The post-synaptic neuron fires a spike when $u_i$ reaches the threshold $u_{th}$, and then $u_i$ is reset to $u_{rest}$. After a spike is fired, the post-synaptic neuron will keep silent in the refractory period $t_{ref}$.

In the following experiments, the initial parameters for the neurons and synapses are $u_i = 13.5\,\text{mv}$, $\tau_n = 30\,\text{ms}$, $u_{th} = 15\,\text{mv}$, $I_{ij} = 0\,\text{mA}$, $I_0 = 13.5\,\text{mA}$ and $t_{ref} = 3\,\text{ms}$, $\tau_s = 3\,\text{ms}$ for the excitatory neurons, $t_{ref} = 2\,\text{ms}$, $\tau_s = 6\,\text{ms}$ for inhibitory neurons, respectively. Besides, Euler integration with a time step of 1 ms is adopted.

## 2.4 State Extraction

The reservoir converts spatio-temporal properties into high dimensional states, and the readout layer extracts the states from the reservoir for linear classification. Each readout neuron extracts the state from a specific neuron in the reservoir, the model of readout neurons is also I&F but it only calculates the membrane potential without firing spikes and without a threshold. The membrane time constant of the readout neurons is $\tau_{read} = 30\,\text{ms}$ and the membrane potential is $u_{read} = 0\,\text{mv}$.

Each readout neuron receives spikes from a single neuron (excitatory or inhibitory), thus the membrane potential only reflects the spike counts and spike time of per-synaptic neurons. The state defined in this work is the membrane potential of readout neurons at the end of each spike train converted from the spatio-temporal data injected into the network, then the membrane potential is set to the initial value.

After all sequences are injected into the network and states are obtained, a linear classifier (Softmax) is employed in this work to evaluate the score of pattern recognition.

# 3   Parameter Optimization with CMA-ES

## 3.1   CMA-ES

It has been found that the final classification performance of the LSM heavily depends on the parameter settings of the neurons and the reservoir connectivity. Optimization of the LSM parameters can be regarded as a non-linear and non-convex optimization function. It is computationally prohibitive to obtain the global minimum of the problem, we therefore aim to get an approximate minimum using a derivative-free optimization method instead. In this work, the covariance matrix adaptation evolution strategy (CMA-ES) is employed to search for the optimal initial parameters of the LSM.

In the CMA-ES, a population of offspring individuals is generated by sampling from a multivariate normal distribution [4]. The offspring $s_k^{(g+1)}$ is generated as follows:

$$s_k^{(g+1)} \sim m^{(g)} + \sigma^{(g)} \mathcal{N}(0, C^{(g)}) \qquad \text{for } k = 1, \ldots, \eta \qquad (6)$$

where $g = 0, 1, 2...$ is the generation number, $\eta$ is the population size. The multivariate normal distribution of the next generation will be recalculated by updating the mean $m^{(g)}$, the covariance matrix $C^{(g)}$ and the step-size $\sigma^{(g)}$. Only the best individuals out of the offspring population will be selected as the parent for the next generation, where $\gamma < \eta$ is the number of the parent population size, and typically $\gamma \approx \eta/2$.

A small offspring size can lead to fast adaptation of $C^{(g)}$ and reliable distribution for CMA-ES. Therefore, CMA-ES is more suited for parameter optimization of computationally expensive problems.

## 3.2   Parameter Selection

There are many parameters in the LSM but only some of them are very sensitive to the input signals. The feature of spike trains encoded from various input sequences needs a unique density of connections to map the data set onto a high-dimensional space with a good linear separability. According to the connection rule, $\lambda$ is a parameter controlling the density of connections and the number of synapses in the reservoir, which directly affect the response intensity of the reservoir to the stimuli. Additionally, the synaptic strength should be adjusted synchronously to adapt density of connections. In this work, all distribution of the synaptic strength is fixed, thus the distribution of the synaptic strength can be decided by the factor of $f$. Thus, $\lambda$ and $f$ will be subject to evolutionary optimization.

As to the neuron level, the fluctuating intervals between two spikes may lead to different membrane potentials in post-synaptic neurons. However, whether these differences can be captured depends on the membrane time constant. A too short membrane time constant may miss some important information, but a too long membrane time constant may record too much noise. Therefore, the membrane time constant $\tau_n$, a scale of the neuron to dispose of temporal information between spikes is also to be optimized.

Thus, only three most critical parameters ($\lambda$, $\tau_n$ and $f$) are to be optimized by the CMA-ES to reduce the computational cost and the parameters are initialized empirically. The objective function of CMA-ES is the test accuracy in each pattern recognition task. The dimension of the search space is thus $n = 3$ and the $\eta = 4 + \lfloor 3 \times log(n) \rfloor = 7$ and $\gamma = \lceil \frac{\eta}{2} \rceil = 3$ are selected by default as recommended in [4]. Finally, the boundary condition is set such that all inputs are positive.

## 4     Pattern Recognition Results

The performance of the proposed framework is investigated with image recognition, speaker recognition and human motion recognition tasks. The simulation tools used in this work is Brian2 [16]. Three independent runs are performed for the CMA-ES. The proposed method is compared with a grid search and some existing LSM models. The range of the grid search is empirically determined and the grids are uniformly distributed with 10 steps in each dimension.

**Table 1.** Comparative results of pattern recognition tasks.

|   | Architectures | MNIST | Jv | KTH |
|---|---|---|---|---|
| 1 | Softmax regression | **92.6%** | - | - |
| 2 | LSM with STDP [2] | - | 90.8% | 66.7% |
| 3 | GRN-BCM [11] | - | - | 88.15% |
| 4 | LSM with grid search | 86.4% | 90.2% | 98.42% |
| 5 | LSM with CMA-ES | 88.5% | **94.8%** | **98.55%** |

Feed-forward SNNs have been used for classification tasks, especially for the data without temporal properties such as image data [18]. However, little work has been reported on the application of the LSM image data classification. The image data used in this work is MNIST dataset, which contains 60000 training data and 10000 testing data, where each data is a $28 \times 28$ matrix. In this work, the $28 \times 28$ matrix is converted to 784 dimensional input. The training data and testing used for parameters searching are 3000 and 500 randomly selected from training data and testing data, respectively. Each dimension is encoded by the square cosine encoding with one encoding neuron and the duration of encoding is 30 ms. The initial parameters are $\lambda = 0.9$, $\tau_n = 0.5$, $f = 1.2$, and $\sigma = 0.1$ for

CMA-ES. The range of the grid search for image recognition is $\tau_n \subseteq (10, 100)$, $\lambda \subseteq (0.25, 2.5)$, $f \subseteq (0.15, 1.5)$.

The dataset adopted for speaker recognition task is taken from [7], which contains 9 male speakers uttered two Japanese vowels (Jv) /ae/ successively. Each utterance produced by a speaker forms sequences and each sequence is composed of 12 features. The number of the sequences is 640 in total including one set of 270 sequences for training and the other set of 370 sequences for testing. Each feature input is encoded by 3 encoding neurons and the duration of each data in the sequences is 10 ms for this classification task. The initial parameters of CMA-ES are $\lambda = 0.5$, $\tau_n = 0.5$, $f = 0.5$, and $\sigma = 0.5$. The range of parameters for speaker recognition is $\tau_n \subseteq (30, 300)$, $\lambda \subseteq (0.2, 1)$ and $f \subseteq (0.5, 1)$.

The human motion recognition needs a high dimensional spatial-temporal video data. For example, the KTH [14] is a data set consisting of 2391 video files. There are six actions including boxing, clapping, walking, waving, and jogging. Each video contains $160 \times 120$ pixels with a length of four seconds in average and is taken by a static camera with $25 fps$ frame rate. For this task, the encoding method is based on Eqs. (1) and (2). To reduce the computational complexity, each frame of the video is downsampled to $32 \times 24$ with the $5 \times 5$ max pooling operation. The firing threshold of encoding neuron is 0.2. The initial parameters of CMA-ES are $\lambda = 0.5$, $\tau_n = 0.5$, $f = 0.5$, and $\sigma = 0.5$. Meanwhile, the parameters of the grid search for human motion recognition are in the range of $\tau_n \subseteq (30, 300)$, $\lambda \subseteq (0.2, 2)$ and $f \subseteq (0.1, 1)$.

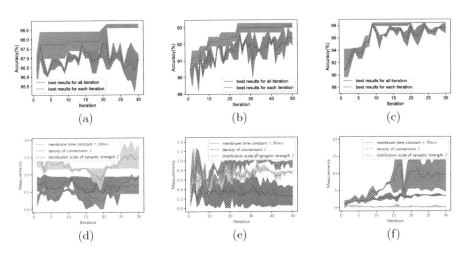

**Fig. 2.** (a) (b) (c) Average classification accuracy of the LSM over the generations. (d) (e) (f) average parameter values of $\tau$, $\lambda$ and $f$ over the generations.

Table 1 presents the pattern recognition results obtained by the five algorithms under comparison in test data. The first algorithm is a softmax regression, which is the most common algorithm for multi-class classification. The softmax regression directly uses 784 dimensional data of MNIST as the input without any hidden layer. The results of the LSM with the STDP are taken from [2], where the STDP used to train the synapses between the excitatory neurons in the reservoir. Recall that GRN-BCM [11] is a feed-forward framework regulated by the BCM plasticity whose parameters are optimized by a CMA-ES. The performance of the proposed framework is better in Jv and KTH but not good enough for MNIST. The reason could be the recurrent connections provide feeble feature extraction on the data without spatio-temporal information.

Figure 2(a) (b) (c) presents the average accuracy, while Fig. 2(d) (e) (f) shows the average evolution paths of $\tau$, $\lambda$ and $f$. The best solution for image recognition appears at about the 20-th generation but the performance heavily fluctuates during the optimization. These results indicate that only a small number of generations are required to find the optimal solutions and the performance of image recognition is definitely sensitive to the selected parameters. To our best knowledge, this is the first work reporting that an LSM is able to achieve similar performance to conventional neural networks without convolution, demonstrating effectiveness to optimize the topology of the LSM and the critical parameter in the neurons. The experiment for speaker recognition dataset runs 50 generations in total, although an acceptable solution appears around the 25-th generation. The evolution path of $\tau$ shows a larger variation range, indicating that $\tau$ has a strong relationship with the spatio-temporal information of Jv. Finally, we can see that the convergence process is more stable than that for image recognition. The convergence of evolution progress for human motion recognition is the most stable and the best solution is obtained within approximately 10 generations. Parameter $\lambda$ is larger than that of the other tasks in the evolution path, which means more connections is required to deal with the spatial-temporal information.

## 5    Conclusion and Discussion

This work aims to optimize three critical parameters in the LSM using an evolution strategy, two specifying the connectivity of the reservoir and one parameter in the spiking neurons. Our empirical results on three complex classification tasks demonstrate the proposed method is computationally acceptable and has enabled the LSM to perform robustly on complex tasks and achieve competitive performance. Our results clearly indicate that the LSM optimized using CMA-ES has achieved better performance on spatio-temporal data than the state-of-the-art LSM models, in particular on the human motion recognition task. Note, however, the performance of the LSM on the image recognition task is not yet as good as that of other learning models. This could be attributed to the fact that recurrent connections do not provide advantages over other feed-forward models on the data without spatio-temporal information.

This connectivity of the reservoir is described globally by one probability distribution. In the future, we are going to exploit the possibility of generating more complex connectivity structures for the reservoir to further improve the learning performance of the LSM.

# References

1. Carlson, K.D., Nageswaran, J.M., Dutt, N., Krichmar, J.L.: An efficient automated parameter tuning framework for spiking neural networks. Front. Neurosci. **8**, 10 (2014)
2. Chrol-Cannon, J., Jin, Y.: Learning structure of sensory inputs with synaptic plasticity leads to interference. Front. Comput. Neurosci. **9**, 103 (2015)
3. Gerstner, W., Kistler, W.M.: Spiking Neuron Models: Single Neurons, Populations, Plasticity. Cambridge University Press, Cambridge (2002)
4. Hansen, N., Ostermeier, A.: Completely derandomized self-adaptation in evolution strategies. Evol. Comput. **9**(2), 159–195 (2001)
5. Jin, Y., Wen, R., Sendhoff, B.: Evolutionary multi-objective optimization of spiking neural networks. In: de Sá, J.M., Alexandre, L.A., Duch, W., Mandic, D. (eds.) ICANN 2007. LNCS, vol. 4668, pp. 370–379. Springer, Heidelberg (2007). https://doi.org/10.1007/978-3-540-74690-4_38
6. Kasabov, N., et al.: Evolving spiking neural networks for personalised modelling, classification and prediction of spatio-temporal patterns with a case study on stroke. Neurocomputing **134**, 269–279 (2014)
7. Kudo, M., Toyama, J., Shimbo, M.: Multidimensional curve classification using passing-through regions. Pattern Recogn. Lett. **20**(11), 1103–1111 (1999)
8. Lake, B.M., Salakhutdinov, R., Tenenbaum, J.B.: Human-level concept learning through probabilistic program induction. Science **350**(6266), 1332–1338 (2015)
9. Maass, W., Natschläger, T., Markram, H.: Real-time computing without stable states: a new framework for neural computation based on perturbations. Neural Comput. **14**(11), 2531–2560 (2002)
10. Marblestone, A.H., Wayne, G., Kording, K.P.: Toward an integration of deep learning and neuroscience. Front. Comput. Neurosci. **10**, 94 (2016)
11. Meng, Y., Jin, Y., Yin, J.: Modeling activity-dependent plasticity in BCM spiking neural networks with application to human behavior recognition. IEEE Trans. Neural Netw. **22**(12), 1952–1966 (2011)
12. Panda, P., Roy, K.: Learning to generate sequences with combination of hebbian and non-hebbian plasticity in recurrent spiking neural networks. Front. Neurosci. **11**, 693 (2017)
13. Schliebs, S., Kasabov, N.: Evolving spiking neural network a survey. Evolving Syst. **4**(2), 87–98 (2013)
14. Schuldt, C., Laptev, I., Caputo, B.: Recognizing human actions: a local SVM approach. In: Proceedings of the 17th International Conference on Pattern Recognition, 2004, vol. 3, pp. 32–36 (2004)
15. Song, S., Miller, K.D., Abbott, L.F.: Competitive hebbian learning through spike-timing-dependent synaptic plasticity. Nat. Neurosci. **3**(9), 919–926 (2000)
16. Stimberg, M., Goodman, D., Benichoux, V., Brette, R.: Equation-oriented specification of neural models for simulations. Front. Neuroinf. **8**, 6 (2014)
17. Wu, Q.X., McGinnity, T.M., Maguire, L.P., Glackin, B., Belatreche, A.: Learning under weight constraints in networks of temporal encoding spiking neurons. Neurocomputing **69**(16), 1912–1922 (2006)
18. Xu, Q., Qi, Y., Yu, H., Shen, J., Tang, H., Pan, G.: CSNN: an augmented spiking based framework with perceptron-inception. In: IJCAI, pp. 1646–1652 (2018)

# An Algebraic Algorithm for Joint Independent Subspace Analysis

Jia-Xing Yang[✉], Xiao-Feng Gong, and Gui-Chen Yu

School of Information and Communication Engineering,
Dalian University of Technology, Dalian 116023, China
jiuyunshikong@mail.dlut.edu.cn

**Abstract.** In this work, we propose an algebraic algorithm called coupled exact joint block decomposition (CE-JBD) for joint independent subspace analysis (JISA), an extension to joint blind source separation. In JISA, tensors admitting coupled rank-$(L_m, L_n, \cdot)$ Block Term Decomposition (BTD) can be constructed using second order statistics of nonstationary signals. And the loading matrices to be estimated will be computed from these tensors via coupled rank-$(L_m, L_n, \cdot)$ BTD based algorithms. However, most of the existing algorithms resort to iterative techniques. They heavily rely on a good starting point. Capable of providing such a point, our proposed CE-JBD, based on coupled rank-$(L_m, L_n, \cdot)$ BTD, achieves JISA only by employing generalized eigenvalue decomposition followed by a clustering step and singular value decomposition. To validate its efficacy, as well as its ability to serve its iterative counterparts, we present some experiment results in the end.

**Keywords:** Coupled rank-$(L_m, L_n, \cdot)$ block term decomposition ·
Second order statistics · Joint independent subspace analysis ·
Coupled exact joint block decomposition

## 1 Introduction

In the past decade, joint blind source separation (JBSS) has been applied to a wide variety of fields, e.g., transformed signals in multiple frequency bins [1], multi-subject functional magnetic resonance imaging [2–4], and multi-subject electrocardiography-hyperscanning [5]. The JBSS is developed from blind source separation (BSS) that traditionally is used to identify the latent statistically-independent sources and loading matrix from an instantaneous mixture. Theoretically, JBSS resolves several BSS problems by exploiting the potential dependence of the one-dimensional sources across different datasets. It turns out that JBSS aligns the corresponding source estimates across datasets [2], which nevertheless is not guaranteed by individual applying BSS.

However, JBSS methods, e.g., generalized non-orthogonal joint diagonalization [6] and double coupled canonical polyadic decomposition [7], fail to handle the inter-dependent datasets where multidimensional sources are within each

© Springer Nature Switzerland AG 2019
H. Lu et al. (Eds.): ISNN 2019, LNCS 11554, pp. 399–408, 2019.
https://doi.org/10.1007/978-3-030-22796-8_42

mixture, e.g., the mixtures in [8]. To solve, joint independent subspace analysis (JISA), as a generalization of JBSS, is recently proposed by Lahat *et al.* in [9]. The JISA identifies independent multidimensional sources within each dataset while keeping their correspondence across different datasets. So far, works for JISA are mainly divided into two groups. The first is about matrix-based methods. In [9], a relative gradient algorithm is proposed to realize JBSS of Gaussian multidimensional components. In [10], the author uses a quasi-Newton algorithm to achieve asymptotically minimal mean square error to identify all of the estimates of loading matrices. In [13], the author presents some sufficient and necessary conditions for the uniqueness and identifiability of JISA and matrix-based coupled block diagonalization algorithms. The second is based on coupled tensor decomposition. In [11], a structured data fusion (SDF) method capable of achieving JISA is studied for coupled and/or structured decompositions of tensors. In [12], two coupled rank-$(L_m, L_n, \cdot)$ block term decomposition (BTD) based algorithms, both based on simultaneous generalized schur decomposition scheme, are proposed.

Coupled tensor decomposition based algorithms are effective for JISA because the tensors able to be constructed within together admit coupled rank-$(L_m, L_n, \cdot)$ BTD, a concept first mentioned and described in [12]. As an emerging member of tensor decomposition, coupled rank-$(L_m, L_n, \cdot)$ BTD is developed on the basis of rank-$(L_m, L_n, \cdot)$ BTD [14]. In comparison to the latter, it decomposes multiple tensors from the perspective of multi-set fusion, making accuracy improvement and identifiability relaxation (see [12] and the references therein). Nevertheless, it is still hard to find an algebraic candidate achieving coupled rank-$(L_m, L_n, \cdot)$ BTD among open literatures.

In this paper, we propose an algebraic algorithm named coupled exact joint block decomposition (CE-JBD), which is on the basis of coupled rank-$(L_m, L_n, \cdot)$ BTD. It evolves from exact joint block decomposition (E-JBD) proposed in [15], which, however, is only applied to ordinary rank-$(L_m, L_n, \cdot)$ BTD. The CE-JBD relies on generalized eigenvalue decomposition (GEVD) [16] followed by a clustering step and singular value decomposition (SVD) to compute all unknown loading matrices, as well as other interesting components (if needed). For comparison, we also implement another two coupled rank-$(L_m, L_n, \cdot)$ BTD based iterative algorithms. The involved algorithms are tested on the coupled tensors constructed by second order statistics (SOS) of the short-time non-stationary binary phase shift key (BPSK) signals.

*Notations*: we represent vector, matrix and tensor as uppercase boldface letter, lowercase boldface letter, and uppercase calligraphic letter, respectively. By denoting the Khatri-Rao product as '$\otimes$', we represent the partition-wise Khatri-Rao product, the column-wise Khatri-Rao product, and the mode-n product as '$\odot$', '$\odot_c$', and '$\times_n$', respectively: $\boldsymbol{A} \odot \boldsymbol{B} \triangleq [\boldsymbol{a}_1 \otimes \boldsymbol{b}_1, \cdots, \boldsymbol{a}_R \otimes \boldsymbol{b}_R]$, $\boldsymbol{A} \odot_c \boldsymbol{B} \triangleq [\boldsymbol{A}_1 \otimes \boldsymbol{B}_1, \cdots, \boldsymbol{A}_R \otimes \boldsymbol{B}_R]$, and $(\boldsymbol{\mathcal{T}} \times_n \boldsymbol{F})_{i_1,\ldots,i_{n-1},j,i_{n+1},\ldots i_N} \triangleq \sum_{i_n=1}^{M} t_{i_1,\ldots,i_N} \boldsymbol{F}_{j,i_n}$, where we suppose that $\boldsymbol{A} = [\boldsymbol{A}_1, \ldots, \boldsymbol{A}_R]$ and $\boldsymbol{B} = [\boldsymbol{B}_1, \ldots, \boldsymbol{B}_R]$ in the partition-wise Khatri-Rao product and $\boldsymbol{A} = [\boldsymbol{a}_1, \ldots, \boldsymbol{a}_R]$ and $\boldsymbol{B} = [\boldsymbol{b}_1, \ldots, \boldsymbol{b}_R]$ in the Khatri-Rao product. We use $\tilde{\boldsymbol{A}}$ to denote the estimate of $\boldsymbol{A}$. We use symbols

'$(\cdot)^{\mathsf{T}}$','$(\cdot)^{\mathsf{H}}$','$(\cdot)^{*}$','$(\cdot)^{\dagger}$', '$E(\cdot)$', and '$||\cdot||_{\mathsf{F}}^2$' to represent transpose, conjugated transpose, conjugate, pseudo-inverse, mathematical expectation, and Forbenius norm, respectively. We use MATLAB notation $\mathcal{T}_{(:,:,k)}$ to represent the $k$-th frontal slice of a third-order tensor $\mathcal{T}$. For a tensor $\mathcal{T} \in \mathbb{C}^{I \times J \times K}$, $(\mathcal{T})_1$, $(\mathcal{T})_2$, and $(\mathcal{T})_3$ are used to denote its mode-1, mode-2, and mode-3 matrix expressions, respectively: $(\mathcal{T})_{1((j-1)K+k,i)} = (\mathcal{T})_{2((i-1)K+k,j)} = (\mathcal{T})_{3((i-1)J+j,k)} = \mathcal{T}_{(i,j,k)}$.

*Definition*: A decomposition of $\mathcal{T} \in \mathbb{C}^{I \times J \times K}$ in a sum of rank-$(L_m, L_n, \cdot)$ terms [14] is a decomposition of the following: $\mathcal{T} = \sum_{r=1}^{R} \mathcal{S}_r \times_1 \mathbf{A}_r \times_2 \mathbf{B}_r, r = 1, ..., R$, where $\mathcal{S}_r \in \mathbb{C}^{L_m \times L_n \times K}$ has mode-1 rank and mode-2 rank equivalent to $L_m$ and $L_n$, respectively. Both $\mathbf{A}_r \in \mathbb{C}^{J \times L_m}$ and $\mathbf{B}_r \in \mathbb{C}^{J \times L_n}$ have full column rank. Note that $\mathbf{A} \triangleq [\mathbf{A}_1, ..., \mathbf{A}_R]$ and $\mathbf{B} \triangleq [\mathbf{B}_1, ..., \mathbf{B}_R]$ are defined as two factor matrices of $\mathcal{T}$. The three matrix representations of $\mathcal{T}$ are written as: $(\mathcal{T})_1 = [(\mathcal{S}_1 \times_2 \mathbf{B}_1)_1, ..., (\mathcal{S}_R \times_2 \mathbf{B}_R)_1]\mathbf{A}^{\mathsf{T}}$, $(\mathcal{T})_2 = [(\mathcal{S}_1 \times_1 \mathbf{B}_1)_2, ..., (\mathcal{S}_R \times_1 \mathbf{B}_R)_2]\mathbf{B}^{\mathsf{T}}$ and $(\mathcal{T})_3 = (\mathbf{A} \odot \mathbf{B})((\mathcal{S}_1)_3^{\mathsf{T}}, ..., (\mathcal{S}_R)_3^{\mathsf{T}})^{\mathsf{T}}$, respectively.

## 2 Problem Formulation

### 2.1 Data Model

Consider a latent signal model where the received $\mathbf{x}^{(m)}(t)$ of size $I^{(m)} \times 1$ is an instantaneous linear mixture as follows:

$$\mathbf{x}^{(m)}(t) = \mathbf{A}^{(m)}\mathbf{s}^{(m)}(t), 1 \le t \le T, 1 \le m \le M, \tag{1}$$

where $T$ and $M$ represent the number of snapshots and the number of datasets, respectively. Matrix $\mathbf{A}^{(m)} \in \mathbb{C}^{I^{(m)} \times L^{(m)}}$ is an invertible matrix. The source vector $\mathbf{s}^{(m)}(t) \in \mathbb{C}^{L^{(m)} \times 1}$ consists of $R$ multidimensional sources, with its $r$th component $\mathbf{s}_r^{(m)}(t) \in \mathbb{C}^{L_m \times 1}$ $(L^{(m)} = RL_m)$. We assume that $\mathbf{s}_{r_1}^{(m)}$ and $\mathbf{s}_{r_2}^{(m)}$ are statistically independent for $1 \le r_1 \ne r_2 \le R$, and that $\mathbf{s}_r^{(m)}$ and $\mathbf{s}_r^{(n)}$ are statically dependent for $1 \le m \ne n \le M$, as illustrated in Fig. 1.

Given the partition pattern of $\mathbf{s}^{(m)}(t)$, we likewise divide $\mathbf{A}^{(m)}$ into $R$ submatrices, i.e., $\mathbf{A}^{(m)} = [\mathbf{A}_1^{(m)}, ..., \mathbf{A}_R^{(m)}]$, with $\mathbf{A}_r^{(m)}$ being size of $I^{(m)} \times L_m$. For any invertible $\mathbf{B}$ of size $L_m \times L_m$, it is unlikely to discriminate $(\mathbf{A}_r^{(m)}, \mathbf{s}_r^{(m)})$ from $(\mathbf{A}_r^{(m)}\mathbf{B}^{-1}, \mathbf{B}\mathbf{s}_r^{(m)})$. In practice, JISA aims at estimating the column space of $\mathbf{A}_r^{(m)}$, namely span($\mathbf{A}_r^{(m)}$). The data model will become that of a standard JBSS when only one-dimensional sources are within each dataset. For convenience, we stack $\mathbf{x}^{(m)}(t)$ as follows:

$$\begin{bmatrix} \mathbf{x}^{(1)}(t) \\ \vdots \\ \mathbf{x}^{(M)}(t) \end{bmatrix} = \begin{bmatrix} \mathbf{A}^{(1)} & 0 & 0 \\ 0 & \ddots & 0 \\ 0 & 0 & \mathbf{A}^{(M)} \end{bmatrix} \begin{bmatrix} \mathbf{s}^{(1)}(t) \\ \vdots \\ \mathbf{s}^{(M)}(t) \end{bmatrix}. \tag{2}$$

It is clear that Eq. (2) can be written in a more compact form:

$$\mathbf{x}(t) = \mathbf{A}\mathbf{s}(t). \tag{3}$$

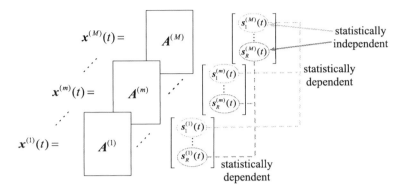

**Fig. 1.** Diagram of the JISA data model. Terms with the same subscript have statistical dependence, while those with different subscripts does not.

In (3), $\boldsymbol{x}(t) = [\boldsymbol{x}^{(1)\mathsf{T}}(t), \ldots, \boldsymbol{x}^{(M)\mathsf{T}}(t)]^{\mathsf{T}}$ and $\boldsymbol{s}(t) = [\boldsymbol{s}^{(1)\mathsf{T}}(t), \ldots, \boldsymbol{s}^{(M)\mathsf{T}}(t)]^{\mathsf{T}}$ are of size $\sum_{m=1}^{M} I^{(m)} \times 1$ and $\sum_{m=1}^{M} L^{(m)} \times 1$, respectively. The $\boldsymbol{A}$ is a block diagonal matrix holding $\boldsymbol{A}^{(m)}$ as its $m$th block-diagonal sub-matrix. Subsequently, we focus on the SOS of $\boldsymbol{x}(t)$ and $E(\boldsymbol{x}(t)\boldsymbol{x}^{\mathsf{H}}(t))$ can be written as:

$$E\{\boldsymbol{x}(t)\boldsymbol{x}^{\mathsf{H}}(t)\} = \boldsymbol{A}E\{\boldsymbol{s}(t)\boldsymbol{s}^{\mathsf{H}}(t)\}\boldsymbol{A}^{\mathsf{H}}, \tag{4}$$

in which $E\{\boldsymbol{s}(t)\boldsymbol{s}^{\mathsf{H}}(t)\}$ can be similarly represented as follows:

$$E\{\boldsymbol{s}(t)\boldsymbol{s}^{H}(t)\} = \begin{bmatrix} \boldsymbol{D}^{(1,1)} & \cdots & \boldsymbol{D}^{(1,M)} \\ \vdots & \ddots & \vdots \\ \boldsymbol{D}^{(M,1)} & \cdots & \boldsymbol{D}^{(M,M)} \end{bmatrix}, \tag{5}$$

where $\boldsymbol{D}^{(m,n)} = 1/T\sum_{t=1}^{T} \boldsymbol{s}^{(m)}(t)\boldsymbol{s}^{(n)\mathsf{H}}(t)$, $1 \leq m, n \leq M$. In light of the aforementioned statistical properties, each $\boldsymbol{D}^{(m,n)}$ is a block-diagonal matrix, with its $r$th diagonal sub-matrix being $1/T\sum_{t=1}^{T} \boldsymbol{s}_r^{(m)}(t)\boldsymbol{s}_r^{\mathsf{H}(n)}(t)$. Let $\boldsymbol{X}^{(m,n)} = \boldsymbol{A}^{(m)}\boldsymbol{D}^{(m,n)}\boldsymbol{A}^{\mathsf{H}(n)}$, such that (4) can be further formulated as:

$$E\{\boldsymbol{x}(t)\boldsymbol{x}^{H}(t)\} = \begin{bmatrix} \boldsymbol{X}^{(1,1)} & \cdots & \boldsymbol{X}^{(1,M)} \\ \vdots & \ddots & \vdots \\ \boldsymbol{X}^{(M,1)} & \cdots & \boldsymbol{X}^{(M,M)} \end{bmatrix}. \tag{6}$$

In (6), $E\{\boldsymbol{x}(t)\boldsymbol{x}^{\mathsf{H}}(t)\}$ comprises $M \times M$ sub-matrices, where the $(m, n)$th is written as $\boldsymbol{X}^{(m,n)}$. After the pattern of $E\{\boldsymbol{x}(t)\boldsymbol{x}^{\mathsf{H}}(t)\}$ is derived, it is convenient for us to relate it with coupled rank-$(L_m, L_n, \cdot)$ BTD in the following.

## 2.2    JISA via Coupled Rank-$(L_m, L_n, \cdot)$ BTD

Besides the two mentioned assumptions, coupled rank-$(L_m, L_n, \cdot)$ BTD based algorithms additionally require the sources to be temporally non-stationary to

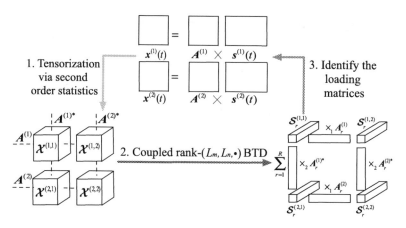

**Fig. 2.** Diagram of $2 \times 2$ sub-tensors admitting coupled rank-$(L_m, L_n, \cdot)$ BTD.

capture the time diversity. We control the time-frame of the received $\boldsymbol{x}(t)$ so that $K$ sample blocks are generated. Then, we stack SOS of these frames into a third-order tensor $\boldsymbol{\mathcal{X}}$ along its third mode, i.e., $\boldsymbol{\mathcal{X}}_{(:,:,k)} = E\{\boldsymbol{x}_k(t)\boldsymbol{x}_k^{\mathsf{H}}(t)\}$ with $1 \leq k \leq K$. Clearly, $\boldsymbol{\mathcal{X}}$ consists of $M \times M$ sub-tensors, with the $(m, n)$th denoted as $\boldsymbol{\mathcal{X}}^{(m,n)}$. Each sub-tensor $\boldsymbol{\mathcal{X}}^{(m,n)}$ (also claimed as tensor $\boldsymbol{\mathcal{X}}^{(m,n)}$ in the following context) admits rank-$(L_m, L_n, \cdot)$ BTD [14] and can be written as the sum of a set of rank-$(L_m, L_n, \cdot)$ terms:

$$\boldsymbol{\mathcal{X}}^{(m,n)} = \sum_{r=1}^{R} \boldsymbol{\mathcal{S}}_r^{(m,n)} \times_1 \boldsymbol{A}_r^{(m)} \times_2 \boldsymbol{A}_r^{(n)*}, \tag{7}$$

in which $\boldsymbol{\mathcal{S}}_r^{(m,n)}$ is of size $L_m \times L_n \times K$ and its $k$th frontal slice $(\boldsymbol{\mathcal{S}}_r^{(m,n)})_{(:,:,k)} = E(s_r^{(m)}(t)s_r^{\mathsf{H}(m)}(t))_k$. Note that the mode-1 rank and mode-2 rank of $(\boldsymbol{\mathcal{S}}_r^{(m,n)})$ equal to $L_m$ and $L_n$, respectively. Each $\boldsymbol{\mathcal{X}}^{(m,n)}$ follows rank-$(L_m, L_n, \cdot)$ BTD [14]. Meanwhile, $\{\boldsymbol{\mathcal{X}}^{(m,n)}, n = 1, ..., M\}$ share $\boldsymbol{A}^{(m)}$ in the first mode, and $\{\boldsymbol{\mathcal{X}}^{(m,n)}, m = 1, ..., M\}$ share $\boldsymbol{A}^{(n)}$ in the second mode. Therefore, we say that $\{\boldsymbol{\mathcal{X}}^{(m,n)}, m, n = 1, ..., M\}$ admit coupled rank-$(L_m, L_n, \cdot)$ BTD (see Fig. 2).

There are some trivial indeterminacies for coupled rank-$(L_m, L_n, \cdot)$ BTD: (i) permutation ambiguity, i.e., the terms in $\boldsymbol{\mathcal{X}}^{(m,n)}$ can be arbitrarily permuted if it is done to all of the other tensors, (ii) rotation ambiguity, i.e., for any invertible $\boldsymbol{Z}_r^{(m)}$ and $\boldsymbol{Z}_r^{(n)}$, it is unlikely to discriminate $\tilde{\boldsymbol{\mathcal{S}}}_r^{(m,n)} = \boldsymbol{\mathcal{S}}_r^{(m,n)} \times_1 \boldsymbol{Z}_r^{(m)-1} \times_2 \boldsymbol{Z}_r^{(n)-1}$, $\tilde{\boldsymbol{A}}^{(m)} = \boldsymbol{A}_r^{(m)}\boldsymbol{Z}_r^{(m)}$, and $\tilde{\boldsymbol{A}}^{(n)} = \boldsymbol{A}_r^{(n)}\boldsymbol{Z}_r^{(n)}$ from $\boldsymbol{\mathcal{S}}_r$, $\boldsymbol{A}_r^{(m)}$, and $\boldsymbol{A}_r^{(n)}$. We say that coupled rank-$(L_m, L_n, \cdot)$ BTD identifies all of the loading matrices uniquely up to the two ambiguities.

## 3    Proposed Algorithm

The primary step of CE-JBD is to perform E-JBD [15] individually on each tensor $\boldsymbol{\mathcal{X}}^{(m,m)}$ to find a relatively accurate loading matrix. Let us first construct $\boldsymbol{X}_{k_1 k_2}^{(m)}$ as follows:

$$X_{k_1 k_2}^{(m)} = \mathcal{X}_{(:,:,k_1)}^{(m,m)} \mathcal{X}_{(:,:,k_2)}^{(m,m)\dagger} = A^{(m)} S_{(:,:,k_1)}^{(m,m)} S_{(:,:,k_2)}^{(m,m)\dagger} A^{(m)\dagger}, \tag{8}$$

In (8), $\mathcal{X}_{(:,:,k_1)}^{(m,m)}$ and $\mathcal{X}_{(:,:,k_2)}^{(m,m)}$ are two significant frontal slice of $\mathcal{X}^{(m,m)}$. We perform GEVD of $X_{k_1 k_2}^{(m)}$ and the derived $L$ eigenvectors, associated to the $L$ most significant eigenvalues, constitute span($A^{(m)}$). We represent span($A^{(m)}$) as $V^{(m)}$, such that $A^{(m)}$ can be further written as:

$$A^{(m)} = V^{(m)} P^{(m)}, \tag{9}$$

where $P^{(m)}$ is a permutation matrix of size $L^{(m)} \times L^{(m)}$, clustering the eigenvalue vectors into corresponding subspaces. We pre-multiply and post-multiply $X_{(:,:,k)}^{(m,m)}$ by $V^{(m)\dagger}$ and $(V^{(m)H})^\dagger$, respectively, and the result is written as:

$$W_k^{(m)} = V^{(m)\dagger} \mathcal{X}_{(:,:,k)}^{(m,m)} (V^{(m)H})^\dagger = P^{(m)} S_{(:,:,k)}^{(m,m)} P^{(m)\dagger}, \tag{10}$$

in which $W_k^{(m)}$ is a matrix with non-zero elements being placed symmetrically. Let $W^{(m)} = 1/K \sum_{k=1}^{K} W_k^{(m)}$ such that $P^{(m)}$ can be identified via the positions of the most significant $L_m$ values at each row of $W^{(m)}$. Here we refer the readers to [15] for more detailed information about E-JBD.

Although all $A_k^{(m)}$ can be identified via E-JBD, their subspaces are not aligned due to its ignoring the latent inter-dependence across different sets. Rather, we select the most accurate loading matrix using the fitting error:

$$\xi^{(m)} = \|(\mathcal{X}^{(m,m)} - \tilde{\mathcal{X}}^{(m,m)})\|_{\mathsf{F}}^2 / \|(\mathcal{X}^{(m,m)})\|_{\mathsf{F}}^2, \tag{11}$$

where a lower $\xi^{(m)}$ means a more accurate estimation. For convenience, without loss of generality, we assume that $\tilde{A}^{(1)}$ is the selected loading matrix in that exchanging indices will not make any difference.

Now that $\tilde{A}^{(1)}$ is obtained, the next step is to use SVD to identify the remaining mixing matrices. It is worth noting that identifying the $m$th mixing matrix requires the estimates of $\{A^{(1)}, ..., A^{(m-1)}\}$ to be computed and the involvement of $\{\mathcal{X}^{(m,n)}, \mathcal{X}^{(n,m)}, n = 1, ..., m-1\}$. Specifically, we write the mode-2 matrix expression [14] of $\mathcal{X}^{(m,n)}$ as:

$$(\mathcal{X}^{(m,n)})_2 = [(S_1^{(m,n)} \times_1 A_1^{(m)})_2, ..., (S_R^{(m,n)} \times_1 A_R^{(m)})_2] A^{(n)H}, \tag{12}$$

and the mode-1 matrix expression [14] of $\mathcal{X}^{(n,m)}$ as:

$$(\mathcal{X}^{(n,m)})_1 = [(S_1^{(n,m)} \times_2 A_1^{(m)*})_1, ..., (S_R^{(n,m)} \times_2 A_R^{(m)*})_1] A^{(n)T}. \tag{13}$$

In (12) and (13), $(S_r^{(m,n)} \times_1 A_r^{(m)})_2$ and $(S_r^{(n,m)*} \times_2 A_r^{(m)})_1$ are the $r$th sub-matrices of $(\mathcal{X}^{(m,n)})_2 (A^{(n)H})^{-1}$ and $(\mathcal{X}^{(n,m)*})_1 (A^{(n)H})^{-1}$, respectively. We post-multiply (12) by $(A^{(n)H})^{-1}$ such that

$$(\mathcal{X}^{(m,n)})_2 (A^{(n)H})^{-1} = [(S_1^{(m,n)} \times_1 A_1^{(m)})_2, ..., (S_R^{(m,n)} \times_1 A_R^{(m)})_2]. \tag{14}$$

We post-multiply (13) by $(A^{(n)T})^{-1}$ and impose conjugate on its both sides. Consequently, (14) can be further expressed as:

$$(\boldsymbol{\mathcal{X}}^{(n,m)*})_1(A^{(n)\mathsf{H}})^{-1} = [(\boldsymbol{\mathcal{S}}_1^{(n,m)*} \times_2 A_1^{(m)})_1, ..., (\boldsymbol{\mathcal{S}}_R^{(n,m)*} \times_2 A_R^{(m)})_1]. \quad (15)$$

Given the three matrix expressions of a sub-tensor, to bare $A_r^{(m)}$, we convert $(\boldsymbol{\mathcal{S}}_r^{(m,n)} \times_1 A_r^{(m)})_2$ and $(\boldsymbol{\mathcal{S}}_r^{(m,n)} \times_1 A_r^{(m)})_1$ into $(\boldsymbol{\mathcal{S}}_r^{(n,m)*} \times_2 A_r^{(m)})_1$ and $(\boldsymbol{\mathcal{S}}_r^{(n,m)*} \times_2 A_r^{(m)})_2$, respectively. Equivalently, we write the transformed results as $(\boldsymbol{\mathcal{S}}_r^{(m,n)})_1 \times A_r^{(m)\mathsf{T}}$ and $(\boldsymbol{\mathcal{S}}_r^{(n,m)*})_2 \times A_r^{(m)\mathsf{T}}$, respectively. We then concatenate $\{(\boldsymbol{\mathcal{S}}_r^{(m,n)})_1 \times A_r^{(m)\mathsf{T}}, (\boldsymbol{\mathcal{S}}_r^{(n,m)*})_2 \times A_r^{(m)\mathsf{T}}, n = 1, ..., m-1\}$ as follows:

$$\boldsymbol{G}_r^{(m)} = \begin{bmatrix} (\boldsymbol{\mathcal{S}}_r^{(m,1)})_1 \times A_r^{(m)\mathsf{T}} \\ \vdots \\ (\boldsymbol{\mathcal{S}}_r^{(m,n)})_1 \times A_r^{(m)\mathsf{T}} \\ (\boldsymbol{\mathcal{S}}_r^{(1,m)*})_2 \times A_r^{(m)\mathsf{T}} \\ \vdots \\ (\boldsymbol{\mathcal{S}}_r^{(n,m)*})_2 \times A_r^{(m)\mathsf{T}} \end{bmatrix} = \begin{bmatrix} (\boldsymbol{\mathcal{S}}_r^{(m,1)})_1 \\ \vdots \\ (\boldsymbol{\mathcal{S}}_r^{(m,n)})_1 \\ (\boldsymbol{\mathcal{S}}_r^{(1,m)*})_2 \\ \vdots \\ (\boldsymbol{\mathcal{S}}_r^{(n,m)*})_2 \end{bmatrix} \times A_r^{(m)\mathsf{T}}, \quad (16)$$

where $\boldsymbol{G}_r^{(m)}$ is of size $\sum_{n=1}^{m-1} 2KL_n \times I^{(m)}$. The subspace of $A_r^{(m)}$ can be obtained via SVD of $\boldsymbol{G}_r^{(m)}$, consisting of the first $L_m$ right singular vectors. Varying $r$ and $m$, we finally identify all subspaces of $\{A_r^{(m)}, r = 1, ..., R, m = 2, ..., M\}$.

## 4  Experiment Results

The experimental setup is as follows: $M = 2$, $I^{(m)} = 10$ $\forall m$, $R = 2$, $T = 160000$, and $L_r^{(m)} = 2$ $\forall r, m$. We generate the source signal $\boldsymbol{S}^{(m)} = [\boldsymbol{S}_1^{(m)\mathsf{T}}, ..., \boldsymbol{S}_R^{(m)\mathsf{T}}]^\mathsf{T}$ of size $L^{(m)} \times T$ using the short-time non-stationary BPSK signals, whose value at instant $t$ is chosen from either $1 + i$ or $1 - i$ with identical possibility. The correlation can be introduced as:

$$\boldsymbol{S}_r(t) = \boldsymbol{H}_r[\boldsymbol{S}_r^{(1)\mathsf{T}}, ..., \boldsymbol{S}_r^{(M)\mathsf{T}}]^\mathsf{T}, 1 \leq r \leq R, \quad (17)$$

in which $\boldsymbol{H}_r \in \mathbb{C}^{\sum_{m=1}^M L_m \times \sum_{m=1}^M L_m}$ is an invertible matrix following Gaussian normal distribution with zero mean and unit variance. We fix overlapping rate $\lambda$ and time-frame to 0.5 and 16000, respectively. The real and imaginary parts of the mixing matrices are drawn from Gaussian normal distribution with zero mean and unit variance. As such, noise-free coupled tensors $\{\boldsymbol{\mathcal{X}}^{(m,n)}, m, n = 1, ..., M\}$ can be generated via Eq. (4)–(7). We add Gaussian noise term $\boldsymbol{\mathcal{N}}^{(m,n)}$ to each $\boldsymbol{\mathcal{X}}^{(m,n)}$ as follows:

$$\boldsymbol{\mathcal{T}}^{(m,n)} = \sigma_s \boldsymbol{\mathcal{X}}^{(m,n)} / \|\boldsymbol{\mathcal{X}}^{(m,n)}\|_\mathsf{F}^2 + \sigma_n \boldsymbol{\mathcal{N}}^{(m,n)} / \|\boldsymbol{\mathcal{N}}^{(m,n)}\|_\mathsf{F}^2, \quad (18)$$

where $\sigma_s$ and $\sigma_n$ denote the signal and noise level, respectively. The signal-to-noise ratio is defined as SNR$= 20 \log_{10}(\sigma_s/\sigma_n)$.

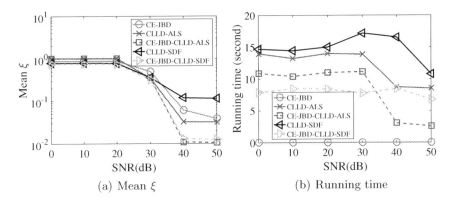

(a) Mean $\xi$          (b) Running time

**Fig. 3.** Comparison of CE-JBD, CLLD-ALS, CLLD-SDF, CE-JBD-CLLD-ALS, and CE-JBD-CLLD-SDF for SNR varying from 0 dB to 50 dB.

There are three involved algorithms, CE-JBD, CLLD-ALS (of which the details will be presented in our future works), and CLLD-SDF [11]. The latter two are iterative algorithms. For CLLD-ALS, it alternately updates each $\boldsymbol{A}^{(m)}$ and $\boldsymbol{\mathcal{S}}^{(m,n)}$, randomly generated at the beginning, to decrease the fitting error:

$$\xi = \sum\nolimits_{m,n=1}^{M}(\|\boldsymbol{T}^{(m,n)} - \tilde{\boldsymbol{T}}^{(m,n)}\|_{\mathsf{F}}^2/\|\boldsymbol{T}^{(m,n)}\|_{\mathsf{F}}^2)/M^2, \qquad (19)$$

where $\xi$ indicates the accuracy of obtained estimates, and a lower $\xi$ means a more successful estimation. During each update, the updating factor matrix is regarded as unknown, denoted as $(\cdot)_{new}$, but the others are fixed, denoted as $(\cdot)_{old}$. In the $i$th iteration, CLLD-ALS sends $\{(\boldsymbol{A}^{(n)})_{old}, 1 \le n \ne m \le M\}$ and $\{X^{(m,n)}, X^{(n,m)}, 1 \le n \ne m \le M\}$ into (12) and (13), respectively, and uses the same technique as (16) to compute $(\boldsymbol{A}^{(m)})_{new}$. We terminate CLLD-ALS when either $|\xi_{new} - \xi_{old}|/\xi_{old} \le 10^{-8}$ or the maximum iteration number $N = 1000$ is satisfied. For CLLD-SDF, we set the tolerance parameters TolFun and TolX in the '*SDF_NLS*' function of Tensorlab [17] to 0.01 and $10^{-5}$, respectively.

We do 200 independent runs of CE-JBD, CLLD-ALS and CLLD-SDF. For comparison, we also initialize CLLD-ALS and CE-JBD with CE-JBD (CE-JBD-CLLD-ALS and CE-JBD-CLLD-SDF), and independently run them for 200 times. The average results of accuracy and CPU running time against SNR are shown in 3(a) and (b), respectively. We can see that if SNR values is sufficiently high, all of the involved algorithms will accomplish JISA, and that in the whole range of SNR from 0dB to 50dB, CE-JBD is faster than CLLD-ALS and CLLD-SDF. In addition, CE-JBD-CLLD-ALS and CE-JBD-CLLD-SDF have improved accuracy and reduced running time compared to CLLD-ALS and CLLD-SDF, respectively. This signifies that CE-JBD is capable of providing a good starting point for its iterative counterparts.

# 5   Conclusion

In this paper, we propose an algebraic algorithm CE-JBD for JISA. The algorithm performs coupled rank-$(L_m, L_n, \cdot)$ BTD on the constructed tensors, of which the pattern is shown in Subsect. 2.2. More precisely, to compute all loading matrices from the coupled tensors, it first finds a relatively accurate one via GEVD followed by a clustering step, and then resorts to SVD to derive the remaining ones. Experiment results demonstrate that CE-JBD is effective for JISA, and that it can provide a good starting point for its iterative counterparts, causing accuracy improvement and running time reduction.

**Acknowledgments.** This research is funded by national natural science foundation of China (Grant nos. 61331019 and 61379012).

# References

1. Kim, T., Attias, H.T., Lee, S.Y.: Blind source separation exploiting higher-order frequency dependencies. IEEE Trans. Audio Speech Lang. Proc. **15**(1), 70–79 (2006)
2. Li, Y.O., Adali, T., Wang, W.: Joint blind source separation by multiset canonical correlation analysis. IEEE Trans. Signal Proc. **57**(10), 3918–3929 (2009)
3. Correa, N.M., Eichele, T., Adali, T.: Multi-set canonical correlation analysis for the fusion of concurrent single trial ERP and functional MRI. Neuroimage **50**(4), 1438–1445 (2010)
4. Correa, N.M., Adali, T., Li, Y.O.: Canonical correlation analysis for data fusion and group inferences. Signal Proc. Mag. IEEE **27**(4), 39–50 (2010)
5. Chatel-Goldman, J., Congedo, M., Phlypo, R.: Joint BSS as a natural analysis framework for EEG-hyperscanning. In: IEEE International Conference on Acoustics, Speech and Signal Processing, pp. 1212–1216. IEEE Press, Vancouver (2013)
6. Gong, X.F., Wang, X.L., Lin, Q.H.: Generalized non-orthogonal joint diagonalization with LU decomposition and successive rotation. IEEE Trans. Signal Proc. **63**(5), 1322–1334 (2015)
7. Gong, X.F., Lin, Q.H., Cong, F.Y., De Lathauwer, L.: Double coupled canonical polyadic decomposition for joint blind source separation. IEEE Trans. Signal Proc. **66**(13), 3475–3490 (2018)
8. Lahat, D., Jutten, C.: Joint independent subspace analysis using second-order statistics. IEEE Trans. Signal Proc. **64**(18), 4891–4904 (2016)
9. Lahat, D., Jutten, C.: Joint blind source separation of multidimensional components: model and algorithm. In: European Signal Processing Conference, pp. 1417–1421. IEEE Press, Lisbonne (2014)
10. Lahat, D., Jutten, C.: Joint independent subspace analysis: a quasi-Newton algorithm. In: Vincent, E., Yeredor, A., Koldovský, Z., Tichavský, P. (eds.) LVA/ICA 2015. LNCS, vol. 9237, pp. 111–118. Springer, Cham (2015). https://doi.org/10.1007/978-3-319-22482-4_13
11. Sorber, L., Barel, M.V., De Lathauwer, L.: Structured data fusion. IEEE J. Sel. Top. Signal Proc. **9**(4), 586–600 (2015)
12. Gong, X.F., Lin, Q.H., et al.: Coupled rank-$(Lm, Ln, \cdot)$ block term decomposition by coupled block simultaneous generalized schur decomposition. In: International Conference on Acoustics, Speech and Signal Processing, pp. 2554–2558. IEEE Press, Shanghai (2016)

408     J.-X. Yang et al.

13. Lahat, D., Jutten, C.: Joint independent subspace analysis: uniqueness and iden-tifiability. IEEE Trans. Signal Proc. **67**(3), 684–699 (2019)
14. De Lathauwer, L.: Decompositions of a higher-order tensor in block terms-part I: lemmas for partitioned matrices. SIAM J. Matrix Anal. Appl. **30**(3), 1022–1032 (2008)
15. Nion, D.: A tensor framework for nonunitary joint block diagonalization. IEEE Trans. Signal Proc. **59**(10), 4585–4594 (2011)
16. Anderson, T.W.: An Introduction to Multivariate Statistical Analysis, New York (2003)
17. Tensorlab 3.0 - numerical optimization strategies for large-scale constrained and coupled matrix/tensor factorization. https://www.tensorlab.net

# MaxEntropy Pursuit
# Variational Inference

Evgenii Egorov[1], Kirill Neklydov[2,3], Ruslan Kostoev[1],
and Evgeny Burnaev[1(✉)]

[1] Skolkovo Institute of Science and Technology, Moscow, Russia
{e.egorov,r.kostoev,e.burnaev}@skoltech.ru
[2] National Research University Higher School of Economics, Moscow, Russia
k.necludov@gmail.com
[3] Samsung AI Center in Moscow, Moscow, Russia

**Abstract.** One of the core problems in variational inference is a choice of approximate posterior distribution. It is crucial to trade-off between efficient inference with simple families as mean-field models and accuracy of inference. We propose a variant of a greedy approximation of the posterior distribution with tractable base learners. Using Max-Entropy approach, we obtain a well-defined optimization problem. We demonstrate the ability of the method to capture complex multimodal posterior via continual learning setting for neural networks.

**Keywords:** Variational inference · Deep learning ·
Maximum Entropy · Bayesian Inference

## 1  Introduction

The posterior distribution evaluation is the primary challenge in Bayesian model construction. Calculating the exact posterior distribution is intractable, and methods like MCMC while being flexible can also be unacceptably expensive. In turn, the variational inference is a method to approximate complicated probability distributions with the simpler ones. Now variational inference is used in semi-supervised classification, drives the most realistic generative models of images, and is a useful tool for analysis of any dynamical system. Inference requires that intractable posterior distributions be approximated by a class of known probability distributions, over which we search for the best representative of the chosen family.

We study the problem of the posterior approximation by a sequentially fitting composition of simple distributions given that one can turn the considered problem to the tractable optimization problem. The structure of the resulting model makes the work with the posterior approximation efficient.

The rest of the paper is organized in the following way. In Sect. 2, we review the variation inference framework. In Sect. 3, we derive the stochastic optimization algorithm for sequential approximation of posterior distribution, named

© Springer Nature Switzerland AG 2019
H. Lu et al. (Eds.): ISNN 2019, LNCS 11554, pp. 409–417, 2019.
https://doi.org/10.1007/978-3-030-22796-8_43

MaxEntropy Pursuit Variational Inference. In Sect. 4, we apply the proposed approach to incremental learning of neural networks. In Sect. 5, we discuss the obtained results and future work.

**Notations.** We denote: the differential entropy of distribution $h$ by $\mathcal{H}[h] := -\int h \log h d\theta$; the inner product between two Lebesgue integrable functions by $\langle f_1, f_2 \rangle := \int f_1 f_2 d\theta$; the full likelihood of the probabilistic model over the dataset $X$ by $L(\theta) := p(X|\theta)p(\theta)$; the posterior distribution by $p(\theta|X) \propto L(\theta)$.

## 2   Variational Inference

We consider the posterior distribution of latent variables $\theta$ given observations $X$:

$$p(\theta|X) = \frac{p(X,\theta)}{\int p(X,\theta)d\theta}.$$

The integral in the denominator is high dimensional, so the normalization is intractable.

The idea of the Variational Inference is to introduce some variational distribution $q_\lambda(\theta)$, and instead of computing the normalization constant we approximate the posterior with the simpler distribution $q$, parametrized by the variational parameter $\lambda$ to get the best matching with $p$.

One of the most common approaches to evaluate proximity between $p$ and $q$ is to use KL-divergence (also known as relative entropy or information gain):

$$D_{KL}(q(\theta)||p(\theta)) = -\int q(\theta) \log \frac{p(\theta)}{q(\theta)} d\theta.$$

KL-divergence is asymmetric $(D_{KL}(q||p) \neq D_{KL}(p||q))$, non-negative and equals to zero iff $q(\theta) = p(\theta)$.

KL-divergence asymmetry provides two different approximation methods: variational inference and expectation propagation (not reviewed in this paper).

Reducing KL-divergence to zero leads to exact matching of distributions, but usually, the variational family $q \in Q$ is not flexible enough for this.

We can formulate minimization of KL-divergence in another way:

$$\log p(X) = \log \int p(X,\theta)d\theta = \log \int \frac{p(X,\theta)q_\lambda(\theta)}{q_\lambda(\theta)} d\theta$$

$$= \log \mathbb{E}_{q_\lambda(\theta)} \left[ \frac{p(X,\theta)}{q_\lambda(\theta)} \right] \geq \mathbb{E}_{q_\lambda(\theta)} \left[ \log \frac{p(X,\theta)}{q_\lambda(\theta)} \right] = \mathcal{F}[q] =: ELBO.$$

ELBO (Evidence Lower Bound) with the KL divergence between the variational distribution and the posterior form the true log marginal probability of the data:

$$\log p(X) = \mathcal{F}(\lambda) + D_{KL}(q_\lambda(\theta)||p(X,\theta)),$$

so the minimization of KL-divergence is equivalent to the maximization of ELBO.

However, optimizing over a parametric variational family of distributions and getting the optimal solution $q^* = \arg\max_{\mathcal{Q}_\lambda} \mathcal{F}[q]$ still leads to the approximation gap [9], equal to $\log p(X) - \mathcal{F}[q^*]$. Many papers showed that the choice of the variational family $\mathcal{Q}_\lambda$ is important for quality of the variational approximation [1, 23, 25].

There are a number of approaches for reducing the approximation gap. Some of them propose to increase the flexibility of the approximation family, e.g. normalizing flows [22] or hierarchical variational models [21]. The other research direction explores the idea of incrementally expanding variational family by the additive mixture of tractable base learners [11, 18]. In [17] they investigate the theoretical justification of such approach from an optimization perspective. In general, the both approaches are able to capture the multimodality and nonstandard posterior shapes. However, it seems that the incremental learning of the posterior approximation is more promising from the applied point of view, as the additive mixture composes the approximation using simple and easy-to-evaluate building blocks.

Here we address several problems with this approach. Firstly, starting from the Maximum Entropy principle [8], we obtain a natural regularized optimization problem, instead of the ad-hoc regularization, proposed in other works. This leads to interesting connections with other fields and allows to use stochastic optimization approaches in contrast to the original boosting approach [11]. We show the ability of the proposed approach to approximate complex posteriors by using Bayesian Neural Networks, which is a data-intensive and challenging task [28].

# 3    Max Entropy Pursuit Variational Inference

In this section we derive algorithm in which problem of the posterior distribution is solved by additive mixture. Each component is obtained sequentially. Each step consists of the two optimization problems: for new component $h$ and for the corresponding mixture weight $\alpha$.

## 3.1    Optimization over New Component $h$

Consider that we given some approximation of the posterior distribution $q_t$. Our goal is to improve accuracy of the approximation in terms of the KL-divergence $D_{KL}[q_t(\theta)||p(\theta|X)]$ by using the additive mixture:

$$q_{t+1} = (1 - \alpha)q_t + \alpha h, \ \alpha \in (0; 1), \ h \in Q.$$

Hence, using Maximum Entropy Approach [8] we can state the following optimization problem:

$$\max_{h \in \mathcal{Q}} \mathcal{H}[h], s.t.$$
$$\mathcal{F}[q_{t+1}] - \mathcal{F}[q_t] > 0. \tag{1}$$

As the optimization problem in Eq. (1) is highly non-linear, we propose to follow the framework based on the Frank-Wolfe algorithm [17,26] and consider the constraint as a functional perturbation.

Expanding the $\mathcal{F}[q_{t+1}]$ term, we get

$$
\mathcal{F}[q_{t+1}] = \int [q_t + \alpha(h - q_t)]\left(\log\frac{L(\theta)}{q_t} - \log\left(1 + \alpha\frac{h - q_t}{q_t}\right)\right) d\theta
$$

$$
= \underbrace{\int q_t \log\frac{L(\theta)}{q_t} d\theta}_{\mathcal{F}[q_t]} + \alpha\int (h - q_t)\left(\log\frac{L(\theta)}{q_t} - \log\left[1 + \alpha\frac{h - q_t}{q_t}\right]\right) d\theta
$$

$$
- \int q_t \log\left(1 + \alpha\frac{h - q_t}{q_t}\right) d\theta.
$$

Using Taylor expansion, we obtain the constraint in the following form:

$$
\mathcal{F}[q_{t+1}] - \mathcal{F}[q_t] = \alpha\left\langle h - q_t, \log\frac{L(\theta)}{q_t}\right\rangle - \alpha^2\int\frac{(h - q_t)^2}{q_t} d\theta + o\left(\alpha\left\|\frac{h - q_t}{q_t}\right\|_2\right).
$$

Considering the first order terms, we get the following optimization problem:

$$
\max_{h \in Q} \mathcal{H}[h] + \lambda\left\langle h, \log\frac{L(\theta)}{q_t}\right\rangle. \tag{2}
$$

We can perform scalable optimization by the doubly stochastic gradient descent [14,24]. The $\lambda > 0$ is the corresponding Lagrange multiplier of the constraint. Exact solution of the dual problem for the optimal $\lambda$ is intractable. Below we provide some analysis of how the solution depends on $\lambda$. It allows us to propose practically useful heuristic to select a value of $\lambda$.

Note, that retaining only the first order terms corresponds to the "functional gradient" of the KL-divergence [11]. However, MaxEntropy approach allows obtaining the natural regularization term. Further, we show that it is critical to obtain a data scalable algorithm and interpret the parameter $\lambda$. Also, in Sect. 4 we discuss whether the first order terms expansion is enough for high dimensional problems.

### 3.2    Analysis of Optimization Problem for $h$

To provide the heuristic rule of choosing the $\lambda$, we optimize in Eq. (2) not over some parametric family $Q$ of base learners $h$, but over all probability densities. As the objective is concave over $h$, we can derive the global optimal of the maximization problem from the first-order conditions:

$$
\frac{\delta}{\delta h}\left[\mathcal{H}[h] + \lambda\left\langle h, \log\frac{L(\theta)}{q_t}\right\rangle\right] + \gamma\left(\int h d\theta - 1\right) = 0,
$$

$$
h^* = \left[\frac{L(\theta)}{q_t}\right]^\lambda \exp(\gamma - 1).
$$

Hence, the optimal new component has the following form:

$$h^* = \frac{1}{Z(\lambda)} \left[ \frac{L(\theta)}{q_t} \right]^\lambda . \tag{3}$$

The solution $h^*$ is intractable, as finding the normalization constant $Z = \int \left[ \frac{L(\theta)}{q_t} \right]^\lambda d\theta$ has the same complexity as solving the original problem. Still, as the global optimum is known, instead of the optimization problem in Eq. (2) we can consider another optimization problem:

$$\min_{h \in Q} D_{KL} \left( h \middle\| \frac{1}{Z(\lambda)} \left[ \frac{L(\theta)}{q_t} \right]^\lambda \right) . \tag{4}$$

The problem (4) is a well-known optimization problem for which there are a lot of black-box variational inference (BBVI) solvers, see e.g. [10,20]. Hence, any practitioner can benefit from our approach without additional significant costs of implementing or reformulating the initial statistical problem. Moreover, we could provide intuition for selecting $\lambda$ by establishing a connection with Renyi divergence [16] thanks to the analyses of the form of (3). Namely, we consider a parametric mapping in the probability density space:

$$T_\lambda : p \to \frac{p^\lambda(\theta)}{\int p^\lambda(\theta)d\theta}, \quad \lambda > 0. \tag{5}$$

Consider a pair of a uniform distribution $U$ and $p : \mathcal{H}[p] > \mathcal{H}[U]$. We can easily prove that

$$D_{KL}(U\|p) > D_{KL}(U\|T_\lambda p), \quad \text{for } \lambda > 1,$$
$$D_{KL}(U\|p) < D_{KL}(U\|T_\lambda p), \quad \text{for } \lambda < 1. \tag{6}$$

Hence, we can state that for $\lambda > 1$ we obtain a mode-seeking solution and for $\lambda < 1$ we get a mass covering solution. Interestingly, in case of the Renyi divergence optimization in [19] they describe the same behavior for different values of $\alpha$. Hence, we can refer to $\lambda$ as the temperature and select some annealing schedule for each step of the optimization process to tune $\lambda$.

Let us consider the corner case, i.e. $\lambda = 1$. Then we can rewrite the objective in (2):

$$\arg\max_{h \in Q} \mathcal{H}[h] + \left\langle h, \log \frac{L(\theta)}{q_t} \right\rangle = \arg\max_{h \in Q} \underbrace{\int h \log \frac{L(\theta)}{h} d\theta}_{\text{term (1)}} - \underbrace{\int h \log q_t d\theta}_{\text{term (2)}} .$$

$$\tag{7}$$

Hence, the term 1 in (7) corresponds to the standard optimization objective in case of variational inference [12]. At the same time the term 2 in (7) plays a role of a penalty for the similarity with the current solution $q_t$.

### 3.3    Optimization over Mixture Weight $\alpha$ Corresponding to $h$

After we obtain the new mixture component $h$ for the current variational approximation $q_t$, we should select the mixture weight $\alpha$ to obtain a new variational approximation as a convex combination:

$$q_{t+1}(\theta) = (1 - \alpha)q_t(\theta) + \alpha h(\theta).$$

Hence, let us state the optimization problem over $\alpha \in (0; 1)$:

$$\min_{\alpha \in (0;1)} D_{KL}((1 - \alpha)q_t(\theta) + \alpha h(\theta)||p(\theta|X)). \tag{8}$$

Using Taylor expansion we can get the approximation for any $f$-divergence [27] by the Pearson Chi-squared divergence:

$$D_f(q||p) \approx f''(1)\chi^2(q||p).$$

Hence, we can re-formulate the approximation problem:

$$\min_{\alpha \in (0;1)} \int \frac{1}{p(\theta|X)} [q_t + \alpha(h - q_t)]^2 d\theta. \tag{9}$$

Consider the gradient and the hessian of the objective in (9) w.r.t. $\alpha$:

$$\nabla_\alpha \int \frac{1}{p}[q_t + \alpha(h - q_t)]^2 d\theta = 2 \int \frac{1}{p(\theta|X)}[q_t + \alpha(h - q_t)](h - q_t)d\theta,$$

$$\nabla_\alpha^2 \int \frac{1}{p(\theta|X)}[q_t + \alpha(h - q_t)]^2 d\theta = 2 \int \frac{(h - q_t)^2}{p(\theta|X)} > 0.$$

As the objective (9) is convex, we can obtain the solution of the optimization problem (9) from the first order condition:

$$\alpha^* = -\frac{\int \frac{1}{p(\theta|X)} q_t(h - q_t)d\theta}{\int \frac{1}{p(\theta|X)}(h - q_t)^2 d\theta} = -\frac{\int \frac{1}{L(\theta)} q_t(h - q_t)d\theta}{\int \frac{1}{L(\theta)}(h - q_t)^2 d\theta}. \tag{10}$$

In practice such estimator has high variance. Estimation for each sample requires the forward pass through the whole dataset, hence the variance can not be reduced by averaging efficiently. Therefore we propose to use the exact solution (10) in case of middle-size datasets and use the stochastic gradient approach with a projection for the objective from Eq. (8) in case of large-scale datasets.

## 4    Neural Network Incremental Learning via Bayesian Inference

Deep neural networks provide the state-of-art solution for the image classification problems. However, as a network is trained to do a specific classification task, it is problematic to incrementally learn any new task. This situation was described as the catastrophically forgetting behaviour of neural networks. However, intuitively we expect the other situation: performance should similar to that when training over the whole dataset in the offline mode [13]. In this section, we show how our approach helps to overcome this limitation.

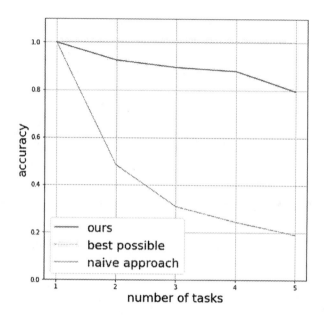

**Fig. 1.** Mean test accuracy for a sequence of models, trained on a sequence of tasks (subsets of the training set). Each next task is equal to a previous task plus some new subset of the initial training data.

**Experimental Setup.** We perform the incremental class learning experiment using the MNIST dataset with the LeNet-5 Convolutional Neural Network (CNN) [15]. The dataset contains grey scale images belonging to 10 classes. We split the dataset in 5 tasks, the first task containing digits '0' and '1', the second task containing digits '2' and '3', and so on. For each task, we perform 10 epoch of training. We compare our incremental posterior approximation of the neural network parameters with a baseline naive continual neural network learning. The size of the test dataset is $10^4$ samples, the total train size for all tasks is $5 \times 10^4$. As the prior distribution on the neural network parameters we use the fully factorized standard normal distribution. The predictive distribution of the model is approximated by an ensemble of the weights sampled from the variational approximation.

**Results.** As result, we find our incremental posterior distribution approximation to maintain higher test accuracy through the whole sequence of tasks, almost matching the performance of a network trained simultaneously on all observed data. Figure 1 shows the test accuracy as new tasks are observed. We conclude from our results that the incremental posterior approximation leads to a drastic increase in performance for incremental learning tasks.

## 5   Conclusion

In this work, we developed an efficient approach for learning complex multimodal posteriors by constructing an additive mixture of simple densities. Following the MaxEntropy approach, we state well defined and tractable optimization problem. Additive mixture allows us to control the complexity of the posterior by simply increasing or decreasing the number of components.

An important avenue of future research is to develop approaches for modeling covariance structure that accurately account for different characteristics of the posterior and that still allow for efficient computations in case of deep neural networks.

Also, we plan to consider various applications of the proposed approximation scheme including uncertainty quantification [2–4] and Bayesian parameter estimation for Gaussian Processes regression [5–7].

**Acknowledgements.** The work was supported by the Russian Science Foundation under Grant 19-41-04109.

## References

1. Burda, Y., Grosse, R., Salakhutdinov, R.: Importance weighted autoencoders. arXiv preprint arXiv:1509.00519 (2015)
2. Burnaev, E., Panin, I.: Adaptive design of experiments for Sobol indices estimation based on quadratic metamodel. In: Gammerman, A., Vovk, V., Papadopoulos, H. (eds.) SLDS 2015. LNCS (LNAI), vol. 9047, pp. 86–95. Springer, Cham (2015). https://doi.org/10.1007/978-3-319-17091-6_4
3. Burnaev, E., Panin, I., Sudret, B.: Effective design for Sobol indices estimation based on polynomial chaos expansions. In: Gammerman, A., Luo, Z., Vega, J., Vovk, V. (eds.) COPA 2016. LNCS (LNAI), vol. 9653, pp. 165–184. Springer, Cham (2016). https://doi.org/10.1007/978-3-319-33395-3_12
4. Burnaev, E., Panin, I., Sudret, B.: Efficient design of experiments for sensitivity analysis based on polynomial chaos expansions. Ann. Math. Artif. Intell. **81**(1), 187–207 (2017)
5. Burnaev, E., Zaytsev, A., Spokoiny, V.: Properties of the posterior distribution of a regression model based on Gaussian random fields. Autom. Remote Control **74**(10), 1645–1655 (2013)
6. Burnaev, E., Zaytsev, A., Spokoiny, V.: The Bernstein-von Mises theorem for regression based on Gaussian processes. Russ. Math. Surv. **68**(5), 954–956 (2013)
7. Burnaev, E., Zaytsev, A., Spokoiny, V.: Properties of the Bayesian parameter estimation of a regression based on Gaussian processes. J. Math. Sci. **203**(6), 789–798 (2014)

8. Caticha, A.: Relative entropy and inductive inference. In: AIP Conference Proceedings, vol. 707, pp. 75–96 (2004)

9. Cremer, C., Li, X., Duvenaud, D.: Inference suboptimality in variational autoencoders. In: International Conference on Machine Learning, pp. 1086–1094 (2018)

10. Duvenaud, D., Adams, R.P.: Black-box stochastic variational inference in five lines of python. In: NIPS Workshop on Black-box Learning and Inference (2015)

11. Guo, F., Wang, X., Fan, K., Broderick, T., Dunson, D.B.: Boosting variational inference. arXiv preprint arXiv:1611.05559 (2016)

12. Hoffman, M.D., Blei, D.M., Wang, C., Paisley, J.: Stochastic variational inference. J. Mach. Learn. Res. **14**(1), 1303–1347 (2013)

13. Kemker, R., McClure, M., Abitino, A., Hayes, T.L., Kanan, C.: Measuring catastrophic forgetting in neural networks. In: Thirty-Second AAAI Conference on Artificial Intelligence (2018)

14. Kingma, D.P., Welling, M.: Auto-encoding variational bayes. arXiv preprint arXiv:1312.6114 (2013)

15. LeCun, Y., Bottou, L., Bengio, Y., Haffner, P., et al.: Gradient-based learning applied to document recognition. Proc. IEEE **86**(11), 2278–2324 (1998)

16. Li, Y., Turner, R.E.: Rényi divergence variational inference. In: Advances in Neural Information Processing Systems, pp. 1073–1081 (2016)

17. Locatello, F., Khanna, R., Ghosh, J., Ratsch, G.: Boosting variational inference: an optimization perspective. In: International Conference on Artificial Intelligence and Statistics, pp. 464–472 (2018)

18. Miller, A.C., Foti, N.J., Adams, R.P.: Variational boosting: iteratively refining posterior approximations. In: Proceedings of the 34th International Conference on Machine Learning-Volume 70, pp. 2420–2429. JMLR. org (2017)

19. Minka, T., et al.: Divergence measures and message passing. Technical report, Microsoft Research (2005)

20. Ranganath, R., Gerrish, S., Blei, D.: Black box variational inference. In: Artificial Intelligence and Statistics, pp. 814–822 (2014)

21. Ranganath, R., Tran, D., Blei, D.: Hierarchical variational models. In: International Conference on Machine Learning, pp. 324–333 (2016)

22. Rezende, D., Mohamed, S.: Variational inference with normalizing flows. In: International Conference on Machine Learning, pp. 1530–1538 (2015)

23. Salimans, T., Kingma, D.P., Welling, M., et al.: Markov chain Monte Carlo and variational inference: bridging the gap. In: ICML, vol. 37, pp. 1218–1226 (2015)

24. Titsias, M., Lázaro-Gredilla, M.: Doubly stochastic variational Bayes for non-conjugate inference. In: International Conference on Machine Learning, pp. 1971–1979 (2014)

25. Tran, D., Ranganath, R., Blei, D.M.: The variational Gaussian process. arXiv preprint arXiv:1511.06499 (2015)

26. Wang, C., Wang, Y., Schapire, R., et al.: Functional Frank-Wolfe boosting for general loss functions. arXiv preprint arXiv:1510.02558 (2015)

27. Wang, D., Liu, H., Liu, Q.: Variational inference with tail-adaptive f-divergence. In: Advances in Neural Information Processing Systems, pp. 5742–5752 (2018)

28. Welling, M., Teh, Y.W.: Bayesian learning via stochastic gradient Langevin dynamics. In: Proceedings of the 28th International Conference on Machine Learning, ICML 2011, pp. 681–688 (2011)

# Moderated Information Sets

Manish Aggarwal[1($\boxtimes$)] and Madasu Hanmandlu[2]

[1] IIM Ahmedabad, Ahmedabad, India
magwal5@gmail.com
[2] MVSR Engineering College, Hyderabad 501510, India
mhmandlu@gmail.com

**Abstract.** In this chapter, the information set theory is extended with
a moderator's information about the elements of the information set.
The role of moderator is to provide a second opinion about the validity
of the information values comprising the information set. The properties
of the proposed moderator information sets (MIS) are also investigated.
Many illustrative examples are included to show the usefulness of MIS
in the real world decision-making.

**Keywords:** Information set · Moderator · Vagueness ·
Multi attribute decision-making

## 1 Introduction

A fuzzy set [1] represents in-exactness, ill-definedness in an information source
value. However, it has its own difficulties [2]. Foremost among them is the interpretation difficulties involved with a membership function. That is, the interpretation of a membership grade is specific to the onlooker agent. For instance,
a membership grade of 0.8 in the fuzzy set *very tall* conveys little information
about the actual height of the person. Also, it means a different value for each
interpreting agent. Inspired by these drawbacks of a membership function, the
information set is proposed in [2]. It basically gives a set of perceived values
specific to the agent, which can be seen as a combination of the information
source value and its evaluation by the agent.

The perceived values are in fact the entropy values that are referred to as
*information value*. These entropy values are easier to interpret and deal with,
as shown in [2]. Since, an information source value is looked upon differently
by different agents, a function to represent the same should be general enough
to represent multiple forms, a few of which could be Gaussian, triangular, or
trapezoidal etc.

Besides, it is especially useful in representing vague or subjective assessment of the decision-maker (DM), in multi criteria decision making (MCDM).
However, when the values are ill-defined (or vague), the agent inevitably faces
confusion. That is, often (s)he may have multiple possible perceptions for an
information source value. This difficulty is addressed in [5], where information

© Springer Nature Switzerland AG 2019
H. Lu et al. (Eds.): ISNN 2019, LNCS 11554, pp. 418–425, 2019.
https://doi.org/10.1007/978-3-030-22796-8_44

set has been extended as hesitant information set (HIS), inspired by the concept of hesitant fuzzy set. A HIS comprises all the perceived values that an agent may have for an information source value.

HIS provides an excellent data structure to take into consideration the hesitancy of the agent in the representation of the perceived values, also referred to as the information values. However, hesitancy is a feature of one's own perception, and it is not uncommon to have the actual information misinterpreted during its presentation. This might happen because of the limited domain knowledge of the agent or due to the lack of the standard terminologies. For instance, a patient may try to convey his/her severity of symptoms in different ways, and the interpreted severity may be different from that intended to be conveyed. This is sought to be corrected by the proposed work.

In this chapter, we intend to validate the information values of an information set through a second opinion about them. We accomplish this objective by taking a moderator's opinion in the form of a partial degree that (partially/fully) endorses the original information values provided by the agent. The partial degree indicates the extent of the moderator's agreement with the information values provided by the agent. We term the proposed data structure as the moderated information set (MIS).

The framework of MIS has the moderator's input indicating the credibility of the information values in an information set. The moderator's information thus provides a kind of check-point for an agent to relook at the original information values of the information set. If the degree given by the moderator is greater than 0.5, it indicates that the information values are reasonably credible depending on how close the moderator's degree is to 1. In contrast, this degree less than 0.5 indicates a level of disagreement between the moderator and the agent.

The proposed MIS attempts to minimize the possibility of potential errors (due to the agent's lack of judgment) by taking account of the moderator's opinion on the same. For example, a patient may tend to misrepresent the actual severity of the symptoms, due to which the agent (the specialist doctor) with limited time and going by the patient's severity values may arrive at an incorrect diagnosis. It may be more judicious to have a junior doctor as the *moderator* to moderate the severity of the symptoms.

With this motivation, we extend the concept of information set with another parameter to represent the degree of the moderator's agreement with the information values provided by the agent. The data structure, so formed, is termed as moderated information set. The rest of the chapter is organized as follows. Section 2 gives the background for the study. In Sect. 3, we introduce the concept of moderated information set. Section 4 gives the conclusions.

## 2 Background

### 2.1 Information Set

In multi attribute decision making (MADM), it is the perceived value that is more at play rather than the actual attribute value. In this regard, the concept of information set provides an interesting formalism to model the perceived values. An information set is a set of entropy values, each of is an information value as perceived by the agent. These perceived values correspond to the given information source values for a particular entity, and are computed through the Hanman-Anirban entropy function [4].

In the context of MADM, let us consider a set of objects $U = \{u_1, \ldots, u_m\}$ that we call as the reference set. Each of the $u_i$s is described by multiple attributes, denoted by $E$. An *information source value* $I_e(u_i)$ denotes the value that attribute $e \in E$ takes for $u_i$. The collection of these information source values is shown as:

$$\mathcal{I}_e = \{I_e(u_i) \mid \forall u_i \in U\}, \tag{1}$$

$\mathcal{I}_e$ in (1) is an information function $\mathcal{I}_e : U \to I_e(u_i)$. For the ease of interpretation, we consider $I_e(u_i)$ to be the normalized in the interval $[0, 1]$.

The set of *normalized* information source values, when put together attribute-wise, generates "soft" classes (concepts). Each of these soft classes corresponds to an attribute $e$, and is denoted as $\mathcal{I}_e$. With this reference, an information set could be seen as a soft class, giving a collection of the agent's perceived information values.

Though the attribute values (information source values) are real crisp values, but they induce have different interpretations for as many individuals. For instance, in the case of MADM, the various values that an attribute takes for the given alternative (option) in the choice set are the information source values. These values for a particular attribute $e$ are denoted by $\mathcal{I}_e$. The information set formalism helps to modify such an actual information source value in light of the agent's specific evaluation of the same to give an *information value*. This evaluation is determined by the Hanman-Anirban entropic gain function [4], performing the role of an *agent*. This gain function is, henceforth, referred to as the *agent*[1]. The perceived value or the information value is nothing but the Hanman-Anirban entropy value.

The information gain function is given as:

$$g_e(u_i) = e^{-\left(a_e(I_e(u_i))^3 + b_e(I_e(u_i))^2 + c_e(I_e(u_i)) + d_e\right)^{\alpha_e}} \tag{2}$$

where $a_e, b_e, c_e, d_e$ and $\alpha_e$ are the adjustable parameters, unique to the given choice set and the attribute $e$, i.e. $\{I_e(u_i)\}, \forall u_i \in U$. These parameters help generate several evaluation functions $g_e(\cdot)$. In this regard, the DM's specific evaluation scheme for a particular attribute is uniquely portrayed by a combination of these parameters. The two moments of distribution of the information source values, i.e. mean and/or standard deviation, could also be one set of the of these parameters. For instance, if we take these parameter set values as: $a_e = 0$, $b_e = 0$, $c_e = \frac{1}{\sigma_X}$, $d_e = -\frac{m_e}{\sigma_e}$ (where $m_e$ refers to the mean of the values of all the values that $e$ takes for the given choice set, i.e. $\frac{1}{n}\sum_i I_e(u_i)$; and $\sigma_e$ refers to the standard deviation of the values in $\mathcal{I}_e$.), then we obtain $g_e(\cdot)$ as a

---

[1] A human agent too perceives his/her environment differently in his/her own way.

generalized Gaussian function, shown as:

$$g_e(u_i) \;=\; e^{-\left(\frac{I_e(u_i)-m_e}{\sigma_e}\right)^{\alpha_e}} \tag{3}$$

where, $\alpha_e$ uniquely determines the DM's evaluation pattern for the attribute $e$. Different shapes of agent, i.e. $g_e(\cdot)$ are obtained for different values of $\alpha$. A few of these shapes obtained with $\alpha_e = 0.5, 1, \ldots, 5$ are shown in Fig. 1. For instance, at $\alpha_e = 2$, a Gaussian function is generated.

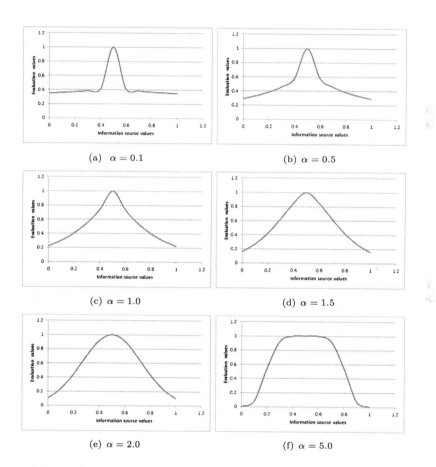

**Fig. 1.** Information gain values of the form $g_e(u_i)$, obtained with the information gain function given in (3).

The uncertainty in the information source values $\{I_e(u_i)\}, \forall u_i \in U$, as evaluated by $g_e(.)$, can be quantified as:

$$Q_e \;=\; \sum_i I_e(u_i) g_e(u_i) \tag{4}$$

where $I_e(u_i)g_e(u_i)$ gives the agent's *information value*, i.e. the perceived value for $I_e(u_i)$. For example, a particular value of an attribute, say *height*, evokes different conceptual perceptions for different agents. For instance, an agent from a geographical region with tall people would have his/her conceptual perceptions about a height value as per his/her own background. If this height value is found to be lesser than his definition of normal, then (s)he may perceive this height as *low*. Similarly a person from a region of short people (like Congo in Africa) may perceive this height to be very tall.

The perceived height is computed as $I_e(u_i)g_e(u_i)$ through the information set formalism, where $e$ refers to height as the attribute, and $I_e(u_i)$ is the specified temperature value.

Information set $\mathcal{S}_e$ consists of these information values of the form shown as:

$$\mathcal{S}_e = \{I_e(u_i)g_e(u_i)\}, \forall u_i \in U \tag{5}$$

Each element of $\mathcal{S}_e$ is expressed as:

$$S_e(u_i) = I_e(u_i)g_e(u_i), \tag{6}$$

which is nothing but the value of the entropy function. The exponential gain function (agent) $g_e(.)$ helps generate several evaluation patterns (agents) through a combination of the parameters. Besides, information set formalism helps to connect the information source values with the corresponding evaluations, specific to the agent. In contrast to this these remain delinked in a fuzzy set.

## 3    Moderated Information Set

The information set is extended as a moderated information set (MIS) by taking the moderator's assessment of the information values provided by the agent. The moderated information values hold a key in representing the information source values in the conceptual framework. Without moderation, the original evaluation remains uncertified in the conventional framework. In this section, we present MIS along with illustrative examples. The properties of MIS are also investigated.

A *moderated information set* $\mathcal{M}$ over soft universe $(U, C)$, is defined by a mapping $F : C \rightarrow \mathcal{S} \times \mathbb{R}$, and $M : C \rightarrow \mathbb{R}$. The moderated information set so formed is expressed as:

$$\mathcal{M} = \{(F(e_i) \mid M(F(e_i)))\}, \forall e_i \in C \tag{7}$$

where $M(F(e_i))$ indicates the moderator's degree of agreement with the information values provided by the agent for an alternative from a set of alternatives, denoted by $C$ $e_i$, and $\mathbf{M} = \{M(F(e_i))\}, \forall e_i \in C$; $F(e_i) \in P(U)$; $M(F(e_i)) \in [0, 1]$. We denote each element of the proposed MIS as:

$$(e_i) = (F(e_i) \mid M(F(e_i))) \tag{8}$$

Each element of the proposed MIS is a tuple of the agent's perceived value along with the moderator's evaluation about the same. This makes it potentially useful in MCDM to take into consideration a second opinion (of the moderator) about the agent's evaluation in terms of the information values.

**Example 3.1.** *Let $U = \{u_1, u_2, u_3, u_4\}$ be a set of patients and the set of symptoms be denoted by $E = \{fever\ (e1),\ loss\text{-}of\text{-}appetite\ (e2)\}$. The conventional information set is then shown as:*

$$\mathcal{S}_i = \{S_{i1}, S_{i2}\}, \tag{9}$$

*where $S_{ij}$ denotes the information value giving the perception of the patient $u_i$ about his/her symptom $e_j$. Let the partial degree given by the moderator about the patient's information values be $M(e_1) = (0.8)$, $M(e_2) = (0.3)$. Then the corresponding moderated information set is:*

$$\mathcal{M} = \{(S_1|\,\mathbf{0.8})\,,\ (S_2|\,\mathbf{0.3})\} \tag{10}$$

The moderator's values are indicated in the bold, and are reflected in the final decision outcome. The main operations on the proposed MIS are given as follows:

**Definition 3.1 (Subset).** *Let $\mathcal{M}^1$ and $\mathcal{M}^2$ be two MISs defined over $(U, C)$. Then $\mathcal{M}^1$ is a subset of $\mathcal{M}^2$, i.e. $\mathcal{M}^1 \subseteq \mathcal{M}^2$ iff*

*i. $M(F^1(e_i)) \leq M(F^2(e_i)),\ \forall e_i \in C$*
*ii. $F^1(e_i) \subseteq F^2(e_i),\ \forall e_i \in C$.*

**Example 3.2.** *Let us relook at $\mathcal{M}$ of Example 3.1. Let $\mathcal{M}^1$ be another MIS. Determine if one of these is a subset of the other. The two MISs are shown as:*

$$\mathcal{M} = \{(S_1\,|\,\mathbf{0.8})\,,\ (S_2\,|\,\mathbf{0.3})\}$$
$$\mathcal{M}^1 = \{(S_1\,|\,\mathbf{0.1})\,,\ (S_2\,|\,\mathbf{0.2})\}$$

*Since, $M(F^1(e_1)) \leq M(F(e_1))$, and $M(F^1(e_3)) \leq M(F(e_3))$; and also $F^1(e_1) \subseteq F(e_1)$, and $F^1(e_3) \subseteq F(e_3)$, therefore $\mathcal{M}^1$ is a subset of $\mathcal{M}$.*

**Definition 3.2 (Complement).** *Let the complement of $\mathcal{M}$ be denoted by $\overline{\mathcal{M}}$, then the complement is defined as*

$$\overline{\mathcal{M}} = \left\{\left(\overline{S}_i\,|\,\overline{M}_i = 1 - M_i\right)\right\},\ \forall e_i \in C \tag{11}$$

The complement of a MIS is thus just opposite to that of the given MIS.

**Example 3.3.** *We reconsider the MIS given in Example 3.1. Its complement is shown as:*

$$\overline{\mathcal{M}} = \{(S_1|\,\mathbf{0.2})\,,\ (S_2|\,\mathbf{0.7})\} \tag{12}$$

**Definition 3.3 (Union).** *The union of $\mathscr{M}^1$ and $\mathscr{M}^2$, denoted by $\mathscr{M}^1 \cup \mathscr{M}^2$, is defined as*

$$\mathscr{M}^1 \cup \mathscr{M}^2 = \left\{\left(S_i^1 \cup S_i^2 \mid M_i^1 \cup M_i^2\right)\right\}, \forall e_i \in C \tag{13}$$

**Definition 3.4 (Intersection).** *The intersection of $\mathscr{M}^1$ and $\mathscr{M}^2$ is given as following:*

$$\mathscr{M}^1 \cap \mathscr{M}^2 = \left\{\left(S_i^1 \cap S_i^2 \mid M_i^1 \cap M_i^2\right)\right\}, \forall e_i \in C \tag{14}$$

**Example 3.4.** *The union and intersection of MISs $\mathscr{M}$ and $\mathscr{M}^1$, as shown in Example 3.2 are presented as following:*

$$\mathscr{M} \cup \mathscr{M}^1 = \left\{\left(S_1 \cup S_1^1 \mid \mathbf{0.8}\right), \left(S_2 \cup S_2^1 \mid \mathbf{0.3}\right)\right\}$$
$$\mathscr{M} \cap \mathscr{M}^1 = \left\{\left(S_1 \cap S_1^1 \mid \mathbf{0.1}\right), \left(S_2 \cap S_2^1 \mid \mathbf{0.2}\right)\right\}$$

**Definition 3.5 (Null Moderated Information Set).** *The null moderated information set defined over $(U, C)$, and denoted as $\mathscr{M}_0$, has all the moderator's assessments for its elements as 0, i.e.:*

$$M_i = 0, \ \forall e_i \in C$$

**Example 3.5.** *The corresponding null moderated information set for MIS considered in Example 3.1 is shown as:*

$$\mathscr{M}_0 = \{(S_1, S_2 \mid \mathbf{0.0})\}$$

**Definition 3.6 (Absolute Confidence Information Set).** *The absolute moderated information set is denoted by $\mathscr{M}_1$, and has the maximum, i.e. 1, moderator's assessments for its elements. It is as following:*

$$M_i = 1, \ \forall e_i \in C$$

**Example 3.6.** *The corresponding absolute moderated information set for the MIS in Example 3.1 is given as:*

$$\mathscr{M}_1 = \{(S_1, S_2 \mid \mathbf{1.0})\}$$

**Definition 3.7. (Empty Moderated Information Set).** *A moderated information set defined over $(U, C)$ is said to be empty if $S = \phi$, and $M = 1$. It is denoted by $\mathscr{M}_\Phi$, and is shown as:*

$$\mathscr{M}_\Phi = \{(\{\} \mid \mathbf{1.0})\} \tag{15}$$

### 3.1   Properties

Let $\mathscr{M}$ be a MIS defined over $(U, C)$, then

(i) $\mathscr{M} \subseteq (\mathscr{M} \cup \mathscr{M})$
(ii) $(\mathscr{M} \cap \mathscr{M}) \subseteq \mathscr{M}$
(iii) $\mathscr{M} \cup \mathscr{M}_0 = \mathscr{M} \cap \mathscr{M}_0 = \mathscr{M}_0$
(iv) $\mathscr{M} \cup \mathscr{M}_1 = \mathscr{M} \cap \mathscr{M}_1 = \mathscr{M}$

# 4   Conclusions

This chapter extends the concept of the recent information set to the moderated information set by introducing a moderator's assessment of the first-hand information values. The usefulness of MIS is demonstrated through several examples. As a future work, the structure of the moderator's input can be tailored to tackle specific decision making problems in different domains. Suitable methods can be thought of to improve upon the original evaluations by the agent, in light of the moderator's inputs. The refined evaluations, so obtained, would be much more useful for arriving at an accurate decision.

# References

1. Zadeh, L.A.: Fuzzy sets. Inf. Control **8**(3), 338–353 (1965)
2. Aggarwal, M., Hanmandlu, M.: Representing uncertainty with information sets. IEEE Trans. Fuzzy Syst. **24**(1), 1–15 (2016)
3. Zimmerman, H.J.: Using fuzzy sets in operational research. Eur. J. Oper. Res. **13**(3), 201–216 (1983)
4. Hanmandlu, M., Das, A.: Content-based image retrieval by information theoretic measure. Defence Sci. J. **61**(5), 415–430 (2011)
5. Aggarwal, M.: Hesitant information sets and application in group decision making. Appl. Soft Comput. **75**, 120–129 (2019)

# Coherence Regularization for Neural Topic Models

Katsiaryna Krasnashchok$^{(\boxtimes)}$ and Aymen Cherif

EURA NOVA, Rue Emilie Francqui 4, 1435 Mont-Saint-Guibert, Belgium
katherine.krasnoschok@euranova.eu
http://euranova.eu

**Abstract.** Neural topic models aim to predict the words of a document given the document itself. In such models perplexity is used as a training criterion, whereas the final quality measure is topic coherence. In this work we introduce a coherence regularization loss that penalizes incoherent topics during training of the model. We analyze our approach using coherence and an additional metric - exclusivity, responsible for the uniqueness of the terms in topics. We argue that this combination of metrics is an adequate indicator of the model quality. Our results indicate the effectiveness of our loss and the potential to be used in the future neural topic models.

**Keywords:** Topic modeling · Neural networks · NPMI ·
Topic coherence

## 1 Introduction

Topic Modeling is an established area of text mining focused on discovering topics in a collection of documents. Generative models like Latent Dirichlet Allocation (LDA) [1] have been long used as a standard in Topic Modeling. With the popularization of the Neural Networks and Deep Learning, several neural topic models have been suggested [2,3], implementing the same generative principles: generating a document given the same document, through the topic/word and document/topic distributions. The advantages of neural topic models include better resource management through the flexibility of training, and natural use of embeddings, a highly effective type of word representation, as opposed to bag-of-words model, traditionally used in LDA. Neural topic models, being unsupervised, use a metric like perplexity as a training criterion, to control the ability of the model to generate documents. However, perplexity does not reflect the human judgment [4] of the topic quality, unlike coherence, commonly used in evaluation of topic models. Therefore, such models cannot guarantee the best quality of the final output. Furthermore, coherence cannot be regarded as a single defining quality metric: being an average between all topics, the final coherence will favor repeating topics, coherent but containing the same words.

© Springer Nature Switzerland AG 2019
H. Lu et al. (Eds.): ISNN 2019, LNCS 11554, pp. 426–433, 2019.
https://doi.org/10.1007/978-3-030-22796-8_45

In this work we present a new regularization loss for a neural topic model, inspired by the coherence measure. Our contributions are the following: (i) we propose a simple and straightforward regularization function for a neural topic model, aimed to control the coherence of the intermediate topics, generated during training; (ii) we propose the usage of a new (to the best of our knowledge) composite metric, combining coherence and a uniqueness measure - *exclusivity* [15], thus solving the coherence's bottleneck of repeating topics; (iii) we introduce new multi-criteria training procedure – a combination of perplexity and aforementioned composite metric, to better control the topic model quality during training.

The paper is organized as follows: in Sect. 2 we outline the related work that has influenced our approach; Sect. 3 details the proposed regularization technique; implementation and experiments are described in Sect. 4, and their results are listed and analyzed in Sect. 5. Finally, Sect. 6 finalizes our findings and sets up a plan for the future work.

# 2 Related Work

In this section we outline the previous research that have shaped up our work.

## 2.1 Regularization for Standard Topic Models

Regularization of topic models is not a novel concept. [5] proposed to modify the LDA model by building a structured prior over words using a covariance matrix, enforcing co-occurring words to appear in the same topics. Another regularizer in form of a Markov Random Field, was presented in [6], with the same idea of incorporating word correlation knowledge into LDA. [7] offered a more efficient method called Sparse Constrained LDA. All of the aforementioned models carry the same idea of adding word co-occurrence information to the LDA algorithm. This concept is transferred to neural topic models in our work.

## 2.2 Neural Topic Models

The success of Neural Networks and Deep Learning in many NLP tasks has lead to the research in Neural Topic Modeling. The early methods use autoencoders [8], a Restricted Boltzmann Machine [9], autoregressive models [10] and Deep Boltzmann Machine [11]. [2] proposed a straightforward way to represent the topic model with a Neural Network. Their model uses word embeddings, which enables the input consisting of ngrams instead of single words. Topically Driven Neural Language Model (TDLM) [3] adopts a topic model to improve the language model. The authors report a state of the art coherence for several datasets, outperforming [2] and in most cases LDA. For this reason, TDLM serves in our work as the baseline and base model, where our regularization loss is applied.

## 2.3    Topic Quality Measures

Topic quality measures have been evolving in the recent years: from using perplexity [12] to various metrics for topic coherence [13]. Normalized Point-wise Mutual Information (NPMI) coherence is one of the most well performing and popular with researchers [13]. Whereas coherence is well-informative, it does not measure redundancy [1], e.g. a model containing N coherent, but identical topics would get a high score, despite being low quality. For this reason, many works conduct a qualitative analysis of the topics. The other way to handle redundancy problem is to add another metric, such as inter-topic similarity from [1], or generality in [14]. In this work we follow the latter approach and employ the exclusivity metric [15] for measuring the "uniqueness" of the topics. We then introduce a new metric – a combination of coherence and exclusivity, in order to give better quantitative assessment of the model quality.

## 3    Coherence Loss

Traditionally, in neural models dropout is used for regularization, to prevent overfitting. For neural topic models, other types of regularizers, besides dropout, are necessary to improve the coherence of the resulting topics. The main loss function of a neural topic model, such as [2,3], is designed for an autoencoder, where the training is led by metrics like perplexity, which, as we know, does not correlate with coherence [4]. From the need to explicitly control the coherence of the topics we have drawn the idea of adding a coherence loss as a new regularization technique.

In this work we use NPMI coherence in the evaluation, the reason being its popularity and superior performance. The definition for NPMI coherence is inferred from the framework by [13]:

$$C_{NPMI} = \sigma^a_{t=1...N}(\sigma^a_{i,j \in \binom{N_t}{2}}(NPMI(w_i, w_j)))  \tag{1}$$

where $\sigma^a$ is an arithmetic mean [13], $N$ is the number of topics, $w_i, w_j \in W_t$ – the set of top $N_t$ terms of each topic and

$$NPMI(w_i, w_j) = \frac{\log \frac{P(w_i, w_j) + \epsilon}{P(w_i) \cdot P(w_j)}}{-\log P(w_i, w_j) + \epsilon}  \tag{2}$$

NPMI values lie in $[-1; 1]$ interval. Probabilities $P$ are estimated based on the word occurrence and co-occurrence matrices. It is assumed that the more the terms of a topic appear together in a corpus, the better they are related, which makes the topic more coherent. The co-occurrence matrix can be built on any corpus, including the training one, though the coherence calculated on a large external corpus (like Wikipedia) is known to perform better [16]. Our proposed coherence loss is inspired by the $C_{NPMI}$ coherence formula (1) and defined as:

$$L_{NPMI} = \alpha * \sigma^a_{t=1...N}(\sigma^a_{i,j \in \binom{N_t}{2}}(\phi_{t,i,j} * (1 - NPMI(w_i, w_j)))),  \tag{3}$$

where $\alpha$ is the loss coefficient, to be decided empirically. The loss definition differs from the coherence formula in two aspects: firstly, we use $(1 - NPMI(w_i, w_j))$ instead of $NPMI(w_i, w_j)$, to represent the coherence penalty, and secondly, we add a topic-specific coefficient $\phi_{t,i,j}$, calculated from topic/word distribution $\phi$ of words $w_i, w_j$: $\phi_{t,i,j} = \phi_{t,i} * \phi_{t,j}$. The value of $\phi_{t,i,j}$ can be considered as the "importance" of an individual coherence loss, since it depends on the position of terms in the topic descriptor: the higher pair of terms in the list will be given more weight for its coherence penalty, while the last words in the descriptor are allowed to be slightly less coherent. We apply a *softmax* over the top $N_t$ $\phi$ values for each topic to avoid very small numbers in loss calculation.

Each step of the training process, our coherence cost is computed and added to the main cost, e.g. cross-entropy for [3], with a coefficient $\alpha$.

## 4    Experiments

Seeing that our goal in this work is to improve coherence, we have chosen the state of the art neural topic model as both our baseline and base model: TDLM from [3], with a straightforward and elegant topic model neural representation. For our experiments we exclude the language model part of TDLM and focus only on the topic model, which also significantly accelerates the training time.

### 4.1    Implementation

For testing the proposed coherence loss, we have used the open-source tensorflow implementation[1] from [3], where we have disabled the language model part, leaving only the topic model. Occurrence and co-occurrence matrices for datasets were computed once and saved. For the lack of sufficient resources, we calculated the matrices based on the training data, instead of using an external bigger corpus. Since the goal is to demonstrate the work of the coherence loss in comparison with the base model, "local" co-occurrence information suits the task well. The $\alpha$ coefficient in (3) is a hyperparameter, of which several values were tested. In our results we report the best performing values.

### 4.2    Datasets

For the sake of comparison, three datasets from [3] have been chosen: IMDB reviews from [17], APNEWS[2] - collection of news from Associated Press, and British National Corpus (BNC)[3] [18]. The training and validation sets were taken directly from the open-source TDLM project, along with the default parameters. Due to difference in model and evaluation, TDLM's coherence values reported in [3] could not be used for comparison, thus all metrics were recalculated.

---

[1] https://github.com/jhlau/topically-driven-language-model.
[2] https://www.ap.org/en-gb/.
[3] http://www.natcorp.ox.ac.uk/.

### 4.3   Evaluation

As mention in the previous sections, using perplexity as a training criterion does not guarantee best coherence. Therefore, we modified the training and evaluation procedures in three different aspects, described below.

**Quality Measures.** We added a new metric, called *exclusivity*: $excl = \frac{|W_u|}{|W|}$, where $|W_u|$ is the number of unique terms and $|W|$ is the total number of terms in topic descriptors. It has been used in [15] as complimentary measure for coherence, to cover the problem of redundancy. Exclusivity is a simple and straightforward variation of the metrics offered by [1,14], and takes values from interval $(0;1]$. The latter property allows to view exclusivity as a measure of "quality of coherence". Naturally, some redundancy is unavoidable in a topic model, yet, the less redundant model is the more coherent one [1]. Therefore, both metrics should be considered for evaluation. In this work we propose a composite measure $Q = C_{NPMI} * excl$, that captures both coherence and exclusivity, providing a fair assessment of the topics. This formula is only used with *positive* coherence values, as negative coherence already indicates low quality model. Eventually, more complex formulas may be used, depending on the importance of both measures, but for this work we focus on the basic definition.

**Multi-criteria Training.** After each epoch, we compute a tuple of metrics, consisting of validation perplexity $ppl$ and $Q$. It is then checked in the following manner: if none of the metrics are improving compared to the previous epochs, the epoch is considered failed and the parameters are restored to the previous epoch, as in [3]. Otherwise, the training continues without changes. This way we make sure that good quality epochs do not get restarted because of worse perplexity.

**Final Evaluation.** We run the model for 20 epochs and average the results from several runs for each epoch. The best coherence and $Q$ values may not be the final ones, therefore per-epoch analysis is performed. We compare three types of values: best coherence, best exclusivity, and finally best $Q$ among epochs.

## 5   Results

We ran the experiments on IMDB, APNEWS and BNC for $N = 50, 100, 150$ topics. For each combination of dataset/$N$ we found the value of $\alpha$ empirically. Table 1 shows the obtained results (with superior values **in bold**). Comparison of coherence indicates that the regularized version improves the metric in the majority of cases, while exclusivity values decrease, to different extents. This observation is justified, since our regularization is based on average coherence. An exclusivity regularizer may be beneficial, which we leave for the future work.

**Table 1.** Topic quality results, maximum values from 20 epochs

| Dataset | IMDB | | | APNEWS | | | BNC | | |
|---|---|---|---|---|---|---|---|---|---|
| $C_{NPMI}$ | | | | | | | | | |
| N ($\alpha$) | 50 (1) | 100 (1) | 150 (1) | 50 (1) | 100 (2) | 150 (2) | 50 (2) | 100 (2) | 150 (2) |
| Baseline | 0.026 | 0.044 | **0.043** | 0.150 | **0.162** | 0.160 | **0.145** | 0.140 | 0.137 |
| Regularized | **0.035** | **0.045** | 0.041 | **0.151** | 0.155 | **0.163** | 0.143 | **0.142** | 0.137 |
| *excl* | | | | | | | | | |
| Baseline | **0.634** | **0.422** | **0.366** | 0.868 | 0.659 | **0.531** | 0.885 | **0.656** | **0.510** |
| Regularized | 0.620 | 0.409 | 0.361 | **0.869** | **0.674** | 0.504 | **0.905** | 0.620 | 0.504 |
| $Q = C_{NPMI} * excl$ | | | | | | | | | |
| Baseline | 0.016 | 0.018 | 0.014 | 0.129 | **0.105** | 0.082 | 0.128 | **0.092** | 0.067 |
| Regularized | **0.021** | 0.018 | 0.014 | **0.130** | 0.103 | 0.082 | **0.129** | 0.088 | **0.069** |

The analysis of the composite metric $Q$ concludes that our regularization loss is mostly beneficial for smaller number of topics. The difference is especially significant with IMDB, our smallest dataset, where regularized model gained a big difference in $C_{NPMI}$ and $Q$, without adding much redundancy to the topics. This indicates that the coherence loss is particularly useful for small and noisy corpora, where it is harder to obtain coherent topics. For a more detailed analysis, Fig. 1 shows the trends of the metrics for IMDB $N = 50$, $\alpha = 1.0$. It is evident that our proposed loss have improved the coherence at each iteration while only slightly decreasing the exclusivity. This trend is persistent through the training process. Moreover, the best quality topics do not correspond to the final epoch, which justifies our per-epoch analysis. It is also worth noticing that loss in perplexity for our regularized model is considerably small, compared to the gain in coherence.

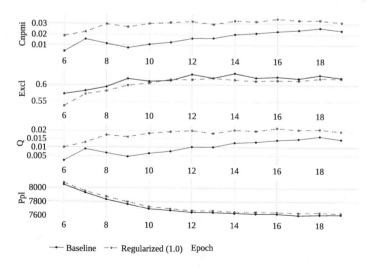

**Fig. 1.** Metrics per epoch of IMDB corpus for 50 topics (from epoch 6)

Other tests show the increase in coherence, but bigger decrease in exclusivity, which results in the trade-off $Q$ metric being equal or (in two cases) worse than the baseline TDLM. We can account that on the resource limitations that allowed us to only test several values of the parameter $\alpha$ for large number of topics. We believe that more extensive experiments will reveal better values, with possible addition of an exclusivity regularization.

Based on the obtained results, we can observe that the proposed coherence loss does indeed improve the topic coherence metric. Moreover, our new composite quality metric $Q$ proved to be a necessary addition to the evaluation process, showing that a better coherence value does not always indicate better model quality, as it may add much redundancy to the topics. Our new quality metric allows to estimate the trade-off between gain in coherence and loss in exclusivity, thus simplifying and improving the comparative analysis of the models.

## 6    Conclusion

In this work we presented new coherence loss, intended as a regularization loss for neural topic models. We have tested our loss on the state of the art model TDLM [3] and demonstrated its ability to increase the coherence of the topics (in most cases), and the overall quality of the topics (in few cases). Our regularization technique is flexible: the loss can be applied to any neural topic model, where a topic/word distribution can be computed during training. Moreover, we introduced a composite metric for the topic quality evaluation, representing the trade-off between topic coherence and the level of redundancy in topics (exclusivity). Finally, we proposed a multi-criteria training procedure, which allowed us to control both perplexity and topic quality metrics during training.

**Future Work.** Testing of our proposed loss revealed the need to add an exclusivity regularization to control the redundancy in topics. This can be achieved by adding an entropy-based loss that would ensure that each word in the vocabulary is assigned to maximum one topic. This addition we leave for the future work. Furthermore, the coherence loss should be tested on other neural topic models, such as NTM [2], where the main loss differs from TDLM. Additionally, future work will include testing more hyperparameter values for large topic numbers, with co-occurence matrices computed on a reference corpus.

**Acknowledgements.** The elaboration of this scientific paper was supported by the Ministry of Economy, Industry, Research, Innovation, IT, Employment and Education of the Region of Wallonia (Belgium), through the funding of the industrial research project Jericho (convention no. 7717).

## References

1. Arora, S., et al.: A practical algorithm for topic modeling with provable guarantees. In: International Conference on Machine Learning, pp. 280–288 (2013)
2. Cao, Z., Li, S., Liu, Y., Li, W., Ji, H.: A novel neural topic model and its supervised extension. In: AAAI, pp. 2210–2216 (2015)

3. Lau, J.H., Baldwin, T., Cohn, T.: Topically driven neural language model. In: Proceedings of the 55th Annual Meeting of the Association for Computational Linguistics, vol. 1: Long Papers. ACL (2017)
4. Chang, J., Gerrish, S., Wang, C., Boyd-Graber, J.L., Blei, D.M.: Reading tea leaves: how humans interpret topic models. In: Advances in Neural Information Processing Systems, pp. 288–296 (2009)
5. Newman, D., Bonilla, E.V., Buntine, W.: Improving topic coherence with regularized topic models. In: Advances in Neural Information Processing Systems, pp. 496–504 (2011)
6. Xie, P., Yang, D., Xing, E.: Incorporating word correlation knowledge into topic modeling. In: Proceedings of the 2015 Conference of the North American Chapter of the Association for Computational Linguistics: Human Language Technologies. ACL (2015)
7. Yang, Y., Downey, D., Boyd-Graber, J.: Efficient methods for incorporating knowledge into topic models. In: Proceedings of the 2015 Conference on Empirical Methods in Natural Language Processing. ACL (2015)
8. Ranzato, M.A., Szummer, M.: Semi-supervised learning of compact document representations with deep networks. In: Proceedings of the 25th International Conference on Machine Learning - ICML 2008. ACM Press (2008)
9. Hinton, G.E., Salakhutdinov, R.R.: Replicated softmax: an undirected topic model. In: Advances in Neural Information Processing Systems, pp. 1607–1614 (2009)
10. Larochelle, H., Lauly, S.: A neural autoregressive topic model. In: Advances in Neural Information Processing Systems, pp. 2708–2716 (2012)
11. Srivastava, N., Salakhutdinov, R., Hinton, G.: Modeling documents with a deep Boltzmann machine. In: Uncertainty in Artificial Intelligence, p. 616. Citeseer (2013)
12. Wallach, H.M., Murray, I., Salakhutdinov, R., Mimno, D.: Evaluation methods for topic models. In: Proceedings of the 26th Annual International Conference on Machine Learning, pp. 1105–1112. ACM (2009)
13. Röder, M., Both, A., Hinneburg, A.: Exploring the space of topic coherence measures. In: Proceedings of the Eighth ACM International Conference on Web Search and Data Mining - WSDM 2015. ACM Press (2015)
14. O'Callaghan, D., Greene, D., Carthy, J., Cunningham, P.: An analysis of the coherence of descriptors in topic modeling. Expert Syst. Appl. **42**(13), 5645–5657 (2015)
15. Krasnashchok, K., Jouili, S.: Improving topic quality by promoting named entities in topic modeling. In: Proceedings of the 56th Annual Meeting of the Association for Computational Linguistics (Volume 2: Short Papers), vol. 2, pp. 247–253 (2018)
16. Newman, D., Lau, J.H., Grieser, K., Baldwin, T.: Automatic evaluation of topic coherence. In: Human Language Technologies: The 2010 Annual Conference of the North American Chapter of the Association for Computational Linguistic, pp. 100–108. ACL (2010)
17. Maas, A.L., Daly, R.E., Pham, P.T., Huang, D., Ng, A.Y., Potts, C.: Learning word vectors for sentiment analysis. In: Proceedings of the 49th Annual Meeting of the Association for Computational Linguistics: Human Language Technologies, vol. 1, pp. 142–150. ACL (2011)
18. BNC Consortium. The British National Corpus, version 3 (BNC XML Edition). Distributed by Oxford University Computing Services on behalf of the BNC Consortium (2007)

# A Solution of Two-Person Zero Sum Differential Games with Incomplete State Information

Kanghao Du[1], Ruizhuo Song[1(✉)], Qinglai Wei[2], and Bo Zhao[3]

[1] University of Science and Technology Beijing, Beijing 100083, China
dukangemail@163.com, ruizhuosong@ustb.edu.cn
[2] Institute of Automation, Chinese Academy of Sciences, Beijing 110016, China
[3] School of Systems Science, Beijing Normal University, Beijing 100875, China
zhaobo@bnu.edu.cn

**Abstract.** This paper introduces a two-player zero-sum differential game with imperfect state information, focusing on the situation with linear system, quadratic cost functional and state measurement contains white Gaussian noise. A solution is put forward in view of a special situation where one controller has available noise-corrupted measurement and the other has only one priori information. In addition, Kalman filter is used for state estimation. In subsequent section, a simulation based on a linear differential game problem is proposed, which is a well corroborate of the theoretical part. Finally concludes the research work on this paper, and points out the need to further expand in the future.

**Keywords:** Differential game · Imperfect state information · Noise-corrupted measurement · Kalman filter

## 1 Introduction

The study of differential games has already caused considerable concern [1–3]. Its origins can be traced back to the literature published by Von Neumann and Morgenston [4] in 1944. Subsequently, Rufus [5], an American game theorist, made an intensive study of this field and appeared the first monograph on differential games in the world in 1965, which laid the foundation for further research work. Most of the research in this field is focused on deterministic differential games with complete information, in which players or controllers clearly know the state of the system at any time. But the practical application is often not so ideal. Hereafter researchers commence the study of nondeterministic differential games that noise (usually zero-mean white Gaussian noise) occurs in system state or state measurement. Y. C. Ho proposes a solution to a class of this problem used variational technique in reference [6] where one player controls the state, and the other can only measure the state with incomplete or noisy measurements. After that, Haurie and Başar [7] consider another kind of problem, where one player

© Springer Nature Switzerland AG 2019
H. Lu et al. (Eds.): ISNN 2019, LNCS 11554, pp. 434–443, 2019.
https://doi.org/10.1007/978-3-030-22796-8_46

has complete information and the other has only noisy measurement of system state. In addition, many researchers [8–10] have used different methods to give various arguments for differential games under various conditions.

When differential game theory was established, it was naturally associated with the optimal control theory [11]. But gradually, it was found that the optimal control method could not be simply applied to differential games [12–14]. Essentially, differential game is a multifaceted optimal control problem, and the optimal control problem can be regarded as a special type of differential game. There are further discussions on this issue in the literature [15]. In recent decades, the theory of differential games has made great progress with the deepening of the research, and the problem of differential games also presents a variety of types, such as deterministic differential game, stochastic differential game [16], pursuit-evasion game, zero-sum game [17], multi-person cooperative differential game [18], non-cooperative differential game [19], leader-follower differential game [20] and other dynamic game types.

In this paper, we assume a nondeterministic differential game in which players can only have available noise-corrupted measurements of system state. Further, attention mainly points to the special case of a linear system, a quadratic cost function and an independent zero-mean white Gaussian noises occurring in state measurement. From the formula of Sect. 2, we can see how they are described. The formal solution to this type of problem in special cases under the premise of Nash equilibrium is given in Sect. 3. Furthermore, the validity of Kalman filter for state estimation in this special case is proved by establishing two propositions in Sect. 4. In the Sect. 5, a simulation is given to verify the theoretical part of the paper. Finally, get a summary of the theoretical study and practical research.

## 2 Problem Statement

The two-player zero-sum differential game problem of imperfect information involving a linear system, a quadratic cost function and an independent zero-mean white Gaussian noises occurring in state measurement. Our purpose here is to give an optimal differential strategy to this problem. Consider a continuous-time linear system described by the vector differential equation

$$\dot{x} = Ax + Bu(t) + Dv(t) \tag{1}$$

where the n-dimensional vector $x(t)$ is the system state; The control vectors $u(t)$ and $v(t)$ are p-dimension and q-dimension, respectively; $A$, $B$, $D$ are matrices of appropriate dimension. In addition, assuming that the system is stabilizable. Consider also a quadratic cost functional or payoff to this problem

$$J_t(u, v) = \int_t^{t_f} r(x, u, v, t)\, dt + L(x(t_f)) \tag{2}$$

where $r(x, u, v, t) = u^T R_1(t) u + v^T R_2(t) v$, $L(x(t_f)) = x^T(t_f) P_f x(t_f)$, and $t_0 \leq t \leq t_f$, $t_f$ is the fixed termination time. Note that $R_1(t)$ is a symmetric positive definite matrix and $R_2(t)$ is a symmetric negative definite matrix;

$P_f$ is a positive semidefinite matrix. Here, controlling $u$ and $v$ select their own actions to maximize or minimize the expectations of the payment function $J$. Controller 1, controlling $u$, has available noise-corrupted measurements of the form

$$z_1(t) = H_1(t)x(t) + w_1(t) \tag{3}$$

while Controller 2, controlling $v$, has no available measurements and, yet, has only one priori information to use

$$z_2(t) = H_2(t)x(t) + w_2(t), H_2(t) \equiv 0 \tag{4}$$

where $H_1(t)$, $H_2(t)$ are matrices of appropriate dimension. Suppose $w_1$ and $w_2$ are zero-mean white Gaussian noise with covariances

$$cov(w_i(t), w_i(\tau)) = W_i(t)\delta(t - \tau), i = 1, 2 \tag{5}$$

where $\delta$ is a unit impulse function, and $W_i(t)$ is symmetric nonnegative positive definite matrix. Suppose that both controllers regard the initial state $x(t_0)$ as a Gaussian random vector unrelated with $w_1(t)$, $w_2(t)$, and has mean $\bar{x}_0$, moreover $cov(x(t_0), x(t_0)) = M_0$.

$Z_i$ is defined as the measurement set of controller $i, i = 1, 2$ on interval $[t_0, t)$

$$Z_i(t) = \{(z_i(s), s)|s \in [t_0, t)\}. \tag{6}$$

For controller 2, $H_2(t) \equiv 0$ and $Z_2(t) \equiv Z_2(t_0)$ for all $t$. Both controllers must choose their own control variables on the basis of their respective measurements.

## 3   The Solution

This two-player zero-sum differential game consists of the optimization of (2) subject to (1), with controller 1 having available noise-corrupted measurements of the state $x(t)$ and controller 2 having only one priori information of the form given in (4). The optimal closed-loop control laws $(u^o, v^o)$ is obtained when the condition

$$J_t(u^o, v) \le J_t(u^o, v^o) \le J_t(u, v^o). \tag{7}$$

is satisfied for all admissible strategies. Then, if (7) is established, there is a saddle point for $J_t(u, v)$, and $u^o, v^o$ are called the optimal strategy [21]. We introduce the estimate of $x(t)$

$$\hat{x}_i(t|t) = E\{x(t)|Y_i(t)\}, i = 1, 2 \tag{8}$$

and $\tilde{x}_i(t|t) = x(t) - \hat{x}_i(t|t), i = 1, 2$. Known from the previous, both state $x(t)$ and measurements $z_i(t)$ are Gaussian random vectors in this paper, also, $\hat{x}_i(t|t)$ is the minimal mean square error estimate of $x(t)$ under given condition $Y_i(t)$. For the closed-loop control laws $(u, v)$, define the value function [22] as

$$\begin{aligned} V(x(t), \tilde{x}_2(t|t), t) \\ = x^T(t)P(t)x(t) - \tilde{x}_2(t|t)^T N(t)\tilde{x}_2(t|t) + b(t) \\ = x^T(t)P(t)x(t) - (x(t) - \hat{x}_2(t|t))^T N(t)(x(t) - \hat{x}_2(t|t)) + b(t). \end{aligned} \tag{9}$$

We can also define this two-player zero-sum differential game as

$$V^o = J_t(u^o, v^o) = \min_u \max_v J_t(u, v)$$

where $J$ is defined in (2) and $V^o$ is the optimal value function. This differential game has a unique solution if either saddle point exists or the *Nash* condition [23]

$$V^o = \min_u \max_v J(u, v) = \max_v \min_u J(u, v)$$

is satisfied. Note that $V(x(t), \tilde{x}_2(t|t), t) = J_t(u, v)$. The Hamilton function is defined as follows

$$
\begin{aligned}
&H(x, \tilde{x}_2, u, v, t) \\
&= V_t(x, \tilde{x}_2, t) + V_x(x, \tilde{x}_2, t)\dot{x} + V_{\tilde{x}_2}(x, \tilde{x}_2, t)\dot{\tilde{x}}_2 + r(x, u, v, t) = 0
\end{aligned}
\tag{10}
$$

where $i = 1, 2$, $V_t = \frac{\partial V}{\partial t}$, $V_x = \frac{\partial V}{\partial x}$, $V_{\tilde{x}_2} = \frac{\partial V}{\partial \tilde{x}_2}$. Note that controlling $u$ and controlling $v$ can only be obtained from their respective measurements. Rewriting system equation (1) as follows

$$\dot{x} = Ax + BE\{u|Y_1\} + DE\{v|Y_2\}. \tag{11}$$

Refer to Eqs. (8) and (11)

$$
\begin{aligned}
\dot{\hat{x}}_2 &= E\{\dot{x}|Y_2\} = E\{Ax + BE\{u|Y_1\} + DE\{v|Y_2\}|Y_2\} \\
&= A\hat{x}_2 + BE\{u|Y_2\} + DE\{v|Y_2\}
\end{aligned}
\tag{12}
$$

$$
\begin{aligned}
\dot{\tilde{x}}_2 &= \dot{x} - \dot{\hat{x}}_2 \\
&= Ax + BE\{u|Y_1\} + DE\{v|Y_2\} - (A\hat{x}_2 + BE\{u|Y_2\} + DE\{v|Y_2\}).
\end{aligned}
\tag{13}
$$

Note that the external expectation of the last term is based on $Y_2$, meanwhile, the internal expectation based on $Y_1$ is also valid, since $Y_2(t) = Y_2(t_0)$ (a priori information available to both controller) is a subset of $Y_1(t)$, so $E\{E\{\bullet|Y_1\}|Y_2\} = E\{\bullet|Y_2\}$. Thus, substitution of (9), (11), and (12) into Eq. (10) gives

$$
\begin{aligned}
H(x, \tilde{x}_2, u, v, t) =\ & x^T \dot{P} x - \tilde{x}_2^T \dot{N} \tilde{x}_2 + \dot{b} + u^T R_1 u + v^T R_2 v \\
& + 2x^T P(Ax + BE\{u|Y_1\} + DE\{v|Y_2\}) \\
& - 2\tilde{x}_2^T N[(Ax + BE\{u|Y_1\} + DE\{v|Y_2\}) \\
& - (A\hat{x}_2 + BE\{u|Y_2\} + DE\{v|Y_2\})]
\end{aligned}
\tag{14}
$$

### 3.1   The Form of Optimal Strategy

According to Bellman's optimal principle [24], the optimal strategy can be obtained as follows

$$u^o = \arg\min_u E\left\{H\left(x, \tilde{x}_2, u, v, t\right) | Y_1\left(t\right)\right\}$$

$$= -\frac{1}{2} R_1^{-1} B^T E\left\{Px - N\tilde{x}_2 | Y_1\left(t\right)\right\} = -R_1^{-1} B^T \left(P\hat{x}_1 - N\left(\hat{x}_1 - \hat{x}_2\right)\right), \quad (15)$$

$$v^o = \arg\max_v E\left\{H_2\left(x, u, v, t\right) | Y_2\left(t\right)\right\}$$

$$= -\frac{1}{2} R_2^{-1} D^T E\left\{Px - N\tilde{x}_2 | Y_2\left(t\right)\right\} = -R_2^{-1} D^T P\hat{x}_2. \quad (16)$$

We can also get

$$E\left\{u | Y_2\right\} = \arg\min_u E\left\{H\left(x, \tilde{x}_2, u, v, t\right) | Y_2\left(t\right)\right\}$$

$$= -R_1^{-1} B^T E\{Px - N\left(x - \hat{x}_2\right) | Y_2\left(t\right)\} = -R_1^{-1} B^T P\hat{x}_2. \quad (17)$$

Substitution of (15), (16) and (17) into (14) then gives, after some algebra,

$$H\left(x, \tilde{x}_2, u^o, v^o, t\right) = x^T\left(\dot{P} + PA + A^T P - PBR_1^{-1} B^T P\right.$$

$$- PDR_2^{-1} D^T P)x - \tilde{x}_2^T\left(\dot{N} + NA - A^T N\right.$$

$$- P\left(BR_1^{-1} B^T + DR_2^{-1} D^T\right) P)\tilde{x}_2 \quad (18)$$

$$+ \tilde{x}_2^T \left(P - N\right) BR_1^{-1} B^T \left(P - N\right) \tilde{x}_2$$

$$+ E\left\{x_1^T \left(P - N\right) BR_1^{-1} B^T \left(P - N\right) \tilde{x}_1 | Y_1\right\} + \dot{b} = 0$$

It then follows immediately that the symmetric matrix $P(t)$ satisfies the *Riccati* equation

$$\dot{P} + PA + A^T P - PBR_1^{-1} B^T P - PDR_2^{-1} D^T P = 0, \quad (19)$$

while the symmetric matrix $N(t)$ satisfies the differential equation

$$\dot{N} + NA - A^T N - P\left(BR_1^{-1} B^T + DR_2^{-1} D^T\right) P$$

$$+ \left(P - N\right) BR_1^{-1} B^T \left(P - N\right) = 0 \quad (20)$$

with the boundary condition $N(T) = 0$. The penultimate term with the expectation based on $Y_1$ of (18) is

$$tr\left[E\left\{\tilde{x}_1 \tilde{x}_1^T | Y_1\right\} \left(P - N\right) BR_1^{-1} B^T \left(P - N\right)\right]$$

$$= tr\left[M\left(P - N\right) BR_1^{-1} B^T \left(P - N\right)\right], \quad (21)$$

where the symmetric matrix $M$ is defined as

$$M(t) = E\left\{\tilde{x}_1(t|t)\tilde{x}_1(t|t)^T | Y_1(t)\right\}, \quad (22)$$

and from Sect. 2, we know the boundary condition $M(t_0) = cov\left(x(t_0), x(t_0)\right) = M_0$. The scalar $b(t)$ can be solved by the differential equation

$$\dot{b} = -tr[M(t)\left(P(t) - N(t)\right) \cdot B(t)R_1^{-1} B^T(t)\left(P(t) - N(t)\right)], \quad (23)$$

with boundary condition $b(T) = 0$.

## 3.2    State Estimation Using Kalman Filter

Combined with (12), (15), (16) and (17), the optimal state estimation of controller 2 , given only one priori information, without available the measurements, is given by

$$\dot{\hat{x}}_2 = A\hat{x}_2 + BE\{u|Y_2\} + DE\{v|Y_2\}$$
$$= A\hat{x}_2 - BR_1^{-1}B^T P\hat{x}_2 - DR_2^{-1}D^T P\hat{x}_2 \tag{24}$$

with initial condition $\dot{\hat{x}}_2(t_0|t_0) = \bar{x}_0$.

For controller 1, the measurement and state estimate are independent of controller 2, thus the optimal estimate based on measurement set $Y_1(t)$ is carried out by kalman filter theory [25], and is designed by a linear dynamical system as follows

$$\dot{\hat{x}}_1(t|t) = A\hat{x}_1(t|t) + Bu(t) + Dv(t) + K(t)\tilde{z}(t) \tag{25}$$
$$\tilde{z}_1(t) = z_1(t) - H_1(t)\hat{x}_1(t|t) \tag{26}$$

with initial state $\dot{\hat{x}}_1(t_0|t_0) = \bar{x}_0$. The fundamental purpose of using $kalman$ filter is to estimate the real state of the system as accurately as possible. This requires that the filtering process be stable, which indicates that the state estimation is effective. Next, we will prove the boundedness and convergence of error estimates.

**Boundedness of the Estimated Error.** If the estimation error $\tilde{x}$ is acceptable, that is, there is no divergence, then the process of state estimation is stable. Note that it has been assumed in (8) that $\hat{x}_1$ is the minimum mean square error of $x$ given $Y_1(t)$.

**Theorem 1.** *If $\hat{x}_1(t|t)$ is the minimal mean square error estimate of $x(t)$ given $Y_1(t)$, then the estimation error $\tilde{x}_1(t|t)$ is certainly bounded and the process of state estimation is certainly stable.*

*Proof.* For any $t \in [t_0, t_f]$, $\hat{x}_1(t|t)$ is the minimal mean square error estimate of $x$ given $Y_1(t)$, so for any estimator $x_1^*$, $E\{\hat{x}_1(t|t) - x(t)\}^2 \le E\{x_1^* - x(t)\}^2$ is valid. Clearly, we can find an admissible $\epsilon$ that makes $E\{\hat{x}_1(t|t) - x(t)\}^2 \le E\{x_1^* - x(t)\}^2 \le \| \epsilon \|^2$ permanent established. According to the nature of norm $\| \tilde{x}_1(t|t) \|^2 = \| E\{\hat{x}_1(t|t) - x(t)\} \|^2 \le \| E\{x_1^* - x(t)\} \|^2 \le \| \epsilon \|^2$. That is to say, the estimation error $\tilde{x}_1(t|t)$ is bounded, which indicates that the process of state estimation is stable. But this stability does not guarantee the convergence directly. Inspired by the definition in (22), to prove that $\tilde{x}_1(t|t)$ convergence only needs to prove $M(t)$ convergence.

According to (1), (3), (25) and (26), we can get, after some algebra, $\dot{\tilde{x}}_1 = A\hat{x}_1 - K(H_1\tilde{x}_1 + w_1)$. Using kalman optimal filtering theory [25], the optimal gain can be displayed as $K = MH_1^T W_1^{-1}$. Thus we can get immediately that $\dot{\tilde{x}}_1 = A\tilde{x}_1 - MH_1^T W_1^{-1}(H_1\tilde{x}_1 + w_1)$. The remaining unknown matrix $M$ is a solution of the differential equation

$$\dot{M} = AM + MA^T - MH_1^T(t)W_1^{-1}(t)H_1(t)M \tag{27}$$

of the *Riccati* type with the boundary condition $M(t_0) = M_0$ (recall (22)).

**Convergence of Solutions to the Differential Equations.** If (27) given a fixed initial time $t_0$ and a nonnegative matrix $M_0$ satisfies a *Lipschitz* condition, it follows at once that there exist a unique solution $M(t) = \Psi(t|M_0, t_0)$. Here, Theorem 2 contained two sufficient conditions is given to conclude $M(t)$ exists for all $t$.

**Theorem 2.** *The solution* $\lim_{t\to\infty} \Psi(t|M_0, t_0) = M^*(t)$ *of* (27) *exists for all $t$ if either*

I. *the system equation* (1) *is uniformly asymptotically stable;*
II *the system equation* (1) *is completely observable for all t.*

*Proof.* Theorem 2 is proved in [26]. Since $M(t)$ has been proved to be convergent, refer to (22), we can easily get the error estimate $\tilde{x}_1$ is also convergent.

## 4    Simulations

In order to verify the correctness of the solution presented above, the following second-order continuous time linear system is considered [27]

$$\dot{x} = \begin{bmatrix} 0 & 1 \\ -1 & -3 \end{bmatrix} x + \begin{bmatrix} 0 \\ 0.6 \end{bmatrix} u + \begin{bmatrix} 1 \\ 4 \end{bmatrix} v, \tag{28}$$

where $x$ is the system state; $u$ and $v$ are control vectors with appropriate dimensions. Before the simulation, we select the following weighting term $R_1 = 1, R_2 = -1, H_1 = [1, 0], W_1 = 2$. First, solve the Eq. (27), where $M$ is a symmetric matrix in the form of $M = [m_{11}, m_{12}; m_{12}, m_{22}]$, and assume its initial matrix is $M_0 = [1, 0; 0, 1]$, iterative interval is $t \in [0, 20]$. Using the Runge-Kutta algorithm, we can see the convergence process of $M$ as shown in the following Fig. 1.

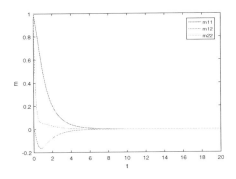

**Fig. 1.** Convergence process of $M$, $W_1 = 2$.

At the end of the iteration, we can get $M = 0$, which indicates that the estimation error $\dot{x}_1$ converges to zero. Therefore, we continue to solve

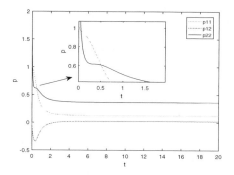

Fig. 2. Convergence of $P$, $W_1 = 2$.

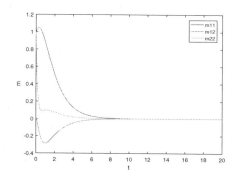

Fig. 3. Convergence of $N$, $W_1 = 2$.

Eqs. (19) and (20). Assuming that the form of the symmetric matrix $P$ is $P = \left[ p_{11}, p_{12}; p_{12}, p_{22} \right]$ and has an initial value $P_0 = \left[ 1, 0; 0, 2 \right]$, the iteration process of $P$ on interval $[0, 20]$ can be obtained as shown in the Fig. 2. Analogously, assuming that the form of the symmetric matrix $N$ is $N = \left[ n_{11}, n_{12}; n_{12}, n_{22} \right]$ and has an initial value $N_0 = \left[ 1, 1; 1, 1 \right]$, the iteration process of $N$ on interval $[0, 30]$ can be obtained as shown in the Fig. 3. Thus, we can get the solutions of symmetric matrices $P$ and $N$ as $P = \begin{bmatrix} 0.1222 & 0.02199 \\ 0.02199 & 0.3648 \end{bmatrix}, N = \begin{bmatrix} 0.1222 & 0.02205 \\ 0.02205 & 0.3643 \end{bmatrix}$.

Next, according to (15) and (16), the optimal control strategy $(u^o, v^o)$ of the differential game can be obtained, since other matrices are already known. Note that the estimates of $\hat{x}_1$ and $\hat{x}_2$ can be obtained in (24) and (25). Further, if only the value of $W_1$ is changed while other conditions stay the same, the results shown in Figs. 2 and 3 can be obtained. From these two graphs, we can see that $W_1$ still converges to steady state when it changes. This means that the Kalman filter used in this paper can remove different levels of noise (Figs. 4 and 5).

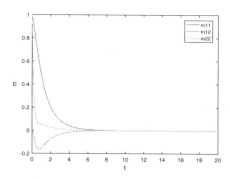

Fig. 4. Convergence of $M$, $W_1 = 4$.

Fig. 5. Convergence of $M$, $W_1 = 20$.

## 5   Conclusions and Next Work

In this paper, we propose an approximate optimal control method for differential games with incomplete information, especially when one controller has noise-corrupted measurement and the other only has a prior experience. Firstly, the formal solution of the optimal control laws $(u^o, v^o)$ is obtained by constructing Hamilton function and value function and using the optimal control theory. Secondly, the effectiveness of state estimation using Kalman filter is analyzed. Finally, the simulation results have demonstrated the validity of the proposed optimal control laws. In this paper, we only give the solution of a special case of nondeterministic two-person differential games. In the next step, it can be considered that noise occurs in two controllers or that noise exists simultaneously in the system state equation.

**Acknowledgment.** This work was supported in part by the National Natural Science Foundation of China under Grants 61873300, 61722312, and in part by the Fundamental Research Funds for the Central Universities under Grant FRF-GF-17-B45.

## References

1. Dong, J., Zhang, X., Jia, X.: Strategies of pursuit-evasion game based on improved potential field and differential game theory for mobile robots (2012)
2. Wang, G., Xiao, H.: Linear quadratic non-zero sum differential games of backward stochastic, differential equations with asymmetric information. Eprint Arxiv arXiv:1407.0430 (2016)
3. Zhang, S.: A mean-field linear-quadratic leader-follower stochastic differential game. In: 2018 Chinese Control and Decision Conference, Shenyang (2018)
4. Von Neumann, J., Morgenston, O.: The Theory of Games and Economic Behaviour. Princeton University Press, Princeton (1944)
5. Rufus, P.I.: Differential Games. A Mathematical Theory with Application to Warfare and Pursuit, Control and Optimization. Wiley, Hoboken (1966)
6. Ho, Y.C.: Optimal terminal maneuver and evasion strategy. SIAM J. Control **4**(3), 421–428 (1966)
7. Haurie, A.: Feedback equilibria in differential games with structural and modal uncertainties. In: Advances in Large Scale Systems, pp. 163–201 (1984)
8. Bensoussan, A., Lions, J.L.: Stochastic differential games with stopping times. Differential games and control theory II. In: Proceeding of Second Kingston Conference, pp. 377–399 (1976)
9. Elliott, R.J., Davis, M.H.A.: Optimal play in a stochastic differential game. SIAM J. Control Optim. **19**(4), 543–554 (1981)
10. Bardi, M., Raghavan, T.E.S., Parthasarathy, T.: Stochastic and differential games: theory and numerical methods. J. Royal Stat. Soc. **50** (1999)
11. Nicholas, W.G.: Stochastic differential games and control theory. Thesis, Virginia Polytechnic Institute and State University, Virginia (1971)
12. Friedman, A.: Stochastic differential equations and applications. Stoch. Diff. Eqn. Appl. **8**(11), 107–146 (2014). Vol. 2
13. Ho, Y.C.: On the minimax principle and zero sum stochastic differential games. In: IEEE Conference on Decision and Control. IEEE Press (1974)

14. Ramachandran, K.: Weak convergence of partially observed zero-sum stochastic differential games. Dyn. Syst. Appl. **4**, 329–340 (1995)
15. Ho, Y., Bryson, A., Baron, S.: Differential games and optimal pursuit-evasion strategies. IEEE Trans. Autom. Control **10**(4), 385–389 (1965)
16. Jacobson, D.H.: Optimal stochastic linear systems with wxponential performance criteria and their relation to deterministic differential games. IEEE Trans. Autom. Control **18**(2), 124–131 (1973)
17. Anderson, C.: Zero-sum game. Nature **356**(6365), 97–97 (1992)
18. Smol, E.R.: New theory of cooperative games. Cybern. Syst. Anal. **41**(5), 767–774 (2005)
19. Chistyakov, S.V.: On noncooperative differential games. Milan J. Math. **79**(2), 357–427 (2011)
20. Shi, J., Wang, G., Jie, X.: Leader-follower stochastic differential game with asymmetric information and applications. In: Leader-Follower Stochastic Differential Game with Asymmetric Information and Applications (2016)
21. Ramachandran, K.M., Tsokos, C.P.: Stochastic differential games: theory and applications. Mol. Microbiol. **35**(5), 961–73 (2012)
22. Rhodes, I., Luenberger, D.: Differential games with imperfect state information. IEEE Trans. Autom. Control **14**(1), 29–38 (1969)
23. Dockner, E.J., Takahashi, H.: On the saddle-point stability for a class of dynamic games. J. Optim. Theory Appl. **67**(2), 247–258 (1990)
24. Bellman, R.: Dynamic programming. Science **153**(3731), 34–37 (1966)
25. Kalman, R.E.: New results in linear filtering and prediction theory. Basic Eng. ASME Trans. Ser. D. **83**(83), 109 (1960)
26. Kalman, R.E., Bertram, J.E.: Control system analysis and design via the second method of Lyapunov: (I) continuous-time systems (II) discrete time systems. IRE Trans. Autom. Control **4**(3), 112–112 (1960)
27. Fu, Y., Chai, T.: Online solution of two-player zero-sum games for continuous-time nonlinear systems with completely unknown dynamics. IEEE Trans. Neural Netw. Learn. Syst. **27**(12), 2577–2587 (2017)

# ACPJS: An Anti-noise Concept Drift Processing Algorithm Based on JS-divergence

Xin Song[1,2(✉)], Shizhen Qin[1], Shaokai Niu[1], and Yan Wang[1]

[1] School of Computer Science and Engineering, Northeastern University, Shenyang, China
bravesong@163.com
[2] State Key Laboratory of Management and Control for Complex Systems, Institute of Automation, Chinese Academy of Sciences, Beijing, China

**Abstract.** Concept drift involving noise is an important research in the field of data mining. Many concept drift detection models are proposed to promote the research of traditional concept drift detection. In this paper, we propose an anti-noise concept drift processing algorithm based on entropy of information, named ACPJS. In ACPJS, the JS-divergence and Hoeffding Bounds are used to set double threshold for concept drift detection and subsequently a horizontal integrated model will be constructed for anti-noise concept drift processing. In the comparison experiments of multiple data sets, the presented algorithm has shown good performance in concept drift detection, anti-noise performance and classification accuracy.

**Keywords:** Concept drift · JS-divergence · Horizontal integrated model

## 1 Introduction

Nowadays, we face with a tremendous number of data from sensor networks, social networks, Web applications, scientific experiments and financial activities, etc. [1]. Therefore, an efficient processing method for a large amount of data has become an urgent need. In the related research, data processing is usually in a dynamic environment, one of the most important challenges in learning from data streams is reacting to concept drift [2]. Due to the environments are often nonstationary, the variables to be predicted may change over time in a processing, called concept drift. Most of the current concept drift detection algorithms are only for specific types of concept drift. Especially in the noise-containing data stream, they perform poorly. In order to ensure the classification accuracy and improve the anti-noise performance of the models, this paper proposes an anti-noise concept drift processing algorithm based on JS-divergence (ACPJS). When the ACPJS detects the concept drift on training sets, the noise detection is added at the same time, which can reduce the interference of the noise data to the classification model and improve the robustness of the classification model. The main contributions of this paper as follows:

© Springer Nature Switzerland AG 2019
H. Lu et al. (Eds.): ISNN 2019, LNCS 11554, pp. 444–453, 2019.
https://doi.org/10.1007/978-3-030-22796-8_47

1. This paper proposes a method based on Hoeffding Bounds and JS-divergence to set double threshold in the detection of noise data and concept drift. The detection performance of this method is better than other methods.
2. The paper proposes a horizontal integration model based on decision tree classification algorithm in the noisy concept drift processing stage. The model can reuse the buffer information, and the concept drift classification score is the dynamic weighted average score of each base classifier. Therefore, the model has higher accuracy and fault tolerance.
3. The paper integrates the detection phase of concept drift and the classification processing phase to form a complete anti-noise concept drift processing model.

The remainder of this paper is organized as follows: Sect. 2 presents related works. In Sect. 3, we elaborate the extraction of noise data, horizontal integration processing framework and concept drift detection process of ACPJS algorithm. Finally, in Sects. 4 and 5, the analysis of experimental results and conclusions are carried out, respectively.

## 2   Related Works

In recent years, with the rapid development of information technology, the amount of data generated in various related fields have also been on the rise. Meanwhile, the speed of data generation is also accelerating. Consequently, the management and mining of real-time data are becoming more and more important. Since Schlimmer et al. [3]. proposed "concept drift" in 1986, many scholars have made tremendous contributions. Since the 1990s, the research on concept drift detection in the field of data stream mining has become a hot topic, and the research results obtained have been widely used [4]. In Reference [5], KL-divergence is used to measure the difference between two probability distributions. However, there are still three limitations:

- It needs to discretize the data to calculate the probability density;
- It can only deal with the concept of drift between the two categories and many categories can only be based on the results of two classes to decide;
- Guided and discrete processes can be time consuming.

Therefore, the improved KL-divergence algorithm, JS-divergence, is used in this paper, which effectively avoids the limitation of the existence of KL-divergence. In Reference [6, 7], two classical concept drift detection algorithms CVFDT and DDM are proposed. The CVFDT introduces a sliding window based on the VFDT [8] algorithm, which effectively improves the problem that the concept drift detection is not good only by using the Heoffding tree. The DDM algorithm achieves better performance in conceptual drift detection by setting two thresholds, warning and drift for the error rate. Both CVFDT and DDM are existing classical algorithms. Many scholars conduct research on them. This paper will compare these two algorithms to explore the improvement of concept drift detection performance. In the research of anti-noise algorithms for data streams, the main methods are integration technology [9, 10], random forests [11–13], unsupervised learning [14–16] and decision trees [17, 18], etc. In order to highlight the performance improvement of the algorithm in the anti-noise performance, this paper will compare with many classic concept drift algorithms using different technologies, such as the algorithms using integrated technology (OzaBoost [19], OzaBag [20], Weighted-bagging [21], ASHT-bagging [22]), the algorithms using decision tree technology, MSRT [23] and the algorithm using the sliding window setting technique, DWSCD [24]. Although there are many new research methods and ideas in the research of concept drift detection in data stream, how to quickly and accurately detect and process concept drift is still a research hotspot in the field of data stream mining. Research still faces many challenges and needs to be explored in greater depth.

# 3   An Anti-noise Concept Drift Processing Algorithm Based on JS-Divergence

## 3.1   ACPJS Algorithm Concept Drift Detection Process

### Extraction of Noise Data

The existence of noise will cause huge interference to the judgment of concept drift, which makes it difficult for the system to distinguish whether there is a concept drift phenomenon or noise data. Therefore, it is necessary to distinguish between the two by certain means. In this paper, the ACPJS uses the Hoeffding Bounds inequality theory to realize the ex-traction of noise data. The theory of Hoeffding Bounds inequality is defined as assuming that $r$ is a real number random variable, its maximum value is R, and the mean value of d independent observation for random variable $r$ is $\bar{r}$, then the probability of $r$'s real average value at least $\bar{r} - \varepsilon$ is $1 - \delta$, that is:

$$P(r \geq \bar{r} - \varepsilon) = 1 - \delta, \varepsilon = \sqrt{R^2 In(1/\delta)/2d} \tag{1}$$

In formula (1), $d$ is the number of data blocks. Hoeffding Bounds inequality reflects the closeness between random variables and their means, to classify the data stream of each data block in the integrated classifier classification accuracy $e$ as a random variable, in the ideal case, the value of the random variable is equal to the average.

$$P\{|e - \bar{e}| \leq \varepsilon\} = 1 - \delta \tag{2}$$

When $e$ can not satisfy Eq. (2), then it proves that there is a concept drift or noise interference. In order to verify whether a concept drift or just noise interference is generated, the algorithm sets two thresholds $T_l = c_1\varepsilon$ and $T_h = c_2\varepsilon$ ($c_1$, $c_2$ are constants and satisfy $c_1 < c_2$). Now suppose $e$ is the misclassification rate on the different data block integration classifiers, the detection process is following:

- The integrated classifier is used to classify the samples in each data block and record the misclassification rate $e_i$ of each data block, and finally calculate the average misclassification rate $\bar{e}$.

- Calculate $\Delta e$ by $\Delta e = |e_i - \bar{e}|$, if $\Delta e \leq T_l$, there is no concept drift occurs, if $\Delta e \geq T_h$, indicated that the concept drift may occur and need to use JS-divergence for further judgment.
- Update $e_i$ for the next detection and repeat the above steps until the end of the training.

### Concept Drift Detection Based on JS-divergence

Before using JS-divergence for concept drift detection, the KDQ tree is used to divide the data, which is beneficial to the uniform distribution of data in the feature space. In concept drift detection, the JS-divergence is used to calculate the data distribution distance between two data blocks. JS-divergence solves the asymmetry problem of KL-divergence with respect to entropy, and can well represent the relationship between two data segments. The formula for JS-divergence is as follows:

$$JSD(P\|Q) = \sum_{x \in X} (p(x) \log \frac{2p(x)}{p(x)+q(x)} + q(x) \log \frac{2q(x)}{p(x)+q(x)}) \qquad (3)$$

Supposed that P and Q in formula (3) are two contiguous blocks of data. When the calculation result is less than the set threshold $\tau$, it is proved that the concept drift is not detected, and conversely, the concept drift is generated. In order to determine the size of the threshold $\tau$, the ACPJS uses the BootStrap to perform multiple trials in the form of put back sampling, and uses JS-divergence to calculate the results each time. Finally, sort from small to large and select top 95% as the confidence interval to find the appropriate threshold.

## 3.2 Concept Drift Processing of ACPJS Algorithm

The ACPJS uses the C4.5 decision tree algorithm to learn the n data blocks in the buffer, and finally uses the obtained n base classifiers to form a horizontal integrated classifier, as shown in Fig. 1. The main technical contributions of this integrated classifier are as follows:

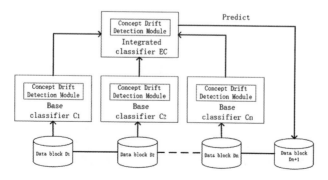

**Fig. 1.** Horizontal integration framework

- The classifier based on decision tree shows certain friendliness for continuous data, discrete data and high dimensional data.
- By using a plurality of base classifiers to integrate the classification model, it is possible to adapt to the instability of the data flow distribution. Moreover, the final processing result is the dynamic weighted average score of each base classifier, and the processing result is more accurate.
- The data in the buffer can be used repeatedly, reducing the running time of the subsequent test data stream block.

The integration model is integrated using dynamic weighting, and the update of C_weigh $t_j$ is updated by Eqs. (4) and (5).

$$MSE_r = \sum_c p(c)(1 - p(c))^2 \tag{4}$$

$$\text{C\_weigh} \, t_j = MSE_r - C\_err_j \tag{5}$$

$MSE_r$ represents the predicted mean square error of the classifier, which is related to the distribution of the category in the data block Sn + 1. $C\_err_j$ represents the error rate of each base classifier. $p(c)$ is the proportion of each category in $S_{n+1}$.

### 3.3    Implementation Process of ACPJS Algorithm

The ACPJS mainly includes the construction of integrated models, the detection of concept drift, the detection of noise data and the integration of model updates.

- Integrated classifier stage: The data needs to be divided into n blocks before the data enters the classifier, and then the n base classifiers are trained with the C4.5 decision tree. Finally, the n base classifiers are combined into an integrated classifier.
- The detection of noise data stage: Import data into integrated classifier and detect conceptual drift or noise data interference according to the methods in Sect. 3.1.
- The detection of concept drift stage: In the case where it is not possible to determine whether a concept drift occurs in the detection of noise data stage, the methods of Sect. 3.2 is used to perform the 2nd detection of the concept drift.
- Integration of model updates stage: Update the integration model by used the methods mentioned in Sect. 3.2.

## 4    Experimental Results and Analysis

### 4.1    Experimental Environment and Parameter Setting

The computer system is Windows 7, the CPU is 2.67 Hz and the memory size is 4G. The test platform of the algorithm is a large-scale online analysis open source platform (massive online analysis, MOA). The basic parameter settings are shown in Table 1.

**Table 1.** Experimental parameter setting

| Parameters | Meaning | Value |
|---|---|---|
| n | Number of base classifiers | 10 |
| $\tau$ | JS-divergence threshold | 0.25 |
| $\delta$ | Degree of confidence | 0.03 |
| c1 | Lower bound coefficient of noise | 2 |
| c2 | Upper bound coefficient of noise | 3 |

## Experiment of Concept Drift Detection

### Data Set

In the experimental evaluation of concept drift detection ability of ACPJS, we chose SEA data set, HyperPlane data set and KDD99 data set for experiments, and compare the experimental results with the classical concept drift detection algorithm CVFDT and DDM. In the experiment, the SEA size is 300k, including 10% noise data and 5 concept drifts. The HyperPlane size is 500k, including 10% noise data and 20 concept drifts; the KDD99 size is 490k and contains 48 concept drifts.

### Experimental Results and Analysis

As shown in Table 2, on the SEA, the ACPJS proposed in this paper accurately detects all the concept drift phenomena, and there is no missed or false detection. In contrast, CVFDT and DDM do not detect all concept drift, and there are certain missed detections. On the KDD99, although the ACPJS has missed detection and misdetection, it has better performance than the other two methods as a whole. On the HyperPlance, a special phenomenon has appeared. Although the ACPJS performs well in the number of false detections and the number of errors, it is not superior to the CVFDT algorithm and the DDM algorithm in the number of missed detections. Two reasonable explanations for this phenomenon as follows:

**Table 2.** Concept drift detection results

| Data set | Algorithm | Number of detection | Number of false positives | Number of omission |
|---|---|---|---|---|
| SEA | ACPJS | 278 | 0 | 0 |
|  | CVFDT | 247 | 2 | 0 |
|  | DDM | 261 | 1 | 0 |
| HyperPlane | ACPJS | 477 | 2 | 2 |
|  | CVFDT | 452 | 6 | 2 |
|  | DDM | 465 | 7 | 1 |
| KDD99 | ACPJS | 508 | 2 | 2 |
|  | CVFDT | 439 | 3 | 4 |
|  | DDM | 472 | 3 | 3 |

- In order to ensure the performance of noise monitoring on various data sets, the ACPJS uses the noise detection threshold with the best average detection effect, so it cannot be fully applied to the HyperPlance data environment;
- The size of the window is a fixed value. When the window is too small, there may be insufficient detection information, which may affect the experimental results.

In view of the above analysis, the ACPJS proposed in this paper shows better performance in the experimental results of concept drift detection. At the same time, it also finds the shortcomings of the proposed ACPJS and the direction of further research in the future.

## 4.2    Experiment of Anti-noise Detection

### Data Set

The anti-noise performance test evaluates the anti-noise performance of the ACPJS in different noisy environments. There are two main data sets used in the experiment: HyperPlane data set with continuous data type and the LED-drift data set with discrete data type. On these two data sets (noise range is 5% –30%), we compare ACPJS with other classic concept drift data stream classification algorithms (including OzaBag, ASHT-bagging, weighted-bagging, OzaBoost, DWCDS and MSRT).

### Experimental Results and Analysis

On the HyperPlane, the comparison of classification accuracy and noise rate of the algorithm is shown in Fig. 2. When the noise ratio is 5%, the classification accuracy of ACPJS is 0.7% lower than the ASHT-bagging. However, when the noise rate is increased to 10%, the classification accuracy of the ACPJS is 1% higher than the highest ASHT-bagging. Finally, the classification accuracy is 6% higher than the ASHT-bagging when the noise rate is increased to 30%. As shown in Fig. 3, On the LED-drift, ACPJS has been significantly better than others from the range of 5% to 30%. In summary, with the increase of noise content, ACPJS can effectively distinguish noise data and concept drift, so that the data stream mining model maintains high classification accuracy, and the model has better anti-noise performance.

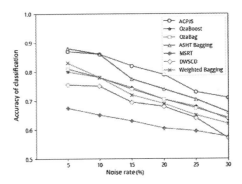

**Fig. 2.** Relationship between classification accuracy and noise rate in HyperPlane

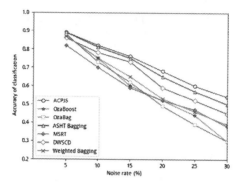

**Fig. 3.** Relationship between classification accuracy and noise rate of LED-drift.

## 4.3    Experiment of Validity Test of Multi-data Set Algorithm

### Data Set

In order to further verify the validity of ACPJS for concept drift detection, this paper compares the ACPJS with the classical OzaBag, OzaBoost, MSRTH and ASHT-bagging algorithms on the multi-objective concept data set. The data set used is 10% noise, the training set size is 400 K and the test set size is 200 k. The two data sets used are as follows:

- The Waveform-drift (WF-drift) dataset is a Waveform dataset with noisy data. The data type of the data set is discrete and the size is 21 dimensions, wherein the 20-dimensional has a concept drift attribute;
- SEA contains four target concepts, namely SEA-1, SEA-2, SEA-3 and SEA-4.

### Experimental Results and Analysis

As shown in Table 3, on the four conceptual SEA datasets, several classical algorithms maintain similar classification accuracy, but ACPJS classification accuracy is higher. On the Waveform-drift (WF-drift), the ACPJS has higher classification accuracy than the other three classical algorithms (OzaBag, OzaBoost and MSRT), and is almost identical to the classification accuracy of ASHT-bagging. In summary, the ACPJS can

**Table 3.** Comparison of classification accuracy with other algorithms

| Data set | ACPJS | OzaBag | OzaBoost | MSRT | ASHT-bagging |
|----------|-------|--------|----------|------|--------------|
| WF-drift | 0.843 | 0.797 | 0.801 | 0.337 | 0.845 |
| LED-drift | 0.751 | 0.425 | 0.740 | 0.675 | 0.742 |
| SEA-1 | 0.896 | 0.874 | 0.855 | 0.768 | 0.865 |
| SEA-2 | 0.865 | 0.855 | 0.862 | 0.732 | 0.824 |
| SEA-3 | 0.867 | 0.838 | 0.837 | 0.710 | 0.857 |
| SEA-4 | 0.831 | 0.818 | 0.820 | 0.718 | 0.795 |

effectively detect the occurrence of concept drift on multi-class data sets with noises and reconstruct the data mining classification model, so that the system maintains a high classification accuracy and the applicability of the algorithm is better.

## 5  Conclusions

This paper proposes an anti-noise concept drift detection algorithm based on JS-divergence for concept drift data stream mining problem with noisy data. The Heoffding Bounds algorithm and JS-divergence are used to detect noise data and concept drift, and the concept drift is processed by using a dynamically weighted horizontal integrated classifier. The ACPJS effectively reduces the false detection rate of the concept drift detection and improves the classification accuracy of the algorithm. At the same time, the algorithm updates the base classifier dynamically, avoiding the waste of resources caused by continuous passive updating. The experimental results show that compared with the existing concept drift detection algorithm, the proposed algorithm has better anti-noise performance, and has a great improvement in concept drift detection rate and classification accuracy.

**Acknowledgment.** The research work was supported by the National Natural Science Foundation of China under Grant No. 61603083, the Fundamental Research Funds of the Central Universities under Grant No. N162304009, N182303036, the Major Project of Science and Technology Research of Hebei University under Grant No. ZD2017303, and open research fund of State Key Laboratory of Management and Control for Complex Systems, Institute of Automation, Chinese Academy of Sciences under grant No. 20180105.

## References

1. Xu, W., Qin, Z., Chang, Y.: Clustering feature decision trees for semi-supervised classification from high-speed data streams. J. Zhejiang Univ. Sci. C **12**(8), 615–628 (2011)
2. Brzezinski, D., Stefanowski, J.: Reacting to different types of concept drift: the accuracy updated ensemble algorithm. IEEE Transact. Neural Netw. Learn. Syst. **25**(1), 81–94 (2014)
3. Schlimmer, J.C., Granger, R.: Incremental learning from noisy data. Mach. Learn. **1**(3), 317–354 (1986)
4. Gama, J., Zliobaite, I., Bifet, A.: A survey on concept drift adaptation. ACM Comput. Surv. **46**(4), 44 (2014)
5. Dasu, T., Krishnan, S., Venkatasubramanian, S.: An information-theoretic approach to detecting changes in multi-dimensional data streams. In: Proceedings of the Symposium on the Interface of Statistics, Computing Science, and Applications, pp. 1–24 (2006)
6. Hulten, G., Spencer, L., Domingos, P.: Mining time-changing data streams. In: Proceedings of the ACM SIGKDD International Conference on Knowledge Discovery and Data Mining (2001)
7. Gama, J., Medas, P., Castillo, G.: Learning with drift detection. SBIA Braz. Symp. Artif. Intell. **3171**(17), 286–295 (2004)
8. Domingos, P., Hulten, G.: Mining high-speed data streams. In: Proceedings of the Sixth ACM SIGKDD International Conference on Knowledge Discovery and Data Mining, pp. 71–80, Boston (2000)

9. Susnjak, T., Barczak, A.L.C., Hawick, K.A.: Adaptive cascade of boosted ensembles for face detection in concept drift. Neural Comput. Appl. **21**(4), 671–682 (2011)
10. Scholz, M., Klinkenberg, R.: Boosting classifiers for drifting concepts. Intell. Data Anal. **11** (1), 3–28 (2007)
11. Liu, A., Lu, J., Liu, F., Zhang, G.: Accumulating regional density dissimilarity for concept drift detection in data streams. Pattern Recognit. **76**, 256–272 (2018)
12. Song, G., Ye, Y., Zhang, H., Xu, X., Lau, R.Y.K., Liu, F.: Dynamic clustering forest: an ensemble framework to efficiently classify textual data stream with concept drift. Inf. Sci. **357**, 125–143 (2016)
13. Rad, R.H., Haeri, M.A.: Hybrid Forest: A Concept Drift Aware Data Stream Mining Algorithm (2019)
14. Benczúr, A.A., Kocsis, L., Pálovics, R.: Reinforcement learning, unsupervised methods, and concept drift in stream learning. In: Sakr, S., Zomaya, A. (eds.) Encyclopedia of Big Data Technologies, pp. 1–8. Springer, Cham (2018). https://doi.org/10.1007/978-3-319-63962-8
15. De Mello, R.F., Vaz, Y., Grossi, C.H., Bifet, A.: On learning guarantees to unsupervised concept drift detection on data streams. Expert Syst. Appl. **117**, 90–102 (2019)
16. Lavaire, J.D., Singh, A., Yousef, M., Singh, S., Yue, X.: Dimensional scalability of supervised and unsupervised concept drift detection: an empirical study. In: IEEE International Conference on Big Data. IEEE (2015)
17. Song, X., He, H., Niu, S., Gao, J.: A data streams analysis strategy based on hoeffding tree with concept drift on hadoop system. In: International Conference on Advanced Cloud Big Data. IEEE (2017)
18. Hulten, G., Spencer, L., Domingos, P.M.: Mining time-changing data streams. In: International Conference on Knowledge Discovery and Data Mining, pp. 97–106. ACM, New York (2001)
19. Oza, N.C.: Online Bagging and Boosting. In: IEEE International Conference on Systems (2006)
20. Bifet, A., Holmes, G., Kirkby, R.B., Pfahringer, B.: MOA: massive online analysis. J. Mach. Learn. Res. **11**(2), 1601–1604 (2010)
21. Kolter, J.Z., Maloof, M.A.: Dynamic weighted majority: a new ensemble method for tracking concept drift. IEEE Computer Society (2003)
22. Bifet, A., Holmes, G., Pfahringer, B., Kirkby, R., Gavalda, R.: New ensemble methods for evolving data streams. In: Knowledge Discovery and Data Mining (2009)
23. Li, P., Hu, X., Wu, X.: Mining concept-drifting data streams with multiple semi-random decision trees. In: Tang, C., Ling, C.X., Zhou, X., Cercone, N.J., Li, X. (eds.) ADMA 2008. LNCS (LNAI), vol. 5139, pp. 733–740. Springer, Heidelberg (2008). https://doi.org/10.1007/978-3-540-88192-6_78
24. Zhu, Q., Hu, X., Zhang, Y., Li, P., Wu, X.: A double-window-based classification algorithm for concept drifting data streams, pp. 639–644 (2010)

# A Collaborative Neurodynamic Approach to Sparse Coding

Hangjun Che[1,3]($\boxtimes$), Jun Wang[1,2,3], and Wei Zhang[4]

[1] Department of Computer Science, City University of Hong Kong,
Kowloon, Hong Kong
hjche2-c@my.cityu.edu.hk, jwang.cs@cityu.edu.hk
[2] School of Data Science, City University of Hong Kong, Kowloon, Hong Kong
[3] Shenzhen Research Institute, City University of Hong Kong, Shenzhen, China
[4] Key Laboratory of Intelligent Information Processing and Control of Chongqing
Municipal Institutions of Higher Education, Chongqing Three Gorges University,
Wanzhou, Chongqing 404100, China
cqec126@126.com

**Abstract.** In this paper, a collaborative neurodynamic approach is proposed for sparse coding. As the formulated sparse coding optimization problem with $l_0$-norm objective function is NP-hard, it is reformulated as a global optimization problem based on an inverted Gaussian function. A group of neurodynamic optimization models is employed to solve the reformulated problem by gradually decreasing the value of the parameter of the inverted Gaussian function. The experimental results show the superior performance of the proposed approach.

**Keywords:** Sparse coding ·
Collaborative neurodynamic optimization · Signal reconstruction

## 1 Introduction

Sparse coding is to find sparse representations or solutions of a given problem. It has various applications such as sparse channel estimation [1], hierarchical data aggregation [27], face recognition [20], medical image processing [17], directions-of-arrival estimation [3], just to name a few. The sparsification of underdetermined systems of linear equations can be formulated as the following constrained $l_0$-minimization problem:

$$\min \quad \|x\|_0 \quad \text{s.t.} \quad Ax = b, \tag{1}$$

This work was supported in part by the Research Grants Council of the Hong Kong Special Administrative Region of China, under Grants 11208517 and 11202318, in part by the National Natural Science Foundation of China under grant 61673330, and in part by International Partnership Program of Chinese Academy of Sciences under Grant GJHZ1849.

H. Lu et al. (Eds.): ISNN 2019, LNCS 11554, pp. 454–462, 2019.
https://doi.org/10.1007/978-3-030-22796-8_48

where $x \in \Re^n$, $A \in \Re^{m \times n}$ with $m < n$, $\|x\|_0$ is defined as the number of non-zero components of $x$. If $\|x\|_0 = k$, then $x$ is called $k$−sparse. In [2], it is proven that if $A$ satisfies the restricted isometry property and $x$ is $k$−sparse, then $x$ can be reconstructed by solving problem (1).

As the $l_0$-norm objective function, problem (1) is NP-hard and computationally intractable. Therefore, many alternative functions are used to replace the $l_0$-norm function. In [12,25,32], $\|x\|_1$ is used to replace $\|x\|_0$, then problem (1) is solved by convex optimization approach. In [7], $l_q$-norm is used as the objective function where $0 < q \le 1$. In [36], the difference of $\|x\|_1$ and $\|x\|_2$ is used as the objective function, and the proposed algorithm $DCAL_{1-2}$ is proven to be almost sure convergent to the global minima of the formulated optimization problem.

In [8,10], the sum of inverted Gaussian functions shows well approximation to $l_0$ norm. In [22], it is proven that the solution is sparse under the RIP condition. Unfortunately, the sum of inverted Gaussian functions is nonconvex which makes the reformulated problem (1) as a global optimization problem.

In recent years, various neurodynamic models are proposed for constrained optimizations such as nonsmooth optimization [15,19], nonconvex optimization [16], variational inequalities and related optimization [18], bilevel optimization [11,24], minimax optimization [13], multiobjective optimization [14,35], distributed optimization [21,33,34], just to name a few. As most single-model neurodynamic approaches would be stuck into local minima at global optimization problems, in [4,5,28,31], collaborative neurodynammic optimization approach (CNO) is proposed for global optimization by employing a group of neurodynamic models with particle swarm optimization (PSO) to reinitialize the neuronal states iteratively. CNO is proven to be almost sure convergent to the global optima [28]. In addition, CNO is applied in model predictive control [29,30], non-negative matrix factorization [6] and emission dispatch [26].

In this paper, problem (1) is formulated as a global optimization problem by using the sum of inverted Gaussian functions to approximate the $l_0$ norm. Collaborative neuroydnamic optimization approach is proposed for solving the reformulated problem. In each model, the value of parameter $\sigma$ is designed to decrease over time until the required sparsification is achieved. The paper is organized as follows. In Sect. 2, preliminaries are reviewed. In Sect. 3, collaborative neurodynamic optimization is described. In Sect. 4, experimental results are discussed. The conclusions are made in Sect. 5.

## 2  Preliminaries

### 2.1  Problem Formulation

According to the definition of $l_0$ norm, $\|x\|_0$ can be equivalently expressed as follows [8]:

$$\|x\|_0 = n - \sum_{i=1}^{n} \delta(x_i), \tag{2}$$

where $\delta(x_i)$ denotes the unit impulse function defined as

$$\delta(x_i) = \begin{cases} 1, & x_i = 0; \\ 0, & x_i \neq 0. \end{cases} \tag{3}$$

To approximate the $l_0$-norm function of (1), following inverted Gaussian function is introduced [10]:

$$g(x_i) = 1 - e^{-x_i^2/\sigma^2}, \tag{4}$$

where $\sigma$ is a positive parameter. Figure 1 shows that the larger value of $\sigma$ makes $g(x_i)$ smoother, but the worse approximation to $l_0$ norm. Conversely, the smaller value of $\sigma$ makes $g(x_i)$ better approximation. For $|x_i| \ll \sigma$, $g(x_i) \approx 0$, and for $|x_i| \gg \sigma$, $g(x_i) \approx 1$. Note that $\lim_{\sigma \to 0} g(x_i) = 1 - \delta(x_i)$ and $\lim_{\sigma \to 0} \sum_{i=1}^{n} g(x_i) = \|x\|_0$.

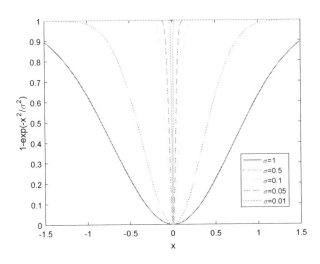

**Fig. 1.** Inverted Gaussian function with various values of $\sigma$.

Let $f_\sigma(x) = n - \sum_{i=1}^{n} e^{-x_i^2/\sigma^2}$, problem (1) is reformulated approximately as the following global optimization problem:

$$\min f_\sigma(x) \quad \text{s.t.} \quad Ax = b. \tag{5}$$

## 2.2  Existing Neurodynamic Model

In this section, a one-layer neurodynamic model [9] is introduced for sparsification:

$$\epsilon \frac{dx}{dt} = -Px - (I - P)\nabla f(x) + q \tag{6}$$

where $\epsilon$ is a time constant, $x$ is the state vector, $I$ is the identity matrix, $P = A^T(AA^T)^{-1}A$, $q = A^T(AA^T)^{-1}b$, and $\nabla f(x) = (\partial f(x)/\partial x_1, ..., \partial f(x)/\partial x_n)^T$ is the gradient of the objective function.

# 3    Collaborative Neurodynamic Optimization Approach

As the minimization of the sum of inverted Gaussian functions is depended on the value of $\sigma$, decreasing $\sigma$ over time is a reasonable strategy to have a sparse solution. Neurodynamic model (6) with an additional variable is described as follows:

$$\epsilon_1 \frac{dx}{dt} = -Px - (I - P)\nabla f_\sigma(x) + q$$
$$\epsilon_2 \frac{d\sigma}{dt} = -(1 - k/\|x\|_0)\sigma, \tag{7}$$

where $\epsilon_1 < \epsilon_2$. $\sigma$ stops decreasing until $\|x\|_0 = k$.

Let $N$ neurodynamic optimization models be employed. The collaborative neurodynamic optimization approach for solving (5) is described as follows:

- *Initialization*: Initialize neuronal states $x_i(0)$ randomly where $i = 1, ..., N$. Set the individual solution $x_i^p(0) = x_i(0)$ and the group best solution $x_g(0) = \arg\min_{x_i}(f(x_i(0)))$. Set the error tolerance $\varepsilon$ and the time constants $\epsilon_1, \epsilon_2$.
- *Main loop*:
  1. Reinitialize the neuronal states based on PSO.
  2. Compute equilibrium points $\bar{x}_i(j)$ based on (9).
  3. Update individual solution $x_i^p(j)$ if $f(\bar{x}_i(j)) < f(x_i^p(j-1))$.
  4. Update group best solution $x_g(j)$ if $f(x_i^p(j)) < f(x_g(j-1))$.
  5. $j = j + 1$.
- *Termination*: Terminate the iteration process if $\|x_g(j) - x_g(j-r)\|_2 < \varepsilon$ where $r$ is a given positive integer.

# 4    Experimental Results

In the experiment, let $N = 3$, $\epsilon_1 = 10^{-4}$ and $\epsilon_2 = 10^{-2}$. The matrix $A$ and the $k$-sparse signal $x$ are generated as follows:

1. Each signal $x$ is randomly generated from $[-10, 10]^n$ where $n - k$ components are randomly chosen and set to 0.
2. $A$ is generated from standard normal distribution $\mathcal{N}(0, 1)$.
3. $b = Ax$.

To measure the reconstruction quality, the relative error is defined as follows:

$$\sqrt{\frac{\sum_{i=1}^{n} |x_i - \hat{x}_i|^2}{\sum_{i=1}^{n} |x_i|^2}} \tag{8}$$

where $\hat{x}$ is the recovered signal. A signal is considered to be recovered successfully if the relative error is smaller than $10^{-2}$ [8].

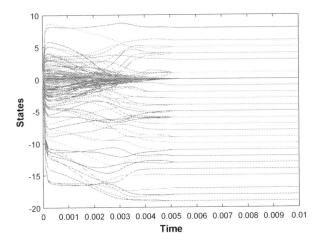

**Fig. 2.** A snapshot of the transient behaviors of states.

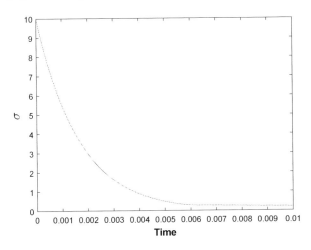

**Fig. 3.** A snapshot of the transient behavior of $\sigma$.

Figure 2 shows a snapshot of the transient behaviors of states where $k = 50$, $m = 90$, $n = 128$. It shows that most of states are convergent to zero. Figures 3 and 4 show respectively a snapshot of transient behavior of $\sigma$ and a snapshot of the transient behavior of $\|x\|_0$ where $k = 50$, $m = 90$, $n = 128$. It shows that the value of $\sigma$ decreases until $\|x\|_0$ achieves required sparsification. Figure 5 shows the transient behaviors of the group best solution where $N = 1, 2, 3, 4$. It shows that the proposed approach with more neurodynamic optimization models needs less iterations to achieve the required sparsification.

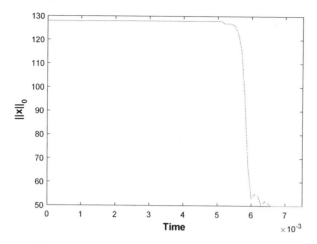

**Fig. 4.** A snapshot of the transient behavior of $\|x\|_0$.

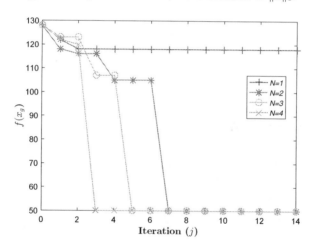

**Fig. 5.** Transient behaviors of the group best solution where $N = 1, 2, 3, 4$.

Furthermore, the proposed approach is used to compare with eight existing sparse signal reconstruction methods including OMP [25], $l_1$-LS [12], CoSaMP [23], $l_q$ [7], SL$_0$ [22], YALL1 [32], DCAL$_{1-2}$ [36] and RNN$_{gy}$ [8]. For each pair $(k, m)$, nine methods are run in 100 trials, Fig. 6 shows that the percentage of achieving sparsification of the proposed approach increases faster than other methods with the increase of measurement under the sparsity level $k = 50$. Figure 7 shows that the percentage of achieving sparsification of the proposed approach decreases slower than other methods with the increase of sparsity level under measurement $m = 80$.

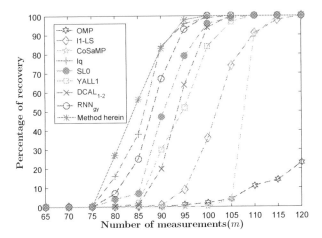

**Fig. 6.** The percentage of recovered signal with relative error smaller than $10^{-2}$ for various $m$ with fixed sparsity level $k = 50$ in 100 trials.

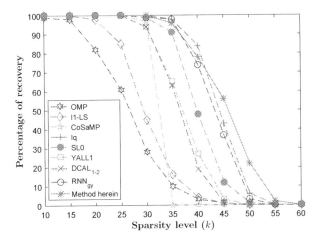

**Fig. 7.** The percentage of recovered signal with relative error smaller than $10^{-2}$ for various $k$ with fixed measurement $m = 80$ in 100 trials.

## 5    Conclusions

The constrained $l_0$-minimization optimization problem is reformulated as a global optimization problem by minimizing an inverted Gaussian function with a tunable parameter. Initialized by using particle swarm optimization repeatedly, a group of neurodynamic models is employed for solving the reformulated problem. The experimental results show the superiority of the proposed approach.

# References

1. Berger, C.R., Wang, Z., Huang, J., Zhou, S.: Application of compressive sensing to sparse channel estimation. IEEE Commun. Mag. **48**(11), 164–174 (2010)
2. Candès, E.J., Romberg, J., Tao, T.: Robust uncertainty principles: Exact signal reconstruction from highly incomplete frequency information. IEEE Trans. Inf. Theory **52**(2), 489–509 (2006)
3. Carlin, M., Rocca, P., Oliveri, G., Viani, F., Massa, A.: Directions-of-arrival estimation through Bayesian compressive sensing strategies. IEEE Trans. Antennas Propag. **61**(7), 3828–3838 (2013)
4. Che, H., Wang, J.: A two-timescale duplex neurodynamic approach to biconvex optimization. IEEE Trans. Neural Netw. Learn. Syst. (2018). https://doi.org/10.1109/TNNLS.2018.2884788
5. Che, H., Wang, J.: A collaborative neurodynamic approach to global and combinatorial optimization. Neural Netw. **114**, 15–27 (2019)
6. Fan, J., Wang, J.: A collective neurodynamic optimization approach to nonnegative matrix factorization. IEEE Trans. Neural Netw. Learn. Syst. **28**(10), 2344–2356 (2017)
7. Foucart, S., Lai, M.J.: Sparsest solutions of underdetermined linear systems via $l_q$-minimization for $0 < q <= 1$. Appl. Comput. Harmon. Anal. **26**(3), 395–407 (2009)
8. Guo, C., Yang, Q.: A neurodynamic optimization method for recovery of compressive sensed signals with globally converged solution approximating to $l_0$ minimization. IEEE Trans. Neural Netw. Learn. Syst. **26**(7), 1363–1374 (2015)
9. Guo, Z., Liu, Q., Wang, J.: A one-layer recurrent neural network for pseudoconvex optimization subject to linear equality constraints. IEEE Trans. Neural Netw. **22**(12), 1892–1900 (2011)
10. Guo, Z., Wang, J.: A neurodynamic optimization approach to constrained sparsity maximization based on alternative objective functions. In: Proceedings of International Joint Conference on Neural Networks (IJCNN), pp. 273–280. IEEE (2010)
11. He, X., Li, C., Huang, T., Li, C., Huang, J.: A recurrent neural network for solving bilevel linear programming problem. IEEE Trans. Neural Netw. Learn. Syst. **25**(4), 824–830 (2014)
12. Kim, S.J., Koh, K., Lustig, M., Boyd, S., Gorinevsky, D.: An interior-point method for large-scale $l_1$-regularized least squares. IEEE J. Sel. Top. Sig. Process. **1**(4), 606–617 (2007)
13. Le, X., Wang, J.: A two-time-scale neurodynamic approach to constrained minimax optimization. IEEE Trans. Neural Netw. Learn. Syst. **28**(3), 620–629 (2017)
14. Leung, M.F., Wang, J.: A collaborative neurodynamic approach to multiobjective optimization. IEEE Trans. Neural Netw. Learn. Syst. **29**(11), 5738–5748 (2018)
15. Li, G., Yan, Z., Wang, J.: A one-layer recurrent neural network for constrained nonsmooth invex optimization. Neural Netw. **50**, 79–89 (2014)
16. Li, G., Yan, Z., Wang, J.: A one-layer recurrent neural network for constrained nonconvex optimization. Neural Netw. **61**, 10–21 (2015)
17. Lingala, S.G., Jacob, M.: Blind compressive sensing dynamic MRI. IEEE Trans. Med. Imaging **32**(6), 1132–1145 (2013)
18. Liu, Q., Huang, T., Wang, J.: One-layer continuous-and discrete-time projection neural networks for solving variational inequalities and related optimization problems. IEEE Trans. Neural Netw. Learn. Syst. **25**(7), 1308–1318 (2014)

19. Liu, Q., Wang, J.: A one-layer projection neural network for nonsmooth optimization subject to linear equalities and bound constraints. IEEE Trans. Neural Netw. Learn. Syst. **24**(5), 812–824 (2013)

20. Liu, Q., Wang, J.: $L_1$-minimization algorithms for sparse signal reconstruction based on a projection neural network. IEEE Trans. Neural Netw. Learn. Syst. **27**(3), 698–707 (2016)

21. Liu, Q., Yang, S., Wang, J.: A collective neurodynamic approach to distributed constrained optimization. IEEE Transactions on Neural Networks and Learning Systems **28**(8), 1747–1758 (2017)

22. Mohimani, H., Babaie-Zadeh, M., Jutten, C.: A fast approach for overcomplete sparse decomposition based on smoothed $l_0$ norm. IEEE Trans. Sig. Process. **57**(1), 289–301 (2009)

23. Needell, D., Tropp, J.A.: Cosamp: Iterative signal recovery from incomplete and inaccurate samples. Appl. Comput. Harmon. Anal. **26**(3), 301–321 (2009)

24. Qin, S., Le, X., Wang, J.: A neurodynamic optimization approach to bilevel quadratic programming. IEEE Trans. Neural Netw. Learn. Syst. **28**(11), 2580–2591 (2017)

25. Tropp, J.A., Gilbert, A.C.: Signal recovery from random measurements via orthogonal matching pursuit. IEEE Trans. Inf. Theory **53**(12), 4655–4666 (2007)

26. Wang, T., He, X., Huang, T., Li, C., Zhang, W.: Collective neurodynamic optimization for economic emission dispatch problem considering valve point effect in microgrid. Neural Netw. **93**, 126–136 (2017)

27. Xu, X., Ansari, R., Khokhar, A., Vasilakos, A.V.: Hierarchical data aggregation using compressive sensing (HDACS) in WSNs. ACM Trans. Sens. Netw. (TOSN) **11**(3), 45 (2015)

28. Yan, Z., Fan, J., Wang, J.: A collective neurodynamic approach to constrained global optimization. IEEE Trans. Neural Netw. Learn. Syst. **28**(5), 1206–1215 (2017)

29. Yan, Z., Le, X., Wang, J.: Tube-based robust model predictive control of nonlinear systems via collective neurodynamic optimization. IEEE Trans. Ind. Electron. **63**(7), 4377–4386 (2016)

30. Yan, Z., Wang, J.: Nonlinear model predictive control based on collective neurodynamic optimization. IEEE Trans. Neural Netw. Learn. Syst. **26**(4), 840–850 (2015)

31. Yan, Z., Wang, J., Li, G.: A collective neurodynamic optimization approach to bound-constrained nonconvex optimization. Neural Netw. **55**, 20–29 (2014)

32. Yang, J., Zhang, Y.: Alternating direction algorithms for $l_1$ problems in compressive sensing. SIAM J. Sci. Comput. **33**(1), 250–278 (2011)

33. Yang, S., Liu, Q., Wang, J.: Distributed optimization based on a multiagent system in the presence of communication delays. IEEE Trans. Syst. Man Cybern.: Syst. **47**(5), 717–728 (2017)

34. Yang, S., Liu, Q., Wang, J.: A multi-agent system with a proportional-integral protocol for distributed constrained optimization. IEEE Trans. Autom. Control. **62**(7), 3461–3467 (2017)

35. Yang, S., Liu, Q., Wang, J.: A collaborative neurodynamic approach to multiple-objective distributed optimization. IEEE IEEE Trans. Neural Netw. Learn. Syst. **29**(4), 981–992 (2018)

36. Yin, P., Lou, Y., He, Q., Xin, J.: Minimization of $l_{1-2}$ for compressed sensing. SIAM J. Sci. Comput. **37**(1), A536–A563 (2015)

# A Kernel Fuzzy C-means Clustering Algorithm Based on Firefly Algorithm

Chunying Cheng[✉] and Chunhua Bao

College of Computer Science and Technology,
Inner Mongolia University for Nationalities, Tongliao 028043, China
chengchunying_80@163.com

**Abstract.** In this paper, the KFCM algorithm selects the Gaussian kernel function, maps the data into the high-dimensional feature space for clustering, and uses the optimal clustering center of the firefly algorithm as the initial value of KFCM, and then processes it through KFCM. Based on class analysis, a kernel fuzzy C-means clustering algorithm based on firefly algorithm (FA-KFCM) is proposed. Numerical experiments results show that FA-KFCM is superior to other algorithms in clustering accuracy and time efficiency.

**Keywords:** Firefly algorithm · Clustering · Kernel fuzzy C-means clustering · The classical UCI data set

## 1 Introduction

The kernel fuzzy C-means clustering algorithm introduces the idea of kernel in the traditional FCM algorithm, solves the problem of linear indivisibility, and realizes effective clustering of various data structures. Currently, it has been widely applied in many fields, but there are still many problems, such as sensitivity to the initial clustering center, long calculation time, diversity of kernel functions and parameter selection. [1] Therefore, we can use the kernel method to improve the accuracy of clustering.

A new swarm intelligence algorithm, the firefly algorithm (FA) [2], was proposed by Yang in 2008. FA is an interdisciplinary research achievement using swarm intelligence and stochastic algorithms, and it is strong and fast. The firefly algorithm has become an increasingly important tool of Swarm Intelligence that has been applied in almost all areas of optimization, as well as engineering practice.

FA has been used in several fields, for example, for solving minimizing the makespan for the permutation flow shop scheduling problem [3], solving the task graph scheduling problem [4] and solving the Job shop scheduling problem [5]. This study applies The FA-KFCM algorithm overcomes the FCM algorithm trapped local optimum and being sensitive to initial value effectively, and enhances the capacity of local search of firefly algorithm. And the validity of the algorithm is verified by Iris, Cmc, Wine and Zoo data in the public dataset.

© Springer Nature Switzerland AG 2019
H. Lu et al. (Eds.): ISNN 2019, LNCS 11554, pp. 463–468, 2019.
https://doi.org/10.1007/978-3-030-22796-8_49

## 2  FA-KFCM Algorithm

In this section, we first describe the firefly algorithm and kernel fuzzy C-means. Then, we combine the firefly algorithm kernel fuzzy C-means with mechanism obtain our FA-KFCM algorithm.

### 2.1  FA Algorithm

Firefly algorithm is based on a physical formula of light intensity $I$ hat decreases with the increase of the square of the distance $r^2$. However, as the distance from the light source increases, the light absorption causes that light becomes weaker and weaker. These phenomena can be associated with the objective function to be optimized.

(1)  All fireflies are unisex.
(2)  Their attractiveness is proportional to their light intensity.
(3)  The light intensity of a firefly is affected or determined by the landscape of the fitness function.

In the standard firefly algorithm, the light intensity $I$ of a firefly representing the solution s is proportional to the value of fitness function $I(x) \propto f(x)$, whilst the light intensity $I(r)$ varies according to the following equation: [6]

$$I(r) = I_0 \times e^{-\gamma r^2} \tag{1}$$

The attractiveness $\beta_0$ of fireflies is proportional to their light intensities $I(r)$. The attraction $\beta$ can be described by Eq. (1)

$$\beta(r) = \beta_0 \times e^{-\gamma r^2} \tag{2}$$

Where $\beta_0$ is the attractiveness at $r = 0$. The light intensity $I$ and attractiveness $\beta$ are in some way synonymous.

The distance between any two fireflies $x_i$ and $x_j$ in the basic firefly algorithm is calculated from the Euclidean distance:

$$r_{ij} = ||x_i - x_j|| = \sqrt{\sum_{k=1}^{n} (x_{i,k} - x_{j,k})^2} \tag{3}$$

Firefly $i$ is moved by the attraction of its bright firefly $j$, The position of the movement can be expressed as:

$$x_i = x_i + \beta_0 \times e^{-\gamma r_{ij}^2} \times (x_j - x_i) + \alpha(rand - 0.5) \tag{4}$$

Where $x_i$, $x_j$ are the positions of fireflies $i$ and $j$ in the solution space; is the step factor, which is the constant on the interval $[0, 1]$; $rand$ is the random factor, which obeys the interval $[0, 1]$ Evenly distributed.

## 2.2   KFCM Algorithm

The kernel function clustering method has a much better performance than the classical clustering algorithm. The kernel clustering method performs better clustering of samples by non-linear mapping of samples of the input space.

If the sets $Y = \{y_j | j = 1, 2, \ldots, M\}$, the objective function of fuzzy C-means is as shown in formula (5)

$$J_b(U, V) = \sum_{j=1}^{M} \sum_{i=1}^{C} u_{ji}^{m} \|y_j - v_i\|^2 \tag{5}$$

Where $u_{ji}$ is

$$\forall j, \sum_{i=1}^{C} u_{ji} = 1; \forall j, i, \mu_{ji} \in [0, 1]; \forall i, \sum_{j=1}^{M} \mu_{ji} > 0 \tag{6}$$

Using Lagrange multiplier method:

$$\bar{J}(U, V, \delta) = J_b(U, V) + \sum_{j=1}^{M} \delta_j (\sum_{i=1}^{C} \mu_{ji} - 1) \tag{7}$$

Introducing nonlinear mapping here $\varphi : y \to \varphi(y)$

$$\|\varphi(y_j) - \varphi(v_i)\| = KFCM(x_j, x_j) + KFCM(v_i, v_i) - 2KFCM(y_j, v_i) \tag{8}$$

The objective function of KFCM is:

$$J_\varphi = \sum_{j=1}^{M} J_j = \sum_{j=1}^{M} \sum_{i=1}^{C} \mu_{ji}^{b} \|\varphi(y_j) - \varphi(y_i)\|^2 \tag{9}$$

Here the kernel function selects the Gaussian function as:

$$KFCM(y, x) = \exp[-(y - x)^2 / \sigma^2] \tag{10}$$

Substituting Eq. (10) into Eq. (8), then Eq. (9) is:

$$J_\varphi = 2 \sum_{j=1}^{M} \sum_{i=1}^{C} \mu_{ji}^{b} [1 - KFCM(y_j, v_i)] \tag{11}$$

For $J_\varphi$, the partial derivatives of $u$ and $v$ are separately obtained as partial derivatives

$$v_j = \frac{\sum_{j=1}^{M} \mu_{ji}^b KFCM(y_j, v_i) y_j}{\sum_{j=1}^{M} \mu_{ji}^b KFCM(y_j, v_i)} \tag{12}$$

$$\mu_{ji} = \frac{(1 - KFCM(x_j, v_i))^{-1/(b-1)}}{\sum_{i=1}^{C} (1 - KFCM(x_j, v_i))^{-1/(b-1)}} \tag{13}$$

Kernel fuzzy C-means first calculates the kernel function by the formula (10), and updates the membership matrix according to the formulas (12) and (13) until the optimal cluster center is found, and then the final clustering result is obtained.

### 2.3    The Proposed FA-KFCM Algorithm

In the FA-KFCM algorithm, the position of each firefly represents a clustering center, expressed in terms of vector $V = \{v_1, v_2, \cdots, v_C\}$, where $v_i$ is the $i$ th cluster center. The light intensity of the firefly is determined by the objective function of the fuzzy clustering. According to the characteristics of the firefly algorithm, the light intensity function of the firefly can be defined as:

$$I(V) = \frac{1}{1 + J_\varphi(U, V)} \tag{14}$$

The specific algorithm steps of the KFCM algorithm based on the FA are:

Step 1: Initialize the light absorption coefficient $\gamma$, randomized parameter $\alpha$, Maximum number of iterations $T_{max}$, Maximum attraction $\beta_0$, C and b.
Step 2: Initialize the position of the firefly $V_1, V_2, \cdots, V_N$.
Step 3: Calculate the $KFCM(y_j, v_i)$.
Step 4: For each firefly, calculate the $U$ according to Eq. (12), calculate the light intensity of each firefly $I(V_j)$ according to Eq. (14).
Step 5: Compare the light intensity of fireflies. If $I(V_i) > I(V_j)$, it means that firefly $j$ is in a good position, attract fireflies $i$ to move to itself, and calculate the attraction and update position according to formula (2) and formula (4).
Step 6: According to (12), update the membership matrix of the fireflies.
Step 7: According to (14), recalculate the light intensity of the fireflies.
Step 8: Repeats steps (5) to (7), until the finds out the number of iterations is found.
Step 9: Output result.

### 2.4    Experimental Evaluation

This article uses MATLAB7.0 as a tool for programming under the Windows 8 operating system. The experiments used Iris, Cmc, Wine and Zoo public dataset in the UCI dataset to test the clustering accuracy and time efficiency of the FA-KFCM algorithm proposed in this paper, and compares it with the references [7] and [8]. The results are shown in Table 1.

**Table 1.** Data experiment sample dataset.

| Dataset | Class | Dimension | Number of samples |
|---------|-------|-----------|-------------------|
| IRIS | 3 | 4 | 180 |
| CMC | 3 | 9 | 1800 |
| WINE | 3 | 13 | 150 |
| ZOO | 3 | 8 | 110 |

In this paper, the number of clusters is 3, fuzzy index $b = 2$, maximum iteration number $T_{max} = 100$, light absorption coefficient $\gamma = 0.9$, Maximum attraction $\beta_0 = 1$, randomized parameter $\alpha = 0.1$. After running the current algorithm 50 times, the results are shown in Table 2.

**Table 2.** Comparison of clustering results of three algorithms on dataset.

| Algorithm | Average number of iterations | | | | The maximum number of iterations | | | | Mean fitness value | | | |
|-----------|------|------|------|------|------|------|------|------|------|------|------|------|
| | Iris | Cmc | Wine | Zoo | Iris | Cmc | Wine | Zoo | Iris | Cmc | Wine | Zoo |
| Ref [7] | 100 | 100 | 100 | 100 | 32 | 37 | 41 | 36 | 89.4 | 5545 | 16487 | 3531 |
| Ref [8] | 100 | 100 | 100 | 100 | 26 | 29 | 34 | 22 | 86.6 | 5378 | 15799 | 3269 |
| FA-KFCM | 100 | 100 | 100 | 100 | 21 | 20 | 25 | 19 | 85.1 | 5217 | 15541 | 3120 |

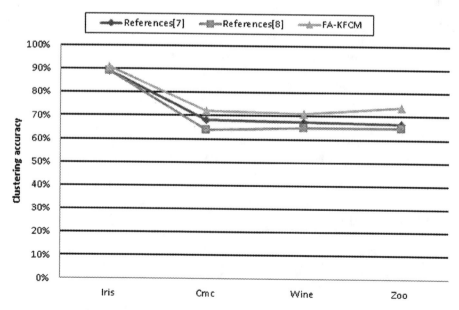

**Fig. 1.** Comparison of clustering precision of three algorithms.

From the comparison results of Table 2, the results obtained by FA-KFCM are obviously better than those of other algorithms and the clustering quality of FA-KFCM is better. For four datasets, the Average number of iterations and the maximum number of iterations obtained by the algorithm in this paper are all the minimum of several clustering algorithms, which further shows that the clustering quality of the algorithm is better.

It could be learned from Fig. 1 that when clustering four data samples, Iris, Cmc, Wine and Zoo, the average clustering accuracy of the FA-KFCM algorithm is better than that of the other two algorithms.

## 3   Conclusion

In this paper, the FA algorithm and KFCM algorithm are analyzed in detail, and the FA-KFCM algorithm is proposed. The algorithm uses the optimal clustering center of the firefly algorithm as the initial value of KFCM, and then processes the clustering analysis through KFCM. In order to verify the validity of the new algorithm, the author used Iris, Cmc, Wine and Zoo data for numerical experiments and compared with the literature algorithm. The experimental results show that the proposed algorithm has better performance and good clustering results.

**Acknowledgments.** This work was supported by Inner Mongolia University for Nationalities Funds of China under Grant No. NMDYB18008.

## References

1. Kim, D.W., Lee, D.: Evaluation of the performance of clustering algorithms in kernel-Induced feature space. Pattern Recogn. **38**, 607–611 (2005)
2. Yang, X.S.: Firefly algorithm. In: Nature-Inspired Metaheuristic Algorithms, pp. 79–90 (2008)
3. Sayadi, M., Ramezanian, R., GhaariNasab, N.: A discrete firefly metaheuristic with local search for makespan minimization in permutation flow shop scheduling problems. Int. J. Ind. Eng. Comput. **1**(1), 1–10 (2010)
4. Hnig, U.: A firefly algorithm-based approach for scheduling task graphs in homogeneous systems. In: Informatics, pp. 24–33 (2010)
5. Khadwilard, A., Chansombat, S., Thepphakorn, T.: Application of firefly algorithm and its parameter setting for job shop scheduling. In: 1st Symposium on Hands-On Research and Development, pp. 1–10 (2011)
6. Yang, X.S.: Optimization and metaheuristic algorithms in engineering. In: Metaheuristic in Water Geotechnical and Transport Engineering, pp. 1–23. Elsevier (2013)
7. Zhongwei, G., Fuyuan, X.: Clustering algorithm based on adaptive chaotic particle swarm optimization. Comput. Eng. Des. **36**(6), 1510–1585 (2015)
8. Kuo, R.J., Syu, Y.J., Zhenyao, C.: Integration of particle swarm optimization and genetic algorithm for dynamic clustering. Inf. Sci. **195**(15), 124–140 (2012)

# A Fuzzy Clustering Algorithm with Multi-medoids for Multi-view Relational Data

Eduardo C. Simões and Francisco de A. T. de Carvalho$^{(\boxtimes)}$

Centro de Informatica, Universidade Federal de Pernambuco, Av. Jornalista Anibal Fernandes s/n - Cidade Universitaria, Recife, PE 50740-560, Brazil
{ecs4,fatc}@cin.ufpe.br

**Abstract.** There is an increasing interest for multi-view clustering due to its ability to manage data from several sources. The majority of multi-view clustering algorithms are suitable to analyse vector data, but much less attention has been given for the analysis of relational data. This paper provides a fuzzy clustering algorithm with multi-medoids for multi-view relational data (MFMMdd). Experiments with real multi-view data sets show the good performance of the MFMMdd in comparison with previous multi-view clustering algorithms for relational data, concerning the quality of the partitions provided by these algorithms.

**Keywords:** Multi-view clustering · Multi-medoids · Relational data

## 1 Introduction

The popularity of multi-view clustering is mainly due to its ability to manage data from several sources [1]. For example, in tumor studies one needs to take into account simultaneously genomic, epigenomic, and proteomic data [2]. The main approaches to cluster multi-view data are concatenation or data fusion, distributed approaches and centralized approaches [3]. In concatenation approaches, the views are previously merged in a single data table, before the application of a classical clustering algorithm. In the distributed approach, classical clustering algorithms are previously applied on the views and the multi-view algorithm provides a consensus partition from the partitions obtained with each individual view. Finally, the centralized approach is able to take into account simultaneously all views aiming to provide a single partition of the data.

The most part of the multi-view clustering algorithms are suitable to analyse vector data, but less attention has been given for analyse relational data described by multiple dissimilarity matrices, expressing the relationships between the samples according to each view [4]. Relational data are needed when the views are not easily described as vector data, as when the views are text or links of a web page. Relational data can be also very useful for confidentiality reasons or when a particular dissimilarity function is suitable for a particular problem.

© Springer Nature Switzerland AG 2019
H. Lu et al. (Eds.): ISNN 2019, LNCS 11554, pp. 469–477, 2019.
https://doi.org/10.1007/978-3-030-22796-8_50

This paper proposes MFMMdd, a multi-view version of the fuzzy clustering algorithm with multi-medoids (FMMdd). In FMMdd, the representative or prototype weight allows each cluster to be represented by multiple objects, which are weighted based on their importance in a cluster [5]. MFMMdd applies a centralized approach and minimizes a suitable objective function aiming to provide a fuzzy partition in a fixed number of clusters, cluster representatives as vectors of prototype weights and a vector of relevance weight whose components provide the importance of each dissimilarity matrix in the clustering task. In comparison with FW4M algorithm [6], a previous multi-view fuzzy clustering algorithm with multi-medoids, MFMMdd has one less parameter to be tuned.

The paper is organized as follows. Section 2 presents the multi-view clustering algorithm MFMMdd. In Sect. 3, experiments with data sets mainly from the UCI machine learning repository shows the good performance of the MFMMdd in comparison with previous multi-view fuzzy clustering algorithms for relational data [4,6,7], concerning the quality of the partitions provided by these algorithms. Section 4 gives the final remarks of the paper.

## 2   Multi-view Relational Fuzzy Clustering Algorithm with Multi-medoids

Let $E = \{e_1, \ldots, e_N\}$ be a set of $N$ objects and let $P$ dissimilarity matrices $\mathbf{R}_p = (r_{ij}^{(p)}) (1 \le p \le P)$, where $r_{ij}^{(p)}$ is the dissimilarity between objects $e_i$ and $e_j (1 \le i, j \le N)$ on dissimilarity matrix $\mathbf{R}_p$.

The Fuzzy Clustering Algorithm With Multi-Medoids for Multi-View Relational Data (hereafter named MFMMdd) aims to provide:

- A fuzzy partition represented by the matrix $\mathbf{U} = (u_{ci}) (1 \le c \le K; 1 \le i \le N)$, where $u_{ci}$ denotes the membership of object $e_i$ in cluster $c$;
- A matrix $\mathbf{V} = (v_{cj}) (1 \le c \le K; 1 \le j \le N)$ of prototype weights of the objects with respect to the clusters [5], where $\mathbf{v}_c = (v_{c1}, \ldots, v_{cN})$ is the vector of prototype weights of the objects with respect to cluster $c$, and $v_{cj}$ is the prototype weight of object $e_j$ with respect to cluster $c$;
- A vector $\boldsymbol{\lambda} = (\lambda_1, \ldots, \lambda_P)$, where $\lambda_p$ is the relevance weight of dissimilaritymatrix $\mathbf{R}_p$. The larger $\lambda_p$ is, the more important dissmilarity matrix $\mathbf{R}_p$ is.

Starting from an initial solution, the matrix of memberships $\mathbf{U}$, the matrix of prototype weights $\mathbf{V}$, and the vector of relevance weights $\boldsymbol{\lambda}$ are obtained interactively in three steps (assignment, representation, and weighting) though the minimization of a proper objective function, here-after named as $J_{MFMMdd}$, that computes the heterogeneity of the fuzzy partition as the sum of the heterogeneity in each fuzzy cluster:

$$J_{MFMMdd}(\mathbf{U}, \mathbf{V}, \boldsymbol{\lambda}) = \sum_{c=1}^{K} \sum_{i=1}^{N} (u_{ci})^m d_{\boldsymbol{\lambda}}(e_i, \mathbf{v}_c) \tag{1}$$

subject to: (i) $\sum_{c=1}^{K} u_{ci} = 1, \forall i$ and $u_{ci} \geq 0, \forall c$ and $i$; (ii) $\sum_{j=1}^{N} v_{cj} = 1, \forall c$ and $v_{cj} \geq 0, \forall c$ and $j$; (iii) $\prod_{p=1}^{P} \lambda_p = 1$ and $\lambda_p > 0, \forall p$; and where

$$d_{\boldsymbol{\lambda}}(e_i, \mathbf{v}_c) = \sum_{p=1}^{P} \lambda_p\, d(e_i, \mathbf{v}_c) = \sum_{p=1}^{P} \lambda_p \left[ \sum_{j=1}^{N} (v_{cj})^n r_{ij}^{(p)} \right] \tag{2}$$

is the adaptive dissimilarity between object $e_i$ and cluster prototype $\mathbf{v}_c$ parameterized by the vector of relevance weights $\boldsymbol{\lambda} = (\lambda_1, \dots, \lambda_P)$. Moreover, $m$ is the traditional parameter that controls the fuzziness of memberships, and according to Ref. [5] the parameter $n$ controls the level of smoothness of the distribution of prototype weights among all the objects in each of the clusters. Still according to Ref. [5], the prototype weight allows each cluster to be represented by multiple objects, which are weighted based on their importance in a cluster.

The minimization of the objective function $J_{MFMMdd}$ aiming to provide the optimal solution for $\mathbf{U}$, $\mathbf{V}$, and $\boldsymbol{\lambda}$ is achieved by using the method of Lagrange multipliers. The Lagrangian function is as follows:

$$L_{MFMMdd}(\mathbf{U}, \mathbf{V}, \boldsymbol{\lambda}) = J_{MFMMdd}(\mathbf{U}, \mathbf{V}, \boldsymbol{\lambda}) - \sum_{i=1}^{N} \alpha_i \left( \sum_{c=1}^{K} u_{ci} - 1 \right)$$
$$- \sum_{c=1}^{K} \beta_c \left( \sum_{j=1}^{N} v_{cj} - 1 \right) - \gamma \left( \prod_{p=1}^{P} \lambda_p \right) \tag{3}$$

where $\alpha_i$, $\beta_c$ and $\gamma$ are the Lagrange multipliers.

During the assignment step, the matrix $\mathbf{V}$ of prototype weights and the vector $\boldsymbol{\lambda}$ of relevance weights are kept fixed. The objective function $J_{MFMMdd}$ is optimized with respect to the memberships. Taking the partial derivatives of $L_{MFMMdd}$ w.r.t $u_{ci}$ and $\alpha_i$, and setting them to zero, we obtain:

$$u_{ci} = \left[ \sum_{f=1}^{K} \left( \frac{d_{\boldsymbol{\lambda}}(e_i, \mathbf{v}_c)}{d_{\boldsymbol{\lambda}}(e_i, \mathbf{v}_f)} \right)^{\frac{1}{m-1}} \right]^{-1} = \left[ \sum_{f=1}^{K} \left( \frac{\sum_{j=1}^{N} (v_{cj})^n \sum_{p=1}^{P} \lambda_p r_{ij}^{(p)}}{\sum_{j=1}^{N} (v_{fj})^n \sum_{p=1}^{P} \lambda_p r_{ij}^{(p)}} \right)^{\frac{1}{m-1}} \right]^{-1} \tag{4}$$

During the representation step, the matrix $\mathbf{U}$ of memberships and the vector $\boldsymbol{\lambda}$ of weights of relevance are kept fixed. The objective function $J_{MFMMdd}$ is optimized with respect to the prototype weights. Taking the partial derivatives of $L_{MFMMdd}$ w.r.t $v_{cj}$ and $\beta_c$, and by setting the partial derivatives to zero, and after some algebra we obtain:

$$v_{cj} = \left[ \sum_{h=1}^{N} \left( \frac{\sum_{i=1}^{N} \sum_{p=1}^{P} (u_{ci})^m (\lambda_p) r_{ij}^{(p)}}{\sum_{i=1}^{N} \sum_{p=1}^{P} (u_{ci})^m (\lambda_p) r_{ih}^{(p)}} \right)^{\frac{1}{n-1}} \right]^{-1} \tag{5}$$

During the weighting step, the matrix $\mathbf{U}$ of memberships and the matrix $\mathbf{V}$ of prototype weights are kept fixed. The objective function $J_{MFMMdd}$ is optimized with respect to the weights of relevance. Taking the partial derivatives of $L_{MFMMdd}$ w.r.t $\lambda_p$ and $\gamma$, and setting them to zero, we obtain:

$$\lambda_p = \frac{\left\{ \prod_{h=1}^{P} \left[ \sum_{c=1}^{K} \sum_{i=1}^{N} (u_{ci})^m \sum_{j=1}^{N} (v_{cj})^n r_{ij}^{(h)} \right] \right\}^{\frac{1}{P}}}{\sum_{c=1}^{K} \sum_{i=1}^{N} (u_{ci})^m \sum_{j=1}^{N} (v_{cj})^n r_{ij}^{(p)}} \qquad (6)$$

These three steps are repeated until the convergence of MFMMdd. Algorithm 1 summarize these steps.

---

**Algorithm 1.** MFMMdd algorithm

---

1: **Input**
2:     $\mathcal{D} = \{\mathbf{R}_1, \ldots, \mathbf{R}_P\}$ (the data set); $K$ (the number of clusters); $T$ (maximum number of iterations); $\epsilon$ (threshold parameter);
3: **Output**
4:     $\mathbf{U}$: the matrix of memberships;
5:     $\mathbf{V}$: the matrix of prototype weights;
6:     $\boldsymbol{\lambda}$: the vector of relevance weights of the dissimilarity matrices.
7: **Initialization**
8:     $t = 0$;
9:     Set $\boldsymbol{\lambda}^{(t)} = (1, \ldots, 1)$;
10:     Randomly initialize the matrix $\mathbf{U}^{(t)} = (u_{ci}^{(t)})\,(1 \leq c \leq K; 1 \leq i \leq N)$ such that $\sum_{c=1}^{K} u_{ci}^{(t)} = 1 \,\forall i$ and $u_{ci}^{(t)} \geq 0 \,\forall c, i$;
11:     Compute the components $v_{cj}^{(t)}(1 \leq c \leq K; 1 \leq j \leq N)$ of the the matrix $\mathbf{V}^{(t)}$ according to Eq. (5);
12:     Compute $J_{MFMMdd}(\mathbf{U}^{(t)}, \mathbf{V}^{(t)}, \boldsymbol{\lambda}^{(t)})$ according to Eq. (1);
13: **repeat**
14:     $t = t + 1$;
15:     **Step 1: assignment**.
16:         Compute the components $u_{ci}^{(t)}(1 \leq c \leq K; 1 \leq i \leq N)$ of the the matrix $\mathbf{U}^{(t)}$ according to Eq. (4);
17:     **Step 2: representation**
18:         Compute the components $v_{cj}^{(t)}(1 \leq c \leq K; 1 \leq j \leq N)$ of the the matrix $\mathbf{V}^{(t)}$ according to Eq. (5);
19:     **Step 3: weighting**
20:         Compute the components $\lambda_p^{(t)}(1 \leq p \leq P$ of the the matrix of relevance weights $\boldsymbol{\lambda}^{(t)}$ according to Eq. (6).
21:     Compute $J_{MFMMdd}(\mathbf{U}^{(t)}, \mathbf{V}^{(t)}, \boldsymbol{\lambda}^{(t)})$ according to Eq. (1);
22: **until**
23:     $|J_{MFMMdd}(\mathbf{U}^{(t)}, \mathbf{V}^{(t)}, \boldsymbol{\lambda}^{(t)}) - J_{MFMMdd}(\mathbf{U}^{(t-1)}, \mathbf{V}^{(t-1)}, \boldsymbol{\lambda}^{(t-1)})| < \epsilon$ or $t > T$;

---

*Remark.* The objective function of the FW4M algorithm of Ref. [6] is as follows:

$$J_{FW4M}(\mathbf{U}, \mathbf{V}, \boldsymbol{\lambda}) = \sum_{c=1}^{K} \sum_{i=1}^{N} \sum_{j=1}^{N} \sum_{p=1}^{P} (u_{ci})^m (v_{cj})^n (\lambda_p)^s r_{ij}^{(p)} \qquad (7)$$

subject to: (i) $\sum_{c=1}^{K} u_{ci} = 1, \forall i$ and $u_{ci} \geq 0, \forall c$ and $i$; (ii) $\sum_{j=1}^{N} v_{cj} = 1, \forall c$ and $v_{cj} \geq 0, \forall c$ and $j$; (iii) $\sum_{p=1}^{P} \lambda_p = 1$ and $\lambda_p \geq 0, \forall p$. According to Ref. [6], the additional parameter $s$ controls the level of smoothness of vector's weights.

## 3  Empirical Results

This section provides a performance comparison between the proposed MFM-Mdd algorithm with CARD-R [4] (a multi-view version of NERF [8]), FW4M [6], and MFCMdd-RWG-P [7] (a multi-view version of FCMdd [9]), state of the art prototype based fuzzy clustering algorithms for multi-view relational data sets.

Six datasets from the UCI Machine learning Repository, and Phonema dataset (http://www.math.univ-toulouse.fr/staph/npfda/npfda-datasets.html), were considered in this study. Table 1, in which $N$ is the number of objects, $C$ is the number of ta priori classes, $nvar$ is the number of variables and $nV$ is the number of views, summarizes these data sets. In can be observed that each variable corresponds to a single view in data sets Glass, Iris, Seeds, Wine. Moreover, in these data sets (except Phonema data set) each view corresponds to a dissimilarity matrix computed with the Euclidean distance. The algorithms were implemented in the C language and performed on the same machine (OS: Windows 10 64-bits, Memory: 12 GB, Processor: Intel Core i7-4790 CPU @ 3.60 GHz).

To compare the time trajectories described by Phonema data set a "cross-sectional longitudinal" dissimilarity [10] was considered. This dissimilarity combines the comparison of the position of each pair of trajectories, and two longitudinal dissimilarities, based on the concepts of velocity and acceleration.

**Table 1.** Summary of the data sets

| Datasets | $N$ | $C$ | $nVar$ | $nV$ |
|---|---|---|---|---|
| Glass | 214 | 6 | 9 | 9 |
| Image segmentation | 2310 | 6 | 19 | 2 |
| Iris | 150 | 3 | 4 | 4 |
| Multiple features | 2000 | 10 | 649 | 6 |
| Phonema | 2000 | 5 | 447 | 3 |
| Seeds | 210 | 3 | 7 | 7 |
| Wine | 178 | 3 | 13 | 13 |

MFMMdd, FW4M, CARD-R and MFCMdd-RWG-P were run on these data sets 100 times, with $K$ (the number of clusters) equal to $C$ (the number of a priori classes). Table 2 shows the parameters used with these algorithms. They were fixed in a unsupervised way, without the use of the labels provided by the a priori partition through a grid search such that the suitable combination of parameters has as values those for which the minimum distance between a couple of representatives falls $<0.1$ [11]. Parameters $T$ and $\epsilon$ were set, respectively, to 100 and $10^{-10}$.

**Table 2.** Selected parameters used on the algorithms

| Datasets | Algorithms | | | |
|---|---|---|---|---|
| | MFMMdd | CARD-R | MFCMdd-RWG-P | FW4M |
| Glass | $m = 1.1$ | $m = 1.3$ | $m = 1.1$ | $m = 1.1$ |
| | $n = 1.1$ | $q = 1.5$ | $q = 3$ | $n = 1.5$ and $s = 1.1$ |
| Image segmentation | $m = 1.1$ | $m = 1.1$ | $m = 1.1$ | $m = 1.1$ |
| | $n = 1.3$ | $q = 1.5$ | $q = 3$ | $n = 1.1$ and $s = 1.1$ |
| Iris | $m = 1.1$ | $m = 1.1$ | $m = 1.1$ | $m = 1.1$ |
| | $n = 1.1$ | $q = 1.5$ | $q = 5$ | $n = 1.1$ and $s = 1.5$ |
| Multiple features | $m = 1.1$ | $m = 1.1$ | $m = 1.1$ | $m = 1.1$ |
| | $n = 1.1$ | $q = 1.1$ | $q = 3$ | $n = 1.1$ and $s = 1.3$ |
| Phonema | $m = 1.05$ | $m = 1.05$ | $m = 1.1$ | $m = 1.05$ |
| | $n = 1.05$ | $q = 1.05$ | $q = 5$ | $n = 1.05$ and $s = 1.1$ |
| Seeds | $m = 1.1$ | $m = 1.1$ | $m = 1.1$ | $m = 1.1$ |
| | $n = 1.1$ | $q = 2.0$ | $q = 5$ | $n = 1.1$ and $s = 1.3$ |
| Wine | $m = 1.1$ | $m = 1.1$ | $m = 1.1$ | $m = 1.1$ |
| | $n = 1.1$ | $q = 1.3$ | $q = 3$ | $n = 1.1$ and $s = 1.3$ |

The quality of the fuzzy partitions given by these algorithms was assessed with the Rand index for a fuzzy partition (Rand-F) [4] and the Hullemeyer index (HUL) [12] Rand-F and HUL indexes are suitable to the comparison between the a priori partitions of the datasets and the fuzzy partitions given by the algorithms. They take their values on the interval $[0, 1]$, where 1 means total agreement between partitions.

Table 3 shows the average and standard deviation of the Rand-F and HUL indexes provided by the algorithms on data sets of Table 1.

It can be observed that MFMMdd presented the best performance and was the most robust according to Rand-F index in 5 out 7 data sets. CARD-R and MFCMdd-RWG-P were the worse each in 3 out 7 data sets. MFCMdd-RWG-P was the less robust in 4 out 7 data sets. Moreover, MVRFMMd and FW4M were the best according to HULL index in 3 out 7 data sets. MFMMdd was also the most robust according to HULL index in 5 out 7 data sets. MFCMdd-RWG-P was the worse (in 3 out 7 data sets) and the less robust (in 5 out 7 data sets).

From the fuzzy partition $\mathbf{U}$ it is obtained a crisp partition $\mathcal{Q} = (Q_1, \ldots, Q_M)$, where the cluster $Q_m(m = 1, \ldots, M)$ is defined as: $Q_m = \{e_i \in E : u_{im} = \max_{h=1}^{K} u_{ih}\}$. To evaluate the quality of the crisp partitions given by the algorithms, the F-measure [13], and the overall error rate of classification (OER) [14] were considered. F-measure and OER indexes are useful to provide a comparison

**Table 3.** Performance of the algorithms: fuzzy partitions

| Data sets | Rand-F | | | | HUL | | | |
|---|---|---|---|---|---|---|---|---|
| | MFMMdd | FW4M | CARD-R | MFCMdd-RWG-P | MFMMdd | FW4M | CARD-R | MFCMdd-RWG-P |
| Glass | **0.7130** ($10^{-7}$) | 0.6515 (0.0066) | 0.4827 ($10^{-5}$) | 0.6827 (0.0299) | **0.5188** ($10^{-6}$) | 0.5038 (0.0090) | 0.2684 (0.0001) | 0.5116 (0.0321) |
| Image segmentation | **0.8504** (**0.0057**) | 0.8383 (0.0221) | 0.8190 (0.0346) | 0.8176 (0.0324) | 0.7493 (**0.0070**) | **0.7590** (0.0347) | 0.7349 (0.0510) | 0.7106 (0.0527) |
| Iris | 0.9330 (0.0021) | **0.9481** ($10^{-6}$) | 0.9443 (0.0201) | 0.8893 (0.0008) | 0.8983 (0.0304) | **0.9220** ($10^{-6}$) | 0.9199 (0.0284) | 0.8370 (0.1182) |
| mFeat | **0.9077** (0.0054) | 0.8699 (**0.0013**) | 0.8591 (0.0028) | 0.9074 (0.0102) | 0.7192 (0.0091) | **0.7961** (**0.0012**) | 0.7901 (0.0033) | 0.7421 (0.0278) |
| Phonema | 0.8203 (**0.0026**) | 0.9001 (0.0152) | **0.9046** (0.0277) | 0.7777 (0.0108) | 0.5550 (**0.0119**) | 0.8312 (0.0237) | **0.8506** (0.0408) | 0.4518 (0.0310) |
| Seeds | **0.8521** ($10^{-7}$) | 0.8401 (0.0004) | 0.8468 ($10^{-6}$) | 0.8471 (0.0544) | **0.7834** ($10^{-7}$) | 0.7679 ($10^{-6}$) | 0.7794 ($10^{-6}$) | 0.7770 (0070) |
| Wine | **0.9047** ($10^{-7}$) | 0.8007 ($10^{-5}$) | 0.7913 (0.0791) | 0.8822 (0.0553) | **0.8565** ($10^{-7}$) | 0.7084 ($10^{-5}$) | 0.7028 (0.1161) | 0.8219 (0.0865) |

between the a priori partitions of the datasets and the crisp partitions. F-measure index has its values on the interval $[0, 1]$, where 1 indicates perfect agreement between partitions. OERC index has its values on the interval $[0, 1]$, it aims to measure the ability of a clustering algorithm to find out the a priori classes present in a data set.

Table 4 shows the average and standard deviation of the F-measure and OER indexes provided by the algorithms on data sets of Table 1.

**Table 4.** Performance of the algorithms: crisp partition

| Data sets | F-measure | | | | OER | | | |
|---|---|---|---|---|---|---|---|---|
| | MFMMdd | FW4M | CARD-R | MFCMdd-RWG-P | MFMMdd | FW4M | CARD-R | MFCMdd-RWG-P |
| Glass | **0.5496** (0.3494) | 0.4917 (0.0164) | 0.3458 (**0.0013**) | 0.5134 (0.0432) | 0.3318 ($10^{-7}$) | 0.5234 (0.0052) | **0.6301** (0.0032) | 0.4223 (0.0460) |
| Image segmentation | **0.6472** (0.2226) | 0.6044 (**0.0441**) | 0.5701 (0.0595) | 0.5725 (0.0645) | 0.3827 (0.1823) | 0.4199 (**0.0397**) | 0.4435 (0.0482) | **0.4495** (0.0610) |
| Iris | 0.9508 (0.0252) | **0.9600** ($10^{-7}$) | 0.9512 (0.0212) | 0.8965 (0.1004) | 0.0495 (0.0285) | **0.0400** ($10^{-9}$) | 0.0496 (0.0291) | 0.1081 (0.1102) |
| mFeat | **0.7675** (0.0284) | 0.4696 (0.0042) | 0.4483 (**0.0041**) | 0.7272 (0.0505) | 0.2476 (0.0380) | 0.5500 (0.0140) | **0.5718** (**0.0079**) | 0.2770 (0.0535) |
| Phonema | 0.8343 (**0.0292**) | 0.8398 (0.0334) | **0.8418** (0.0506) | 0.7847 (0.0565) | 0.1661 (0.0375) | **0.1659** (0.0403) | 0.1678 (0.0586) | 0.2251 (**0.0006**) |
| Seeds | **0.8747** ($10^{-9}$) | 0.8654 ($10^{-7}$) | 0.8656 ($10^{-7}$) | 0.8654 (0.0002) | **0.1238** ($10^{-8}$) | 0.1333 ($10^{-8}$) | 0.1333 ($10^{-8}$) | 0.1333 ($10^{-8}$) |
| Wine | **0.9545** ($10^{-9}$) | 0.8331 ($10^{-7}$) | 0.8039 (0.0954) | 0.9199 (0.0769) | **0.0449** ($10^{-9}$) | 0.1629 ($10^{-7}$) | 0.1921 (0.0930) | 0.0793 (0.0773) |

It can be observed that MFMMdd presented the best performance according to F-measure and OER indexes in 5 out 7 data sets. It was also the most robust according to F-measure and OER indexes in 3 out 7 data sets. Moreover, CARD-R was the worse according to F-measure in 3 out 7 data sets, and MFCMdd-

RWG-P was the less robust according to F-measure in 4 out 7 data sets. Besides, CARD-R and MFCMdd-RWG-P were the worse and the less robust according to OER index in 3 out of 7 data sets.

## 4    Final Remarks and Conclusions

This paper presented MFMMdd, a multi-view version of the fuzzy clustering algorithm with multi-medoids (FMMdd). MFMMdd minimizes a suitable objective function aiming to provide a fuzzy partition in a fixed number of clusters, cluster representatives as vectors of prototype weights and a vector of relevance weights of each dissimilarity matrix in the clustering task.

Experiments with multi-view data sets showed the performance of the proposed algorithm. In the majority of the data sets considered, MFMMdd provided fuzzy and crisp partitions of better quality than previous fuzzy clustering algorithms for multi-view relational data. MFMMdd was also the most robust. Finally, a significant practical advantage of MFMMdd regarding FW4M is that the former algorithm has one less parameter to be tuned. This is because the restriction of type product to one is used in MFMMdd to calculate the weights of the importance of the dissimilarity matrices.

**Acknowledgments.** The authors would like to thank the anonymous referees for their careful revision, and the Conselho Nacional de Desenvolvimento Cientifico e Tecnologico - CNPq (303187/2013-1) for partially support this work.

## References

1. Yang, Y., Wang, H.: Multi-view clustering: a survey. Big Data Min. Anal. **1**(2), 83–107 (2018)
2. Shenl, R., Olshen, A.B., Ladanyi, M.: Integrative clustering of multiple genomic data types using a joint latent variable model with application to breast and lung cancer sub-type analysis. Bioinformatics **25**, 2906–2912 (2009)
3. Cleuziou, G., Exbrayat, M., Martin, L., Sublemontier, J.-H.: CoFKM: a centralized method for multiple-view clustering. In: Ninth IEEE International Conference on Data Mining, pp. 752–757. IEEE Press, New York (2009)
4. Frigui, H., Hwang, C., Rhee, F.C.-H.: Clustering and aggregation of relational data with applications to image database categorization. Pattern Recogn. **40**, 3053–3068 (2007)
5. Mei, J.-P., Chen, L.: Fuzzy relational clustering around medoids: a unified view. Fuzzy Sets Syst. **183**, 44–56 (2011)
6. Gao, Y., Sun, C., Qi, H., Wang, S.: Fuzzy clustering based on weighted multi-medoids and multi-matrices. Int. J. Adv. Comput. Technol. **6**, 61–70 (2014)
7. de Carvalho, F.A.T., Lechevallier, Y., de Melo, F.M.: Relational partitioning fuzzy clustering algorithms based on multiple dissimilarity matrices. Fuzzy Sets Syst. **215**, 1–28 (2013)
8. Hathaway, R., Bezdek, J.: Nerf c-means: non-Euclidean relational fuzzy clustering. Pattern Recogn. **27**, 429–437 (1994)

9. Krishnapuram, R., Joshi, A., Yi, L.: A fuzzy relative of the k-medoids algorithm with application to web document and snippet clustering. In: IEEE International Fuzzy Systems Conference, pp. 1281–1286 (1999)
10. D'Urso, P.: Dissimilarity measures for time trajectories. J. Ital. Stat. Soc. **1–3**, 53–83 (2000)
11. Schwaemelle, V., Norregaard, O.: A simple and fast method to determine the parameters for fuzzy c-means cluster analysis. Bioinformatics **26**, 2841–2848 (2010)
12. Huellermeier, E., Henzgen, S., Senge, R.: Comparing fuzzy partitions: a generalization of the rand index and related measures. IEEE Trans. Fuzzy Syst. **20**, 546–556 (2012)
13. Manning, C.D., Raghavan, P., Schuetze, H.: Introduction to Information Retrieval. Cambridge University Press, Cambridge (2008)
14. Breiman, L., Friedman, J., Stone, C.J., Olshen, R.A.: Classification and Regression Trees. Chapman and Hall/CRC, Boca Raton (1984)

# Author Index

Printed in the United States
By Bookmasters